T0389080

Passive Vibration Control of Structures

Passive Vibration Control of Structures

Suhasini Madhekar and Vasant Matsagar

CRC Press is an imprint of the
Taylor & Francis Group, an **informa** business

First edition published 2022
by CRC Press
6000 Broken Sound Parkway NW, Suite 300, Boca Raton, FL 33487-2742

and by CRC Press
2 Park Square, Milton Park, Abingdon, Oxon, OX14 4RN

Library of Congress Cataloging-in-Publication Data

Names: Madhekar, Suhasini, author. | Matsagar, Vasant, author.
Title: Passive vibration control of structures / Suhasini Madhekar and Vasant Matsagar.
Description: First edition. | Boca Raton, FL : CRC Press, 2022. | Includes bibliographical references and index. | Summary: "Research in structural vibration control deals with catastrophic failures of civil structures during natural hazards. Therefore, the focus of the proposed book will be on theory of dynamic response control of structures by using different kinds of passive vibration control devices. The strategies used for controlling displacement, velocity, and acceleration of structures such as buildings and bridges under the action of dynamic loads emanating from earthquake, wind, wave, etc. will be detailed. The book will contain explanation of mathematical modelling of civil structures equipped with the passive controllers, their dynamic analysis and design, and assessment of the results"-- Provided by publisher.
Identifiers: LCCN 2021047603 (print) | LCCN 2021047604 (ebook) | ISBN 9781138035140 (hbk) | ISBN 9781032163475 (pbk) | ISBN 9781315269269 (ebk)
Subjects: LCSH: Structural control (Engineering) | Damping (Mechanics) | Vibration. | Earthquake resistant design.
Classification: LCC TA658.4 .M34 2022 (print) | LCC TA658.4 (ebook) | DDC 624.1/7--dc23/eng/20211117
LC record available at https://lccn.loc.gov/2021047603
LC ebook record available at https://lccn.loc.gov/2021047604

ISBN: 978-1-138-03514-0 (hbk)
ISBN: 978-1-032-16347-5 (pbk)
ISBN: 978-1-315-26926-9 (ebk)

DOI: 10.1201/9781315269269

Typeset in Times
by KnowledgeWorks Global Ltd.

Contents

Photos and Attribution

Sr. No.	Photo No.	Page No. of Book	Attribution
1	1.1	24	(Courtesy of B. F. Spencer)
2	2.1	47	(Courtesy: Indian Institute of Technology (IIT) Roorkee)
3	2.2	48	(Courtesy, NWTF, IIT Kanpur)
4	3.1	68	Inoue-hiro, CC BY-SA 3.0 https://tinyurl.com/4jpw37ak, via Wikimedia Commons
5	3.2	69	(https://tinyurl.com/2p8jxrc6)
6	3.3	69	Ben P L from Provo, USA, CC BY-SA 2.0 https://tinyurl.com/5fn26ey7, via Wikimedia Commons
7	3.4	70	(https://tinyurl.com/27cft8ev)
8	3.5	70	Cristiano Tomás, CC BY-SA 4.0 https://tinyurl.com/4j3ffu2d, via Wikimedia Commons
9	3.6	70	(https://tinyurl.com/2p8apdpm)
10	3.7	71	(https://tinyurl.com/jzu5kaj3)
11	3.8	71	(https://tinyurl.com/jzu5kaj3)
12	4.1	103	(https://tinyurl.com/2p93r7m6)
13	4.2	103	(https://tinyurl.com/2z4vhnm8)
14	4.3	104	(https://tinyurl.com/yuk2d8vr)
15	4.4	110	Jeffmock, Public domain, via Wikimedia Commons
16	4.5	110	Ivan Andreevich, Public domain, via Wikimedia Commons
17	4.6	111	I, Cumulus Clouds, CC BY-SA 3.0 https://tinyurl.com/2p9fvbzk, via Wikimedia Commons
18	4.7	111	(https://tinyurl.com/48vdjztu)
19	4.8	114	Caroline Culler, CC BY-SA 3.0 https://tinyurl.com/4jpw37ak, via Wikimedia Commons
20	4.9	114	(https://tinyurl.com/3br564nv)
21	4.10	115	(https://tinyurl.com/maybtx2f)
22	5.1	154	Wladyslaw, CC BY 3.0 https://tinyurl.com/4kc8e3d4, via Wikimedia Commons
23	5.2	155	J o, CC BY-SA 3.0 https://tinyurl.com/yhhsvdbw, via Wikimedia Commons
24	5.3	156	Johan Burati, Public domain, via Wikimedia Commons
25	5.4	157	I, Tomtheman5, CC BY-SA 3.0 https://tinyurl.com/yhhsvdbw, via Wikimedia Commons
26	5.5	158	Σ64, CC BY 4.0 https://tinyurl.com/3k4b49ke, via Wikimedia Commons
27	5.6	159	Anthony Santiago101, CC BY-SA 4.0 https://tinyurl.com/yzs2ab7d, via Wikimedia Commons
28	5.7	160	Armand du Plessis, CC BY 3.0 https://tinyurl.com/4kc8e3d4, via Wikimedia Commons
29	5.8	161	Vijay B. Barot, CC BY-SA 4.0 https://tinyurl.com/4j3ffu2d, via Wikimedia Commons

Sr. No.	Photo No.	Page No. of Book	Attribution
30	5.9	174	Bernd in Japan, CC BY-SA 3.0 https://tinyurl.com/2p9fvbzk, via Wikimedia Commons
31	5.10	175	J o, CC BY-SA 3.0 https://tinyurl.com/4jpw37ak, via Wikimedia Commons
32	5.11	176	GG001213, CC0, via Wikimedia Commons
33	6.1	189	(Courtesy of Tripat Pall)
34	6.2	192	(https://www.cwejournal.org)
35	6.3	193	(Courtesy of Tripat Pall)
36	6.4	193	(Courtesy of Tripat Pall)
37	6.5	194	Ben Miller, CC BY 2.0 https://tinyurl.com/2p9337dc, via Wikimedia Commons
38	6.6	195	Jamie Lantzy, CC BY-SA 3.0 https://tinyurl.com/4jpw37ak, via Wikimedia Commons
39	6.7	195	(Courtesy of Tripat Pall)
40	6.8	196	(Courtesy of Tripat Pall)
41	6.9	196	(Courtesy of Tripat Pall)
42	6.10 (a) and (b)	197	(Courtesy of Tripat Pall)
43	7.1 (a)	220	(https://tinyurl.com/bdf6kvcx)
44	7.1 (b) & (c)	220	(https://tinyurl.com/2p9x5jyy)
45	7.2	220	(https://www.cwejournal.org)
46	7.3	222	(https://tinyurl.com/58hhhswm)
47	7.4	223	(https://tinyurl.com/58hhhswm)
48	7.5	229	(Courtesy of Shervin Maleki)
49	7.6	230	(Courtesy of Shervin Maleki)
50	7.7	232	(https://tinyurl.com/mwpnct5r)
51	7.8	236	(Courtesy of Hong-Nan Li)
52	8.1	246	(Courtesy of Nigel Priestley)
53	8.2	250	(https://tinyurl.com/2p8supkx)
54	8.3	251	(https://tinyurl.com/2p8supkx)
55	8.4	251	(https://tinyurl.com/3p6468ce)
56	8.5	252	(https://tinyurl.com/2p9as4st)
57	8.6	252	(https://tinyurl.com/yfyrpfju)
58	9.1	261	(https://tinyurl.com/33df296b)
59	11.1	361	(https://tinyurl.com/u38yxcj7)

Preface

The writing of this book has been both a joy and a challenge. Decades of our combined experience in teaching and research has given us the foundation to bring complex theoretical concepts of structural dynamics and structural vibration control within easy reach of a wide spectrum of readers including postgraduate students, research scholars, teaching fraternity, and practicing professionals, in plain composition and easy-to-read language.

Over the centuries, mankind has had to deal with many natural hazards such as cyclones, floods, wind, droughts, volcanic eruptions, etc., which cause huge losses in terms of lives lost and property destroyed. The one most feared among the hazards is an earthquake. This fear stems principally from the totally unpredictable nature of this hazard, its suddenness, and the colossal destruction of life and property that it causes. Building an earthquake-proof structure is only an ideal, and one cannot guarantee complete safety of a structure against a random base excitation, whose characteristics can at best be an informed guess! In structural vibration control, the primary aim remains minimizing damage and save lives. Conventionally, earthquake-resistant design is based on a combination of strength and ductility. For small, frequent disturbances due to dynamic loads, the structures are expected to remain in an elastic state. However, in reality, a traditionally designed structure does not respond linearly when subjected to a severe dynamic loading.

Considering the dynamic nature of environmental disturbances, new and innovative concepts of structural protection have been adopted universally. Though active and semi-active control technologies could have an increasingly important role in structural design in the future, we have restricted the scope of our book to passive control systems, as they have been popularly used and have bright future for worldwide applications. Passive structural control is a promising workable technology and a fast-expanding field.

The passive structural protective systems are broadly divided into two major groups, viz., seismic isolation, using elastomeric bearings, sliding systems, rolling-base isolation devices, and their combination, and passive dampers that include fluid viscous dampers, friction dampers, metallic dampers, tuned dampers, etc. Some of these devices can be effective against multiple natural hazards. The technique of seismic isolation is now widely used in many countries and is a well-established and mature technology.

This book is our small contribution toward achieving good quality in design of structure with vibration abatement systems by clearly understanding the concepts of structural vibration control. It introduces basic concepts of passive energy dissipation and discusses different passive devices. Numerous illustrative examples on structures such as buildings and bridges are designed and solved.

Every effort has been made to ensure the correctness of various devices and examples covered in this book. In spite of this, some errors and misprints could still be found. The authors will be grateful to users of this book for informing them of any error that they may find. The authors would also welcome any suggestions and comments offered. Due to limitation on number of pages, we could not include many recently developed devices; it is planned to shortly launch the second edition of the book; your suggestions and input on the new devices are therefore solicited for.

Success is a combination of patience, persistence, and perspiration. Finally, we have been able to realize our dream of publishing a book on the topic of our deep interest by taking continuous efforts. In this voyage, many individuals have sailed with us and have extended invaluable help and support. It is our pleasure to acknowledge them all. This book draws on information contained in several publications and technical articles available in the public domain. These have been an invaluable source for developing the concepts covered in this book. We wish to express our gratitude to Professor Satish Nagarajaiah of Rice University for his kind words of appreciation in his foreword to this book.

It is our great pleasure to acknowledge the significant contributions made to this monograph by a number of research scholars at Indian Institute of Technology (IIT) Delhi. Mr Daniel Habtamu Zelleke carefully read the complete manuscript and ensured its technical correctness. We are grateful to him for his perseverance and untiring efforts. Drs Sandip Kumar Saha, Said Elias, Aruna Gupta-Rawat, Pravin Jagtap, Tathagata Roy, and several other students and colleagues have contributed enormously in developing the contents in this book; names of all of them could not be accommodated due to space constraints.

We wish to thank Mr Ketan Bajad, who efficiently typed early drafts of this manuscript and helped in exhaustive literature review. We wish to recognize the determined efforts and patience of Mr Shubhankar N. Moye for the artful illustrations that contributed to the effectiveness of the text.

We would like to express our gratitude to our respective families for their unstinted support, sacrifices, and encouragement without which this publication would not have been possible.

Suhasini N. Madhekar and Vasant A. Matsagar

About the Authors

Suhasini N. Madhekar is currently working as Professor of Applied Mechanics in the Department of Civil Engineering at College of Engineering Pune (COEP). She obtained her doctorate degree from Indian Institute of Technology (IIT) Bombay. The subject of her doctoral research was "Seismic Control of Benchmark Highway Bridge." Her PhD thesis is published in a form of book by LAP LAMBERT Academic Publishing GmbH and Company, Germany. Her research interests include structural dynamics and vibration control, bridge engineering, and protection of structures from earthquake. She is a supervisor of students for doctoral research. She is FIE (India) and life member of ISTE, ISSE, Indian ASTR, IIBE, and ICI.

Prof. Madhekar has published a book, *Seismic Design of RC Buildings - Theory and Practice*, coauthored by Mr Sharad Manohar. The book is of international repute, published by Springer. The book was awarded "Best book of the year - Useful to consultants," in 2015, by the Association of Consulting Civil Engineers, India. She has published international journal papers, conference papers, and book chapters. Prof. Madhekar has delivered invited lectures in different educational institutes and is invited as examiner for evaluation of masters and PhD theses by different universities. Prof. Madhekar is recipient of the Excellence in Teaching Award at College of Engineering Pune, the "COEP Star Award." She works as a member of board of studies and as a member of research and review committees at different universities in India.

Vasant A. Matsagar is currently serving as a Dogra Chair Professor in the Department of Civil Engineering at Indian Institute of Technology (IIT) Delhi. He obtained his doctorate degree from the Indian Institute of Technology (IIT) Bombay in 2005 in the area of seismic base isolation of structures. He performed his postdoctoral research at the Lawrence Technological University (LTU), United States, in the area of carbon fiber reinforced polymers (CFRP) in bridge structures for about four years. His current research interests include structural dynamics and vibration control; multi-hazard protection of structures from earthquake, blast, fire, and wind; finite element methods; fiber-reinforced polymers (FRP) in prestressed concrete structures; and bridge engineering. He has guided students at both undergraduate (UG) and postgraduate (PG) levels in their bachelors and masters projects and doctoral research. Besides student guidance, he is actively engaged in sponsored research and consultancy projects at national and international levels. He has published hundreds of international journal papers, other hundreds of international conference manuscripts, a book, six edited proceedings, and has filed for two patents. He is also involved in teaching courses in structural engineering, e.g., structural analysis, finite element methods, numerical methods, structural stability, structural dynamics, design of steel and concrete structures to name a few. He has received the Excellence in Teaching Award. He has organized several short- and long-term courses as part of a quality improvement program (QIP) and continuing education program (CEP) and delivered invited lectures in different educational and research organizations. He is Fellow of Institution of

Engineers, IEI (India), Indian Society of Earthquake Technology (ISET), and member of several professional societies.

Prof. Matsagar is the recipient of numerous national and international awards including "CDRI Fellowship" by the Coalition for Disaster Resilient Infrastructure (CDRI) in 2021; ASEM-DUO-India Professor Fellowship Award in 2020; Erasmus+ Fellowship in 2020; "Humboldt Research Fellowship" in 2017; "Erasmus Mundus Award" in 2013; "DST Young Scientist Award" by the Department of Science and Technology (DST) in 2012; "DAAD Awards" by the Deutscher Akademischer Austausch Dienst (DAAD) in 2009 and 2012; "DAE Young Scientist Award" by the Department of Atomic Energy (DAE) in 2011; "IBC Award for Excellence in Built Environment" by the Indian Buildings Congress (IBC) in 2010; "IEI Young Engineer Award" by the Institution of Engineers (India) in 2009; and "Outstanding Young Faculty Fellowship" by the Indian Institute of Technology (IIT) Delhi in 2009. He has also been appointed as "DAAD Research Ambassador" by the German Academic Exchange Program since the academic session 2010. He is serving as Chief Editor and member of editorials boards of several indexed journals of repute. He is currently serving on code development committees in the Bureau of Indian Standards (BIS) and International Organization for Standardization (ISO).

1 Structural Vibration Control Using Passive Devices

1.1 INTRODUCTION

In recent years, innovative strategies for enhancing structural functionality and safety against natural and manmade hazards have been largely researched and developed. Undesirable time-varying motions of dynamically sensitive structures, such as vibratory platforms, bridges, towers, water tanks and buildings, are mainly caused by uncertain random excitations. In a variety of applications, it is necessary to attenuate these disturbances effectively. Structures, depending on their dynamic properties, could be vulnerable to damages when subjected to earthquake- and wind-induced vibrations. In addition to structural safety, excessive structural vibrations affect occupants' comfort under large environmental excitations. Despite the presence of intensive efforts aimed toward ensuring the safety of structures under dynamic excitation, such as wind and earthquake, conventionally designed structures are found to be still vulnerable to strong wind or severe earthquake excitations. Such structures have limited capacities of load resistance and energy dissipation; and they primarily rely on (i) their stiffness to resist earthquake induced forces and (ii) their small inherent material damping to dissipate dynamic energy. It could be possible to increase the strength and energy dissipation capacity (i.e., ductility) of structures in order to enhance their capacities to withstand stronger dynamic excitations. However, the use of high-strength and ductile construction materials increases the cost of the project. Further, increasing strength by providing larger member cross-sections leads to larger force demand on the members, subsequently requiring even greater strength. Also, an attempt to increase the energy dissipation capacity of structures by introducing ductile members/components leads to increased damage to the structural and nonstructural components, which hampers the post-hazard functionality of the structure and increases maintenance and rehabilitation cost significantly.

1.2 CONTROL OF STRUCTURAL VIBRATIONS

Structural vibrations caused by earthquake or wind can be controlled by various means, such as by modifying rigidities, masses, damping or shape and by providing passive or active counter forces. Vibration control plays a vital role in increasing the safety and performance requirements of structures subjected to dynamic loads. Structural vibrations can be controlled effectively by implementing appropriate control strategies and installing energy-dissipating devices into the structures. The structure then relies on its own strength to withstand dynamic forces, and on the control devices to dissipate energy. Vibration control techniques are mainly categorized as passive, active, semi-active, and hybrid systems. Thus, the structural control systems (also known as earthquake protection systems) passively, actively, or semi-actively suppress the oscillations in the structure. The basic objective of structural control is to reduce response of the structure under dynamic loading, so that: (i) the damage induced to the structure remains minimal and (ii) the vibration limits of desired serviceability criteria are satisfied. The control of the vibration can be achieved by: (i) reducing magnitude input to the structure from ground acceleration through the use of base isolation systems, (ii) dissipation of the energy by providing dampers in the structure, and (iii) avoiding resonance by altering the dynamic properties of the structure by using stiffness-control devices.

DOI: 10.1201/9781315269269-1

Practically, structural control systems have been strongly recommended and applied to, (i) the structures subjected to unusual excitations, such as extreme winds and severe earthquakes, (ii) structures with critical functions and high-safety requirements, e.g., hospitals, community buildings, fire stations, bridges, water tanks, and nuclear power plants; and (iii) structures demanding serviceability considerations, such as towers, tall buildings, long-span span bridges, and other flexible structures with high time period of vibration.

In earthquake engineering, two basic ideas have been established; the first approach requires increased strength of the structure to withstand the earthquake effects, whereas the second approach gains from structural flexibility, which generally results in fewer seismic demands. In the latter approach, the horizontal decoupling of a structure achieved through the insertion of bearings at foundation level transfers it to a lower frequency range, where the seismic energy acting on the structure is beyond that of resonance and dissipates the energy through damping; the technique is known as "base isolation." Seismic base isolation strategy is now broadly accepted for the protection of strategically important and sensitive structures and seismic performance up-gradation of existing structures.

Base isolation systems prevent the transmission of earthquake ground vibrations to the structure; whereas, control devices and control systems apply a control force to serve as an extra damping mechanism by means of devices such as dampers, tendons, and bracings. Active, semi-active, and hybrid control systems are integrated with real-time processing evaluators, controllers, and sensors installed within the structure. They act simultaneously with the hazardous excitation to provide improved structural behavior. Compared to active systems, a semi-active system is less adaptive, because its force-generating capacity is limited by its passive device component.

Merits of Structural Control Systems

The primary benefits of the implementation of structural vibration control strategies are presented as follows.

1. Energy dissipation devices (EDDs) are installed in structures to reduce excessive structural vibrations, increase human comfort, and prevent catastrophic structural failure due to strong winds, severe earthquakes, and such other disturbances.
2. The vibration control systems withstand stronger loadings and severe environments, thereby enhancing structural safety and serviceability and further improve the performance of structures subjected to dynamic loading.
3. Control devices and systems utilize the energy-absorption capability of materials by viscosity and/or nonlinear characteristics, such as yielding.
4. Control systems add damping to the structure and alter its dynamic properties. The time period of the structure is shifted far away from the predominant frequencies of earthquake ground motions.
5. Adding damping increases the structural energy dissipation capacity, and varying structural stiffness avoids resonance to external excitation, consequently, reducing structural response.
6. Control systems can be effectively used to retrofit historic buildings and structures.

Mathematical Modeling of Control Systems

The governing differential equation of a single degree of freedom (SDOF) structure, installed with control systems, under seismic excitation can be expressed as

$$\left(m+m_c\right)\ddot{u}(t)+c\dot{u}(t)+ku(t)+F_c(t)=-\left(m+m_c\right)\ddot{u}_g(t) \tag{1.1}$$

where m_c is the mass of the control device, which is generally negligible compared to the mass of the structure m, and $F_c(t)$ is the control force generated by the control device, which highly depends on the type of the control strategy and the control device. In the case of active and semi-active strategies, $F_c(t)$ modifies the structural properties so that it can respond efficiently to the dynamic loading.

A typical linear model of $F_c(t)$ can be expressed as

$$F_c(t) = c_c\dot{u}(t) + k_c u(t) \tag{1.2}$$

where c_c and k_c, respectively, are the damping constant and the stiffness of the control device. Therefore, Equation (1.1) can be written as

$$(m+m_c)\ddot{u}(t)+(c+c_c)\dot{u}(t)+(k+k_c)u(t)=-(m+m_c)\ddot{u}_g(t) \tag{1.3}$$

The control devices for the selected control strategy are designed to increase the structural damping (c) and to avoid resonance by altering stiffness (k) or mass (m) of the structure, which will result in improved structural response. If the control force primarily adds damping to the structure, much less control effort is required. Altering the system's mass or stiffness requires much more control energy and is less practical than adding damping to large civil engineering structures. Accordingly, the response control systems are designed mainly to add damping to the system, with only minor modifications to the mass or stiffness of the structure.

Passive vibration control systems use passive EDDs. The control force is produced by the motion of the structure. Since c_c and k_c of the passively controlled systems are nonadjustable device properties, the control systems are less adaptable to excitations and thus are referred to as passive.

Active control systems use external power to generate the control force, which is determined by a control algorithm with a measured structure response. The parameters of control algorithm (c_c) and (k_c) can be adjusted by feedback gain within the actuator capacity. Thus, active control systems are adaptive systems that make structures fully smart to environmental excitations. In reality, structures generally have very small damping (~5%), and the resonant component makes major contributions to the response of such lightly damped systems under wide-banded excitation. If the control effort is to alter the system's stiffness or mass, the control force must be approximately equal to the magnitude of the restoring inertia force. Consequently, a large number of force-generating devices are required.

Hybrid control systems commonly refer to a combination of active control system and passive control system or base isolation system. Since a portion of the control objective is accomplished by the passive system, less active control effort is required. This requires a smaller active force-generating device and less power resource.

With constant research and development, significant improvements have been made in structural vibration control technologies. Modeling of control system and developing appropriate control algorithms are done to improve its applicability to large civil structures characterized by severe uncertainties. This is followed by the development of a control device, capable of generating the large force required for seismic response control. Further, experimental characterization by shake table tests is essential or field measurements are conducted to verify the effectiveness of control devices and proposed control algorithms.

Electrorheological (ER) or magnetorheological (MR) materials, piezoelectric layers, shape memory alloys, and optical fiber sensors are being studied and used for various applications. They are used to develop sensors, dampers, and structural members with embedded smart material layers for sensing and actuation. Structural components with smart materials, dampers, and sensors are employed in civil engineering structures such that they are capable of responding spontaneously to seismic excitations in order to minimize undesired effects. For applicability of reliable and robust active, semi-active, and hybrid control systems to full-scale structures; system integration, observer-controller techniques for a sensing system with acceleration measurements, sufficient number of state sensors, and the force-generating capacity of actuators must be ensured. Effectiveness of control system can be maximized by optimally placing the control devices, considering the safety and stability of the overall system.

1.2.1 Response of Structures under Earthquake Excitations

Earthquakes are natural phenomena caused by the internal dynamics of the earth crust and are the most terrifying and deadly events in nature. Most of the world seismicity is concentrated along the plate boundaries (inter-plate earthquakes). However, a significant number of earthquakes, including some large and damaging ones, also occur within the plates (intra-plate earthquakes). The energy released during earthquakes has the capacity of powerfully shaking entire sections of the earth's surface. Earthquakes have a great potential to destroy buildings, nuclear power plants, lifeline structures like bridges, water tanks, hospitals, and infrastructures. Despite the technological advances of the twentieth century, seismic crises continue to cause devastations with thousands of victims and great economic and social losses, originated from poor seismic behavior of structures. This is evident from many earthquakes around the world and those in India, such as 1993 Killari earthquake in Maharashtra and 2001 Bhuj earthquake in Gujarat. Japan is exposed to some of the world's most extreme natural hazards and is considered to be one of the most prepared countries in terms of earthquake and tsunami defense. The March 11, 2011 earthquake occurred on the megathrust at the Japan Trench, which triggered a tsunami over 15-m-high in places along the Sanriku, Miyagi, Joban, and Kanto coasts. The indirect damages of structures during earthquakes are sometimes far greater than the damages due to the earthquake itself, such as outbreak of fire, explosion, rock fall, landslide, avalanches, and tsunamis, i.e., multi-hazards, which further increase the amount of losses. At the same time, the technological development of recent years, sustainable growth, and the maintenance of welfare of the society demand a need for construction of increasingly complex structures. Therefore, design of structures for dynamic loads calls for special attention. In this context, knowledge of the seismic hazard and the study of the behavior of structures subjected to seismic action are of vital importance. Major dynamic actions are caused on structures by both; wind and earthquakes. However, the design for wind forces and earthquake effects is distinctly different. The intuitive philosophy of structural design uses force as the basis, which is consistent in wind design, wherein the building is subjected to a pressure on its exposed surface area. Hence, wind loading is force-type loading. In earthquake resistant design, the structure is subjected to random ground motion, inducing inertia forces in it, which, in turn, build stresses. Thus, earthquake is displacement-type loading. The demand on the building is force, as in force-type loading imposed by wind pressure, and displacement, as in displacement-type loading imposed by earthquake shaking. A major earthquake can lead to a tsunami; a sea wave triggered by large-scale upward movement of the sea floor. Shallow cracks can form during an earthquake or the ground may rupture engulfing structures, and mountainous regions may experience debris avalanches. A closed body of water, like in water tanks, can experience intensive sloshing. Low-density saturated sands of relatively uniform size can suddenly lose their shear strength and capacity to support loads, leading to liquefaction. The damage that a structure can sustain depends on a unique combination of geological and structural factors. The geological factors relate to the nature of soil underneath the foundation, depth to bedrock, ground water table, and land features. Structural factors include degree of confinement of concrete, slenderness of the building and its components, and accuracy of its design and quality of construction. Thus, seismic vulnerability of structures varies as a function of construction materials and type of earthquake resistant system employed. In some cases, a pattern of damage, which is common to different structural members, such as shear failure may occur in reinforced concrete (RC) beams and columns. Local buckling may affect steel beams, columns, and bracings. Services of lifelines and infrastructures, such as transportation systems (highways and railways), bridges, ports, and airports, are vital to the communities and the functioning of urban and industrial regions. Damage to lifelines imposes devastating economic effects on the community.

Response of Buildings under Earthquake Excitations

During the past earthquakes, typically the residential houses of non-engineered construction were damaged. These constructions still dominate in severe seismic zones of India. Primary attributes, which dictate the satisfactory or otherwise performance of a building under earthquake impact, are its strength, stiffness, ductility, and geometric configuration. A building must possess (i) sufficient strength to withstand forces and moments induced by inertial effects under a seismic environment, (ii) adequate stiffness to contain drift within prescribed limits, (iii) appropriate ductile capacity at specified locations to withstand inelastic excursions, and (iv) a sound structural configuration.

Pounding between closely spaced building structures can be a serious hazard in seismically active zones. It can have worse damage, as adjacent buildings with different dynamic characteristics vibrate out of phase, and there can be insufficient separation distance or energy dissipation system to accommodate their relative motions. The highly congested building system in many metropolitan cities constitutes a major concern for seismic pounding damage. Earlier studies on the effect of pounding on structural response have revealed that when floors of two buildings collide, the response of the building with the smaller mass (for equal stiffnesses), and to a lesser extent that of the more rigid one (for equal masses), is amplified, depending on their time period of vibration and mass or stiffness ratios. Unequal height can create problems to the lower building if it is stiffer and may induce large inter-story shear forces in the taller building. Pounding between two buildings can be of two types, (i) slab-to-slab pounding, and (ii) slab-to-column pounding. Pounding effects diminish with increasing separation gap.

The earthquake of Magnitude 6.6 rocked the Uttarkashi region, India, on October 20, 1991. The earthquake caused enormous destruction of rubble masonry residential houses. Telecommunication and power supply were badly affected due to damaged telephone and electric poles. Many school and health buildings were also damaged. The Latur (Killari) Earthquake, India, of Magnitude 6.4 occurred on September 30, 1993. The earthquake caused strong ground shaking in the region of Latur, Osmanabad, Solapur, Gulbarga, and Bidar. There was heavy damage in a localized area of 15 km close to Killari. The random rubble stone houses in mud mortar were totally destroyed. The heavy roofs and thick walls with low shear strength and no tensile strength were the main reasons for the failure. The Bhuj earthquake of Magnitude 6.9 occurred on January 26, 2001 in India, which caused widespread damage to a variety of buildings. For the first time in India, a large number of urban RC frame buildings, including the multistory buildings at Bhuj, Ahmedabad and Gandhinagar, were severely damaged or collapsed. Many newly constructed apartments and office blocks, with and without lifts, collapsed, which were situated far away from the epicenter. Most of these types of construction was of stilt type, i.e., soft story construction, wherein, either very few or no infill walls were provided in the ground floor, as it was utilized for parking. The damage in RC framed buildings was mostly due to failure of infill, or failure of columns and beams, or beam-column joints. The columns damaged by cracking or buckling due to excessive bending combined with gravity load. The columns buckle significantly when they are slender and the spacing of lateral ties in the column is large. Severe cracking occurs near the rigid joints of frame due to shearing action, which may lead to complete collapse. The past experience reveals that the well-designed and well-constructed RC framed buildings, following the design and detailing codes of practice, perform well during earthquakes.

The Bhuj earthquake again showed that stone houses are most vulnerable to earthquakes as it was observed in Uttarkashi, Killari, and Chamoli earthquakes in India. Widespread damage was also observed at the interface of stone or brick masonry infill and RC frame. In most of the cases diagonal cracks appeared in the stone or brick infill.

Response of Bridges under Earthquake Excitations

Recent devastating earthquakes have demonstrated that bridges are the most vulnerable components of the transportation systems. It is observed that the earthquake damages are mainly due to excessive displacement of bridge deck and bearings and due to the shear failure of piers. Bridges are prone to large bearing displacements and deck accelerations, because of their low structural redundancy and small inherent damping. Disruption in the transportation network due to partial or total collapse of bridges after a major earthquake would seriously hamper the relief and rehabilitation operations. Highway crossings are critical infrastructure as they serve as major access and evacuation routes during and after catastrophic events. Seismic design of bridges assumes a great significance since they come under the category of lifeline structures. The Northridge earthquake (January 17, 1994) in the United States caused extensive damage and collapse of several conventionally designed highway bridges. Exactly one year later (January 17, 1995), a devastating earthquake struck the city of Kobe, Japan. This event was a massive disaster that overshadowed the Northridge earthquake. Reported damage to highway bridges included damages to approach slabs, abutments, bearings, pile caps, and decks. The collapse scenarios in both the near-field earthquakes involved severe column flexural and shear failures due to their nonlinear behavior and unseating at in-span and abutment bearings due to excessive displacements. Total collapse or severe damage of bridges was observed in Indian earthquakes, such as the 1993 Latur, the 1997 Jabalpur, the 1999 Chamoli, and the most recent, the 2001 Bhuj earthquake. The vast destruction and economic loss in such earthquakes underscore the importance of finding more rational and substantiated solutions for protection of bridges from severe earthquake attacks.

Response of Liquid Storage Tanks under Earthquake Excitations

Liquid storage tanks normally possess low energy-dissipating capacity than conventional buildings. During lateral seismic excitation, tanks are subjected to hydrodynamic forces. These two aspects are documented by most seismic codes on liquid storage tanks and, accordingly, the provisions specify higher seismic forces than buildings and require modeling of hydrodynamic forces in analysis. Liquid-containing tanks are used in water distribution systems and in industries for storing toxic and flammable liquids. For water supply, most of the villages in the Killari region had RC-elevated water reservoirs, consisting of container supported on moment-resisting frames. The tank capacities ranged from forty thousand to 2 lakh liter. Most of these tanks survived the earthquake with little or no damage in the September 29, 1993, Magnitude 6.4 Killari, Maharashtra Earthquake in India. One tank that suffered a complete collapse came straight down, burying the remains of six supporting columns directly under the bottom dome of the tank. Evidence of a circumferential displacement of about 0.5 m suggests that torsional vibrations were the primary cause of the damage. The main causes of failures of liquid storage tanks due to earthquake ground motion are: (i) buckling of the container shell precipitated by excessive axial compression due to overall bending or beam-like action of the structure; (ii) damage to roof, caused by liquid sloshing; (iii) failure of piping and other accessories, due to their inability to accommodate the flexural deformation of the shells; (iv) damage to the supporting structures, such as stretching of ties, buckling of struts, and tearing, warping, and rupture of gusset plates in steel tanks at end connections; and (v) foundation liquefaction and differential settlement. In the case of shaft-supported elevated tanks, the shell buckling is typically characterized by diamond shape buckling or elephant foot bulging, which appears at a short distance above the base. The large-scale damage to liquid storage tanks during the 1960 Chilean earthquake initiated extensive research on seismic analysis of tanks. Since then, codes of practice have undergone significant changes. The performance of tanks during the 1964 Alaska, the 1973 Hanson, the 1979 Imperial Valley, the 1994 Northridge, and the 2001 Bhuj earthquakes have helped in identifying and improving deficiencies in codes of practices.

Seismic analysis of liquid-containing tanks differs from buildings in two ways: (i) during seismic excitation, liquid inside the tank exerts hydrodynamic forces on tank walls and base, and

(ii) liquid-containing tanks are usually less ductile and have low redundancy as compared to buildings. Traditionally, hydrodynamic forces in a tank-liquid system are evaluated using mechanical analogue in the form of spring-mass system, which simulate the impulsive and convective mode of vibration of a tank-fluid system. Due to low ductility and redundancy, lateral design seismic forces for tanks are usually higher than those for buildings with equivalent dynamic characteristics. This is achieved by specifying lower values of response modification factor. Since tanks have higher utility and damage consequences, codes specify a higher importance factor for liquid-containing tanks, which further increases design seismic forces for tanks.

1.2.2 Response of Structures under Wind Excitations

Wind flow is generated due to atmospheric pressure differentials and manifests itself into various forms, such as, gales and monsoon winds, cyclones/hurricanes/typhoons, tornados, thunderstorms, and localized storms. Speed of strong winds is of about 60–70 km/h; whereas that of gale winds is about 80 km/h and higher. Both types of winds cause serious destruction in the civil engineering structures, usually in tall buildings, bridges, chimneys, and towers. Wind applies lateral pressure along the height of the structure. The design of structures for wind loading is governed by intensity of wind speed, severity of aerodynamic effects, and the dynamic response of the structural systems. For flexible and slender structures, such as sky scrapers and cable-stayed and suspension bridges, consideration of aerodynamic effects in the architectural and structural design is mandatory. For the purpose of structural design, the wind environment is described in meteorological terms, by specifying the types of storm in the region of interest, e.g., hurricanes, thunderstorms, tornadoes, and in micro-meteorological terms, i.e., fluctuations in wind speeds, turbulent flow fluctuations on the surface roughness of structure, and the height of structure. Severe windstorms happen almost everywhere in the world, causing equivalent damages to that of the destructive earthquakes. There are several different phenomena giving rise to dynamic response of structures in wind, such as buffeting, vortex shedding, galloping, and flutter. Slender structures are sensitive to dynamic response, in the direction of wind, as a consequence of turbulence buffeting. Transverse (crosswind) response is more likely to arise from vortex shedding or galloping but may also result from excitation by turbulence buffeting. Flutter is a coupled motion, often being a combination of bending and torsion, and can result in instability of the structure.

Response of Buildings under Wind Excitations

The increasing trend of high-rise construction can be attributed to population growth, scarcity of land, and the consequent increase in land prices, particularly in metropolitan cities. The past few decades have witnessed a tremendous growth of tall buildings all over the world, with diversity in shapes and forms, due to the demand for aesthetics and elegant architectural design. Asymmetrical shape of buildings makes them extremely sensitive to lateral loads like earthquake and wind. For buildings, flutter and galloping are generally not critical. An important problem associated with wind-induced motion of buildings is concerned with human response to wind-induced vibrations and perception of motion of structure under wind excitation. Therefore, for tall buildings, limit state of serviceability generally governs the design compared to the limit state of strength. The characteristics of wind pressure acting on a structure are a function of features of the approaching wind, geometry and shape of the structure, and the geometry and proximity of the structures on the windward side of the structure under consideration. Wind pressures are highly fluctuating and vary with location, partly as a result of the gustiness of the wind, and also because of local vortex shedding at the sharp corners and boundaries of the structure. For dynamically wind sensitive structures, the frequent fluctuating pressures can further result in fatigue damage.

Control of deflection and drift is imperative for tall buildings for limiting damage and cracking of nonstructural members such as facades, internal partitions, and false ceilings. An additional norm that requires careful consideration in wind-sensitive tall buildings is the control of sway accelerations under serviceability conditions. Acceptance criteria for vibrations in buildings are frequently expressed in terms of acceleration limits for a one-year/five-year return period wind speed. They are based on human tolerance to vibration discomfort in the upper stories of buildings. These limits also depend on the sway frequencies of structure. Wind response is relatively sensitive to both mass and stiffness, and structural accelerations can be reduced by increasing either or both of the parameters. However, this is in conflict with earthquake design optimization, where loads are minimized in buildings by reducing both the mass and stiffness. Increasing the damping in structures results in reduction in the wind and earthquake responses. The general design requirements for structural strength and serviceability are particularly important in the case of tall building design, because significant dynamic response can result from buffeting and crosswind excitations. Serviceability with respect to perception of residents to the lateral vibration response can become the governing design criteria, necessitating the implementation of damping systems in order to reduce these vibrations to tolerable levels. The economics of constructing tall buildings are greatly affected by wind as their height increases. To counteract wind loads and keep building motions within comfortable limits require robust structural control systems that increase the cost of the project. The acting loads and structural motions are often subjected to dynamic amplification in the along-wind and cross-wind directions. These effects largely depend on shape of the structure. Hence the aerodynamics of shape is considered at a very early stage in the design of sky scrapers. Wind loads on the curtain walls tend to increase with height, primarily due to the fact that wind speeds increase with the height of structures.

The approach of adding supplementary damping to the structure is beneficial, as it reduces the structural motions by increasing the ability of the structure to dissipate energy. A wide range of supplementary damping systems are implemented for wind response mitigation. They are broadly categorized as distributed damping systems and mass damping systems. Distributed damping systems are located in regions of highest relative movement between the structural elements. Therefore, their locations highly depend upon the type of structural system. The most effective location for a mass damping system, such as tuned sloshing damper (TSD), tuned liquid column damper (TLCD), tuned mass damper (TMD), or active mass damper (AMD), is close to the location of peak amplitude of vibration. Therefore, they are typically located adjacent to the top of the structure. The optimum choice of supplementary damping systems for any given project depends on a variety of factors, such as the amount of damping required, space consideration at their proposed locations, the number of modes of vibration to be considered in the analysis, and accessibility for maintenance.

Response of Bridges under Wind Excitations

Suspension bridges and cable-stayed bridges are very sensitive to dynamic wind loads, because their decks are very thin compared to the heavy decks of conventional prestressed concrete bridges. The well-known structural collapse due to wind is that of the Tacoma Narrows Bridge, USA, in 1940. It collapsed just after four months of service due to aeroelastic flutter. Within a short duration after its construction, it was found to buckle dangerously and sway along its length in windy conditions. Even with the normal winds, the bridge was undergoing noticeable undulations. On the day of collapse, the bridge experienced winds of about 70 km/h. The center stay was torsionally vibrating at a frequency of 36 cycles/min. Over the next hour, the torsional vibration amplitude built up and the motion changed from rhythmically rising and falling to a two-wave twisting. Despite all these motions, the center part of the bridge along the length remained stationary, while its other two portions on each side twisted in opposite directions. Due to alternate sagging and hogging of span members, the towers holding them were pulled toward them. Further, visible and predominant cracks developed before the entire bridge crashed down into the river. Further investigations showed that the bridge collapsed due to a coupled torsional and flexural mode of oscillation, which caused warping of the thin deck.

Difference between Seismic and Wind Excitations

Wind loading at all times competes with the seismic excitation while considering the load combinations in design of structures. Table 1.1 presents major differences between seismic and wind loading.

TABLE 1.1
Difference between Seismic and Wind Excitations

Seismic Excitation	Wind Load
The loading imposed on structure by shaking due to earthquake ground motion is of displacement-type.	The loading imposed by wind on a structure is of force-type.
When a structure is subjected to earthquake action, it is said to be base-excited.	When a structure is subjected to wind forces, it is said to be mass-excited.
The motion of a structure during earthquake is cyclic about its neutral position. The stresses in the structure undergo a large number of complete reversals during the small duration of earthquake.	Reversal of stresses occurs only when the direction of wind reverses, which happens only over a large duration of time.
Seismic force depends on the mass of the structure and its distribution. The force acts at the center of mass at each floor level.	Wind force depends on the exposed surface area of the structure.
Seismic force is distributed among the interior and exterior frames.	Wind force acts primarily on the exposed (exterior) frames and reduces to interior frames due to shielding effect.
The performance of a lighter structure during earthquake is better, as it attracts lesser forces. The structural performance is not influenced by the exposed area.	A structure with higher mass resists the wind load effectively. The structure with lesser exposed surface area performs better, as it attracts smaller wind force.
The shear force is maximum at the base. The story shear decreases toward the top, as height of structure increases. This is due to reduction in cumulative seismic weight at the upper story levels.	The wind force increases as the height of structure increases, for the same exposed area.
The inertia is the main factor that induces seismic forces in a structure.	Inertia has negligible impact in the generation of wind forces.
The soil type on which the structure stands affects its performance during earthquake.	The soil type does not affect the performance of structures during wind.
The performance of a structure during earthquake can be improved by providing base isolators, which prevent the transmission of seismic forces to the structure.	The performance of a structure to wind loads can be improved by altering the shape of the structure; by providing curved boundaries.
Seismic events do not develop any suction effect on the structure.	During wind excitations, negative pressure can act on a structure due to suction effect.
The story displacements are large at upper floors. The displacement profile is parabolic.	The story displacements at upper floors are generally small, compared to seismic forces. The displacement profile is linear.
Earthquake excitations can be artificially generated in laboratories on a shake table.	Wind excitations can be artificially generated in wind tunnel tests.
Seismic forces depend on the focus of the earthquake and soil profiles through which the seismic waves travel.	Wind forces depend on terrain category and topography of location of the structure.
Prediction of earthquakes is only probabilistic; they occur without any warning. Duration of earthquakes varies from a few seconds to a few minutes.	The formation of windstorms can be predicted accurately. The public is warned well in advance before the storms/cyclones hit. The duration of extreme wind varies from a few minutes to even hours (cyclones).
The area affected by occurrence of earthquakes is very large; in km^2.	The area affected by wind is comparatively low, except in case of cyclones.

1.2.3 Response of Structures under Wave Excitations

Waves are random in nature and their motion is unsteady. They get scattered depending on the size of the structure hit by them. They are nonlinear with complex interactions between frequency components, and their spectra are broad-banded. Wave loading is considered to be the most important type of all the environmental loads to be dealt with in the design and operation of offshore structures and bridges near coastal regions. The loads due to waves very often dominate and dictate the design of these structures. Interaction of surface waves with offshore structures and bridge decks is a complex problem, involving fluid-structure interaction, wave breaking, and overtopping. In hostile areas and deeper waters, the significant wave height can frequently exceed 2 m. The most likely largest wave height in 100 years can be more than 30 m. The mean wave period can vary between 15 and 20 s in extreme weather situations. Wind-generated waves consist of a large number of wavelets of different heights, time periods, and directions, superimposed on one another. The interactive effect between the loads due to water current and wind are significant. Determination of wave-induced forces requires solution of two separate, but interrelated problems. The sea state is essentially computed using an idealization of the wave surface profile and the wave kinematics. For structures in deeper waters, because of their compliancy and the flexibility introduced in the design by geometry and structural design, dynamic effects on the structural members becomes predominantly important. Tsunami, storm waves and storm surge are known to be the ultimate agents of such failures. Under severe conditions, such as storms or hurricanes, water level may rise due to storm surge, allowing for larger waves to approach the offshore structures and piers and decks of coastal bridges. The storm surge and approaching waves may ultimately become large enough to cause banging on these structures. An understanding of the failure mechanism and possible solutions requires thorough knowledge of the destructive loads on the structure. In recent years, remarkable progress has been made on this subject, resulting in some new findings about the failure mechanism and the destructive wave loads.

Response of Offshore Platforms to Wave Excitations

Offshore structures are used for multipurpose operations like oil exploration, drilling, and production. These structures have advanced from very stiff and relatively shallow water structures in the past, to very flexible deep-water structures in the recent years. Offshore structural members are normally designed for significant wave height. Floating structures are influenced by major disturbing forces like waves and wind, while the restoring force is provided by gravity as variable buoyancy. In order to tap untouched resources of oil in deep waters or in hostile environment, there is a growing need for improving the design and construction practice of drilling and production platforms with special prominence on their safety. It is essential that these platforms can withstand the action of dynamic forces due to wind, waves, and earthquakes, for normal operating and extreme conditions. Conventional gravity type platforms in deeper water approach natural frequencies that are within the range of occurrence of everyday wave heights resulting in excessive fatigue-related damage. There are many flexible vertical members in offshore platforms like mooring lines and risers that are exposed to waves. With statistical methods, most probable maximum forces during the lifetime of the structure are calculated using linear wave theory. Statistical approach is used in analyzing the fatigue strength of members and to estimate the dynamic response of offshore platforms.

Response of Coastal Bridges to Wave Excitations

Recent natural extreme events, such as Hurricane Ike in the US (2008), Tohoku tsunami in Japan (2011), Typhoon Haiyan in Southeast Asia (2013), and cyclones in India, Cyclone Phailin (2013, wind speed 215 km/h), Cyclone Gaja (2018, wind speed 280 km/h), Cyclone Fani (2019, wind speed 215 km/h), and Cyclone Amphan (2020, wind speed 240 km/h), have caused significant damage to decks of coastal bridges and other structures in coastal region. The failure of the structure occurs when wave-induced loads on decks of coastal bridges exceed the bridge capacity, resulting in partial removal/dislocation or a complete collapse of bridge decks. Possible submergence of the deck and entrapment of air pockets between girders can increase destructive forces and increase the complexities of the problem. Coastal bridges, located near sea shore in shallow waters, are typically designed to avoid interaction of water waves with bridge decks under normal conditions. The decks are usually held in position by their self-weight and structural connections, generating vertical and horizontal resisting forces against external loads.

1.2.4 Need for Structural Response Control

A structure is engineered to perform its functions to withstand and transmit static and dynamic loads acting on it without experiencing any permanent damage during its designed service life. A typical design of a structure must ensure safety (strength and stability) and serviceability (stiffness) of the structure as a whole, and its components, under predicted static and dynamic loads. Civil engineering structures, such as buildings, bridges, water tanks, and towers, may vibrate severely or even collapse when subjected to strong wind or earthquake excitations. Designing earthquake-resistant structures has always remained a challenge for structural engineers. Despite intensive effort toward wind- and earthquake-resistant designs in code development and construction, structures are found to be still vulnerable to strong winds or earthquake excitations.

The conventional approach is to design structures with sufficient strength to withstand loads and the ability to deform in a ductile manner. However, the conventionally designed structures have inadequate capacity to resist dynamic loads and to deform in a ductile manner, because these structures are lightly damped, i.e., they rely on their inherently small (~5%) material damping to dissipate dynamic energy. Traditionally designed structures completely depend on their own stiffness to resist dynamic forces. These structures are passive and are not able to adapt to frequently changing, random wind and earthquake excitations. For a structure to withstand stronger excitations, an increase in its strength and ductility is required. Increasing strength by providing larger cross-sections of selected members of an indeterminate structure essentially attracts more demand force on these members, subsequently requiring even greater strength. Thus, providing larger cross sections is an ineffective approach. Furthermore, inherent damping for common construction materials, such as RC and steel, cannot be increased in any way.

Recent devastating earthquakes around the world have underscored the tremendous importance of understanding the way in which civil engineering structures respond during such dynamic events. In addition to earthquakes, strong winds can also result in unprecedented devastation along their path. These natural events clearly demonstrate how vulnerable the structures are to the environmental forces. Today, one of the main challenges in structural engineering is to develop innovative design concepts to protect civil structures in a better way, including their material contents and human occupants, from the hazards of strong wind and earthquakes. Conventionally, structures have been designed to resist natural hazards through a combination of strength, deformability, and energy absorption. The prevention of the devastating effects resulting from these natural hazards is frequently attained by allowing certain structural damage. Alternatively, some types of structural protective systems may be implemented to mitigate the damaging effects of the environmental forces. The control systems work by absorbing or reflecting a portion of the input energy that would otherwise be transmitted to the structure itself.

Advanced research efforts to make improvements in the conventional design approach lead to the discovery of smart materials and adaptive control systems that can automatically adjust themselves to environmental changes. When systems with smart materials and control devices are installed in a structure, it can monitor itself and adjust to the environment. Structure installed with smart devices has the ability to sense any change in the environment or system, diagnose any problem at critical locations, store and process measured data, and command appropriate action to improve its performance. In addition, it also has ability to preserve structural integrity, safety, and serviceability. With smart structure technology, control devices and control systems can be added to a structure to increase its seismic and wind-resistant capacity. The structure then relies on its own strength to withstand dynamic forces and on the control devices or control systems for energy dissipation.

Smart structure technology is gaining popularity as an attractive alternative to enhance structural safety and serviceability, as it can significantly improve the performance of structural systems subjected to dynamic loads. When constructing a new structure or retrofitting an existing one for safety and serviceability requirements, implementing smart structure systems is a highly recommended option. Currently, vibration control systems have been applied to (i) the structures, subjected to unusual excitations, such as extreme winds or severe earthquakes, (ii) the structures with critical functions and high-safety requirements, such as hospitals, fire stations, and nuclear power plants, (iii) the structures requiring serviceability considerations, such as towers, elevated water reservoirs, tall buildings, long-span roofs, bridges, and flexible structures.

Response control systems are classified into passive control, active control, semi-active control, and hybrid control. Structural damping can be increased by introducing either passive or active or semi-active control devices in structures. Although passive supplemental damping has been widely used, there are several advantages in the incorporation of active and semi-active systems for controlling response of structures. An active and semi-active control systems can respond to several modal frequencies, while a frequency-dependent passive control system is designed only for the predominant modal frequency of the structure. Additionally, active and semi-active control strategies may help to meet different performance requirements to increase serviceability levels under strong winds and, simultaneously, avoid structural damage under large earthquakes. In Japan, there has been rapid progress in research and development in the construction of mechanical industries and at universities, because when active control devices are installed, the structure is expected to exceed the performance limitations of conventional earthquake-resistant structures. As a result, since 1989, several active and semi-active control systems have already been applied to actual buildings in Japan. The 1995 Kobe earthquake opened a door to positive development of semi-actively controlled structures against large earthquakes. Semi-active control research affects both conventional earthquake-resistant structures and passively controlled structures. Active, semi-active and hybrid structural control systems are a natural evolution of passive control technologies. The possible use of active control systems and some combinations of passive and active systems as a means of structural protection against dynamic loads has received considerable attention in the recent years. Active, semi-active, and hybrid control systems are force-delivery devices integrated with real-time processing evaluators/controllers and sensors within the structure. They act simultaneously with the hazardous excitation to provide enhanced structural behavior for improved service and safety. Research to date has also reached the stage where active control systems have been installed in full-scale structures for seismic hazard mitigation.

Base isolation systems cut off the energy transmission of earthquake ground motions to the structure. Control devices or systems apply a control force to serve as an extra damping mechanism by means of devices such as mass dampers, drivers, tendons, and bracings. These systems utilize the energy absorption capability of materials by viscosity and/or nonlinear characteristics, such as yielding, to shift the frequency of the structure away from the predominant frequency of earthquake ground motions. A smart structure can use either base isolation devices or a control system, or a combination of both, to safeguard it from dynamic excitations.

Control systems add damping to the structure and alter its dynamic properties. Adding damping increases the energy dissipation capacity of structure, and altering structural stiffness can avoid resonance to external excitation. If the control objective is to alter the mass or stiffness of the structure, the control force must be approximately equal to the magnitude of the restoring (inertia) force. This approach requires employment of a large number of control devices in the structure. However, if the control force primarily adds damping to the system, the control effort required is significantly less. For huge civil engineering structures, adding damping to the system is more practical than altering its properties. Thus, seismic and wind response control systems are usually designed to add damping to the system, with only minor modifications to its mass or stiffness.

1.3 CLASSIFICATION OF STRUCTURAL VIBRATION CONTROL STRATEGIES

Maintaining the structural integrity becomes particularly important when the structures are subjected to dynamic loading. The structural control techniques have been believed to be one of the most promising technologies for earthquake- and wind-resistant design of structures. Several types of structural control theories and devices have been researched, developed and introduced to large-scale civil engineering structures to effectively mitigate the large structural responses and control the undesirable vibrations of structures in earthquake- and wind-prone regions. The structural control systems are categorized as passive, active, semi-active, and hybrid, based on their working principle, operating strategy, dissipative nature, energy-absorbing capacity, and reliability. These systems passively, actively, or semi-actively suppress the vibrations in structures. The additional protective systems work by absorbing or dissipating a substantial part of the input energy that would otherwise be transmitted to the structure itself, thereby reducing the structural response considerably. Figure 1.1 presents different structural control systems.

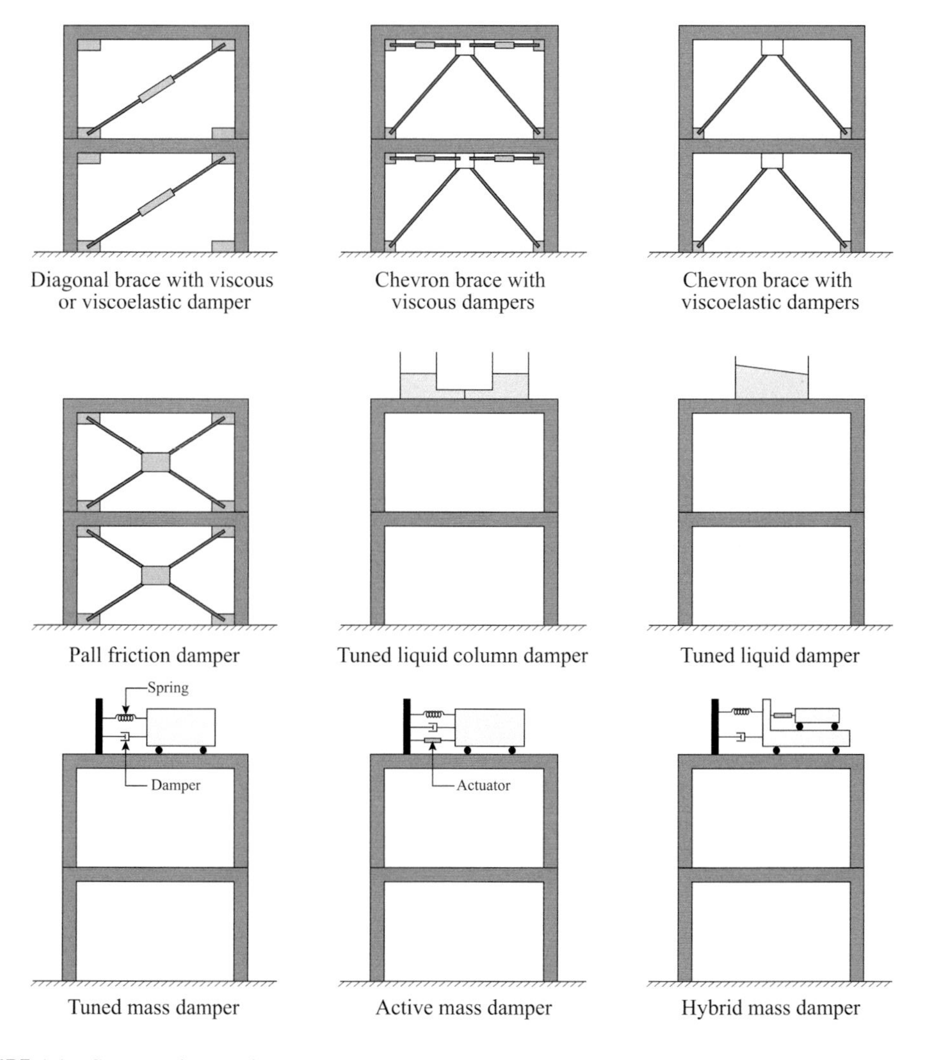

FIGURE 1.1 Structural control systems.

1.3.1 Passive Vibration Control Strategies

Passive vibration control strategies help in suppressing the vibration of structures by: (i) dissipating the energy from the excitation, e.g., viscous damper, (ii) altering the properties of the structure, e.g., base isolation, and (iii) counterbalancing the vibration of the structure, e.g., TMD. Passive control devices dissipate a significant amount of the seismic input energy and thus reduce the demand on the structure. Depending on the construction features, they also increase the stiffness and strength of the structure to which they are attached. The devices utilize the motion of the structure for developing control force. The control force is developed at the location of the device itself, by utilizing the movement of the structure. The amount of control force developed is a function of response (displacement or velocity) of the structure to dynamic forces at that location. Since their stiffness and damping are nonadjustable device properties, the systems are not adaptable to excitations and thus are referred to as "passive." The control is achieved solely by the action of the stiffness of components embedded within the system and damping. In general, such systems are characterized by their capability to enhance energy dissipation in the structural systems to which they are installed.

1.3.1.1 State-of-the-Art Review

For the protection of structures against environmental disturbances such as seismic, wind excitation, and human-made disturbances like traffic or heavy machinery loads, several passive control systems are currently in practice. The term "passive" indicates the inertness of these devices to the external power source, as these devices function independently without any external power supply or input energy. Passive EDDs generally operate on principles such as frictional sliding, yielding of metals, phase transformation in metals, viscoelastic (VE) action in rubber-like materials, shearing of viscous fluid, orificing of fluid, and sliding friction. Passive control systems are the simplest kind of structural control systems used to control the large deformations of structure against the extreme natural hazards. Examples of the primary passive control systems include seismic (base) isolation systems, steel bracings, viscous dampers, viscoelastic dampers (VEDs), friction dampers (FD), TMDs, and mechanical energy dissipaters. Major developments in the theory, design, specification, and installation of these systems have permitted significant applications of these devices to buildings, bridges, industrial plants, nuclear power plants, and other structures. Applications are almost in all of the seismically active countries of the world, however, principally in Italy, Japan, New Zealand, and the United States. Noteworthy advantages have been demonstrated when retrofitting existing structures and designing high-performance and lifeline structures such as hospitals, emergency response facilities, defense installations, and critical bridges.

Based on the performance of energy dissipation or absorption, the passive control devices are classified as:

a. Energy Dissipating Devices
    i. Fluid viscous damper (FVD)
    ii. Viscoelastic damper (VED)
    iii. Friction damper (FD)
b. Base isolation systems
    i. Laminated rubber bearing
    ii. Lead rubber bearing (N–Z system)
    iii. High-damping rubber bearing (HDRB)
    iv. Pure-friction system (PF)
    v. Friction pendulum system (FPS)
    vi. Resilient-friction base isolator
    vii. Electric de France system
    viii. Cylindrical rod
    ix. Elliptical rolling rod
c. Tuned Mass Damper (TMD)

1.3.1.2 Salient Features, Advantages, and Limitations

The basic features of the passive control system are (i) it increases the capacity of energy dissipation of the structure, (ii) it does not require any external power supply, and (iii) there is no adaptability to various external loads.

Basic Principles of Passive Control

Consider an SDOF structural model, consisting of a mass m, supported by springs with total linear elastic stiffness k, and a damper with damping coefficient c. This system is then subjected to an earthquake ground acceleration $\ddot{u}_g(t)$. The excited model responds with a lateral displacement $u(t)$ relative to the ground. The system responds to the equation of motion

$$m\ddot{u}(t)+c\dot{u}(t)+ku(t)=-m\ddot{u}_g(t). \tag{1.4}$$

When a generic passive energy dissipation (PED) element is added into the SDOF model, as shown in Figure 1.2, the equation of motion for the passively controlled SDOF model becomes

$$m\ddot{u}(t)+c\dot{u}(t)+ku(t)+\Gamma u(t)=-\left(m+\bar{m}\right)\ddot{u}_g(t) \tag{1.5}$$

where $\bar{m}$ is the mass of the PED element and $\Gamma u(t)$ is the force corresponding to the added device. Γ represents a generic integrodifferential operator. The specific form of $\Gamma u(t)$ is highly dependent upon the type of PED controller used, which must be known for solving Equation (1.5). The structural behavior is modified by addition of the term $\Gamma u(t)$, so that it can respond more favorably to the earthquake ground motion. The system properties (m and k) remain constant as the load is being applied. A passive structure with added PED elements is also a passive structure.

Figure 1.3 shows the block diagram of a passive control system. Typically, there are two types of PED systems: (i) displacement dependent systems, e.g., Metallic-yielding and FDs, and (ii) velocity-dependent systems, e.g., FVDs and VEDs.

Advantages of Passive Control

Passive control systems encompass a range of materials and devices for enhancing damping, stiffness and strength and can be used for both natural hazard mitigation and rehabilitation of aging or deficient structures. A fully passive control is simple to implement, since it involves only providing

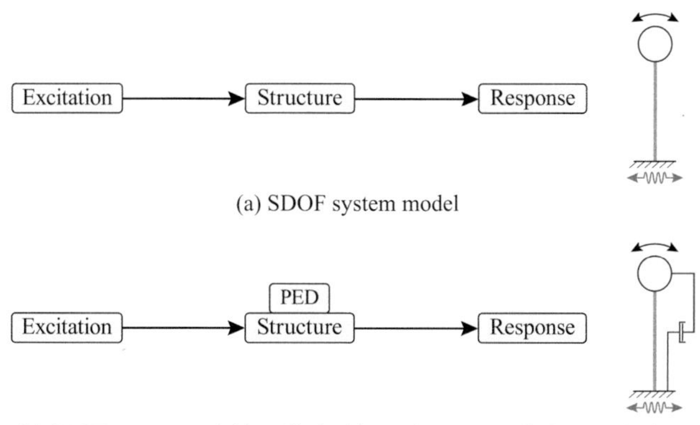

FIGURE 1.2 Passive control of SDOF system.

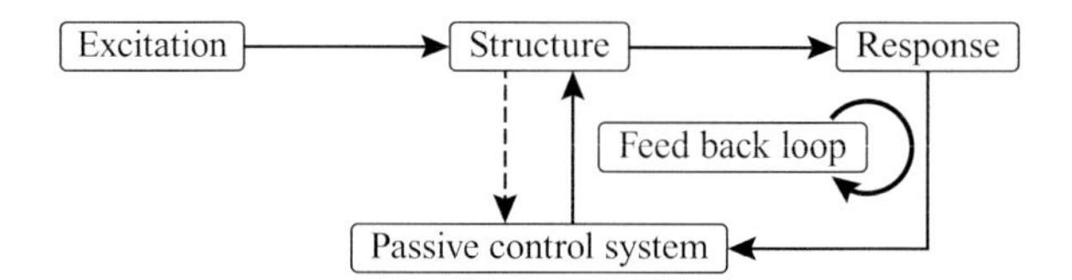

FIGURE 1.3 Block diagram of passive control system.

stiffness and damping, without adding any energy to the structural system. The systems are popular worldwide, because of their following noteworthy advantages.

- A passive control system requires no energy input or interaction with outside source.
- It has energy dissipation capabilities, thereby reducing the level of energy that can be transmitted to the structure.
- The base shear as well as the displacement response of structure is improved.
- The devices are relatively easy to fabricate, easy to install, lightweight, and inexpensive.
- Performance-based design of structures using the passive control techniques can effectively mitigate the loss of life and property, under an earthquake event.
- The devices are complementary for new constructions as well as retrofitting applications and post-earthquake rehabilitation works.

Limitations of Passive Control

Passive control methods are strongly dependent to exact setting and must be specifically designed for each structure, as these devices are incapable of adapting to structural changes and operative parameters. The dependence of passive devices on the local information to develop control force, and their inability to modify their own structural properties, reduce their effectiveness. Furthermore, base isolation systems are less effective for high-rise structures because tall structures typically have longer time period.

Passive control strategies typically include base isolation and supplemental dampers. The different types of base isolation systems include elastomeric bearings, lead rubber bearings (LRB), and high damping rubber bearings (HDRBs). Friction pendulum system (FPS), variable friction pendulum system (VFPS), variable curvature friction pendulum system (VCFPS), and variable frequency pendulum isolator (VFPI) are the examples of sliding friction isolators. Dampers inherently have a few advantages over the base isolation technique as they are easy to install and maintain. In a multistory structure, a multilevel control strategy controls the response of the structure more efficiently. This objective can be achieved with the help of dampers. The different types of dampers used for energy dissipation are metallic yield dampers, friction dampers, viscous fluid dampers, viscoelastic dampers, tuned mass dampers, and tuned liquid dampers. The following sections deal with different types of major passive control systems used globally.

1.3.1.3 Base Isolation

Base isolation is a well-established application of the passive control strategy, which works on the principle of period shift and enhancement of energy dissipation. It increases flexibility of structures, attracts smaller seismic force, and increases structural damping. A structure mounted on a material with low lateral stiffness, such as rubber, achieves a flexible base so as to increase its natural period by increasing the superstructure flexibility. The increase in flexibility typically results in the deflection of a major portion of the earthquake energy, reducing accelerations in the superstructure while increasing the displacement across the isolation level. Thus, during an earthquake, the flexible base filters out high frequencies from the ground motion and prevents the structure from being damaged or from collapsing. Therefore, base isolation is an effective tool for providing seismic protection for low-rise and mid-rise buildings because these types of buildings are characterized as having high frequencies. The design specifications for base isolation have been officially established in many international codes, such as, ASCE 7-16, EN 1998-1, FEMA P-1050, and IRC: SP:114.

Base isolators are generally classified as one of the two major types, elastomeric-type and sliding-type isolation bearings. Elastomeric-type bearings are typically composed of rubber and steel plates, while sliding-type bearings rely upon friction between specially treated surfaces of an assembly unit. Some types of isolation bearings combine characteristics of elastomeric and sliding bearings, but these two basic types are dominantly used in building structures. Several applications of seismic isolation systems have been reported worldwide. Application of base isolation to structures commenced in the 1960s. Many successful examples of installation and performance of base isolation systems are found in the United States. Since the first base isolated building, the Foothill Community Law and Justice Center, located at Rancho Cucamonga, California, was constructed in the United States, base isolation technology has been successfully utilized to isolate different types of building lateral-force-resisting systems, such as steel braced frames, RC shear walls, reinforced or unreinforced masonry walls, and even the use of elastomeric isolators, made from scrap automobile tire rubber pads, for low-cost housings. In Japan, the West Building of the Ministry of Post and Telecommunications and the Research Building of the Matsumura Construction Company showed excellent performance during the 1995 Kobe earthquake. In India, the devastating Bhuj earthquake, in 2001, killed over 175 people in the local hospital in Bhuj (420 km west of state capital Gandhinagar). After this incident, the new 300-bed hospital, employed with lead rubber (NZ) base isolation systems was constructed in 2003.

1.3.1.4 Bracing System

Bracing is a very effective global upgrading strategy to enhance the global stiffness and strength of steel frames and RC-steel composite frames. It can increase the energy absorption capacity of structures and/or decrease the demand imposed by earthquake-induced forces. Structures with increased energy dissipation may safely resist forces and deformations caused by strong ground motions. Generally, global modifications to the structural system are conceived such that the design demands, often denoted by target displacement, on the existing structural and nonstructural components are less than their capacities. Lower demands may reduce the risk of brittle failures in the structure and/or avoid the interruption of its functionality. The attainment of global structural ductility is achieved within the design capacity by forcing inelasticity to occur within dissipative zones and ensuring that all other members and connections behave linearly. Sometimes braces can be aesthetically unpleasant where they alter the original architectural features of the building. In addition, they transmit very high forces to the structural connections and foundations. Therefore, the connections and foundations are required to be strengthened frequently. Several configurations of braced frames may be used for seismic strengthening. Figure 1.1 shows the commonly employed chevron bracing system. The other most common types are concentric braced frames (CBFs), eccentric braced frames (EBFs), and knee-braced frames. Common configurations for CBFs encompass V and inverted-V bracings and K, X, and diagonal bracings. However, V bracings are not advised for seismic retrofitting because of the possibility of damage in the midspan of beam. Under horizontal forces, the compressed braces may buckle, thus reducing their load-bearing capacity abruptly.

1.3.1.5 Viscous Damper

Viscous damper utilizes the fluid viscous properties to absorb seismic energy. It is a velocity-dependent device and operates on the principle of fluid flow through an orifice. The use of viscous fluid for shock and vibration mitigation is familiar to heavy industry and the military. A straightforward design is achieved with classical dashpot and dissipation occurs by converting kinetic energy to heat as a piston moves and deforms a thick, highly viscous fluid. As shown in Figure 1.4, the response acceleration decreases, with increase in damping provided by FVD.

1.3.1.6 Viscoelastic Damper

VEDs utilize the action of solids to enhance the performance of structures subjected to environmental excitations. The dampers utilize high damping from VE materials to dissipate energy through shear deformation. Such materials include rubber, polymers, and glassy substances. A typical VE damper consists of layers of VE materials bonded to steel plates as shown in Figure 1.5. Shear

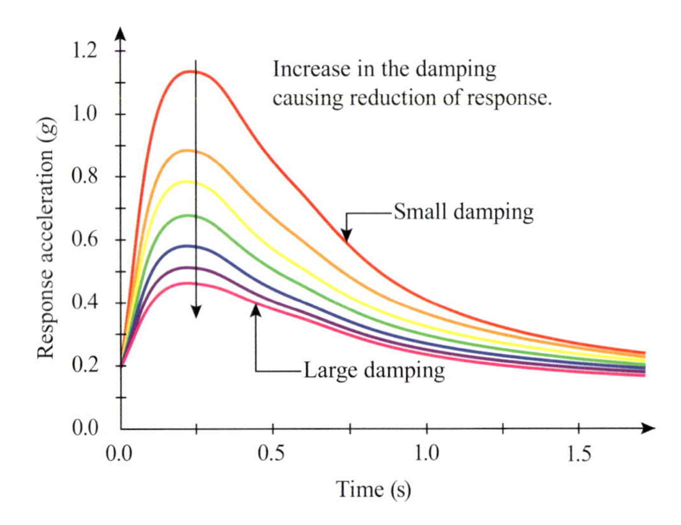

FIGURE 1.4 Effect of damping on response acceleration.

deformation occurs, and energy is dissipated when the structural vibration induces relative motion between the outer steel flanges and the center plate.

1.3.1.7 Mass Damper

In tuned mass, the frequency of the damper is tuned to a specific frequency of the structure so that when that frequency is excited, the damper resonates out of phase with the structural motion. Typically, springs and dampers are inserted between the TMD mass and the structure, and this system exerts out-of-phase lateral forces that counteract the vibration response of the parent structure.

TMDs are generally installed at the top floor of buildings to control the responses of building to natural excitations. TMDs may also be installed in other structures, such as, flexible bridges to control vibrations induced due to wind and traffic. The tuning frequency of TMD is the most important parameter; i.e., the TMD is designed such that its frequency is set equal to the fundamental frequency of the structure. TMDs, in their simplest form, consist of an auxiliary mass m_d, spring k_d, and dashpot c_d. The TMD system is attached to the main structure, usually on the top of the structure as shown in Figure 1.6(a). Figure 1.6(b) presents the basic mechanism of an undamped dynamic vibration absorber, which comprises a small mass m_d, and a spring with stiffness k_d. The absorber is attached to the main structure, with mass m_s and spring stiffness k_s.

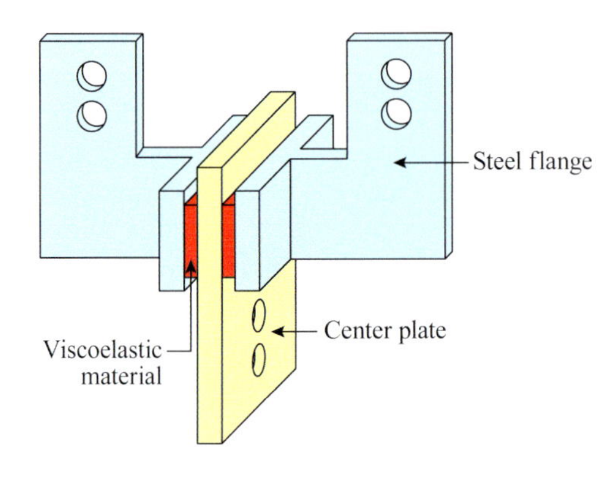

FIGURE 1.5 Schematic representation of viscoelastic damper.

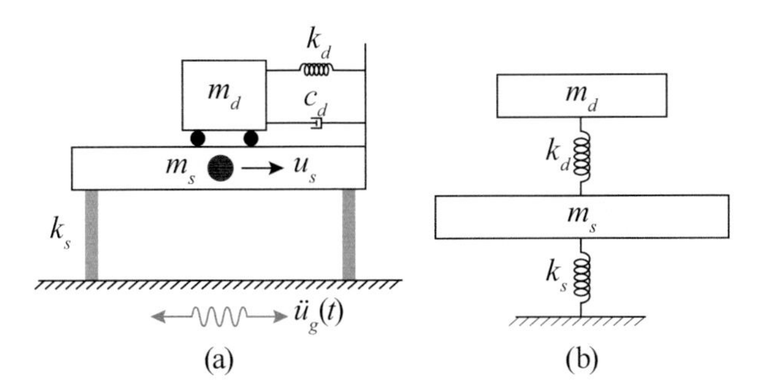

FIGURE 1.6 Tuned mass damper: (a) tuned mass damper installed on structure and (b) undamped dynamic vibration absorber.

1.3.1.8 Friction Damper

Friction is an effective, reliable, economical, and widely applied mechanism to dissipate kinetic energy by converting it to heat. The friction mechanism for energy dissipation is utilized to develop dampers for structural vibration suppression. To achieve this required friction, the damper must have two solid bodies that slide relative to each other. FD works on the principle of sliding of plates on each other and dissipates the energy as heat. Schematic of Pall FD is shown in Figure 1.1.

The working of each damper can be represented by a typical force-deformation loop. Figure 1.7 shows schematic representation of the force-deformation behavior for a viscous damper, viscoelastic damper (VED), and friction damper (FD).

1.3.1.9 Metallic Damper

Inelastic deformation of metallic materials is an effective mechanism for energy dissipation. The traditional seismic-resistant design of structures depends on post-yield ductility of structural members to dissipate earthquake input energy. This concept led to the idea of installing separate metallic hysteretic devices in a structure to absorb seismic energy. Several kinds of metallic yield devices have been developed and implemented.

1.3.1.10 Re-Centering Devices

Re-centering is the capability of any device to return to its original position and configuration after the occurrence of dynamic event. It is the most important aspect for the devices, which possess low restoring capability. Lack of re-centering capability may lead to major damages and collapse of structures, during strong earthquake ground motions. On removal of the applied load, residual deformation of re-centering devices is insignificant, and they retain their indigenous configuration. Devices that fall under re-centering capability group are pressurized fluid damper, preloaded spring friction damper (FD), self-centering seismic-resistant system, and shape memory alloy damper.

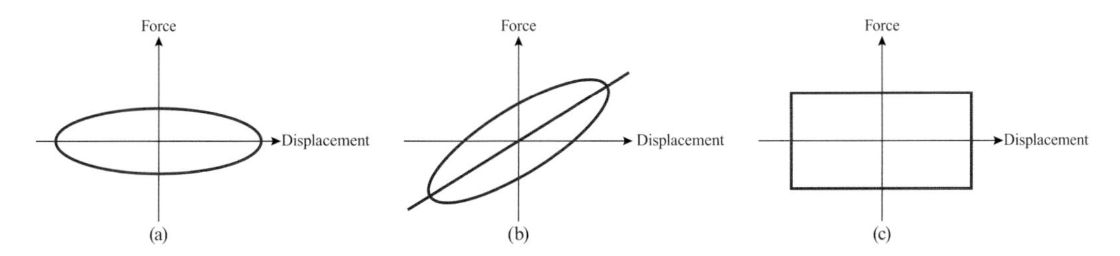

FIGURE 1.7 Schematic representations of the force-deformation behaviors of dampers: (a) viscous damper, (b) viscoelastic damper, and (c) friction damper.

1.3.1.11 Wire Rope Isolator

Wire rope isolators (WRIs) are typically used for shock and vibration isolation in civil constructions and military industry. WRI, also known as wire rope springs or metal cable springs, is used to isolate machinery and equipment from external shocks and vibrations. They are basically type of spring dampers, consisting of stranded stainless-steel wire ropes held between rugged metal retainer bars. The spring function is provided by the elasticity, inherent in the flexing of the cable, made into a loop. The damping function is the result of the relative friction between the individual wires of cables and strands.

1.3.2 Active Control Strategies

Active control strategies employ actuators that enable application of variable control action and introduction of additional control force. Therefore, active control strategies have considerable potential, particularly for structural applications where weight is a major concern, e.g., aerospace vehicles and long-span structures, such as bridges. However, active control strategies require significant energy to operate and may prove to be relatively less reliable as compared to passive and semi-active strategies. The first step for implementing active control is to decide the location of force actuators, which provide the capability for controlling the displacement response in a desired manner. Controllability is a critical issue for active control, which relates to the sensitivity of the response to the control force magnitude. An active structural control system has the ability to determine the current state of the structure. It then decides a set of actions based on the states that change the current state to a more desirable one. All these actions are carried out in a controlled manner and in a very short period of time. Given the distribution, the desired force magnitudes are established using a control algorithm, and instructions are then sent to the actuators to deliver the required magnitude of forces.

1.3.2.1 Salient Features of Active Control Strategies

An active control system typically requires a large power source for operation of electrohydraulic or electromechanical actuators, which supply control forces to the structure in a prescribed manner. In this way, the active control systems are force-delivery devices integrated with real-time processing controllers and sensors within the structure. They react simultaneously with the hazardous excitation to provide enhanced structural behavior for improved service and safety. One of the key decisions in motion-based design is the selection of the appropriate control strategy that uses the measured structural response to calculate the control signal that is appropriate to the actuator.

Structures using active control systems employ external power to generate the control force, which is directly applied to the structure to reduce its response. The control force is determined by a control algorithm with a measured structure response, and its damping and stiffness parameters can be adjusted by feedback gain within the actuator capacity. Thus, the active control systems are adaptive systems that make structures fully smart to environmental excitations. Figure 1.8 shows the block diagram for active control system, which describes its working principle.

The basic configuration of an active control system includes following three types of elements.

i. Sensors
   The sensors can be located at the base of the structure to measure external excitation or can be strategically installed on the structure to measure the response quantities. The control device measures the system response variables, such as displacements, velocities, and accelerations, and accordingly generates control forces. In this system, the signals sent to control actuators are a function of response of the structure, measured with physical sensors.
ii. Actuators
   Actuators, also known as dampers, produce the required control forces in response to the control signals received from the controller.

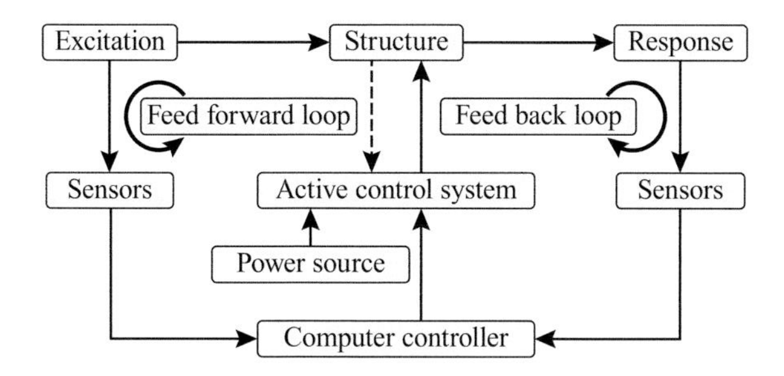

FIGURE 1.8 Block diagram for active control system.

iii. Controller
The controller (a computer) in a structure system receives measurements from sensors, analyzes them, and generates necessary control signals (also called control commands) to drive the actuator on the basis of a predetermined control algorithm. Thus, the controller with a predetermined control algorithm is an information processor that produces actuation signals by a feedback. Control forces are developed based on feedback from sensors that measure the excitation and/or the response of the structure. The feedback from the structural response is measured at locations remote from the location of the active control system. The recorded measurements from the response and/or excitation are monitored by a controller, which determines the appropriate control signal for operation of the actuators. Seismic response control of large-size civil engineering structures demands sizable actuators capable of generating a huge control force. Electrohydraulic actuators, and pulse generators, can be used for this purpose, and they both use external power sources. Therefore, an active control system is designed mainly to increase structural damping with minor modifications of structural stiffness.

1.3.2.2 Advantages of Active Control Strategies

The advantages of active vibration control strategies are presented as follows.

- Active control systems have the advantages of delivering enhanced control effectiveness, adaptability to ground motion, selectivity of control objectives, and applicability to various excitation mechanisms.
- The systems can adapt to different loading conditions and can control different vibration modes of the structure. Thus, they provide enhanced effectiveness in the response control.
- The systems are applicable to multi-hazard mitigation situations, for motion control against both, strong wind and severe earthquakes. Building vibrations are substantially reduced, consequently, increasing the comfort of occupants.
- The active control devices can reduce the structural response more effectively than passive devices due to usage of feedback and feed-forward control loops.

1.3.2.3 Limitations of Active Control Strategies

The limitations of active vibration control strategies are presented as follows.

- External power is required for functioning of active control systems.
- Custom design is always required for the control algorithm.
- The availability of reliable hardware and software systems at a reasonable cost is a critical issue.
- Integrated verification, testing, and validation of the complete control system with proposed control algorithm prior to their implementation to real structures is essential in order

to convince the designers that this technology provides a worthwhile method for reducing structural motions. This is a time-consuming and costly process.

- Active control systems must prove robust, as it is difficult to predict the possible loading conditions and the true structural response.
- While earthquakes threaten major destruction, they are rare events, which may occur once in a hundred years or more. This means that the active control systems are seldom turned on. The control systems must, however, be maintained to a level that, they are reliable and fully operational in the event of an earthquake. To maintain the best condition of the system during this long term is uneconomical.
- The systems are vulnerable to power failure, which is always a possibility during a strong earthquake. Effectiveness of active control systems is satisfactory but reliability is poor, because working of the entire system depends upon power supply, which itself is vulnerable to disasters.
- Loads induced by earthquakes are very difficult to predict and, therefore, present a major challenge to structural engineers. The level of power, energy, and the force required for active control of large-scale civil structures under extreme loading conditions cannot be economically achieved with the present technology.
- The reliability of the system when targeting the improvement of structural safety is a major concern. Active control employs electrical control units and electrically driven actuators. The reliability of these devices may not satisfy the requirement in large civil engineering structures.

1.3.2.4 Applications of Active Control Techniques

Research related to the active control of a variety of structural systems, such as bridges, dams, and lifelines, under different loading conditions such as wind and waves, has been conducted. However, the majority of the work in active control for civil engineering applications to date has been geared toward the structural control of buildings during earthquakes. Active control can be utilized to supplement passive control under service loading. A variety of active control mechanisms have been suggested by researchers, including the active tendon system, the active bracing system, the active TMD/driver, and the active aerodynamic appendage mechanism. The frequently used active control devices are AMD/driver and active tendon systems. Active control was proposed about 50 years ago to improve the response of buildings under extreme events, such as severe earthquake ground motions. The improvement of comfort of habitants during strong winds is a major problem when constructing a high-rise structure in a windy region. Traditional method to deal with this problem is to stiffen the structure and shift the structural characteristics out of frequency range of excitation wind. However, this design requires a huge quantity of structural materials and, therefore, is not cost-effective. Compared to stiffening the structural member, adding extra damping with mass damper system, with a minor modification of stiffness, is a far more inexpensive measure to improve the structural characteristics. The generation of control forces by electrohydraulic actuators requires large power sources, which are on the order of tens of kilowatts for small structures and may reach several megawatts for large structures.

Active Mass Driver (AMD)

There are two mass damper systems: TMD and AMD. The major shortcoming of passive TMD is that it is tuned to a single frequency, i.e., the fundamental frequency of the structure, which needs to be controlled. However, usually the frequencies to be controlled by wind excitation are three: the two translational movements and rotation. An AMD can control a wider frequency band. AMD system can provide additional damping to different modes by a single system and, therefore, is considered as a more economical measure to tackle this problem. AMDs are an adaptation of the passive TMD, in several existing buildings in the United States and Japan. A single AMD is usually installed at the top of the structure. A single damper (active or passive) placed at the top of a

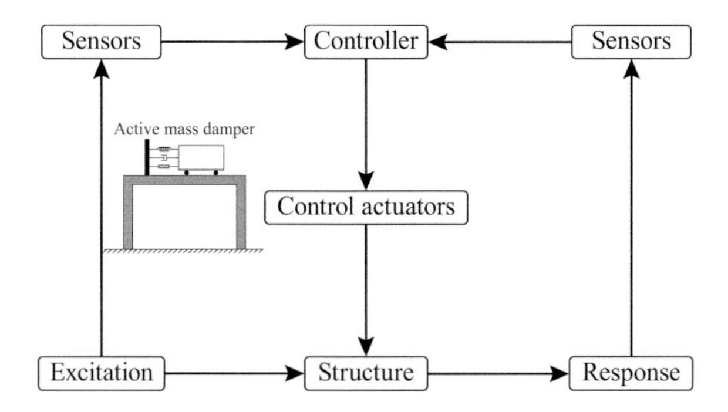

FIGURE 1.9 Control block diagram for a structure installed with AMD system.

building is generally effective in controlling only the fundamental period of the structure, since the frequency of control is governed by the placement of the actuators. Thus, in reality, there is little advantage to using an active rather than a passive mass damper, as far as frequency bandwidth is concerned.

Figure 1.9 shows a schematic representation and functioning of a structure installed with AMD system. AMDs, like passive mass dampers, may be either a pendulum-type mechanism or a sliding mass-type mechanism. One advantage to AMD system is that a smaller mass is required; about 1–2%, as opposed to 2–5% of the structural mass for passive TMD. In addition, the mass is generally variable; it is moved by hydraulic actuators to counteract undesirable motions. Most AMD systems are designed to control torsion as well as motions in the transverse directions of the building.

Full-scale implementation of active control systems to real structures started in 1989, with the first active structural control system, i.e., an AMD, implemented in an 11-story building by the Kajima Corporation in Japan, with the objective of controlling the building response under strong wind and frequent earthquake excitation. Two active mass drivers (AMDs) were installed in the Kyobashi Seiwa building, Tokyo, shown in Photo 1.1. The primary AMD is used for transverse motion control and has a mass of 4 ton, while the secondary AMD has a mass of 1 ton, and is employed to reduce torsional motion.

An AMD is installed in the Takenaka experimental building in addition to the active tendon control system, for direct comparison of the two systems. Figure 1.10 shows schematic diagram of active tendon system, and Figure 1.11 shows the active bracing system with hydraulic actuator.

The AMD system in the Takenaka building is a biaxial pendulum type system. The mass is about 6 ton, which is 1% of the structural mass. The system is capable of generating control force of 10 kN, with a stroke of +1.0 m. During a moderate earthquake in Tokyo in 1992, the maximum relative displacement measured at the top of the building was 6.3 mm, as compared to the estimated uncontrolled displacement of 21.6 mm. Both the AMD and the active tendon control systems are found to reduce the displacements by 20–40% on an average; however, the active tendon control system is more effective in reducing accelerations and in controlling higher order vibration modes.

1.3.3 Semi-Active Control Strategies

Semi-active control systems add adaptive mechanisms to passive systems to adjust their force-generating behavior, thus, often being viewed as controllable passive systems with the adaptability of active systems. A semi-active control system generally originates from a passive control system, which is subsequently modified to allow for the adjustment of mechanical properties, for improved effectiveness. Semi-active control devices combine the best features of both; passive and active control systems.

PHOTO 1.1 Kyobashi Seiwa building installed with two AMD systems. (Courtesy of B. F. Spencer.)

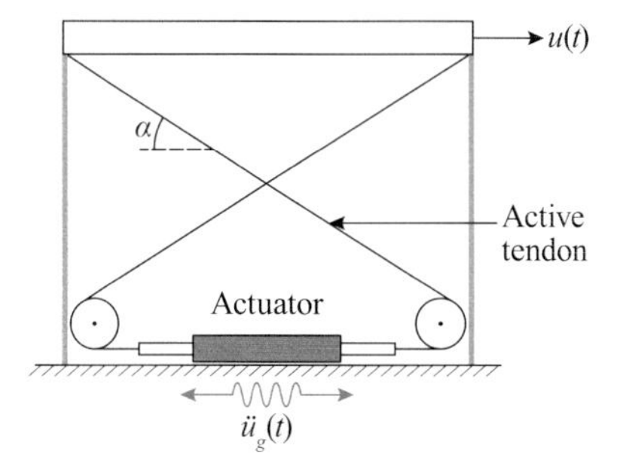

FIGURE 1.10 Schematic diagram of active tendon system.

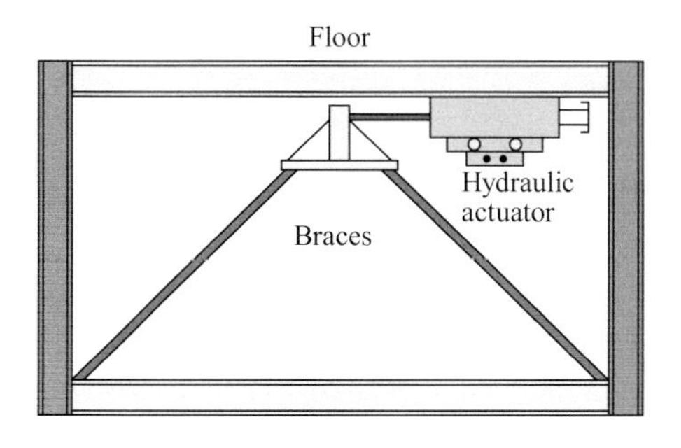

FIGURE 1.11 Active bracing system with hydraulic actuator.

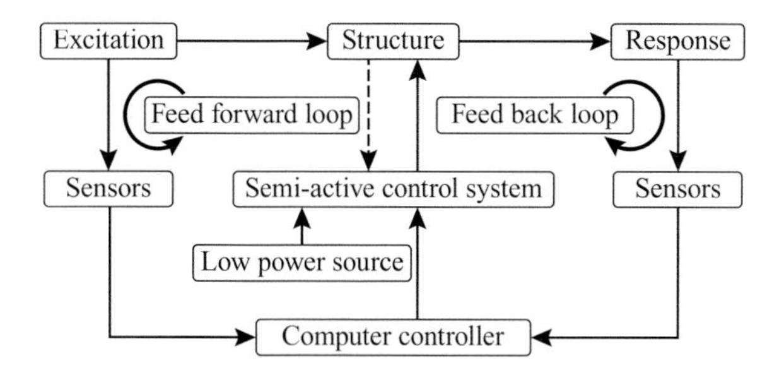

FIGURE 1.12 Block diagram of semi-active control system.

1.3.3.1 Salient Features of Semi-Active Control Strategies

A semi-active control device does not inject mechanical energy into the controlled structural system (i.e., into the structure and the control device) but has properties that can be controlled to optimally reduce the responses of the structure. Consequently, in contrast to active control devices, semi-active control devices do not have the potential to destabilize (in the bounded input/bounded output sense) the structural system; thus, allowing for the possibility of effective response reduction during a wide array of dynamic loading conditions. Extensive studies have indicated that appropriately implemented semi-active systems perform significantly better than passive devices and have the potential to achieve the majority of the performance of fully active systems. Figure 1.12 shows block diagram of semi-active control system. It consists of sensors, a control computer, a control actuator, and a passive damping device. The sensors measure the excitation and/or structural response. The control computer processes the measurement, and generates a control signal for the actuator. Then the actuator acts to adjust the behavior of the passive device. Notably, the actuator is used to control the behavior of the semi-active device, instead of directly providing force onto the structure. Therefore, the external energy required for a semi-active control system is usually orders of magnitude smaller than that for active systems. Therefore, many semi-active systems can operate on battery power. This is one of the most attractive features of semi-active control systems, because during seismic events the main power source to the structure may fail but such battery-operated systems can still work in this situation. The characteristics of the semi-active device can be changed in a very short time. Therefore, the device is always optimized for the actual working conditions. For example, properties can be modified by changing the magnetic field around an MR fluid or by turning a variable orifice. Such operations require a very small amount of energy, typically provided by a small battery.

Figure 1.13 presents the working principle of a typical semi-active control system. Control forces are developed based on feedback from sensors that measure the excitation and/or the response of the structure. The feedback from the structural response is measured at remote locations from

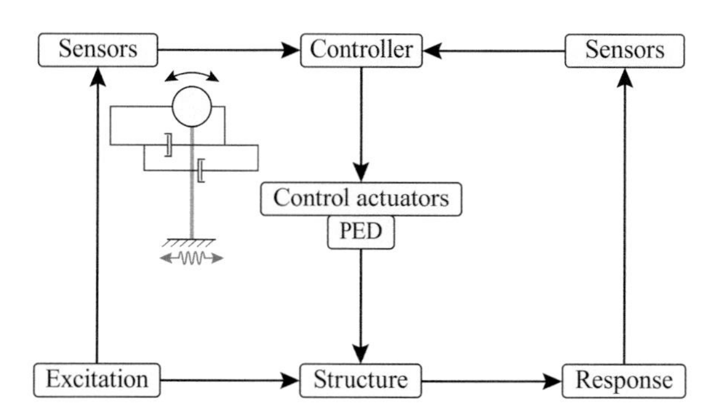

FIGURE 1.13 Working principle of semi-active control system.

the location of installation of the semi-active control system. The mechanical properties of these systems are adjusted based on feedback from the excitation and/or from the measured structural response. As in an active control system, a controller monitors the feedback measurements and generates an appropriate command signal for the semi-active devices; however, the control forces are developed as a result of the motion of the structure itself as in the case of the passive control system. The control forces are developed through appropriate (based on a predetermined control algorithm) adjustment of the mechanical properties of the semi-active control system. Furthermore, the control forces in many semi-active control systems primarily act to oppose the motion of the structural system and therefore promote the global stability of the structure.

1.3.3.2 Advantages of Semi-Active Control Strategies

The advantages of semi-active vibration control strategies are presented as follows.

- Semi-active devices require a relatively small amount of external energy and apply the control force in such a way that the resulting motion is always stable.
- These devices offer the adaptability of active control devices without requiring the associated large power sources.
- The properties of a semi-active control device can be adjusted in real time; however, it cannot add energy into the controlled system and does not create any kind of instability in the system by pumping energy into the controlled structure.
- Many semi-active systems can operate on battery power, which is advantageous during seismic events, when the main power source to the structure may fail.

1.3.3.3 Limitations of Semi-Active Control Strategies

The limitations of semi-active vibration control strategies are presented as follows.

- Like an active control, the implementation of a semi-active control system requires an online control algorithm in order to decide the controllable characteristic parameter of the damper (e.g., the slip force for a variable FD).
- A semi-active damper may perform very differently, when different control algorithms are employed.
- Compared to active control systems, a semi-active control system is less adaptive because its force-generating capacity is restricted by its passive device component.

1.3.3.4 Applications of Semi-Active Control Techniques

Different types of semi-active control devices are variable orifice damper, variable stiffness damper, variable FD, adjustable tuned liquid damper, controllable fluid damper, pseudo negative stiffness damper, piezoelectric FD, and semi-active FVD. A class of semi-active devices, consisting of controllable fluids in a fixed-orifice damper, uses either Electrorheological (ER) fluids or Magnetorheological (MR) fluids.

Semi-active Damper

Semi-active dampers are a natural evolution of passive energy dissipating technology, as they incorporate adaptive systems to improve their effectiveness, and intelligence. They are frequently referred to as controllable or intelligent/smart dampers. Their adaptive system gathers information about the excitation, structural response, and then adjusts the damper behavior on the basis of this information to enhance its performance.

Semi-active Controllable Fluid Damper

Semi-active controllable fluid dampers contain no moving parts other than the piston, which makes them simple and potentially very reliable. The essential characteristics of controllable fluids is their

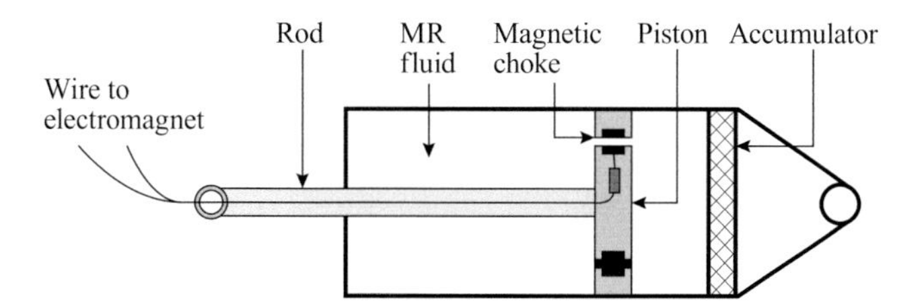

FIGURE 1.14 Schematic details of MR damper.

ability to reversibly change from a free-flowing, linear viscous fluid state to a semi-solid state, with a controllable yield strength, in milliseconds, when exposed to an electric field (for ER fluids), known as ER dampers, or magnetic field (for MR fluids), known as MR dampers. MR fluids typically consist of micron-sized, magnetically polarizable particles dispersed in a carrier medium such as mineral or silicone oil. When a magnetic field is applied to the fluid, particle chains form, the fluid becomes a semi-solid and exhibits visco-plastic behavior. Transition to rheological equilibrium can be achieved in a few milliseconds, allowing construction of devices with high bandwidth. Figure 1.14 shows schematic details of MR damper.

Variable Friction Damper

Dampers utilizing dry friction represent one of the major device categories. An FD is usually classified as one of the displacement-dependent EDDs, because the control force developed by it is independent of the velocity and frequency content of excitations. An FD is activated and starts dissipating energy, only if the friction force exerted on its friction interface exceeds the maximum friction force (slip force); otherwise, an inactivated damper is similar to a regular bracing. The advantages of using FDs over the other types of EDDs are (i) the materials are less affected by the degradation due to aging, (ii) the materials are insensitive to the change of ambient temperature, (iii) there is no material yielding and replacement problems after a large earthquake, and (iv) there is no problem of fluid leakage. For a passive FD, the slip force of the damper is a preset fixed value. Hence, for a particular earthquake event, the damper is activated only within the few cycles of the structural motion in which the exerted force exceeds this fixed value. Therefore, the energy dissipation capacity of the damper is not fully utilized. Another disadvantage of using passive FDs is that during an earthquake, the status of the damper may be frequently switched between slip and stick states, which can result in an unwanted high-frequency structural response. Semi-active variable FDs improve the performance of passive FDs. By regulating the normal contact force (clamping force) applied on the friction interface, a variable FD can adjust its slip load in real time, in response to the structural motion and external excitations, so that the damper may be activated in earthquakes with arbitrary intensities. Variable FDs generate control forces through surface friction and by controlling the slippage of the device. These devices have been proposed to reduce inter-story drifts of seismically excited buildings. When properly confined, piezoelectric materials can generate a large stress and respond quickly to the control voltage with very low electric current.

Adjustable Tuned Liquid Damper

Adjustable tuned liquid damper is a semi-active control device that utilizes the motion of sloshing fluid or of column of fluid for structural response mitigation. These adjustable tuned liquid dampers are based on passive TSDs and TLCDs. As in a TMD, the TSD uses the sloshing liquid in a tank to prevent the resonance of the structural system. Similarly, in a TLCD, the moving mass is a column of liquid that is driven by the vibration of the structure. Several semi-active adjustable tuned liquid dampers have been recently proposed, as they have considerable potential to improve upon the performance of passive tuned liquid dampers, in reducing structural responses.

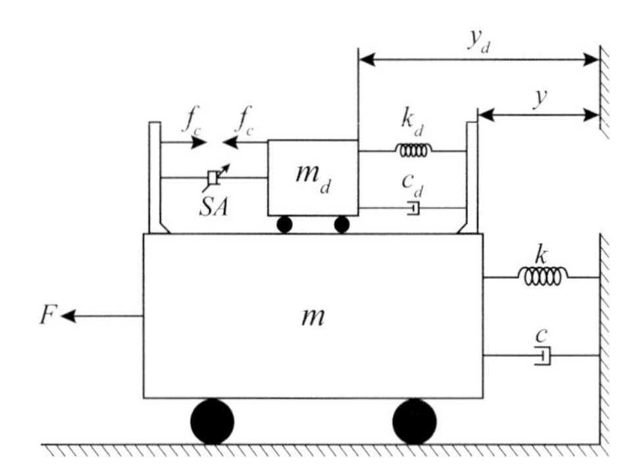

FIGURE 1.15 Mathematical model of semi-active tuned mass damper.

Semi-active TMD

The semi-active TMD system consists of a TMD and an actuator that are installed on top of the main structure, as shown in Figure 1.15. The semi-active TMD has a mass (m_d), damping (c_d), and stiffness (k_d), whereas the main structure is represented by a mass (m), damping (c), and stiffness (k). The actuator, denoted by SA, generates the control force f_c. The control force f_c generated by the actuator contributes to the mitigation of the undesirable structural vibration induced by the excitation (F).

Semi-active Vibration Absorber

An effective method to achieve a controllable damping device is to use a variable orifice valve to adjust the flow of a hydraulic damper. This concept led to the development of semi-active vibration absorber (SAVA), also called semi-active hydraulic damper. As shown schematically in Figure 1.16, the device provides adjustments of both damping and stiffness. The damping capacity is generated from the viscous fluid, and the stiffness is adjusted by the opening of the flow valve. If the valve is closed, the SAVA works as a spring providing stiffness. If the valve is open, the fluid can easily flow through the tube and provides little stiffness to the structure.

1.3.4 Hybrid Control Strategies

A hybrid structural control system is a combination of passive and active control techniques. With two control techniques in synchronized operation, hybrid control system possesses the advantages

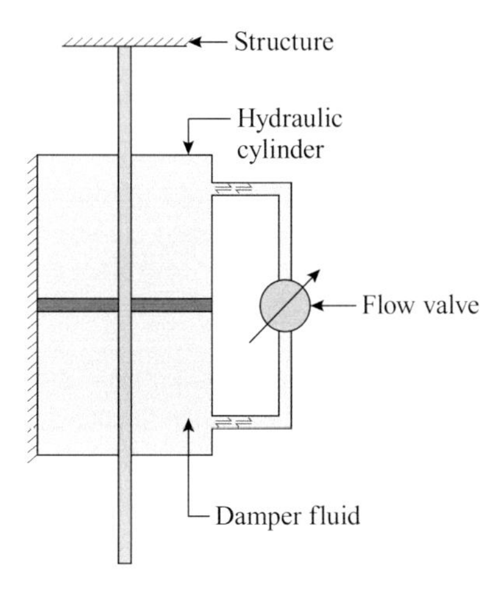

FIGURE 1.16 Schematic of semi-active vibration absorber.

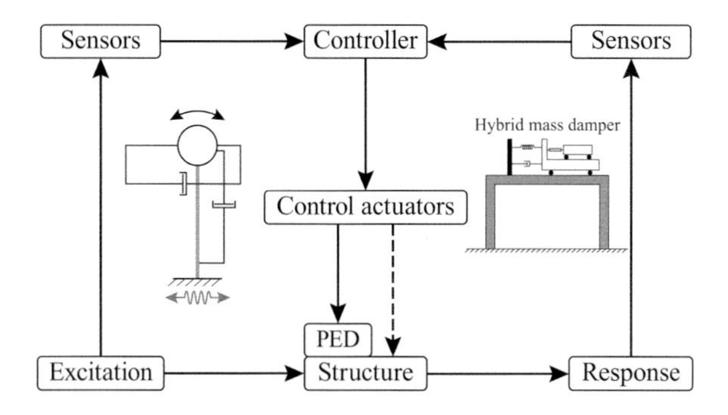

FIGURE 1.17 Schematic of hybrid control system.

of both techniques and alleviates some of the restrictions and limitations that exist, when each system is functioning alone. Operating both systems together enhances the robustness of the passive system and reduces the energy requirements of the active system. As a result, hybrid control systems surpass passive and active systems, and higher levels of structural performance can be achieved.

1.3.4.1 Salient Features of Hybrid Control Strategies

Hybrid control system consists of combined passive and active devices or combined passive and semi-active devices. Figure 1.17 shows schematic of a typical hybrid control system.

1.3.4.2 Advantages of Hybrid Control Strategies

The advantages of hybrid control strategies are presented as follows.

- In the case of a power failure, the passive components of the control system still offer some protection, unlike a fully active control system.
- Since a part of the control objective is accomplished by the passive system, a less active control effort, implying a smaller active force-generating device, and less power resource, are required. Thus, advantages of both systems are fully utilized.
- When both control techniques work together, reliability is ensured by the passive system and the capacity is powered by the active system. The resulting hybrid control system can be more reliable and economical than a fully active system.

1.3.4.3 Limitations of Hybrid Control Strategies

Although hybrid control strategies are advantageous because they offer combined benefits of various types of control strategies, they have some disadvantages. The primary disadvantages are:

- The implementation of hybrid control strategies is relatively more complex technically because hybrid control techniques involve multiple control methods in a single structure.
- The implementation of hybrid control strategies requires specialized professionals with in-depth knowledge about the interactions between the different control strategies and their combined effect on the behavior of the structure.

1.3.4.4 Applications of Hybrid Control Techniques

The implementation of hybrid systems is typically done through two main approaches: the hybrid mass damper (HMD) and the hybrid seismic isolation system. Examples of hybrid control devices include the combination of VEDs as passive part and hydraulic actuators as active part, or using base isolators as passive part and active TMDs as active part. Most of the work on hybrid control in the United States is focused on combining base isolation with some form of active control to limit excessive displacements.

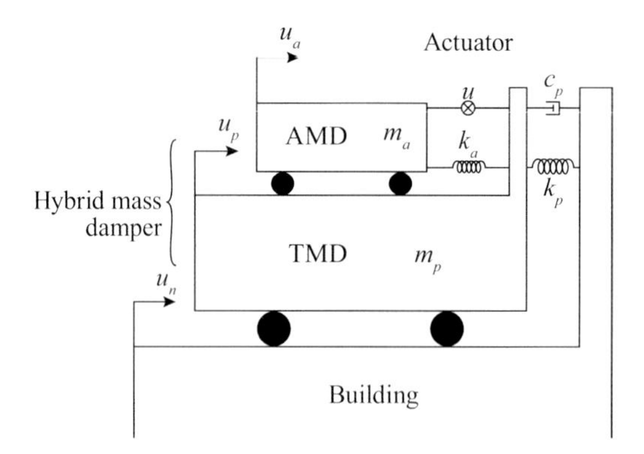

FIGURE 1.18 Hybrid mass damper.

Hybrid Mass Damper

The HMD is the most common control device employed in full-scale civil engineering applications. It combines a passive TMD, and an active control actuator, or an AMD is added to a TMD as shown in Figure 1.18, to enhance the effectiveness of the system in reducing structural vibrations under different loadings. An AMD is attached to a TMD instead of to the structure, as shown in Figure 1.18, so that the AMD can be small; its mass is 10–15% of that for the TMD. The ability of this device to suppress structural responses relies mainly on the natural motion of the TMD. The forces from the control actuator are employed to increase the efficiency of the HMD and to increase its robustness to changes in the dynamic characteristics of the structure. Thus, the TMD is tuned to the fundamental mode of the structure, and the AMD is designed to improve the control effectiveness for higher modes of the structure. The energy and forces required to operate a typical HMD are far less than those associated with a fully active system of comparable performance. Figure 1.19 shows HMD, wherein actuator and TMD are arranged in series.

An example of such an application is the HMD system installed in the Sendagaya INTES building in Tokyo in 1991. The HMD was installed on the top, at 11th floor, consisting of two masses to control the transverse and torsional motions of the structure; while hydraulic actuators provide the active control capabilities. The masses are supported by multistage rubber bearings, intended for reducing the control energy consumed in the HMD and insuring smooth movements of the mass. Sufficient data were obtained for evaluation of the HMD performance when the building was subjected to strong wind, with peak instantaneous wind speed of 30.6 m/s. The response at the fundamental mode was reduced by 18% and 28% for translation and torsion, respectively. Variations of such an HMD configuration include multistep pendulum HMDs, which have been installed in, the Yokohama Landmark Tower in Yokohama, the then tallest building in Japan, and in the TC Tower in Kaohsiung, Taiwan.

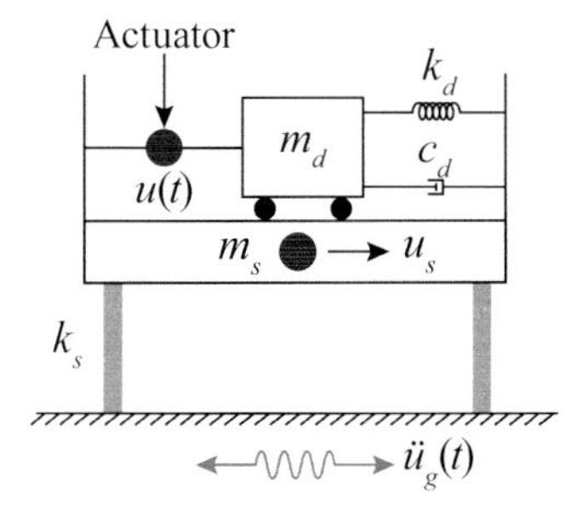

FIGURE 1.19 Hybrid mass damper – actuator and TMD in series.

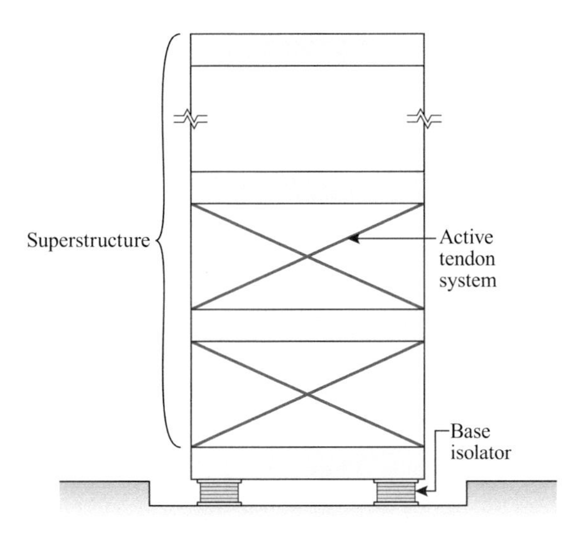

FIGURE 1.20 Hybrid system with base isolation and actuators.

Hybrid Seismic Isolation System

A hybrid seismic isolation system consists of a passive base isolation system implemented in combination with another control device, which is used to enhance the effectiveness of the base isolation system. The hybrid isolation system shown in Figure 1.20 consists of: (i) an active tendon system installed on superstructure and (ii) a base isolation system between the foundation and the superstructure.

Though widely used worldwide, base isolation systems are limited in their ability to adapt to changing demands for structural response reduction. In the case of base isolation using linear springs the active controller is designed by the conventional linear optimal control theory. Many base isolation systems exhibit strong nonlinear behavior, and in these cases, nonlinear robust control laws have been developed and several small-scale hybrid control tests have been performed. With

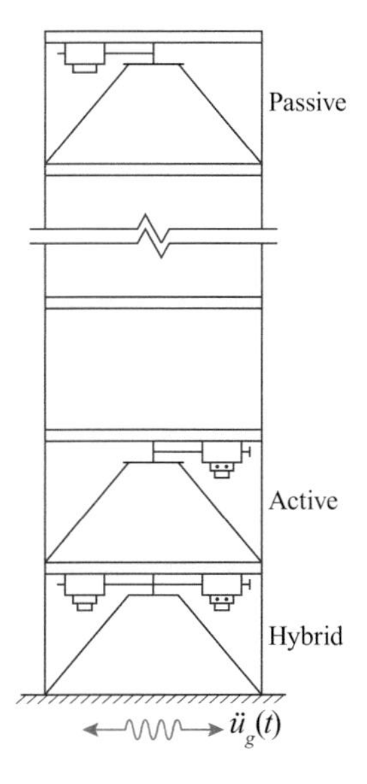

FIGURE 1.21 Configuration of hybrid bracing system and control devices.

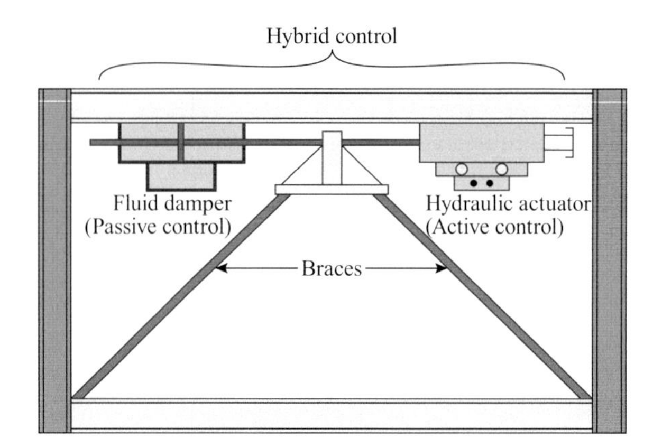

FIGURE 1.22 Installation of hybrid bracing system and control devices.

the addition of an active control device to a base isolated structure, a higher level of performance can be potentially achieved without a substantial increase in the cost, which is very appealing from a practical viewpoint. Since base isolation by itself can reduce the inter-story drift and the absolute acceleration of the structure at the expense of large absolute base displacement, its combination with active control is able to achieve both low inter-story drift and, at the same time, limit the maximum base displacement with a single set of control forces.

Hybrid Damper-Actuator Bracing Control

Hybrid damper-actuator bracing control system mounted by K-braces on the structure is shown in Figures 1.21 and 1.22. Hydraulic actuators are employed as the active device for the system owing to their powerful force-generating capacity, and liquid mass damper, spring dampers, and FVDs can be installed as passive devices.

REFERENCE

Christenson, R. E. (2001). Semiactive Control of Civil Structures for Natural Hazard Mitigation: Analytical and Experimental Studies. PhD Dissertation. University of Notre Dame.

2 Dynamic Loading on Structures and Structural Response

2.1 SELECTION OF EARTHQUAKE GROUND MOTION RECORDS

Earthquake engineering practice is increasingly using nonlinear response history analysis to demonstrate the performance of structures under dynamic loads. This rigorous method of dynamic analysis requires selection and scaling of ground motions appropriate to the design hazard levels. Seismic evaluation of existing structures and of proposed design of new structures is typically based on nonlinear static (pushover) analysis procedures. However, nonlinear response history analysis is now increasingly used, wherein the seismic demands of structures are determined for several ground motions. Procedures for selecting and scaling ground motion records for a site-specific hazard are described in building codes and have been the subject of research in the recent years.

For most of the applications in which the effects of earthquakes on structures are considered, seismic actions are represented in the form of pseudo-acceleration or displacement response spectra. However, many situations exist, for which the specification of structural actions through a response spectrum is deemed insufficient. The code for seismic design of building requires that nonlinear time history analysis should be adopted in the design of high-rise buildings, lifeline structures, and such other important constructions. In this context, the process of selection of appropriate ground motion records to be used in the analysis, and their modification (scaling), plays a vital role. This can lead to responses with low dispersion in nonlinear structural analysis, which in turn results in much more trust in the safety of the designed structure. In such cases, the structural response for a given earthquake is estimated by subjecting the structure to a number of compatible acceleration time histories. Earthquake ground motion variability is one of the major sources of uncertainty in the assessment of the seismic performance of structures. The selection of appropriate ground motion and the method of scaling largely affect the nonlinear response of structures. The problem of selecting a suite of earthquake accelerograms for time domain analysis is of practical and academic interest. Hence, there is a strong need for a clear guidance on appropriate ground motion selection methods. Research in this field has led to numerous approaches for compiling suites of accelerograms that may be used to robustly estimate the median structural response.

Appropriately selecting records considering the hazard conditions for a given site helps to increase the accuracy of the nonlinear analysis. It should be ensured that (i) seismic characteristics of the selected records are the same as those of the site, (ii) the records can provide most critical seismic state for the structure, and (iii) the response resulting from structural analysis, which has been performed using these records, indicates lowest dispersion amount in definite parameters of the response.

During the spectrum of the 1940 Imperial Valley, El Centro, California, accelerogram has large accelerations at periods of 0.1–1 s. In this regard, the design earthquake motions for many seismic areas of the world are similar to what is recorded at El Centro, California, or scaling of this ground motion. Typical earthquake accelerations have dominant periods in this range. Therefore, structures with natural period of vibration in the range of 0.1–1 s are likely to experience resonance under earthquake excitation. Maximum ground accelerations often occur in the range of 0.2–0.6 s. However, deep deposits of soft soil, as in old lake bed zone of Mexico City, may amplify low-frequency earthquake motions, as observed in the Mexico City earthquake of 1985. Such possibilities

DOI: 10.1201/9781315269269-2

reinforce the importance of careful selection of ground motion data that are representative of the site characteristics.

Ground Motion Parameters

To compare the effect of near-fault and far-fault earthquake ground motions, seismic parameters, such as peak ground acceleration (PGA), peak ground velocity (PGV), Arias intensity (AI), cumulative absolute velocity (CAV), and root mean square (RMS) of acceleration, are crucial.

- Earthquake magnitude and epicentral distance
  The magnitude of an earthquake is a measure of energy released during the earthquake event. The energy released can be an effective factor on selecting ground motion of the regions, where past seismic data is not available. Epicentral distance is related to the soil strata, through which the seismic waves travel, affecting their speed.
- Peak Ground Acceleration (PGA)
  PGA is a measure of earthquake acceleration on ground. The peak horizontal ground acceleration is most commonly used in design codes and considered the most common intensity parameter used in design under earthquake hazard. In an earthquake, damage to buildings and infrastructure is related more closely to the characteristics of ground motion, rather than the magnitude of the earthquake. For moderate earthquakes, PGA is usually the best determinant of damage, whereas in severe earthquakes, damage is more often correlated with PGV.

Intensity-Based Scaling Methods

The current performance-based design and evaluation methodologies prefer intensity-based methods to scale ground motions. Intensity-based scaling methods preserve the original nonstationary content of earthquake time history and only modify its amplitude. The primary objective of these methods is to provide scale factors for a small number of ground motion records, so that nonlinear response history analysis of the structure for these scaled records is an accurate representation of the realistic seismic effect. The intensity-based methods are efficient and provide an accurate estimate in the median value of the engineering demand parameters, by minimizing the record-to-record variations in the engineering design parameters. Scaling the ground motions to match a target value of PGA is the original approach to the problem. However, it produces inaccurate estimates with large dispersion in engineering design parameter values.

- Arias Intensity (AI)
  The AI is a measure of earthquake intensity related to the energy content of the recorded signal. It is specified by Equation (2.1).

$$I = \frac{\pi}{2g}\int_{0}^{T_d}\left[\ddot{u}_g(t)\right]^2 dt \tag{2.1}$$

  where g is the gravitational acceleration, $\ddot{u}_g(t)$ is the recorded signature of the earthquake, and T_d is the duration of earthquake.
- Cumulative Absolute Velocity (CAV)
  CAV is an instrumental intensity measure, defined as the integral of the absolute value $|\ddot{u}_g(t)|$ of the acceleration time series, presented mathematically by Equation (2.2), wherein T_{max} is the total duration of the earthquake.

$$\text{CAV} = \int_{0}^{T_{max}} |\ddot{u}_g(t)|\, dt \tag{2.2}$$

The key feature of CAV over peak response parameters is that it includes the cumulative effects of ground motion over the entire duration of earthquake.

- Root Mean Square (RMS) Acceleration
  The RMS acceleration is defined as

$$a_{rms} = \sqrt{\frac{1}{t_d} \int_{t_d} \left[\ddot{u}_g(t) \right]^2 dt} \tag{2.3}$$

where t_d is the duration of strong motion. This index accounts for the effects of amplitude and frequency content of a strong motion record and is directly proportional to the square root of the gradient of the specified interval of AI.

Spectral Shape of Ground Motion

For structures with plan and/or vertical irregularities, or for structures without independent orthogonal lateral load-resisting systems, where three-dimensional analysis need to be performed, ground motions should consist of appropriate horizontal components. Before scaling ground motions, it is required to define the hazard conditions associated with a given site either through deterministic or probabilistic site-specific hazard analysis or alternatively from seismic hazard maps, such as the United States Geological Survey (USGS) seismic hazard maps. Spectral shape, i.e., the response spectrum normalized by PGA, defines ground motion demand characteristic on structural systems. Therefore, in selecting candidate records for nonlinear analysis, it is important to carefully identify the records that have spectral shapes close to each other. The spectral shape of ground motion depends on the following parameters, which are based on the seismic hazard conditions.

- Magnitude dependence: In general, events with larger magnitude yield wider response spectra. In order to find the degree of influence of magnitude on response spectral shape, average spectral shapes of earthquakes, lying in certain specified range, are considered.
- Distance dependence: Predominant period shifts to higher values with increase in the distance from the fault for a given earthquake.
- Soil condition dependence: Predominant period of spectral shape at a rock site is generally lower than that of a soil site.

Spectral Matching Techniques

Spectral matching techniques modify the frequency content of the record to match its response spectrum to the target spectrum. Site- and structure-specific ground motion selection methods often involve selecting a set of ground motions with response spectra matching a site-specific target response spectrum. Target spectrum, including code spectrum, uniform hazard spectrum, and condition mean spectrum, represents the demand of structural design. The code spectrum is the statistical average of response spectrum of multiple records. For general structures, target spectrum is used to determine the earthquake effect in equivalent static analysis. Selecting a set of ground motions to match only a target response spectrum is computationally inexpensive since it can be done by selecting time histories with response spectra individually deviating the least from the target response spectrum. The deviation can be measured using the sum of squared differences between the response spectrum of the record and the target response spectrum. The suitability of a particular ground motion can only be determined in the context of the complete set of ground motions, in which it might be included.

Uniform Hazard Spectrum

Uniform hazard spectrum is defined as the locus of points such that the spectral acceleration value at each period has an exceedance probability equal to the specified target probability, from probability seismic hazard analysis. Thus, it is a set of response spectrum envelope with the same probability of exceedance at all period points, defined by means of the seismic hazard at the site where the structure is supposed to be located. Uniform hazard spectrum is not the representative of the spectra of any individual ground motion. In time history analysis of structures, using uniform hazard spectrum as the target spectrum may result into conservative response compared to that with real ground motion recordings. Seismic provisions in present building codes and standards include guidelines for design of buildings and other structures using response spectrum analysis.

There are generally an intractably large number of possible ground motion sets, and identifying the best set is a computationally expensive combinatorial optimization problem. No automated procedures are available in the literature to select ground motions that match the mean and variance of the response spectrum. In India, there is currently no consensus in the earthquake engineering community on how to appropriately select and scale earthquake ground motions for code-based design and seismic performance assessment of buildings using nonlinear response history analysis. One common state-of-the-art practice in performance-based seismic design is incremental dynamic analysis that scales the same suite of ground motions up and down to cover a wide range of ground motion intensity levels. Ground motion selection provides the necessary link between the seismic hazard and structural response, the important components in performance-based earthquake engineering.

2.1.1 Near-Fault Earthquake Ground Motions

Ground motions resulting from an earthquake mirror the characteristics of the seismic source such as the rupture process, the source-to-site travel path, and local site conditions. Therefore, the features of ground motions in the vicinity of an active fault are significantly different from the far-fault ones that severely affect the damage potential of these earthquakes. Forward-directivity effect and fling-step effect are two well-known characteristics of near-fault ground motions. These characteristics of near-fault earthquake records make structural responses to be considerably different from those expected in far-fault earthquakes. In the near-fault zone, the ground motions may be distinguished by short-duration impulsive motions, permanent ground displacement, and high-frequency content, which have attracted much attention as the critical factors in the design of structures located near the active faults. High-amplitude, long-period velocity pulses are normally observed in the records obtained within about 20 km from the fault rupture. However, there is no universal definition for it over which a site may be classified as in near- or far-fault. A useful criterion to define the near-fault zone is related to the comparison of the source dimension with the source to site distance. For the near-fault ground motions, the fault-normal components having the pulse-like signals are often more remarkable than the fault-parallel components. In the near-fault region, the short travel distance of the seismic waves does not allow enough time for the high-frequency content to be damped out of the record as is normally observed in far-fault records. Near-fault records are characterized by a large high-energy pulse. The energy value of near-fault records is several times larger than the far-fault records. Further, the damage potential of records indicates that near-fault records are typically more destructive than far-fault records.

2.1.1.1 Forward-Directivity

Forward-directivity is the direction of rupture propagation relative to the site. Forward-directivity pulses can be distinguished in velocity time history. It occurs where the fault rupture propagates with a velocity close to the shear-wave velocity. Displacement associated with such a shear-wave velocity is largest in the fault-normal direction for strike-slip faults. The ground motion records affected by forward-directivity show unusual features in the signal resulting in low-frequency cycle pulses in the velocity time-history. In the forward directivity zone, the velocity record is characterized by pulse-type motion of long duration. Due to the radiation pattern of the fault, forward-directivity pulses are mainly oriented in the fault-normal direction. The fault-normal component is of higher PGA than the fault-parallel component at the same recording station. Nevertheless, the fault-parallel direction may also contain strong pulses.

2.1.1.2 Fling-Step Effect

The fling-step is the static component of the near-fault ground motion and is characterized by a ramp-like step in the displacement time history and a one-sided large-amplitude pulse in the velocity time history. The fling-step motion is a result of the development of static permanent ground displacements due to tectonic deformation associated with fault rupture. Due to inaccessibility to various near-fault ground motions database with fling-step effect, the fling records are generally selected from the 1999 Chi-Chi earthquake (M 7.6) (Taiwan), which includes significant permanent tectonic displacement records.

2.1.2 Far-Fault Earthquake Ground Motions

Far-fault long-period motions are characterized by the presence of later-arriving surface waves, consisting of long-period components, which are manifested in the form of sinusoidal waves in a velocity time series. The features of long-period ground motions are of significant concern to engineering communities, largely due to resonance-induced responses of long-period structures to far-fault long-period ground motions. It is well documented that the primary causation of resonance-induced responses during the Michoacán earthquake, 1985, may be attributed to far-fault long-period motions; structures in Mexico City, located approximately 400 km away from the epicenter, were damaged to a more significant degree than those adjacent to near-faults. Subsequent events, such as the 2003 Tokachi, the 2008 Wen Chuan, and the 2011 Tohoku earthquake, have permitted researchers to recognize the vulnerability of structures with an intrinsic long-period motions to far-fault long-period motions, even if their amplitude is relatively small at distant sedimentary layers. Such long-period waves arrive later than body waves which feature high-frequency components, thus extending the duration of the shaking process and further accounting for the increased amplitudes.

2.1.3 Scaling of Earthquake Ground Motions

Sometimes modifications to existing ground motion time histories are required in order to better-fit the specified spectral shape, including the use of the amplitude-scaling approach to scale the amplitude of time histories to achieve an averaged fit to a target spectrum or spectrum matching approaches to adjust the ground motion time histories either in the time domain or in the frequency domain, so that the modified response spectrum is closely matched to the design spectrum. Three different modification methods are used for scaling of earthquake ground motion to minimize the scatter in the results of nonlinear dynamic analysis.

2.1.3.1 Scaling Based on ASCE 7-16 (2017)

In this method, scaling of the acceleration response spectrum of ground motion is performed on appropriate ground acceleration histories that are obtained from records of events having magnitudes, fault distances, and source mechanisms that are consistent with those that control the maximum considered earthquake at the site. The ground motions are scaled such that the average value of the 5% damped response spectra for the suite of motions is not less than the design response spectrum of the site in the neighborhood of the considered structural period. According to ASCE 7-16 (2017), a pair of recorded horizontal components of an earthquake should be used for the analysis. For each pair of the horizontal components, Square-Root-of-Sum-of-Squares (SRSS) acceleration response spectrum must be calculated for a damping ratio of 5%. The scale factor is a numerical coefficient multiplied to the SRSS spectrum such that the resulting modified spectrum just touches the design spectrum in the period range of $0.2T$–$1.5T$, where T is the fundamental period of the structure subjected to dynamic analysis. The same scale factor is also multiplied to the time histories of the ground motion. When several earthquakes are to be used, a common scale factor is calculated for the average spectrum of SRSS spectra of different earthquakes.

2.1.3.2 Scaling Based on Conditional Mean Spectrum

The second modification method is based on the conditional mean spectrum (CMS). The CMS provides the expected, i.e., mean response spectrum, conditioned on occurrence of a target spectral acceleration value at the period of interest. Each ground motion is scaled, so that the average response spectrum over the periods of interest is equal to the average of the target spectrum over the same periods. The CMS is a rather new and much effective concept in scaling of ground motions. In this method, first, a CMS is constructed, and then it is used as a target spectrum to modify individual ground motion. To construct a CMS, a key parameter ε is used as the spectral distance between the record under consideration and the target spectrum at a certain period. Modification of ground motion based on the CMS is implemented by calculating a scale factor as the ratio of the average of the target spectrum at the considered periods to the average of the response spectrum at the same periods, presented by Equation (2.4).

$$\text{Scale factor} = \frac{\sum_{j=1}^{n} S_{a,CMS}\left(T_j\right)}{\sum_{j=1}^{n} S_a\left(T_j\right)} \tag{2.4}$$

Such a modification makes the response spectra have a good resemblance with the CMS. In Equation (2.4), $\sum_{j=1}^{n} S_{a,CMS}\left(T_j\right)$ is the sum of spectral accelerations of the CMS, and $\sum_{j=1}^{n} S_a\left(T_j\right)$ is the sum of spectral accelerations of the response spectrum for the period interval considered. In this method, each record will have a specific scale factor.

2.1.3.3 Scaling Based on Uniform Design Method

The third modification procedure of ground motion records is called the uniform design method (UDM). In this method, a scale factor is produced for each different earthquake by tuning the response spectrum such that it results in a designed structure having a same period as the code-based designed structure. UDM is simple and has a superior behavior regarding dispersion in statistical responses. The UDM introduces a certain scale factor for each different earthquake as follows. First, the structure under study is designed for the original response spectrum of the considered earthquake. The fundamental period of this structure is denoted as T_1^e. The same structure is again designed for the code-based design spectrum. To obtain a uniform design under different earthquakes, and under the design spectrum, the scale factor specified in Equation 2.5 is introduced for each ground motion time history.

$$\text{Scale factor} = \left(\frac{T_1^e}{T_1^{code}}\right)^2 \tag{2.5}$$

where T_1^e and T_1^{code} are the fundamental periods of the structure designed under the earthquake response spectrum and the design spectrum, respectively.

2.1.4 Synthetic Earthquake Ground Motions

Real-time histories can reflect more information on earthquakes, compared to the simulated ground motions. Structural assessment via dynamic analysis requires some characterization of the seismic input which reflects the seismic hazard as well as the soil conditions at the site. Following are the commonly used three types of seismic signals:

i. Artificial waveforms, which are generated recurring to the random vibration theory to match some target spectral shape
ii. Simulated accelerograms, obtained via modeling of the seismological source, that account for travel path and site effects
iii. Ground motion records from real events

The lack of real ground motion records and the need to have seismic input closely representing a specific scenario to match, for example, in terms of a target spectral shape or a magnitude-distance pair, have supported the use of artificial records, by means of which several signals with nominal characteristics of interest can be produced. The recently increasing accessibility to accelerograms recorded during real ground shaking via digital seismic networks has pointed the attention of researchers to investigate the issues related to the use of real records for seismic structural design and assessment. One source where a large collection of ground motion records can be obtained is the Pacific Earthquake Engineering Research Center (PEER) Strong Motion Database of Berkeley.

Some types of artificial accelerograms have shown shortcomings in being a realistic representation of possible ground motion. When selection of real records for seismic design of structures is concerned, the current state of best engineering practice is based on the uniform hazard spectrum. Actual strong motion records may be used if they are applicable to the geologic and tectonic setting of the site under consideration. When actual records are not available, the synthesis of ground motions using appropriate source, path, and site model becomes a good alternative.

The stochastic method is a popular seismological tool to simulate earthquake ground motions for areas that lack real strong ground motion records. Using the Monte Carlo simulation, a suite of synthetic time histories can be generated to account for the random nature of ground motions.

The most common methods of generating artificial time histories include the following:

i. Modification of real-time histories
ii. Generation of time histories in time domain
iii. Generation of time histories in frequency domain
iv. Generation of synthetic ground motions via Green's function methods

2.1.5 Structural Response under Near-Fault and Far-Fault Earthquakes

The near-fault pulses possess immense damage potential specifically for intermediate and long-period structures and may lead to higher seismic demands. Therefore, these must be taken into account for design or retrofit of structures in the near-fault zone. Thus, it is absolutely essential to provide quantitative knowledge to consider the salient effects of near-fault earthquakes on the seismic performance of various elastic and inelastic systems and to develop appropriate design guidelines. The traveling wave effect of pulse-type records causes highly nonuniform distribution of ductility demands over the height of the structure. Near-fault earthquakes may contain large-amplitude long-period pulses in addition to high-frequency content. However, structures designed according to current procedures may be vulnerable to the high-amplitude, long-period velocity pulse-type ground motions. In comparison, the displacement demand in structures subjected to near-fault earthquake records is higher than those subjected to far-fault records. Long-period structures, such as high-rise buildings and base-isolated structures, are more affected due to the long duration and narrow band nature of far-fault excitation. The average structural response under the impact of the near-fault earthquake is greater than those, under the effect of a far-fault earthquake. As compared to long-period structures, short-period structures are less affected by the long-period velocity pulse.

2.2 WIND LOADING

Structural innovations and lightweight construction technology have considerably reduced the stiffness, mass, and damping characteristics of modern buildings. A modern skyscraper, with lightweight partition walls, and high-strength materials, is more prone to wind motion problems than the early-day high-rise buildings, which had the weight advantage of masonry partitions, heavy stone facades, and massive structural members. Wind produces three types of effects on structures: static, dynamic, and aerodynamic. Following are three common forms of wind-induced motion:

i. Galloping
   Galloping is the term used for self-excited transverse oscillations of structures due to the development of aerodynamic forces which are in phase with the motion. It is characterized by the progressively increasing amplitude of transverse vibration with increase in wind speed. Noncircular cross sections are more susceptible to galloping.
ii. Flutter
   Flutter is the self-excited unstable oscillatory motion of a structure due to coupling between aerodynamic force and elastic deformation of the structure. It is the most common form of oscillatory motion due to the combined effect of bending and torsion, generally leading to structural failure. Long-span suspension bridge decks or members of a structure with large value of $\left(\frac{d}{t}\right)$ are prone to low-speed flutter; where d is the depth of structure or a structural member parallel to wind stream, and t is the least lateral dimension of the structure or a structural member.
iii. Ovalling
   Thin-walled structures with one or both ends open, such as oil storage tanks, and natural draught cooling towers, in which the ratio of the diameter of minimum lateral dimension to the wall thickness is of the order of 100 or more are prone to ovalling oscillations. These oscillations are characterized by periodic radial deformation of hollow structures.

The dynamic wind component causes oscillation of structure. The oscillations are generated due to the following reasons.

i. Gust
   The wind velocity at any location varies considerably with time. Wind loads associated with gustiness or turbulence change rapidly and even abruptly, creating effects much larger than those, if the same loads were applied gradually. In addition to a steady wind, there are effects of gusts which last for a few seconds and yield a more realistic assessment of wind load. The intensity of gust depends on its duration. The peak gust is likely to be observed over an average time of 3–15 s depending on the location and size of the structure. Wind loads therefore need to be studied as if they were dynamic in nature. The severity of wind load depends on how fast it varies and also on the dynamic characteristics of the structure. Therefore, whether the pressure on a building created by a wind gust, which may first increase and then decrease, is considered static or dynamic depends to a large extent on the dynamic response of the structure to which it is applied. The action of a wind gust depends on how long it takes the gust to reach its maximum intensity and decrease again, and also on the time period of the building itself. If the wind gust reaches its maximum value and vanishes in a time much shorter than the period of the building, its effects are considered dynamic. On the other hand, the gust can be considered equivalent static load if the wind load increases and vanishes in a time much longer than the fundamental period for the building.
ii. Vortex shedding
   High speed intensity of wind and low stiffness of the building make the building susceptible to vortex-induced vibration. Vortex shedding is the phenomenon by virtue of which

vortices are shed from one side to the other alternatively, due to strong rushing wind, generating zone of low-pressure, resulting into the fluctuating force acting at right angles to the direction of the wind. This type of shedding gives rise to structural vibrations in the flow direction as well as in the transverse direction. When vortex shedding occurs, there is an impulse in the along-wind direction and also in the transverse direction. The transverse impulses are, however, applied alternately to the left and then to the right. The frequency of transverse impulse is precisely half of that of the along-wind impulse. Vortex shedding is predominant in the cross-wind direction than in the windward direction. The frequency of shedding depends mainly on the shape and size of the structure, velocity of flow, and to a lesser degree on surface roughness and turbulence of flow. During vortex shedding, deflection increases at the end of each swing of the structure. If the damping characteristics are small, the vortex shedding can cause building displacements far beyond those predicted on the basis of static analysis. When the wind speed is such that the shedding frequency becomes approximately equal to the natural frequency of the structure, a resonance condition is created. After the structure begins to resonate, further increase in wind speed by a few percent does not change the shedding frequency because the shedding is now controlled by the natural frequency of the structure. The vortex shedding frequency gets locked in with the natural frequency. When the wind speed increases significantly above that causing the lock-in phenomenon, the frequency of shedding is again controlled by the wind speed. The structure vibrates with the resonant frequency only in the lock-in range. For wind speeds either below or above this range, the vortex shedding will not be critical.

When wind acts on a structure, bluff body forces and moments are generated in three mutually perpendicular directions, viz. three translations and three rotations, as shown in Figure 2.1. For civil engineering structures, such as buildings, the force and moment corresponding to the vertical axis (lift and yawing moment) are of little significance. Therefore, in structural design, the flow of wind is simplified and is considered only in two orthogonal directions, i.e., consisting the along-wind and transverse wind.

Along-Wind Direction of Flow

Along-wind is the term used to refer drag forces. For wind blowing along the windward side, dynamic response of structure is due to the buffeting effect caused by air turbulence. Along the windward direction, two components of wind act on the structure; (i) the mean component due to the mean speed of the wind and (ii) the fluctuating component due to the phenomenon of vortex shedding. At low wind speeds, since the vortex shedding occurs at the same instant on either side of the building, the building does not have tendency to vibrate in the transverse direction. It is therefore subjected to along-wind oscillations, parallel to the wind direction. Figure 2.2 shows the schematic of along-wind and across-wind directions.

Cross-Wind Direction of Flow

Transverse wind is the term used to describe cross-wind. Dynamic response of the cross-wind, i.e., wind flowing perpendicular to the wind-direction, is much more adverse than the wind along the windward direction. The cross-wind response causing motion in a plane perpendicular to the

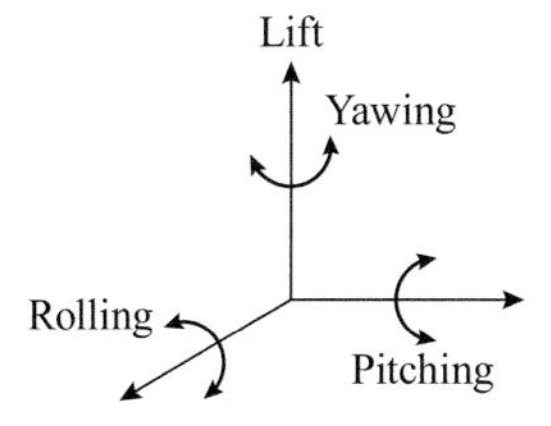

FIGURE 2.1 Direction of bluff body forces and moments.

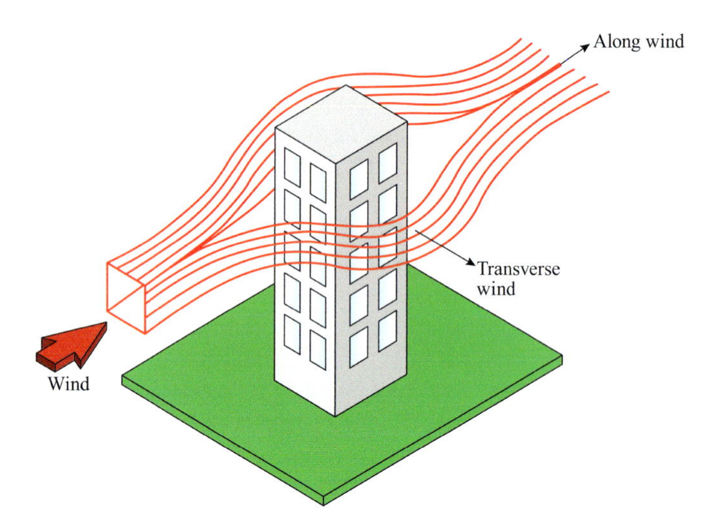

FIGURE 2.2 Along and across wind directions.

direction of wind typically dominates over the along-wind response for tall buildings. In a prismatic building subjected to a smooth wind flow, the originally parallel upwind streamlines are displaced on either side of the building due to boundary layer separation. This results in spiral vortices being shed periodically from the sides into the downstream flow of wind, creating a low-pressure zone due to shedding of eddies, called the wake.

2.2.1 Wind Loading on Buildings

Exposed area of a structure is the key factor for computing wind loading. Wind forces act on exterior/exposed face of structure. The impact of wind forces at the interior face of structure is less, depending on the shielding effect. Due to the increasing need for residential/office space in urban environment, there is a continuous demand for infrastructure like tall buildings with innovative and complex architectural shapes, large-span cable-stayed bridges, communication towers, and such other structures. As the height of the building increases, wind velocity goes on increasing, thereby increasing the wind pressure toward the top. Wind being random both in time and space, for evaluation of wind loads for design of tall buildings and structures, it is necessary to understand the random nature of wind and its effects on structures, viz. aerodynamic pressures/forces, spatial correlations, dynamic amplification, and vortex-induced vibrations. Due to scarcity of land, the structures are commonly closely spaced. This leads to wind-induced interference effects on the primary structure. Therefore, while designing tall buildings for wind loads, it is necessary to consider the presence of neighboring buildings. In designing a building for wind effects, the influence of neighboring buildings and land configuration on the sway response of the building can be substantial. The sway at the top of a tall building caused by wind may be of concern to those occupying its top floors. Further, if the building has a tendency to twist, its occupants may get an illusory sense that the world outside is moving, creating symptoms of vertigo and disorientation. Further, wind creates turbulent air flow of varying nature, with increase in the height of the structure. For example, building located in the free vicinity will have maximum velocity at lower height as compared to the taller building surrounded by structures on all sides blocking the wind.

2.2.1.1 Designing a Building for Wind

In designing a building for wind, it is important to consider (i) its strength and stability; (ii) fatigue in structural members and connections caused by fluctuating wind loads; (iii) excessive lateral deflection (sway) that may cause cracking of internal partitions and external cladding, misalignment of mechanical systems, and possible permanent deformations of nonstructural elements; (iv) frequency and amplitude of sway that can cause discomfort to occupants of tall, flexible buildings; (v) possible buffeting that may increase the magnitude of wind velocities on neighboring buildings; (vi) annoying acoustical disturbances; and (viii) resonance of building oscillations with vibrations of elevator hoist ropes.

2.2.1.2 Mitigation of Wind Forces on Tall Buildings

Wind pressure and wind forces acting on a building can be effectively controlled by employing any suitable method, such as (i) moderating the shape of the building by rounding the corners, providing a spiral, tapered, and setback shapes that significantly reduce the wind forces; (ii) constructing buildings with increased height and lesser width, employing different cross-sections at specified height; (iii) providing large number, and large size openings, so as to allow free passage of wind; (iv) dissipating wind energy by providing additional damping by implementing structural control strategies.

2.2.2 Wind Loading on Bridges

Dynamic effect of wind in the design of long-span bridges needs a very detailed investigation. Like tall buildings, dynamic effect is due to vortex-shedding and the fluctuating wind velocities generated around the bridge due to air turbulence. There is a long history of wind-related problems that arise with different types and different components of bridges. The dynamic response to wind remains a vital and complex consideration in the design of long-span bridges. While it is the response of the road deck that has received the most attention of researchers, other components of bridges are also prone to wind-induced vibrations and have caused severe problems. There are only a small number of well-documented aerodynamic phenomena that can give rise to the dynamic response of elastic structures having slender bluff aerodynamic shapes. When bridge is subjected to the winds with small velocity fluctuations, two types of wind-induced vibrations of bridge occur: vortex-induced vibrations and self-excited vibrations. The response of both types of vibrations is greatly affected by aerodynamic forces and vibratory motion of the bridge because of the aeroelastic effect, which is an interaction between air and the elastic bridge components. Vortex-induced vibration is a result of separate wind flows across the bridge, at the wind speed lower than that of initial wind speed of the self-excited vibrations. The dominant motion causing vortex-induced vibrations reacts in vertical direction and also in wind direction causing torsional modes of vibration. However, the maximum amplitude of vortex-induced vibration is reduced if there is damping provided in the bridge. Due to vibration in the bridge, self-excited aerodynamics forces are induced, causing self-excited vibration in the bridge.

In a bridge, generally galloping occurs in cross-wind direction and is in the vertical direction, whereas flutter causes twisting. Once the self-excited vibration occurs at initial wind speed, the response amplitude increases, with increase in the wind speed. The design wind speed is kept low, compared to the allowable design wind speed, to prevent self-excited vibrations.

The most renowned bridge collapse due to winds is the Tacoma Narrows suspension bridge linking the Olympic Peninsula with the rest of the state of Washington. It was completed and opened to traffic on July 1, 1940. Its 853-m main suspension span was then the third longest in the world. The bridge became famous for its serious wind-induced problems that began to occur soon after it was opened to traffic. Just after four months, the bridge deck oscillated through large displacements in the vertical vibration modes at a wind velocity of about 68 km/h. The motion subsequently changed to a torsional mode. Finally, some key structural members and cables were overstressed, and the main span collapsed. This failure revealed the importance of consideration of the wind-induced dynamic response in flexible bridges with longer spans and cable-supported bridges. Tacoma narrows bridge-experienced resonance from vortex shedding. A small amount of twist in the bridge triggered vortices and zone of low pressure at certain locations that amplified the twisting motion. Torsional aeroelastic flutter eventually induced large stresses in the suspension cables, leading to failure. The failure could have been prevented if the deck was less flexible. Bridge with large steel plates on either side creates some strange interactions with the wind. Therefore, a truss-type deck, through which wind can flow, is a better option.

2.2.2.1 Mitigation of Wind Effects on Bridges

To avoid flutter, a gap can be introduced at the center of the bridge deck so as to equalize the pressure on either side. Alternatively, the bridge deck can be made more aerodynamic, to avoid formation of vortices, that applies push and pull on the structure. Considering the local terrain features of the bridge site, the wind data is analyzed to obtain the maximum wind velocity, dominant direction of wind, turbulence intensity, and the complete wind spectrum. The design period represents a level of failure risk and not the service life of bridge. Wind velocity for flutter (stability) design is usually about 20% larger than that for buffeting (strength) design, although the design period for the former is several orders higher. Once wind characteristics and design velocity are available, a wind tunnel/analytical investigation is conducted.

2.2.2.2 Design Considerations

The aerodynamic behavior of bridges depends mainly on the following key factors.

i. Structural form
   Suspension bridges, cable-stayed bridges, arch bridges, and truss bridges, due to the increase of rigidity in this order, generally have aerodynamic behavior from worst to best. A truss-stiffened section blocks less wind, and hence, it is more favorable than a plate girder-stiffened one. However, a truss-stiffened bridge generally has limited stiffness in torsion.
ii. Stiffness
   For long-span bridges, it is not economical to provide larger sections to increase its stiffness. However, changing the boundary conditions, such as deck and tower connections in cable-stayed bridges, may significantly improve the stiffness. Cable-stayed bridges with A-shaped or inverted Y-shaped towers have higher torsional frequency than the bridges of H-shaped towers.
iii. Type of cross section
   A streamlined section blocks less wind and therefore has better aerodynamic behavior than a bluff section. Small changes in section details may significantly affect the aerodynamic behavior. While designing bridges for dynamic wind loading, major design parameters are usually determined in the preliminary design stage, and then the aerodynamic behavior is evaluated. If the response of the bridge under aerodynamic excitation is not up to the mark, the common way to improve its behavior is to change the deck section. For example, changing the solid parapet to a half-opened parapet or making some venting slots on the bridge deck may significantly improve the aerodynamic behavior.
iv. Damping
   Concrete bridges have higher damping ratios than steel bridges. Consequently, steel bridges have more wind-induced problems than concrete bridges. An increase in damping by installing structural control devices can reduce aerodynamic vibration significantly.

2.2.2.3 Construction Safety

The most common construction method for long-span bridges is segmental (staged) construction, such as balanced cantilever construction of cable-stayed bridges, tie-back construction of arch bridges, and float-in construction of suspended spans. The staged constructions result in different structural configurations during the construction time. Some construction stages have lower stiffness and natural frequency than the completed bridges. Hence, construction stages could often be more critical in terms of either strength of structural members or aerodynamic instability. In the balanced cantilever construction of cable-stayed bridges, three stages are usually identified as critical: the tower before completion, completed tower, and the stage with a longest cantilever arm. Wind tunnel testing is usually recommended and conducted to ensure safety.

2.2.2.4 Cable Vibrations

A common wind-induced problem in long-span bridges is cable vibration. There are a number of types of wind-induced vibrations in cables, individually or as a group, such as vortex excitation, wake galloping, excitation of a cable by imposed movement of its extremities, rain/wind- and ice/wind-induced vibrations, and buffeting of cables in strong turbulent winds. While the causes of cable vibrations are different from each other and the theoretical solutions complicated, following common measures for mitigating cable vibrations can be adopted.

i. Increasing damping
Increasing damping is an effective way of controlling all kinds of cable vibrations. The cables are usually flexible and inherently low in damping; an addition of external damping, usually at the cable ends, can dramatically reduce the vibrations.
ii. Increasing natural frequency
The natural frequency of cable depends on the cable length, the tension force, and the mass. Since the cable force and the mass are determined from the structural analysis and design, commonly the cable length is reduced by using spacers or cross cables.
iii. Changing cable parameters
A change in the cable shape characteristics by increasing the surface roughness or adding protrusions to the cable surface reduces the rain/wind- and ice/wind-induced vibrations.
iv. Implementation of vibration control devices
Vibration control devices could be implemented to attenuate the excessive cable vibration caused by dynamic excitations.

2.2.3 Wind Loading on Chimneys

Tall chimneys and cooling towers are highly susceptible to dynamic wind action. Pollution regulations have forced to increase the height of chimney, and use of high-strength material, thus permitting smaller diameter of shell. This makes the chimneys slender and sensitive to wind-induced vibrations. Along-wind effect on chimneys is due to the buffeting action. For accurate evaluation of along-wind load, chimney model is prepared as a bluff body, and turbulent wind flow test is performed. Most of the codes use equivalent static method for analysis, also known as gust factor method. For reducing effect of wind on chimney, its shape can be modified, i.e., instead of circular chimney of constant cross section, tapered chimneys with low drag and lift coefficient is a better option.

2.2.4 Wind Loading on Towers

Communication towers are required for television, radio, Internet, and mobile phones. Two types of commonly used towers are self-supporting tower, which acts like a cantilever fixed at the base, and guyed tower, which is laterally supported by cables, attached at different levels of the tower along its height. Effect of wind on tower depends on its shape. Failure occurs when the frequency of wind excitation approaches the frequency of the tower. Supplementary damping systems can be installed to effectively control the dynamic response of towers under the effect of wind load.

The dynamic response of towers to wind is quite similar in nature to that of tall buildings. However, the higher modes of vibrations in tower are more likely to be significant in the resonant dynamic response. Because of the smaller cross-wind dimension of tower, the velocity at which the vortex shedding frequency coincides with the first mode vibration frequency is usually much lower than that for tall buildings. Many observation towers and communication towers have circular/square/symmetrical plan. Other cross sections may require wind tunnel tests to determine drag coefficient. The mean or time-averaged drag force per unit height, and hence bending moments, can be calculated using an appropriate sectional drag coefficient with a wind speed suitable to the height, using an appropriate expression for mean wind speed profile.

2.2.5 Wind Tunnel Test

For tall and exclusive-shaped structures, wind tunnel test is recommended. In wind tunnel test, boundary layer conditions are prevailed and the small-scale model of the prototype is tested to simulate the effect of natural wind loading acting on the structure. In order to replicate the practical effect of the topographical environment surrounding the structure, provisions are made in wind tunnel to produce identical environment. The scale of the model is decided such that it gives accurate representation of the features of the original structure. The first wind tunnel machine was made in 1893 by Kernot in Melbourne, Australia, to assess wind forces on buildings. It was called as blowing machine and is shown in Figure 2.3.

Kernot used the blowing machine to study wind forces of various models like cubes, pyramids, cylinders, and roof of various pitches. Kernot's application is now termed as open-test or open-circuit section.

Wind tunnel tests are now very frequently conducted for aeronautical applications, high-rise buildings, towers, chimneys, and cable-supported bridges. Due to enormous modernization in construction industries in metropolitan cities, large size towers are being constructed. For esthetic purpose, construction of unusual shapes of buildings is now very common. Such structures are known as code-exceeding buildings. Along-wind and across-wind response of structures are important from design point of view; however, the appropriate clauses are not covered in detail in standards and codes. Geometry, surrounding conditions and terrain category, primarily influences the response of structures to wind loads. Hence, for unconventional geometry and height, wind tunnel test is carried out for computing wind response of structures. Different international codes recommend wind tunnel tests, depending upon the slenderness ratio, height of structure, and the wind speed. The test results are precise and are considered for structural analysis and design.

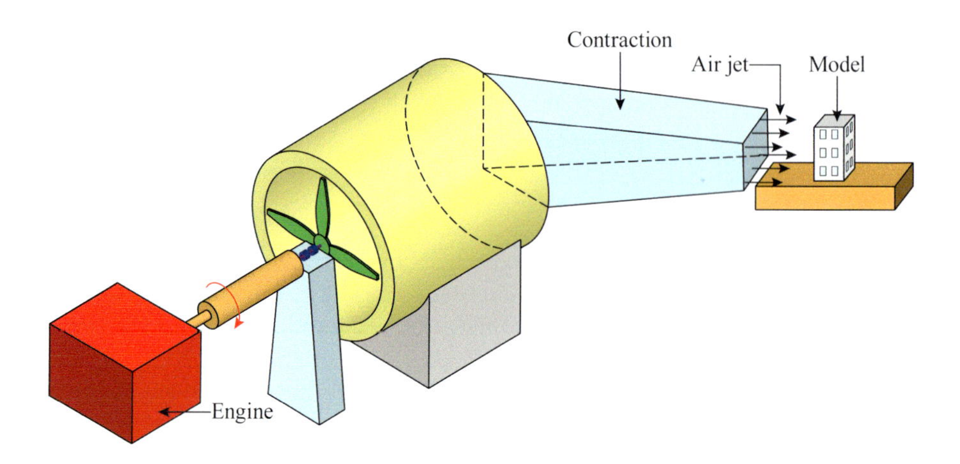

FIGURE 2.3 The Kernot's blowing machine.

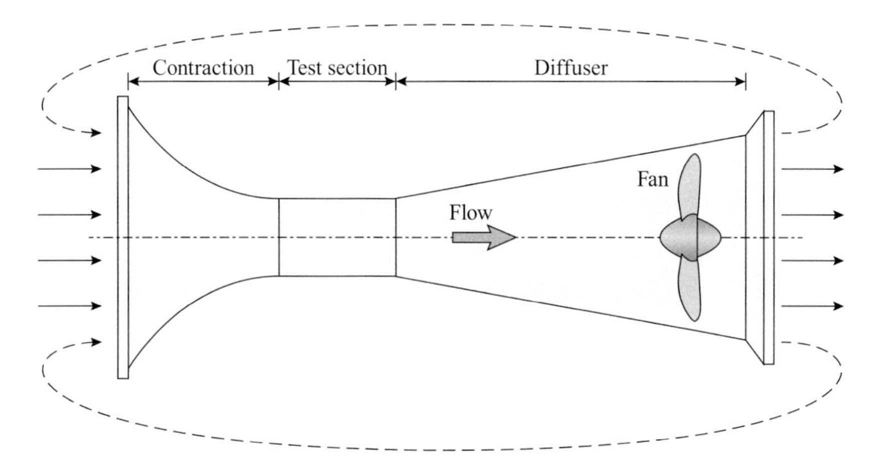

FIGURE 2.4 Open-circuit wind tunnel with small test section.

2.2.5.1 Types of Wind Tunnel Tests

The two basic types of wind tunnels are open-circuit and the closed-circuit.

Open-circuit wind tunnel is also known as Eiffel tunnel. The cost of construction for open-circuit wind tunnel is less, and it requires small area for installation of unit, compared to the closed-circuit type. However, this tunnel produces a lot of noise during operation. Open-circuit wind tunnel fan exhausts the air flow from around the room, and air is allowed to pass through testing section. Figure 2.4 shows an open-circuit wind tunnel with small test section. Photo 2.1 shows open-circuit wind tunnel with large testing section at IIT Roorkee, India.

Figure 2.5 shows open-circuit wind tunnel at IBHS Research Center, South Carolina. It is a full-scale tunnel for testing single- and double-story residential and commercial structures.

Closed-circuit wind tunnel is also known as Prandtl tunnel. The first use of closed-circuit wind tunnel was in Germany. In closed-circuit tunnel, air flow is transmitted constantly through the vanes. The wind tunnel fan exhausts the air flow from the vanes of tunnel. Air passes through test section, and from the outlet of test section, it is carried to the vanes. Closed-circuit wind tunnel is costly, as it contains vanes and ducts. Compared to open-circuit type, large area is required for installation of unit, and the testing operation is not noisy. Because of hotter working conditions, it is essential to employ heat exchanger or active cooling system. Figure 2.6 shows closed-circuit wind tunnel with small testing section. Further, Photo 2.2 shows closed-circuit wind tunnel test set up with large testing section, at the national wind tunnel facility at IIT Kanpur, India.

PHOTO 2.1 Open-circuit wind tunnel with large testing section, IIT Roorkee, India. (Courtesy: Indian Institute of Technology (IIT) Roorkee.)

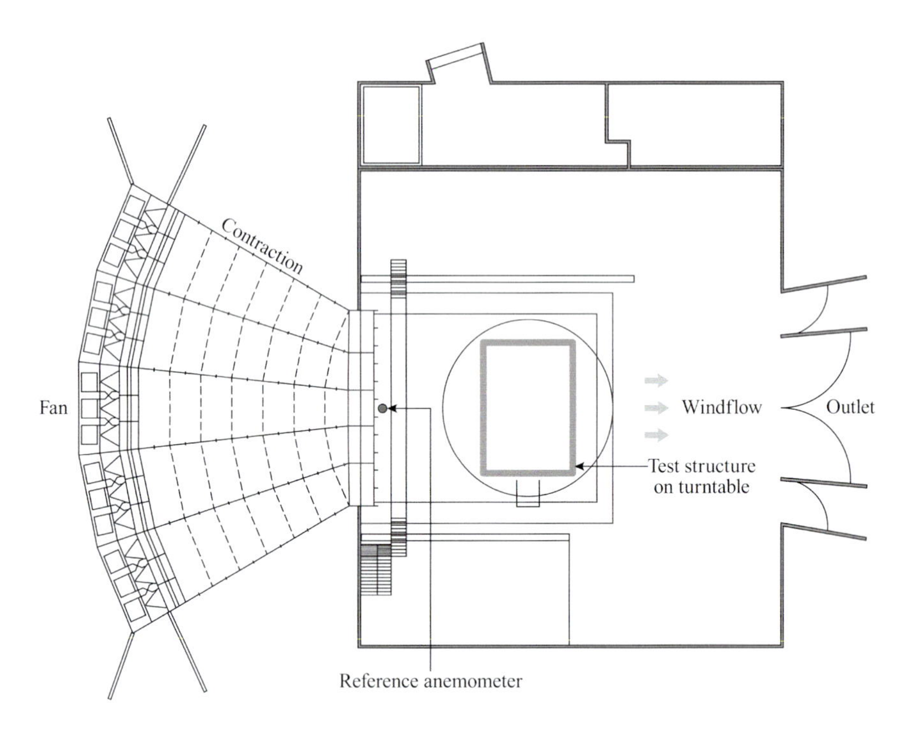

FIGURE 2.5 IBHS Research Center, South Carolina: plan view of open-circuit wind tunnel. (Cope et al. 2012)

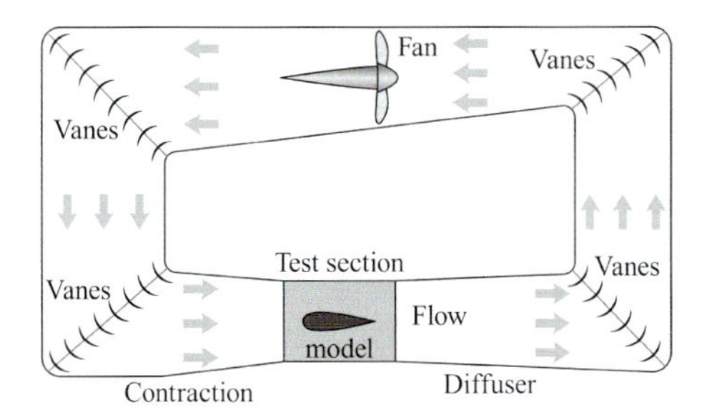

FIGURE 2.6 Closed-circuit wind tunnel with small testing section.

PHOTO 2.2 Closed-circuit wind tunnel with large testing section, IIT Kanpur, India. (Courtesy: NWTF, Indian Institute of Technology (IIT) Kanpur.)

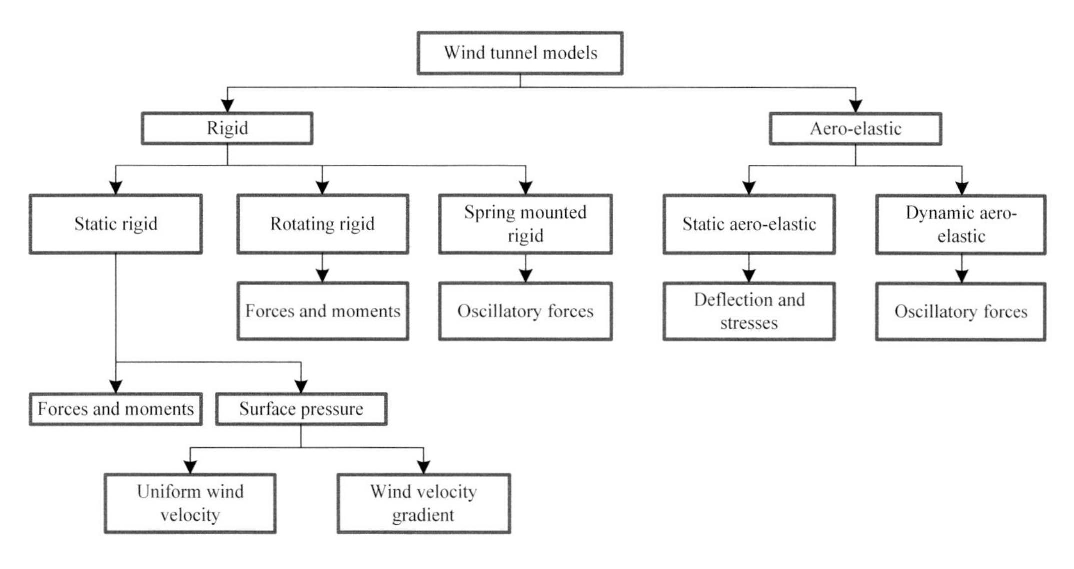

FIGURE 2.7 Classification of wind tunnel models.

Model test may be required to ascertain static and dynamic loads for a structure of unique shape with different features/components, located in unusual terrain condition like hilly area, or high wind valley. Figure 2.7 shows classification of wind tunnel models.

A small-scale replica of real structure is called as the model, and the real structure is known as prototype. The model and the prototype are exactly similar. The model and prototype have geometrical, kinematic, and dynamic similarities. Wind tunnel test on model gives comprehensive information about the performance of the real structure. Figure 2.8 shows the similarities between the prototype and model; m and p correspond to model and prototype, respectively.

2.3 WAVE LOADING

Wave loading is considered to be one of the most important environmental loads acting on offshore structures. The loads due to waves very often dominate and dictate the design of offshore structures and bridges located in the coastal zones. Figure 2.9 shows marine structures with mobile and fixed

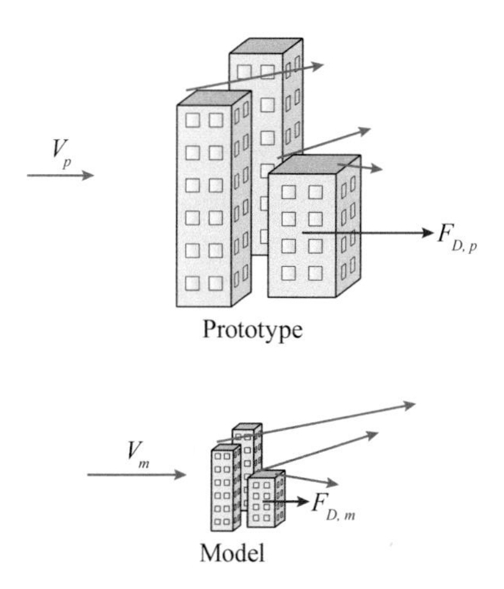

FIGURE 2.8 Similarities between prototype and model.

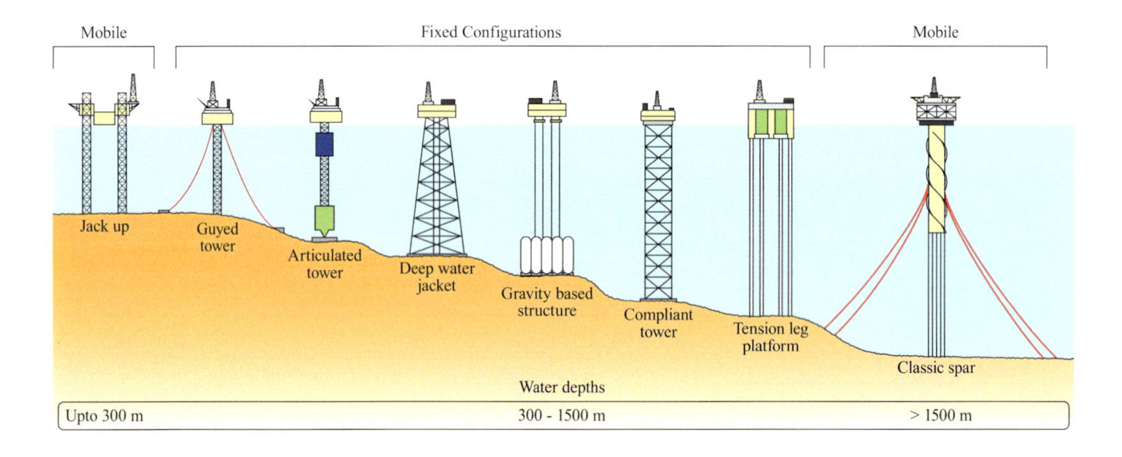

FIGURE 2.9 Marine structures with mobile and fixed configurations.

configurations. The type of structure depends upon the purpose and depth of water at a particular location.

Wave load effects are required for design checks of ultimate, accidental, and fatigue limit states of offshore structures. Ultimate and accidental limit states often govern, depending on extreme loading conditions. For structures in extratropical climates, fatigue may also be an important design criterion and requires an estimate of the number of stress ranges at different loading magnitudes for the service life of the structure. It is reported that the main contribution to fatigue damage is caused by load effects that are of the order of 15–25% of the extreme load effects. Wave load effects for design are commonly determined by quasi-static analysis methods when the structural modes have time periods in the lower range of wave periods, while an intrinsically dynamic analysis approach is required for structures with natural period above the time period of the wave. Figure 2.10 shows bending modes of different offshore structures, depending upon the wave load period.

Besides considering wave loads with period equal to the wave period in the range of 2–20 s, the presence of certain nonlinear features of loading may cause steady-state loads with a period which is a fraction or a multiple of the wave period. Such steady-state loads, as well as wave impact loads and other transient loads, may cause inertia and damping effects which need to be accounted for by using a proper dynamic analysis methodology, for platforms with natural periods below the wave period. For situations where dynamic effects need to be considered, the long-term stochastic character of the wave loading and its effects are important.

2.3.1 Wave-Induced Dynamic Forces

Wind-generated waves consist of a large number of wavelets of different heights, periods, and directions superimposed on one another. Although sea waves are not regular, regular waves can closely model some swell conditions. They also provide the basic components in irregular waves and are commonly used to establish wave conditions for design. Determination of wave forces requires solution of two separate but interrelated problems.

2.3.1.1 Computation of the Sea State

The sea state is computed using an idealization of the wave surface profile and the wave kinematics. The wave kinematics include water particle velocity and variation, in terms of its velocity and acceleration along the depth at any special interval at any specific time. For locating the sea state for designing the system, it is required to select and idealize the wave surface profile and then use any specific theory to compute the wave kinematics or water particle kinematics.

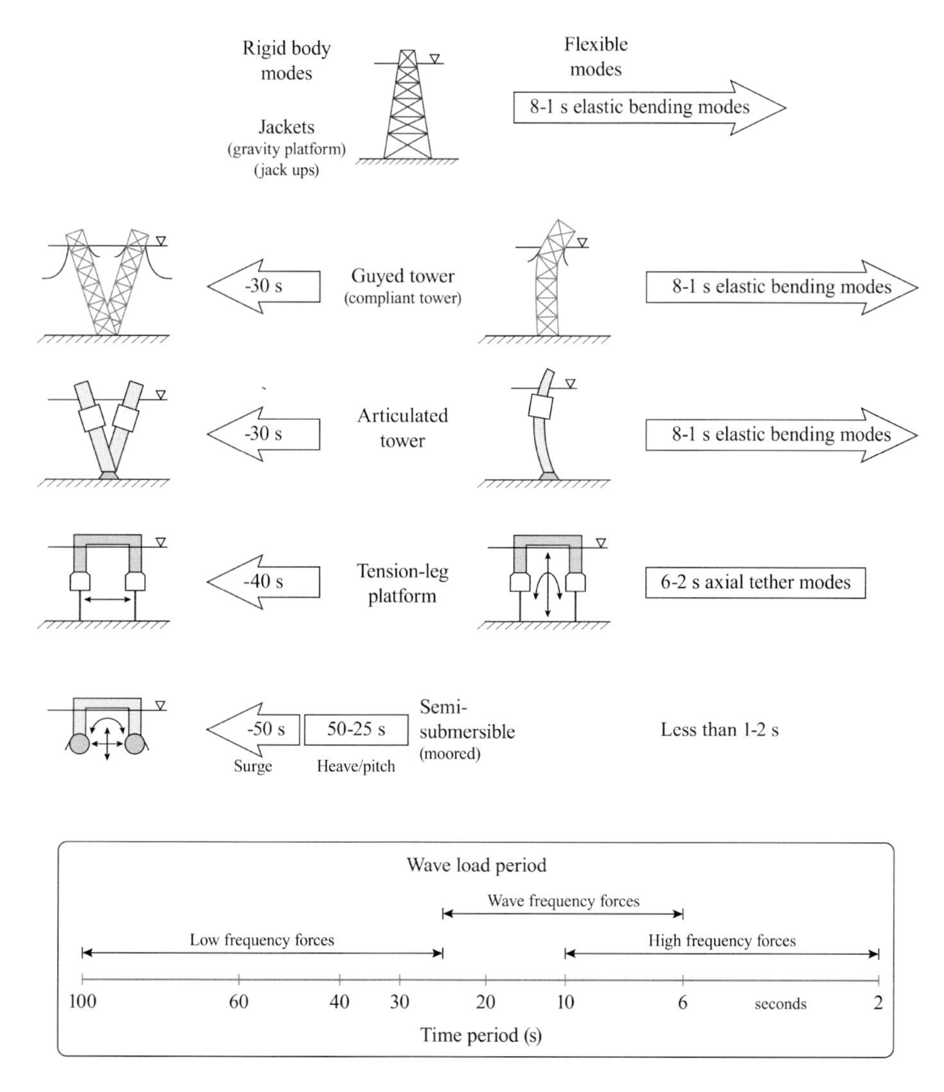

FIGURE 2.10 Natural periods of marine structure and wave excitation period.

2.3.1.2 Computation of the Wave Forces

To estimate wave forces individually on each member and then on the whole structure, following two analyses are commonly used.

a. Single-design wave analysis
   A single-design wave is a regular wave that has well-defined wave height and period. The forces due to this wave are calculated using a higher-order wave theory. Usually a 100-year wave, i.e., the maximum wave height for the return period of 100 years, is selected as a single-design wave for which forces are computed. In such cases, no dynamic behavior of the structure is considered. The limitation of this methodology is that it can be successfully applied to shallow water structures up to the depth of 200 m.
b. Random wave analysis
   Random wave analysis depends on the statistical methodologies. Statistical analysis is done on the basis of wave scatter diagram. For every offshore location, the wave scatter

diagram is available in the literature. Appropriate wave spectra are defined to perform the analysis in frequency domain. These spectra are used to generate a random wave, if dynamic analysis for extreme wave is required for deep water structures.

2.3.2 Dynamic Response of Offshore Structures

For offshore structures in deeper waters, such as those shown in Figure 2.9, because of their compliancy and flexibility introduced in the design by geometry and structural design, dynamic effects on these structures are predominant.

2.3.3 Morison's Equation for Modeling of Wave Forces

Wave theories describe wave kinematics, which is a variation of the velocity and acceleration along the depth, along the spatial values, with respect to time. The wave theories assume that the waves have long crests and can be described by two-dimensional flow field. The waves are characterized by wave height and wave period, which depend on the water depth, wave number, circular frequency, and cyclic frequency. The computation of the forces exerted by waves on a cylindrical object depends on the ratio of the wave length to the member diameter. When this ratio is large (>5), the member does not significantly modify the incident wave. The wave force can then be computed as the sum of a drag force and an inertia force, as shown in Equation (2.6). It is a very classical equation given by Morison:

$$\{F\}=\{F_D\}+\{F_I\}=C_D\left(\frac{w}{2g}\right)AU|U|+C_M\left(\frac{w}{g}\right)V\frac{\partial U}{\partial t} \tag{2.6}$$

where $\{F\}$ is the hydrodynamic force vector, $\{F_D\}$ is the drag force vector, and $\{F_I\}$ is the inertia force vector. Furthermore, C_D is the drag coefficient, w is weight density of water, g is gravitational acceleration, A is the projected area normal to the cylinder axis per unit length, V is the displaced volume of water per unit length, D is the effective diameter of cylindrical member including marine growth, U is component of the velocity vector (due to wave and/or current), C_M is the inertia coefficient, and $\left(\frac{\partial U}{\partial t}\right)$ is the component of the local acceleration vector of water. The limitation of Morison's equation is that it can be applied only when $\left(\frac{D}{L}<0.2\right)$, where D is the member diameter and L is the wavelength, i.e., Morison's equation can be used to compute the forces on slender and hydrodynamically transparent bodies. When $\left(\frac{D}{L}>0.2\right)$, they are called large-volume bodies. In this case, diffraction theory is used for computation of wave forces. A crucial issue in applying Morison's equation is the determination of C_D and C_M. Extensive data from laboratory experiments indicate a general range of the drag coefficient is $0.6 \le C_D \le 1.2$, and that of the inertia coefficient is $1.2 \le C_M \le 2$, depending upon flow conditions and surface roughness. According to Morison's equation, the drag force is nonlinear. This nonlinear formulation is used in designing the offshore structures. For determining the transfer function which is required for frequency domain calculations, the drag force must be linearized. Therefore, frequency domain solutions are appropriate for fatigue life calculations, for which the forces due to the operational level waves are determined by linear inertia term. The nonlinear formation and hence the time domain solutions are required for dynamic analysis of deep-water structures under extreme storm waves, for which the drag portion of the force is dominant.

2.3.4 Pierson-Moskowitz Wave Height Spectrum

One of the primary loading on offshore structures, such as offshore jacket platforms and tension-leg platforms, is the surface waves. The surface wave is usually produced due to wind, and the wave force is modeled as a zero mean ergodic Gaussian process. To represent such waves, the Pierson-Muskowitz wave height spectrum, which is given in Equation (2.7), can be used.

$$S_{hh}(\omega) = \frac{\alpha g^2}{\omega^5} exp\left[-\beta\left(\frac{g}{\omega u'}\right)^4\right] -\infty < \omega < \infty \tag{2.7}$$

where α and β are dimensionless constants equal to 8.1×10^{-3} and 0.74, respectively, $u' = 6.8471\sqrt{H_s}$ is the mean wind velocity, and H_s is the significant wave height.

2.3.5 Governing Equation of Motion for Wave Loading

Excitation acting on a structure is due to wave loading. The mass, damping, and stiffness matrices are made up by contributions from the structure, water, and soil depending on the support conditions of the structure. Equations of motion may be formulated in the time or frequency domain. The choice of formulation depends especially on possible frequency dependence and on nonlinearities of the dynamic properties.

Mass is contributed by structural and contained mass as well as the added hydrodynamical mass.

For a slender cylinder, the hydrodynamical mass is usually taken to be that of the displaced water. The added mass for large-volume structures like caissons of floating gravity structures, and floating bridges, has to be determined by potential theory for the relevant modes of behavior.

The added mass is frequency-dependent.

Damping may be contributed by the structure, water, and soil (rock) and is subject to significant uncertainties. If the soil or rock is activated during vibration, it will contribute radiation and hysteretic (material) damping. Soil damping, especially in rocking motion, is frequency-dependent. If a nonlinear soil model is used, the hysteric damping must be implicitly included in the analysis.

Stiffness is contributed by the structure, water, and the soil (rock). The contribution to the stiffness by water is due to hydrostatic effects. Linear elastic structural models are usually applied, except for possible catenary mooring lines. Water provides buoyancy that influences the stiffness of the bridge supported by pontoons but would be negligible for bottom-supported platforms. The soil is of importance for bottom-supported platforms and may be modeled by equivalent linear properties or by a more refined nonlinear material model. Even if soil stiffness properties are frequency-dependent, the low frequency of water loading implies that the dynamic stiffness is close to the static values.

Figure 2.11 shows simplified dynamic model for tower platform. The $P-\Delta$ effect for platforms with large displacement could be taken into account by linearized negative springs. The equations

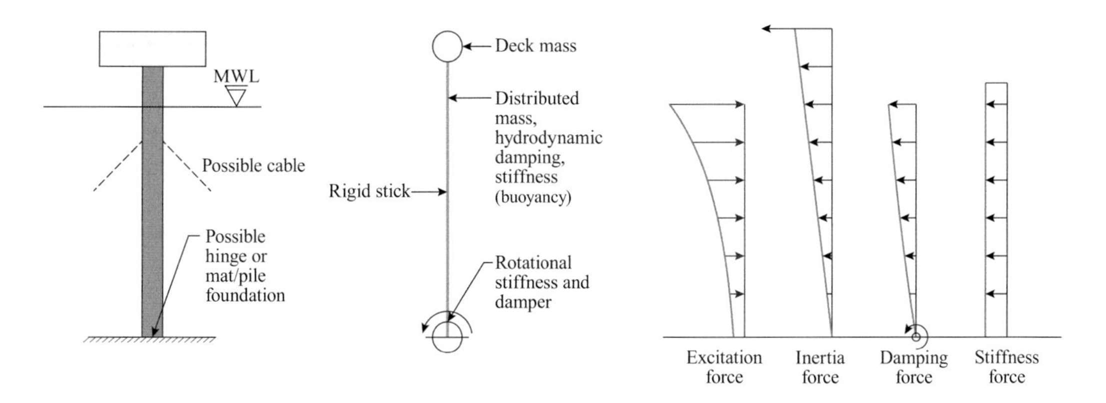

FIGURE 2.11 Simplified dynamic model for of tower platforms.

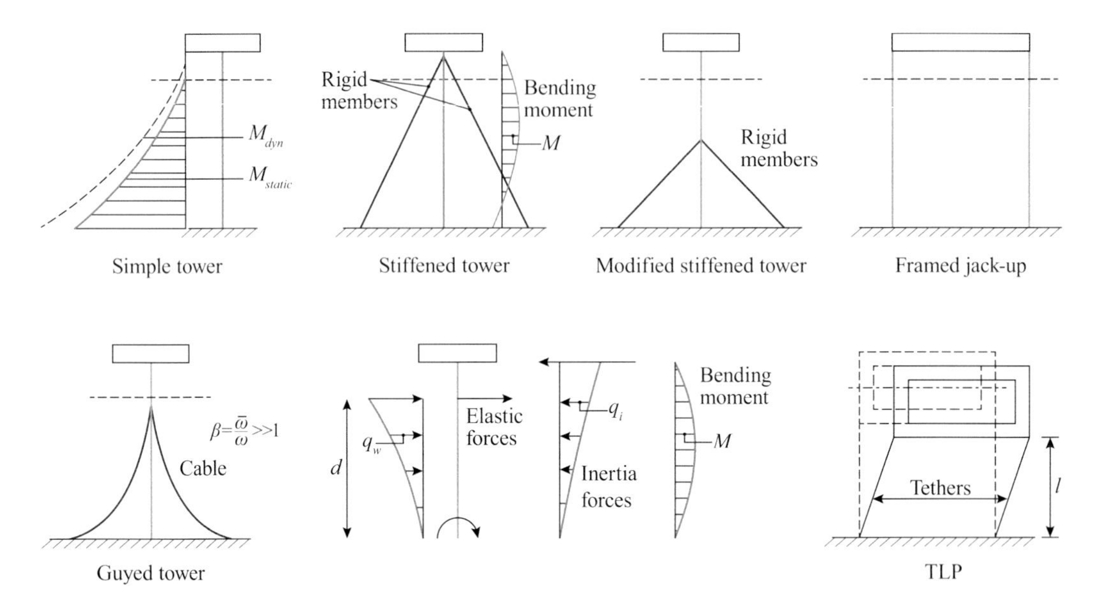

FIGURE 2.12 Types of towers and forces acting.

of motion may be formulated in the time or frequency domain. The choice of formulation depends on possible frequency dependence and nonlinearities of the dynamic properties. Figure 2.12 shows different types of towers and the forces acting on them.

As a cantilever beam, the fixed tower experiences a significant overturning moment and shear force due to waves. Also, the fundamental period of vibration increases with increasing water depth and approaches the range of wave periods associated with significant energy. This fact implies that the response will be dynamically amplified to an increasing extent with increasing water depth. A better platform design for deep water is therefore to stiffen the tower by rigid inclined members, which forms a triangular truss. The bending moment in the central column is then reduced, as the tower essentially becomes a beam, supported at both ends. However, the inclined members also need to be sized adequately. Since these members are subjected to significant lateral loads, their design may not be very cost-effective. The maximum moment in the central tower is significantly reduced, and the truss is less exposed to lateral loads, when it is located at a larger water depth.

2.3.6 Structural Response under Wave Loads

Structures exposed to wave loading experience considerable forces resulting from dynamic pressure variations and water particles motion. The forces are much complex and higher in magnitude compared to the wind forces. Therefore, the pertinent effects shall be considered carefully in the response determination, and appropriate response time history analysis method may be implemented to quantify the response quantities.

REFERENCES

ASCE 7-16. (2017). Minimum Design Loads and Associated Criteria for Buildings and Other Structures. American Society of Civil Engineers.

Cope, A. D., Crandell, J. H., Johnston, D., Kochkin, V., Liu, Z., Stevig, L., Reinhold, T. A. (2012). Wind Loads on Components of Multi-Layer Wall Systems with Air-Permeable Exterior Cladding. ATC & SEI Conference on Advances in Hurricane Engineering: Learning from Our Past, Miami, Florida.

3 Bracing Systems

3.1 INTRODUCTION

Dynamic loads can impart a significantly greater effect toward the response of a structure compared to static loading. As such, the properties of the structure, such as lateral stiffness and damping, play a pivotal role in achieving efficient structural performance against dynamic loads, such as typhoons, earthquakes, blasts, and many others. To avoid structural failure and minimize the excessive lateral movement of the structure, part of the energy exerted by the dynamic loading needs to be dissipated. One of the most efficient methods to achieve this objective is providing structural bracings, which work by providing lateral stiffness and stability to the structure, especially for the multistory and high-rise buildings. Bracing systems subsequently increase the lateral load resistance of the structure and reduce the internal forces in the primary structural system through an appropriate arrangement of members. By incorporating bracings, a stiffer and more economical structure can be achieved. In such systems, the beams and columns forming the primary structural system could typically be designed to resist vertical loads, whereas the bracing systems resist lateral loads. In high-rise structures, bracing frames are provided by arranging the beams and columns in an orthogonal form. Braced frames may be considered cantilevered vertical trusses resisting lateral loads, primarily through the axial stiffness of columns and braces. Since the lateral loads are reversible, braces are subjected to alternate cycles of compression and tension. Hence, they are most often designed for the more severe case, i.e., compression. The diagonal bracing members work as web members resisting the horizontal shear in axial compression or tension, depending on their orientation and direction of inclination. The beams act axially when the system is an entirely triangular truss. They undergo bending only when the braces are eccentrically connected to them. Common types of bracings include the buckling-restrained brace (BRB), concentrically braced frame (CBF), and eccentrically braced frame (EBF). Figure 3.1 presents different structural vibration control approaches, including bracing systems.

3.2 BUCKLING-RESISTANT BRACES (BRBs)

During severe earthquake events, the braces may be subjected to repeated cycles of lateral loading and stresses beyond their elastic limit. In such cases, the braces may yield in tension and buckle in compression. The buckling of braces in compression prevents the utilization of the full capacity of the braces in compression. Thus, the bracing ability of the conventional steel is greatly hampered. The tendency of buckling is influenced by the section property, the compressive force, and the unbraced length of the steel core. The buckling of the steel core results in a severe reduction in the capacity of the braces to resist the earthquake actions and dissipate the energy. To overcome this problem, the concept of buckling-resistant braces (BRBs) was proposed in order to obviate the buckling of the braces and to make the bracing system robust in both tension and compression. A BRB is a steel brace that does not buckle in compression but instead yields in tension as well as in compression. A buckling-restrained braced frame (BRBF) is a CBF that incorporates a buckling-restrained brace. Figures 3.2 and 3.3 illustrate the behavioral difference between the conventionally braced frame and buckling-restrained braced frame under lateral load.

Figures 3.4 and 3.5, respectively, portray the performance of conventional brace and buckling-restrained brace under axial compression. The conventional steel brace is prone to buckling before the compression stress reaches the yield point, whereas the BRB is not susceptible to such buckling because of the lateral support against buckling, as shown in Figure 3.5.

DOI: 10.1201/9781315269269-3

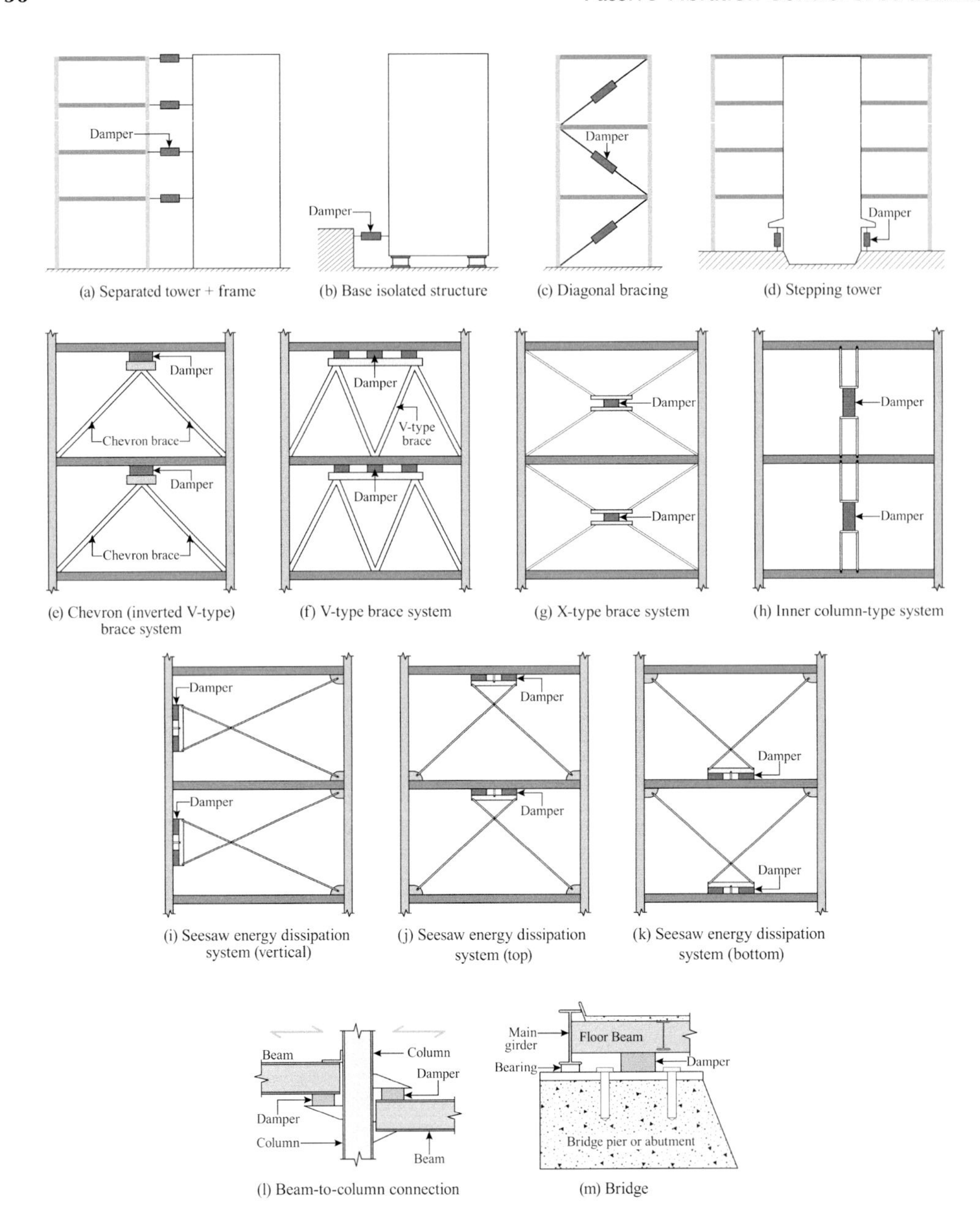

FIGURE 3.1 Different techniques of structural vibration control.

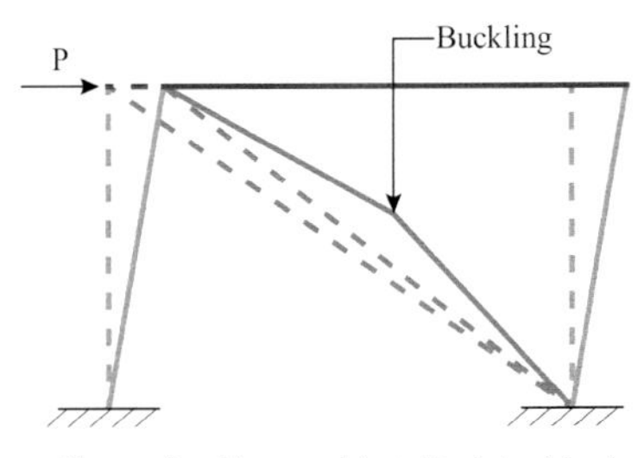

FIGURE 3.2 Behavior of CBF under lateral load.

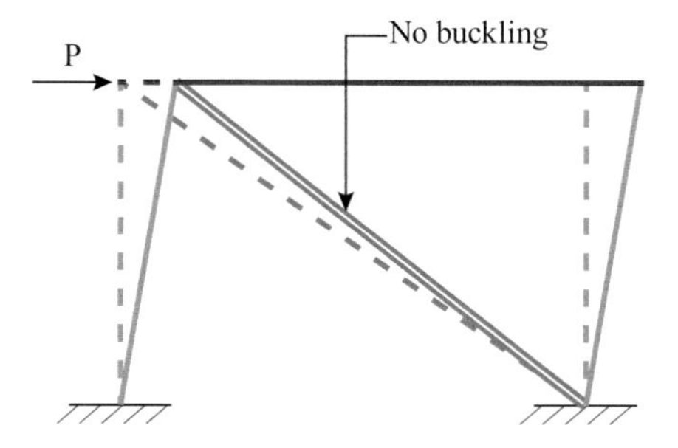

FIGURE 3.3 Behavior of BRBF under lateral load.

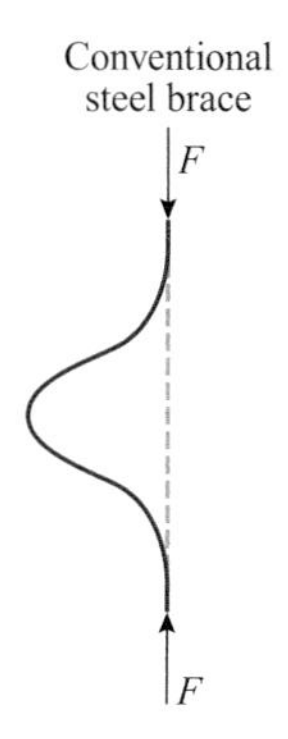

FIGURE 3.4 Behavior of conventional brace under axial compression.

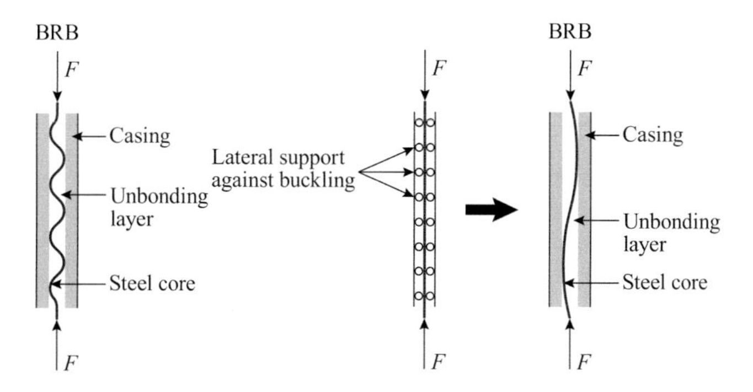

FIGURE 3.5 Behavior of BRB under axial compression.

History of Buckling-Restrained Braces

The concept of buckling-restrained braces, wherein flat steel braces are sandwiched between precast concrete panels, was first proposed in Japan. The concrete panels are used to provide lateral restraint. After that, innovative BRBs were investigated in Japan during the early period of the 1980s. Nippon Steel Corporation, in collaboration with Professor Wada of the Tokyo Institute of Technology, manufactured the first BRB system in Japan using unbonded braces. The system was designed to have a wider section at the end as compared to the reduced section in the middle portion. It was just like the shape of a typical human bone. Further, efforts were made to debond the steel core and concrete to make the brace capable of resisting the horizontal loading and preventing the

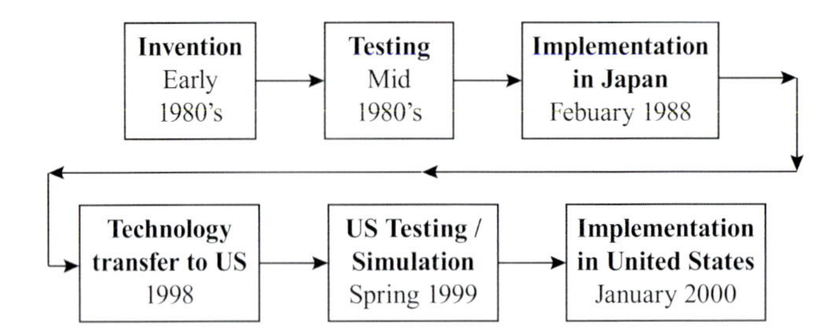

FIGURE 3.6 Timeline of BRB development.

brace from buckling. In 1998, the technology was transferred to the United States, followed by the production of BRBs in 1999. However, the design context of BRB of Japan and the United States are different from each other. In Japan, the BRBs are used along with moment-resisting frames (MRFs), wherein the BRBs work as hysteretic dampers to control the response of the MRF. This combination of BRB and MRF adds significant stiffness to the structure even after the BRBs yield. While in the United States, BRB frames alone provide sufficient strength and stiffness. With this design approach, BRB found its applications in the United States. The first building employed with BRBs is at the University of California, Davis. To date, thousands of buildings are installed with BRBs as a lateral load-resisting system all over the world. Figure 3.6 presents the flowchart of stages of development of BRB.

Anatomy of Buckling-Restrained Braces

The three principal components of a BRB are a steel core, bond-preventing layer, and the outer casing. The schematic of BRB is shown in Figure 3.7, whereas Figure 3.8 shows the yielding core, the restraining unit, and the complete BRB. Furthermore, Figure 3.9 displays the longitudinal section of BRB showing the steel core, concrete casing, and the air gap.

BRBs are made from different components, which serve a specific purpose. The detailed descriptions of the typical components of BRBs, i.e., the steel core, the bond-preventing layer, and the casing, are presented subsequently.

i. Steel core

   The restrained yielding segment is a mild steel rectangular or cruciform cross-section designed to yield under cyclic loading and exhibit high ductility. The restrained non-yielding stiff segment is an extension of the restrained yielding segment. It is the rigid length portion at both ends of the steel core. As the cross-sectional area of the non-yielding segment increases at both ends, the steel core remains elastic, whereas the decreased area in the middle portion transmits plasticity. The unrestrained non-yielding segment is

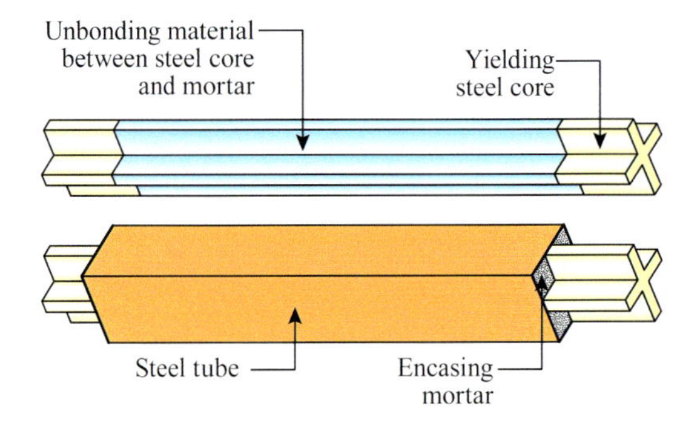

FIGURE 3.7 Principal components of BRB.

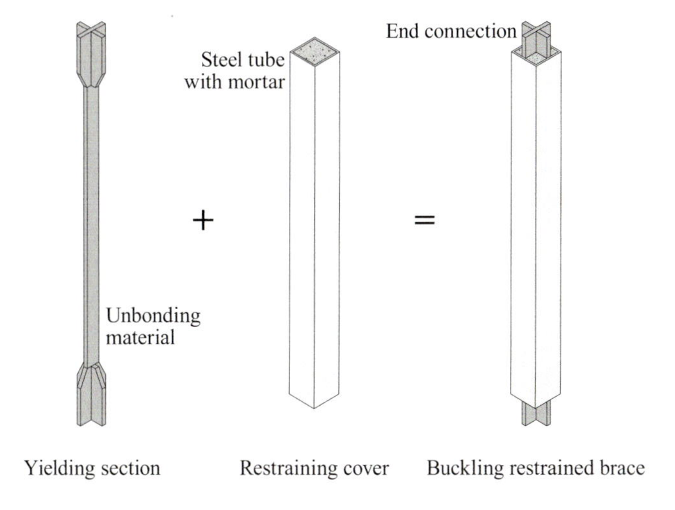

FIGURE 3.8 Assembly of the BRB components.

an extension of the restrained non-yielding segment projecting from the concrete-filled steel tube and designed as a bolted connection or a pinned connection for field erection. Figure 3.10(a) shows the restrained yielding segment, restrained non-yielding segment, and the unrestrained non-yielding segment of BRB. The cross-sectional view of the restrained yielding segment is shown in Figure 3.10(b).

To date, many types of BRB sections have been developed and implemented in engineering practice, such as a rectangular or cruciform section of the yielding steel core members and a concrete-filled tube section as the restraining outer steel casing. Figure 3.11 shows the different types of cross-sections of BRB. Figure 3.11(a) shows the BRB configurations in which filler material is used, and Figure 3.11(b) shows other BRB configurations in which no filler material is used and, hence, no debonding material is required to be provided.

Ductility is strongly affected by the material type and geometry of the yielding steel core segment. The core is encased by a sleeve that is filled with concrete, thus preventing the steel from buckling under compression. The steel core is divided into three segments, i.e., the yielding section, transition segment, and core extension, as shown in Figure 3.12. The middle portion is designed to yield under tensile and compressive loads. Figure 3.13 shows further details of the three zones.

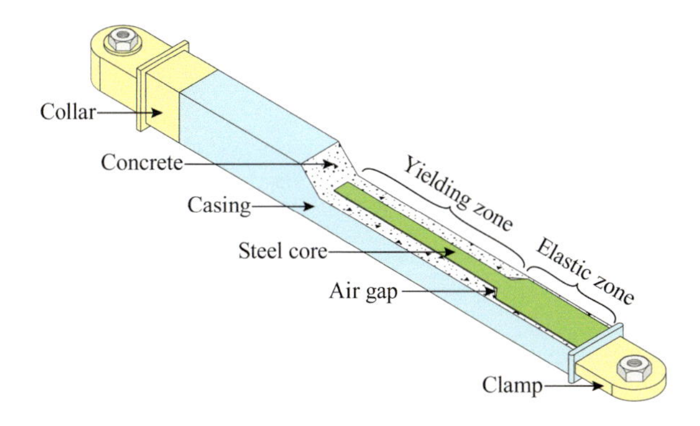

FIGURE 3.9 Longitudinal section of BRB.

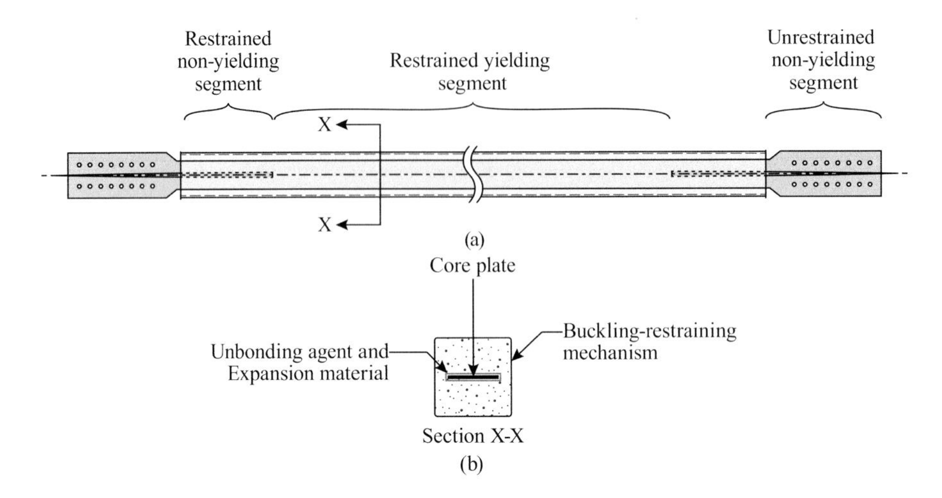

FIGURE 3.10 Buckling-restrained brace: (a) three main segments of BRB and (b) cross-sectional view of BRB.

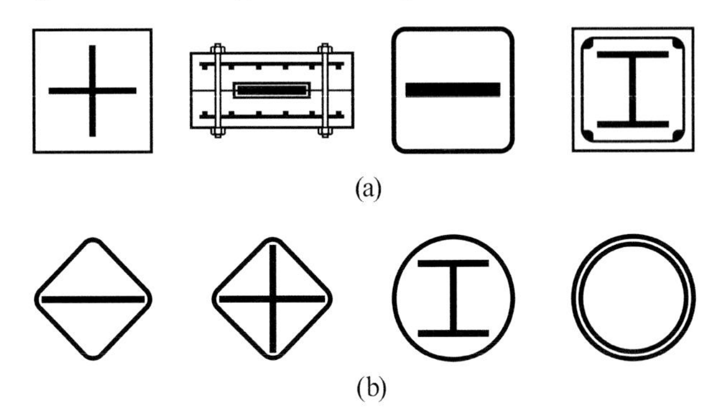

FIGURE 3.11 Different cross-sections of buckling-restrained braces: (a) with filled material and (b) without filled material.

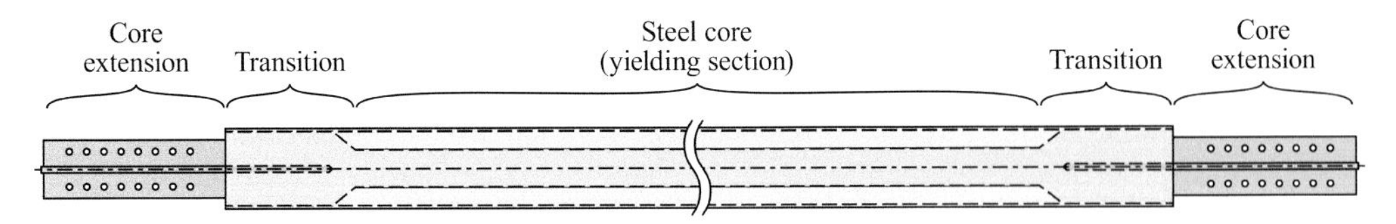

FIGURE 3.12 Segments of steel core.

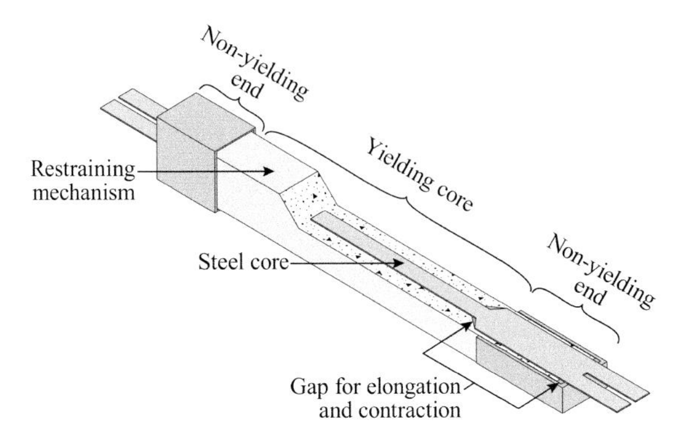

FIGURE 3.13 Non-yielding end and yielding core of BRB.

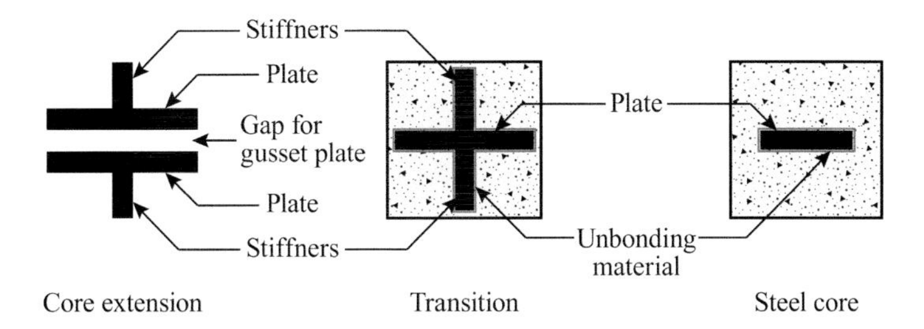

FIGURE 3.14 Cross-sectional view of the steel core at different locations.

ii. Bond-preventing layer
The bond-preventing material/layer is provided at the interface of the steel core and the surrounding concrete, which prevents the adhesion between the two materials. The layer of debonding material carries out the decoupling of the steel casing from the core and, hence, allows the steel core to be designed so that it can provide resistance to any applied load or developed axial forces in the bracing. Hence, buckling-restrained braces are also termed as unbonded braces. Bond-preventing material consists of special coating material, such as polyethylene, silicone grease, or mastic tip. Further, a small gap is provided between the inner steel core and the surrounding concrete to cater to Poisson's expansion of the core in compression and reduce the frictional resistance. The materials in the slip layer must be carefully selected, and the geometry must be carefully designed to allow relative movement between the steel core element and the surrounding concrete, which inhibits local buckling of the steel as it yields in compression. Figure 3.14 shows the cross-sectional view of the steel core and the transition segment. The BRB core extension assembly is attached to the gusset plate. The reinforced core section provided in the transition segment exits the sleeve during tensile yielding, thus ensuring that the reduced section remains inside the concrete. The reduced cross-section of the steel core yields under tensile and compressive loads.

iii. Casing
The concrete and the steel tube encasement provide sufficient flexural strength and stiffness, which prevent the global flexural buckling of the brace. This also allows the steel core to undergo fully reversed axial yield cycles in compression and tension without stiffness degradation and strength deterioration and to dissipate energy through stable tension-compression yield cycles. Global buckling of the core member is restrained as the buckling mode shifts to a pattern with many buckling waves, as shown in Figure 3.5. Buckling-restraining mechanism typically consists of mortar and hollow section steel casing. There are extensions provided at the end of the core to allow bolting of BRB to the gusset plate, as seen in Figure 3.12.

All-Steel Buckling-Restrained Brace

In an all-steel BRB, the inner core is sandwiched by a restraining system made entirely of steel components. All-steel BRB is becoming popular as it provides several advantages, such as lightweight, cost-saving, ease in maintenance and inspection, and faster fabrication and production. Extensive experimental studies have demonstrated the potential of the all-steel BRB system for a satisfactory ductile seismic response. Figure 3.15 shows the schematic representation of components and assembly of all-steel BRB.

Hybrid BRB – BTB and SLBRB

Steel CBFs suffer from strength and stiffness degradation due to the compression buckling of conventional braces during seismic events. Low-cyclic fatigue fractures associated with these buckling-type

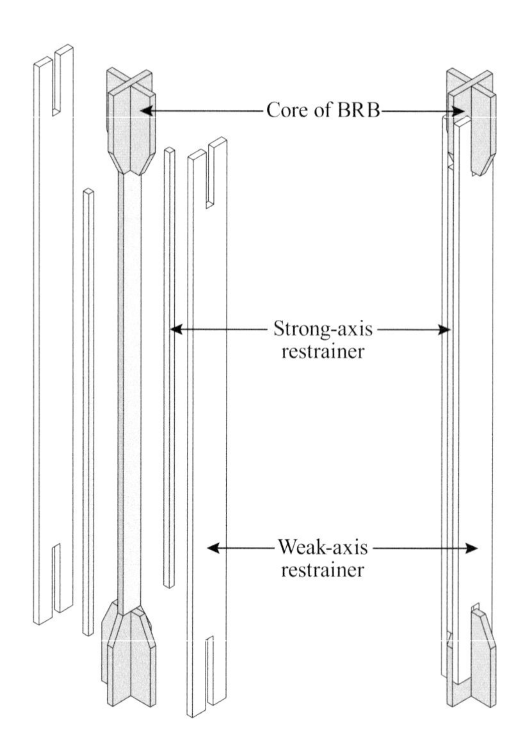

FIGURE 3.15 Schematic representation of components and assembly of all-steel BRB.

braces (BTBs) also reduce the overall ductility of CBFs. In contrast, buckling-restrained braces (BRBs) show symmetric hysteretic response resulting in improved energy dissipation and superior ductility prior to fracture. However, the relatively smaller brace stiffness and the preferred pinned connections in BRBFs make them more flexible, resulting in the increased inter-story drift demand and residual drift response. This may result in severe damages, adversely affecting the retrofitting and reusability of the structure after an earthquake. Therefore, the design of BRBFs is likely to be governed by drift limit, i.e., limit state of serviceability, rather than the strength requirement, and limit state of collapse. This problem is aggravated if the non-moment-resisting (pinned) beam-to-column connections are used instead of moment-resisting (rigid) connections in BRBFs.

This limitation of BRB can be overcome by a hybrid system using BTB and short-length buckling-restrained brace (SLBRB), installed in series. BTB–SLBRB is a composite brace having a short-length buckling-restrained yielding segment and an elastic component in series for the remaining major length. The conventional BTB helps transmit the axial force from the bracing-bent beam to the SLBRB segment and is designed to remain in the elastic range without any premature instability and yielding during the entire loading regime. Since the length of the SLBRB core is much smaller as compared to that in the conventionally used BRBs, the axial stiffness of the hybrid brace is increased, resulting in an increase in the lateral stiffness of the braced frame. Figure 3.16(a) shows the schematic representation of a hybrid (combined) brace system in which an SLBRB is connected in series with a conventional elastic BTB.

This type of hybrid brace system can provide the balanced hysteresis response under cyclic loading, which increases the ductility demand in the SLBRB segments of the hybrid brace in proportion to the decrease in the yield lengths. Figure 3.16(b) shows a typical hysteresis loop of Hybrid BRB – BTB and SLBRB. The conventional BRBs usually have yielding core lengths in the range of 60–70% of their work-point-to-work-point lengths. In contrast, the core lengths of SLBRBs vary in the range of 20–30% of their work-point-to-work-point lengths. Work point (WP) is the point of connection of BRB to the frame, as shown in Figure 3.20. The connection between the SLBRB and

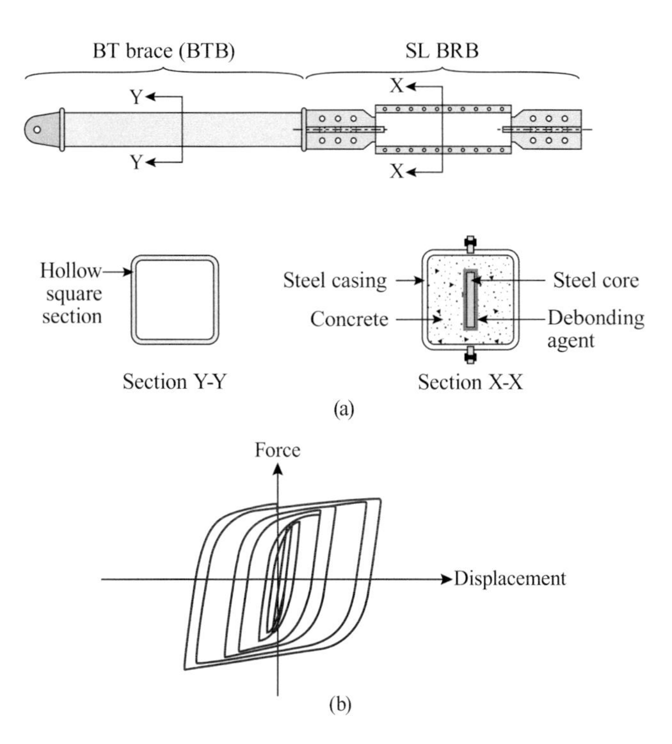

FIGURE 3.16 Hybrid BRB – BTB and SLBRB: (a) components of Hybrid BRB – BTB and SLBRB and (b) hysteresis loop.

BTB components should be rigid and should be designed to safely transfer the axial loads between these components without any failure or instability. Also, there should be minimum force lag in the connection regions to minimize the secondary effects on the brace systems. The drift ratios (ratio of the relative lateral deformation to corresponding story height) directly account for the extent of damage experienced by the bracing system due to the corresponding ground shaking. Thus, it can limit the amount of damage and make retrofitting/reusability of the structure viable. The SLBRB can undergo balanced cyclic yielding during the inelastic action, and the elastic segment in series can provide enhanced axial brace stiffness. The increase in lateral stiffness of the braced frame can reduce the drift demands. In order to make the proportion of yielding length of SLBRBs, the increase in ductility demand should be quantified to develop a robust design procedure for hybrid bracing frames. Past studies have reported that maximum energy dissipation occurs when 25–35% of the total length of conventional BRBs acts as yielding length. The quantification of the energy dissipated by the SLBRB gives the seismic proofing ability of the system against the expected seismic risk.

Connections of BRB

During an earthquake event, connections of the braces play a significant role in dissipating lateral forces acting on the beams and columns. This is because the braces are attached to the gusset plates at the end of the steel core, which in turn is connected to the beams and columns. Also, the type of connection defines the ultimate performance of the system as it affects the mechanism of force transferred to the braces. Typically, three types of connections are used for connecting the gusset plate to the beams and columns: viz., welded, bolted, and pinned connection, as shown in Figure 3.17.

In welded connection, the brace is fully welded to the gusset plate, as shown in Figure 3.17(c). In bolted connection, the gusset plate is directly bolted to the brace using fasteners, as seen in Figure 3.17(a). The large size of the gusset plate is required to accommodate the bolts, resulting in a short core length generating large strains. In another way, the gusset plate and braces are connected

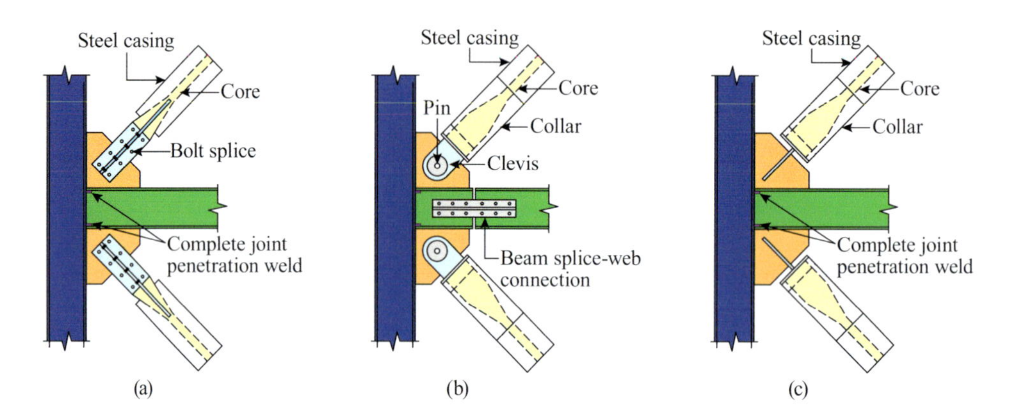

FIGURE 3.17 Types of connections of BRB: (a) bolted connection, (b) pinned connection, and (c) welded connection.

by a pin, as in Figure 3.17(b). The pinned connection allows rotation of braces. The core length of BRB is longer in the case of pinned connections, and thus smaller strain is developed compared to the bolted connections.

Bracing Configurations

While deciding the frame configuration and orientation of BRB, the designer has more freedom compared to other traditional braced frames. This is because BRBs prevent buckling of braces and thus, reducing its further consequences. Various configurations of BRB frame are shown in Figure 3.18, viz., inverted V-bracing (chevron), V-bracing, diagonal bracing (same direction and zigzag pattern), and multistory X-bracing.

In multistory buildings, stacked inverted V-bracing (Chevron) and V-bracing configurations are commonly used. However, the frame beam needs to be sized to have a proper division of vertical and axial loads acting on it. Multistory X-bracing is widely used because it transfers both the vertical and lateral loads to the frame beam and, thus, ensures proper distribution of yielding at multiple levels of the structure. Also, diagonal bracings oriented in the same direction and zigzag pattern are used for minimizing the axial loads in the beam in the frame.

Hysteretic Behavior of BRB

The yielding of BRB in compression and tension causes BRB frames to exhibit ductile cyclic behavior with significant energy dissipation. Typical cyclic force-deformation behavior of BRB and the

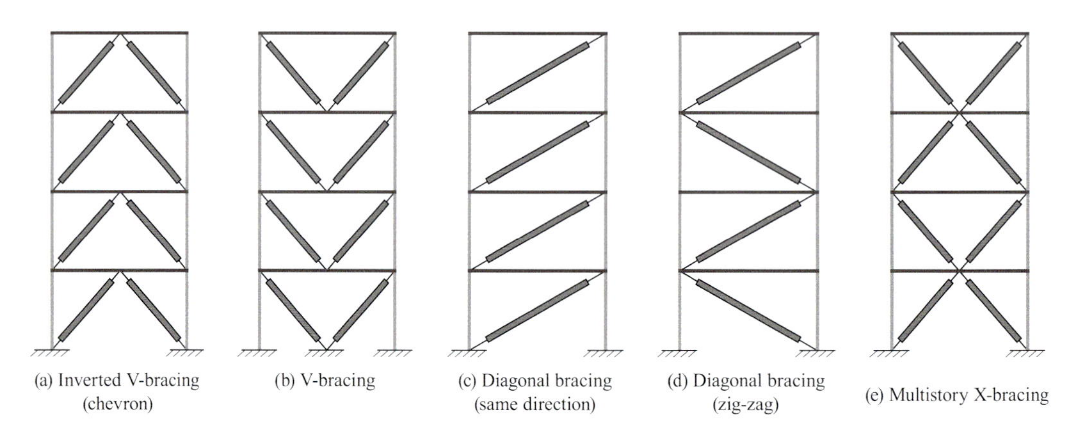

FIGURE 3.18 Various configurations of BRB frame.

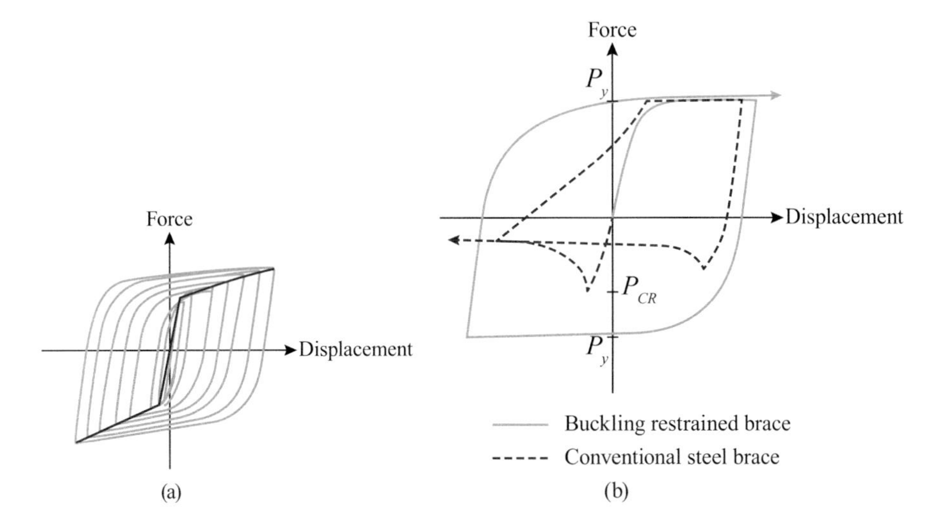

FIGURE 3.19 Cyclic force-deformation behavior of: (a) BRB and (b) conventional steel brace in comparison with that of BRB.

typical backbone curve are shown in Figure 3.19(a). The comparison of the hysteretic behaviors of BRB-restrained frame and CBF is shown in Figure 3.19(b), which shows the evolution of cyclic behavior and the significant strain-hardening response of BRB. BRBs exhibit combined isotropic and kinematic hardening, and they are typically slightly stronger in compression than in tension due to Poisson's expansion and friction at the interface between the core and the restraining mechanism. The energy dissipation capacity of rate-dependent passive energy dissipation devices depends on the velocity across the device, as in the case of the fluid viscous dampers or viscoelastic dampers. Rate-independent devices dissipate the energy through displacement in the device, e.g., metallic-yielding braces and buckling-restrained braces.

Advantages of BRB

Buckling-restrained braces have gained considerable popularity in new constructions and in retrofitting work of structures, as compared to the conventional steel braces, due to their following advantages.

1. BRB has the capability to yield both in tension as well as in compression. The conventional steel-braced system tends to buckle as it is weak in compression. The main advantage of BRBs is that they can restrain buckling effectively, and their hysteretic behavior is more stable than other bracing systems. Traditionally, the CBF system is an effective lateral force-resisting system. However, the literature reports its inferior performance during several past earthquakes. CBF has unsymmetrical tension-compression yielding capacity, as seen in Figure 3.19(b), leading to buckling of bracing members and consequently reducing the strength and stiffness of the structure. Buckling-restrained bracings provide an alternative to the CBF system. They provide the rigidity required to satisfy structural drift limits and stable inelastic behavior. Moreover, BRBs provide substantial energy absorption capability with symmetric hysteretic behavior in tension and compression, as shown in Figure 3.19(b).
2. The yielding of BRBs is always restricted to the steel core, and thus, there is increased control of performance. The designer can make the best use of the mechanical properties of a BRB such as strength, ductility, and stiffness, as the cross-sectional area of its steel core, yield stress, and yield length can be defined separately and varied as per the project design demand. According to the project needs, the designer can monitor the specification of BRB. Such freedom is not available in the case of conventional brace design.

3. Due to the stable hysteretic behavior of BRB, designers can specify the area of steel core, allowing them to meet the capacity demand of each story where BRBs are installed and, hence, the damages in the weak stories can be effectively controlled.
4. As BRBs are effective in restraining lateral buckling, non-structural elements alongside are also free from damage.
5. Compared to the conventional steel-braced system, the erection of BRB is faster, resulting in cost and time-saving. They can be easily incorporated into the structural system using welded, bolted, or pinned connections to gusset plates.
6. When used for retrofitting work, strengthening of structural members and foundation is not required.

Disadvantages of BRB

The major disadvantage of BRB is the absence of a restoring mechanism. Because the yielding core is typically hidden inside the casing, it is usually challenging to detect damages if the brace gets damaged during an earthquake event. Their considerable weight and large size make BRBs difficult to install in the existing structures. Furthermore, skilled labor is required for the installation of BRBs owing to very detailed and complex connections.

Mathematical Modeling of BRB

Andrews et al. (2009) proposed the mathematical model of the prototype system installed with BRB, shown in Figure 3.20. The mathematical model of BRB is shown in Figure 3.21.

The equation of motion for the system, assuming no coupling between the structure and the ground, is

$$\ddot{u}(t)+2\xi_0\omega_0\dot{u}(t)+\frac{P}{m}cos\theta=-\ddot{u}_g(t) \tag{3.1}$$

where ξ_0 is the modal damping ratio of the structure, P is the axial force in BRB and ω_0 is the elastic angular (natural) frequency of the structure given by

$$\omega_0=\sqrt{\frac{K_{WP}cos^2\theta}{m}} \tag{3.2}$$

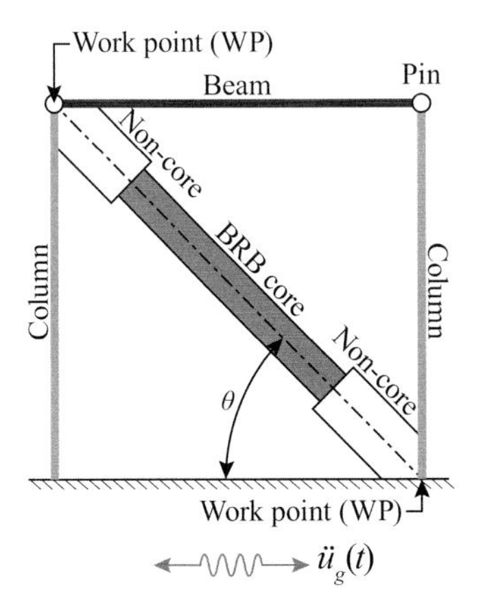

FIGURE 3.20 Prototype structural system installed with BRB.

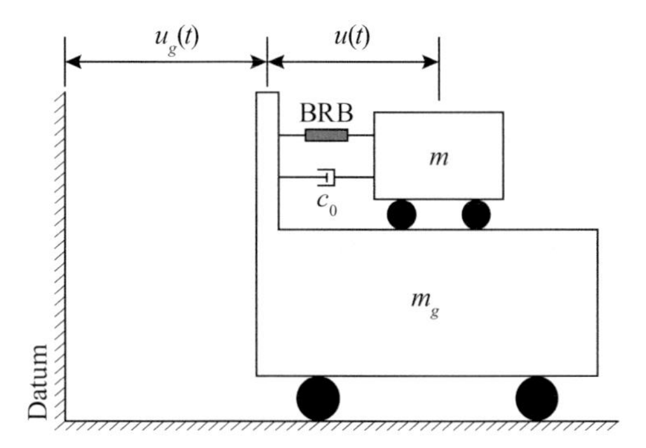

FIGURE 3.21 Mathematical model of BRB.

WPs are the points at which the BRB is diagonally connected to the frame, as seen in Figure 3.20. K_{WP} is the elastic stiffness of the BRB between the WPs given by

$$K_{WP} = \frac{1}{\dfrac{L_c}{A_c E} + 2\left(\dfrac{L_{nc}}{A_{nc} E}\right)} \tag{3.3}$$

where A_c is the core area, L_c is the core length, A_{nc} is the non-core area, L_{nc} is the non-core length, P_y is the core yield force, E is the modulus of elasticity, m is mass of the structure, and θ is the brace angle. These parameters are shown in Figure 3.22.

A Bouc-Wen model (Wen 1976) is used to model the hysteretic response of BRB. The axial force in BRB is given by

$$P = \alpha K_{WP} u + (1-\alpha) K_{WP} u_y z \tag{3.4}$$

where α is the ratio of post-yield to pre-yield stiffness, u is the relative displacement between the WPs in the direction of the BRB, u_y is the BRB WP deformation at initial yielding of the BRB $\left(u_y = P_y / K_{WP}\right)$, and z is the auxiliary variable describing the inelastic response, evolved by the non-linear differential Equation (3.5).

$$u_y \dot{z} + \gamma_{BW} |\dot{u}| z |z|^{n-1} + \beta_{BW} \dot{u} |z|^n - \dot{u} = 0 \tag{3.5}$$

In Equation (3.5), γ_{BW} and β_{BW} are model parameters that affect the shape of the hysteretic response, and n is a model parameter that controls the smoothness of the transition from pre-yield

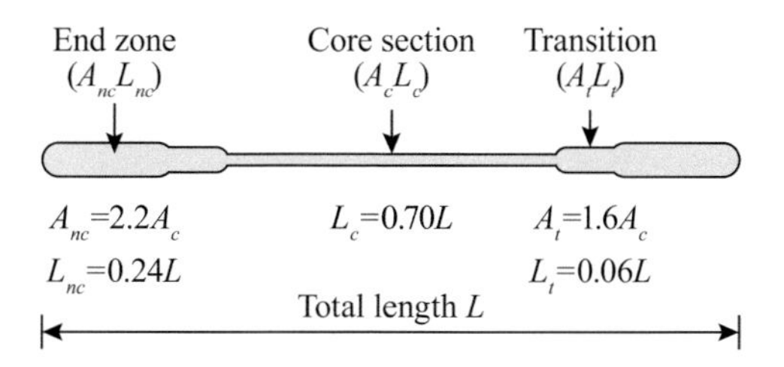

FIGURE 3.22 Parameters of BRB.

to post-yield behavior. The final form of the equation of motion for $n = 1$ is obtained as in Equation (3.6).

$$\ddot{u}(t)+2\xi_0\omega_0\dot{u}(t)+\frac{cos\theta}{m}\left[\alpha K_{WP}u(t)cos\theta+(1-\alpha)K_{WP}u_y z\right]+\ddot{u}_g(t)=0 \tag{3.6}$$

$$u_y\dot{z}+\gamma_{BW}\left|\dot{u}(t)cos\theta\right|z+\beta_{BW}\dot{u}(t)cos\theta\left|z\right|-\dot{u}(t)cos\theta=0 \tag{3.7}$$

In Equations (3.6) and (3.7), $\ddot{u}_g(t)$ is the ground acceleration, m is the system mass, m_g is the mass of ground, $u(t)$ is the relative displacement of mass, $\ddot{u}(t)$ is the absolute acceleration of mass, and c_0 is the damping coefficient of the system.

Worldwide Applications of BRB

BRBs have been successfully employed in various types of structures, such as high-rise buildings, bridges, stadiums, and several other structures. Some of the notable applications of buckling-restrained braces in the world are mentioned below.

- The prominent example of employing the BRB system in Japan is the Osaka International Convention Centre, shown in Photo 3.1. It is a steel structure, rectangular in plan with length and width of 95 m and 59 m, respectively. The center has a total of 13 floors and additional three basement floors, and the total floor area is about 67,545 m^2. Though the center is located in a region of high seismicity and has high steel weight, incorporating BRBs enabled the building to achieve considerable seismic resistance.
- In 2000, the United States found its first application of BRB in the three-story plant and Environmental Sciences Replacement Facility, located at the University of California, Davis campus. The floor area of the C-shaped steel building is about 11,600 m^2. For restraining seismic force, the EBF system was selected initially. For using BRB frames as an alternate system, pushover analysis was performed to compare the performances of EBF, BRB, and CBF systems. It was reported that BRB frames performed well as compared to both EBF and CBF. Photo 3.2 shows the BRBs installed in the building at the UC Berkeley campus.

PHOTO 3.1 Osaka International Convention Centre, Japan.

PHOTO 3.2 BRB system installed at UC Berkeley campus. (Comerio et al. 2006)

- BRBs were used in the seismic rehabilitation of Wallace F. Bennett Federal Building, located in downtown Salt Lake City, Utah. The building was incorporated with 344 BRBs, as shown in Photo 3.3 and Photo 3.4. The 8-story building has a total floor area of 28,000 m^2. Employing BRBs led to an attractive, economical, and high seismic performance of the building.
- BRBs are installed in the Kaiser Santa Clara Medical Centre Project, California (CA), the United States, shown in Photo 3.5. The building is employed with ten bays of BRBs at each floor in the N-S and E-W directions.
- One of the typical retrofit strategies for non-ductile moment frames is to install BRBs as bracing elements along the frames. Photo 3.6 shows early applications of BRBs in multi-story buildings in Japan.

PHOTO 3.3 Wallace F. Bennett Federal Building, Utah.

PHOTO 3.4 Buckling restrained frame system added to the existing building.

PHOTO 3.5 Kaiser Santa Clara Medical Centre Project, California, United States.

PHOTO 3.6 Early application of BRBs in Japan. (Takeuchi and Wada 2017)

PHOTO 3.7 Retrofitted communication tower, Japan. (Takeuchi and Wada 2017)

- BRBs were used in Japan for a retrofit program of a communication tower constructed in 1970. Many such towers have been constructed on the roof of the buildings. They suffer from the risk of collapsing during severe earthquakes due to amplification by the supporting buildings. The damages in the towers are caused by partial stress concentration at critical points on the members. The most effective retrofit method is to replace the damaged members with stable energy dissipation members like BRBs, wherein the plastic strain is evenly distributed among the braces. Photo 3.7 shows the retrofitted communication tower, and Photo 3.8 displays the newly installed BRBs, shown in red.

PHOTO 3.8 Seismic retrofit of tower with BRB. (Takeuchi and Wada 2017.)

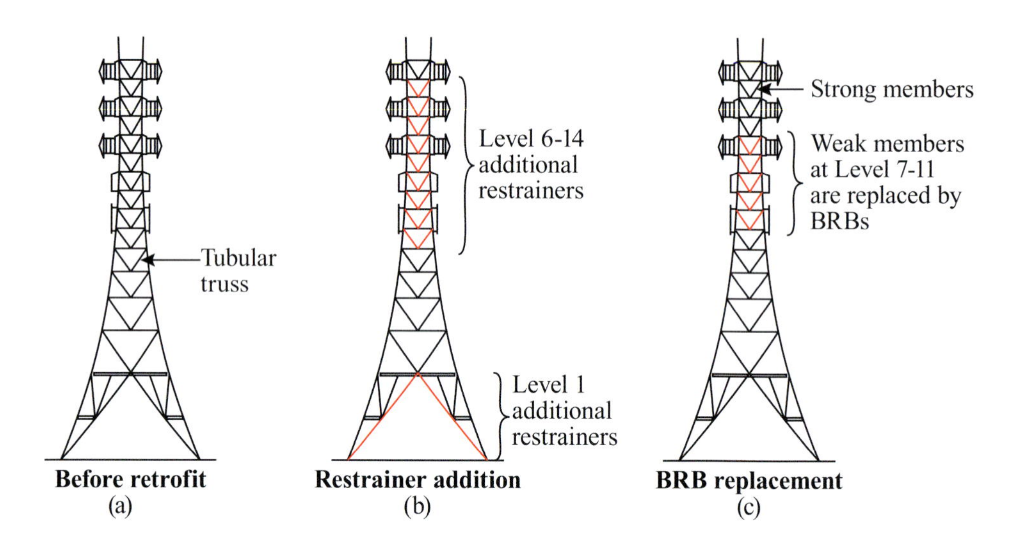

FIGURE 3.23 Retrofitting options for the communication tower, Japan: (a) before retrofit, (b) restrainer addition, and (c) BRB replacement.

Figure 3.23(a) shows the tower before retrofitting, whereas Figures 3.23(b) and (c) show the two options considered for retrofitting. In the alternative presented in Figure 3.23(b), provision of restrainer was considered, whereas weak members are replaced by BRBs (indicated in red) in the option presented in Figure 3.23(c). The option involving replacement with BRBs showed minimum acceleration response throughout the height, the most superior performance, and the lowest construction cost. Hence, the retrofitting work of 50 such towers in Japan was completed using 20 BRBs, in just six days. This proved to be one of the most practical and cost-effective retrofit methods.

3.3 DIAGONAL BRACING

The simplest type of bracing used for resisting lateral loads is diagonal bracing, as shown in Figure 3.18(c) and (d). A single diagonal bracing is provided when the bay width is small. The number of diagonal bracing members inserted in the structure determines the overall efficiency of the braced frame. In the single diagonal braced frame, when a lateral load is applied to the braced frame, the diagonal braces are subjected to compression while the horizontal web member (beam) acts as a tension member to maintain the frame structure in equilibrium. Bracings oriented in both directions in the X-bracing type configuration, as shown in Figure 3.18(e), act as tension and compression members during reversible cycles of dynamic loading. Broadly, two bays of single diagonal braced frames are preferred over a single bay of V brace or inverted V brace. The slenderness ratio of the braces influences their energy absorption capacity. The braces with a smaller slenderness ratio can dissipate a higher amount of energy.

While using diagonal bracing, the designer should check for the possibility of compression buckling due to the vertical component of the brace force at the tensile yield strength. Considering all tension braces framing to the same column as having yielded at the floor levels above the floor level being checked is the most conservative approach in the design of the diagonal bracing system. Figure 3.24(a) shows the yielding of braces in tension, whereas Figure 3.24(b) shows the brace buckling in compression. The latest development in the diagonally braced frame is the Mega X-Bracing (Diagrid). It reduces the lateral drift substantially, minimizes shear deformation, provides maximum resistance against torsion, and possesses greater structural flexibility than other

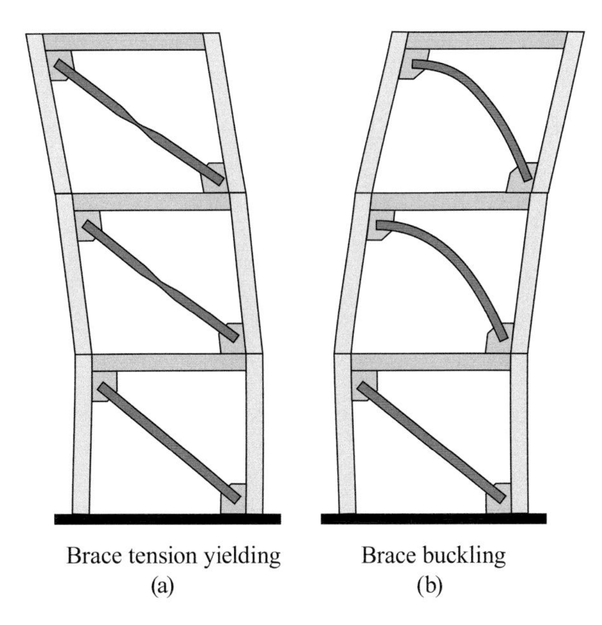

FIGURE 3.24 Yielding and buckling of braces: (a) brace tension yielding and (b) brace buckling under compression.

bracing systems. Further, there is a substantial saving in the quantity of steel. The diagrid can be configured with some adjustment in modules and angle(s) of diagonals to meet architectural and structural requirements.

3.4 CONCENTRICALLY BRACED FRAMES

MRFs are traditionally assumed to contribute higher ductility to the structure compared to CBFs. However, MRF is much more expensive than CBF. CBF uses simple connections and a smaller cross-sectional area of beam and column and can be approximately modeled as a vertical truss.

The CBFs are commonly classified into three major types: viz., diagonally braced frames, V braced frames, and inverted V, i.e., Chevron braced frames. These structures provide stability and lateral stiffness, and they resist the lateral seismic acceleration primarily through the axial forces (tension and compression) and deformation of the braces, beams, and columns. The characteristic property of CBF is a low plastic distribution capacity that tends to cause plastic deformation in some stories. The seismic performance of the CBF is affected by the slenderness and width-to-thickness ratio of the braces. CBFs generally consist of a lateral load-resisting structural system formed by combining one or more diagonal braces like a concentric truss system. In concentric frames, center lines of all members such as beams, girders, and columns intersect each other at a common joint. Various configurations of CBF include chevron bracing, split chevron bracing, X bracing, single diagonal bracing, V bracing, split V bracing, K bracing, and many others.

The typical sequence of failures in bracing systems is shown in Figure 3.25. Buckling and yielding in braces take place in compression and tension, respectively. However, the buckling capacity of braces in compression is very small compared to the tension yield force. For successive buckling cycles, the buckling capacity is decreased further. Thus, in order to maintain similarity in lateral resistance in tension and compression along both directions, X bracings or chevron bracings are employed. Owing to the buckling in compression, axially loaded members fail to regain their original strength. Therefore, structural design codes impose more precise and harsh design requirements on bracings.

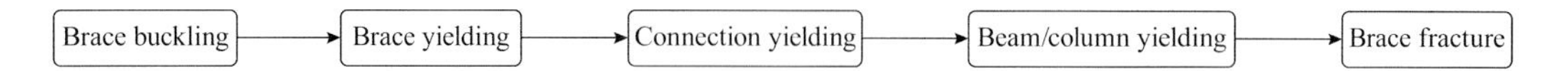

FIGURE 3.25 Sequence of failure of bracing system.

CBFs increase the natural frequency of structure by imparting additional stiffness with minimal weight, thus lowering the lateral story drift. The beam, column, and brace are assumed to be pin connected at a common joint, and hence, there is no bending moment and shear force to be considered between the frame members. Therefore, the axial compression in the column is considerably increased. Figure 3.26 shows preferred damage locations in concentric steel braces.

Seismic Considerations for CBFs

Moment-resistant frames possess considerable energy dissipation characteristics but are relatively flexible when sized from strength considerations alone. These characteristics can result in a less-favorable seismic response, such as low drift capacity and higher accelerations. Any reasonable pattern of braces with singly or multiply braced bays can be designed so that shear is resisted at every story. CBFs are excellent from strength and stiffness considerations and are therefore used widely either by themselves or in conjunction with MRFs for resisting lateral loads caused by the wind and earthquake. They form a vertical concentric truss system, where the axes of the members are aligned concentrically at the joints. The CBF is a common structural steel or composite system that can be employed in structures located in seismically active zones. In the CBF, mechanism forms at the intersection of the beam-column-brace, where the tension brace yields, and compression brace buckles. Figure 3.27(a) shows permitted configurations of CBFs for resisting seismic forces, the special CBF (SCBF), and Figure 3.27(b) shows the configurations that are not permitted.

AISC-Seismic (seismic provisions for structural steel buildings) permits the seismic design of braced frames using either an ordinary concentric braced frame (OCBF) or an SCBF. The difference between the two is in the detailing of the connections. The basic design concept for an OCBF is to limit its response to an elastic behavior, whereas some prescriptive requirements for SCBF are intended to enable them to respond to seismic forces with greater ductility. Both the V and inverted V braced frames, often referred to as chevron braces, have been poor performers during past earthquakes because of buckling of braces and excessive flexure of the beam at midspan where the braces intersect the beam. Buildings with single or multistory X-braces or V braces with zipper columns are deemed better performers, and hence, should be considered for braced frame configurations in high seismic zones. Braced frames with single diagonals are also permitted by AISC-Seismic. However, there is a heavy penalty since the braces must be designed to resist 100% of the seismic force in compression unless multiple single-diagonal braces are provided along a given brace frame line. A preferred but difficult-to-achieve behavior in an SCBF is the in-plane buckling of the brace. Typically, a brace would buckle out-of-plane rather than buckle in the plane of the braced frame. The in-plane buckling is inhibited because the placement of braces in a flat position is generally not permitted for architectural reasons, and the presence of infill metal studs above and below the

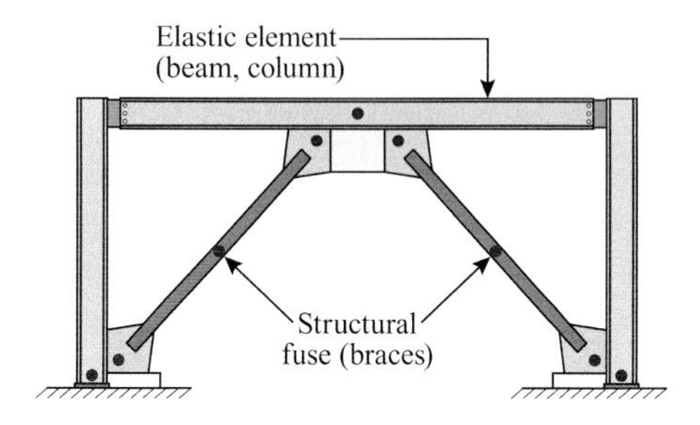

FIGURE 3.26 Preferred locations of damage in CBF.

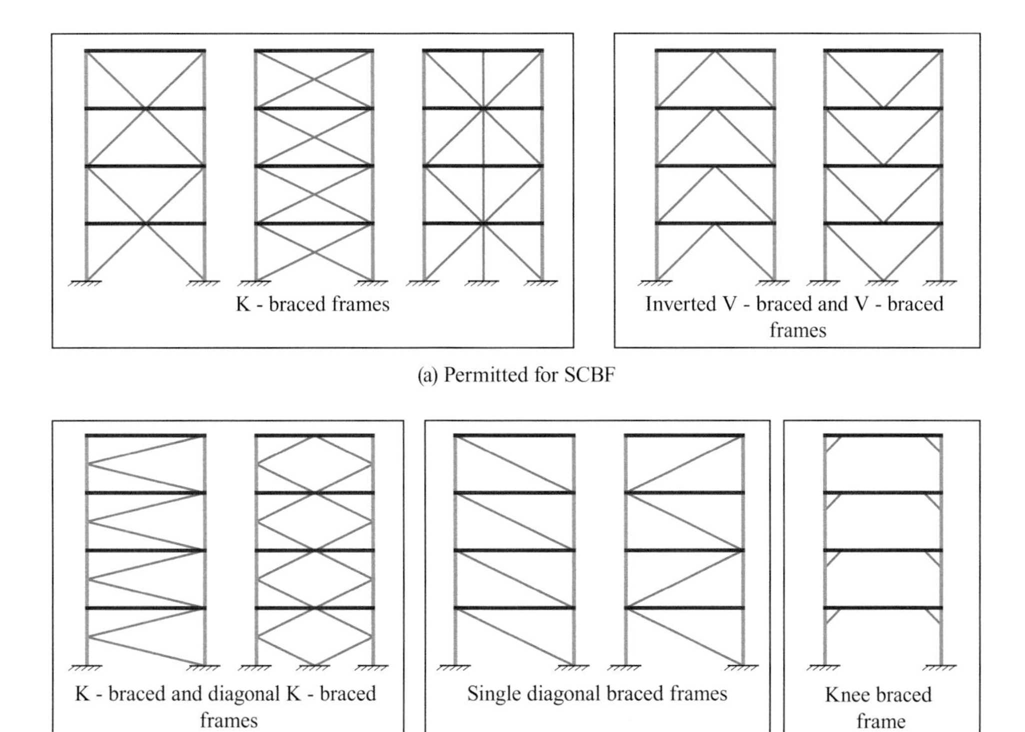

FIGURE 3.27 Configurations of concentrically braced frames from seismic considerations: (a) permitted for SCBF and (b) not permitted for SCBF.

braces adds considerable in-plane stiffness to the braces. Therefore, AISC-Seismic permits out-of-plane buckling of braces, provided an uninterrupted yield line can develop in gusset plates at each end of the brace connection. In detailing the gusset plate of SCBF, the potential restraint that occurs due to the floor slab must be considered. To keep the gusset plate as small as possible, the area of the gusset plate, where the yield line occurs, may be isolated from the slab to allow the yield line to extend below the concrete surface. Beams or columns of braced frames should not be interrupted at the brace intersections to ensure out-of-plane stability of the bracing system at those locations. However, mere continuity of columns or beams at the brace intersections may not be sufficient to provide the required stability. The typical practice is to provide perpendicular framing that engages a diaphragm to provide out-of-plane strength, stiffness, and resistance to lateral torsional buckling of beams. SCBFs are expected to achieve trilinear hysteretic behavior in a large earthquake by going through three ranges of displacement of its brace: the elastic range, the post-buckling range, and the tensile yielding range. The basic design concept for an SCBF is to mitigate brittle modes of failure by controlling its behavior through better detailing. Therefore, connections are designed to develop the yield capacity of the brace.

Special requirements apply to the design of chevron braced frames. Because these braces meet at the midspan of beams, the vertical force resulting from the unequal compression and tensile strengths of these braces can have a considerable impact on the cyclic behavior of the frame. That vertical force introduces flexure in the beam and possibly a plastic hinge, producing a plastic collapse mechanism. Therefore, beams in chevron braced frames must be continuous between the columns. Seismic provisions require that beams in chevron braced frames be capable of resisting their tributary gravity loads, neglecting the presence of the braces. Further, each beam in an SCBF should be designed to resist a maximum unbalanced vertical load calculated using full-yield strength or the brace in tension and 30% of the brace buckling strength in compression. To prevent instability of a beam bottom flange at the intersection of the braces in a chevron braced frame, the top and bottom flanges of beams in both SCBF and OCBF must be designed to resist a lateral force equal to 2% of the nominal beam flange strength. This requirement is best met by the addition of a beam perpendicular to the chevron braced frame. CBFs are expected to undergo inelastic response during large earthquakes. Specially designed diagonal braces in these frames can sustain plastic deformations and dissipate hysteretic energy in a stable manner through successive cycles of buckling in compression and yielding in tension. Therefore, the preferred design strategy is to ensure that plastic deformations only occur in braces, thus preventing damages to the columns, beams, and connections. This allows the structure to survive strong earthquakes without losing resistance to gravity loads.

The plastic hinge that forms at the midspan of a buckled brace, as shown in Figure 3.27, may develop large plastic rotations that could lead to local buckling and rapid loss of compressive capacity and energy dissipation characteristic during repeated cycles of inelastic deformations. Locally buckled braces can also suffer low-cycle fatigue and fracture after a few cycles of severe inelastic deformations, especially when braces are cold-formed rectangular hollow sections. For these reasons, braces in SCBFs must satisfy the width-to-thickness ratio limits for compact sections. For OCBFs, braces can be compact or non-compact, but not slender. When a brace is in tension, net section fracture and block shear rupture at the end of the brace must be avoided. Likewise, the brace connections to beams and columns must be stronger than the braces themselves. Connections must also be able to resist the forces due to the buckling of the brace. Beams and columns in braced frames must be designed to remain elastic when braces have reached their maximum tension or compression capacity to preclude inelastic response in all components except the braces.

3.5 ECCENTRICALLY BRACED FRAMES

The main idea in the design of an EBF is to integrate the advantages of both the MRF and CBF lateral load-resisting systems into a single structural system. The EBF system originated from Japan in the 1970s to achieve a structure with high elastic stiffness and high energy dissipation during severe earthquakes. Eccentric bracing is a unique structural system that attempts to combine the strength and stiffness of a CBF with the inelastic behavior and energy dissipation characteristics of an MRF. In an eccentric bracing system, the connection of the diagonal brace is deliberately offset from the connection between the beam and the column. By keeping the beam to brace connections close to the columns, the stiffness of the system can be made very close to that of concentric bracing. This offset or eccentricity promotes the formation of energy-absorbing hinges in the portion of the beam between the two connections. Figure 3.28 presents the comparison of ductility and stiffness of MRF, CBF, and EBF.

EBFs are made of two short diagonal braces connecting the column to the middle span of the beam with a short segment of the beam to increase the lateral load resistance to seismic forces. EBFs provide architectural functionally, similar to the chevron braced frames. The lateral stiffness of eccentric braces is lower than that of the CBFs, especially the diagonal bracing, because the eccentric braces have a more ductile characteristic. The safety of the occupants in the building is ensured, as eccentric braces delay the structural response toward the earthquake by dissipating a large amount of energy. It gives sufficient time to the occupants to escape from the building.

The role of web members in resisting shear can be demonstrated by following the path of the horizontal shear down the braced bent. Consider the braced frames, shown in Figures 3.29(a)–(e), that are subjected to an external lateral force at the top level. In Figure 3.29(a), the diagonal brace in each story is under compression, causing the beams to be in axial tension. Therefore, the shortening of the diagonal and the extension of the beams give rise to the shear deformation of the bent. In Figure 3.29(b), the forces in the braces connecting to each beam are in equilibrium horizontally,

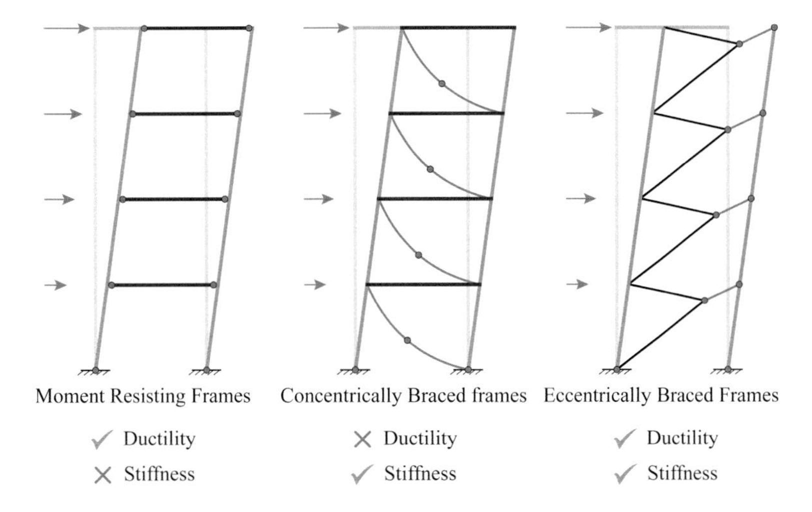

FIGURE 3.28 Comparison of ductility and stiffness of MRF, CBF, and EBF.

with the beam carrying an insignificant axial load. In the frame depicted in Figure 3.29(c), half of each beam is under compression, whereas the other half is subjected to tension. The braces in Figure 3.29(d) are alternately in compression and tension while the beams remain essentially unstressed. Furthermore, the end parts of the beams in Figure 3.29(e) are in compression and tension, and the entire beam is subjected to double curvature bending.

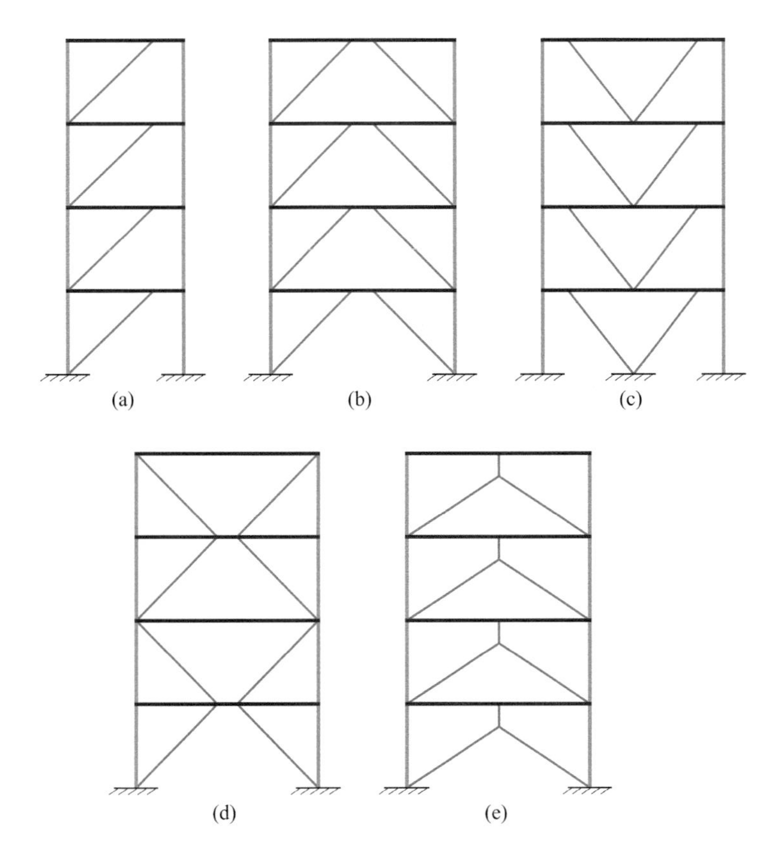

FIGURE 3.29 Some possible bracing arrangements.

Link Elements

EBF is a framing system in which the axial forces induced in the braces are transferred either to a column or another brace through shear and bending in a small horizontal segment in between the connections of two bracing members called link element. Yielding is concentrated only at link segments, and all other members of the frame are proportioned to remain essentially elastic. The links in EBFs act as structural fuses to dissipate earthquake-induced energy in a stable manner and control plastic deformations. A link needs to be properly detailed such that it has adequate strength and stable energy dissipation characteristics to achieve the aforementioned objectives. All other structural components such as beam segments other than the link, braces, columns, and connections are proportioned following capacity design provisions to remain essentially elastic during the design earthquake. There are many variations in terms of the placement of the link beam in the EBF. Different placements or locations of the link in the structure lead to diverse structural response due to the impact of lateral load. The primary function of the link beam is energy dissipation under dynamic excitation. The link is used to control plastic deformation and maintain structural stability through seismic energy dissipation due to its ductile behavior. Moreover, the resistance to lateral movement in the eccentric braces is through the bending of the beam and column. The strength, stability, and ductility of the EBF are influenced by the length of the horizontal links.

A shorter horizontal length of links leads to higher shear-yielding efficiency, as the shear in the story directly influences the shear in the links. It is reported that the short links have better performance in terms of rotation capacities compared to long and intermediate-length links. Further, the short links have greater strength and exhibit ductile performance under several cyclic loading cycles. In shear yielding, the web of beam yields along the entire length. Hence, a sufficient number of web stiffeners must be provided to control inelastic web buckling. The ultimate failure mode is web fracture. Bending may occur in the longer links that contribute more to the flexural yielding. The longer links provide the architecture freedom, especially in the placement of the window and door openings. Flexural yielding occurs near the beam ends. Hence, width-to-thickness limits of web stiffeners are important to control local flange and web deformation. If an intermediate length of the horizontal links is used in the eccentrically braced system, a combination of shear and flexural yielding would be experienced. If damaged during the seismic excitation, it is time-consuming and expensive to replace or repair the horizontal link beam.

Figures 3.30 and 3.31 show a single-story frame with a link element located at the corner and center, respectively. It is seen that yielding occurs within the links, whereas all other members remain essentially elastic. Figures 3.32(a) and (b) show the three-story frame with link elements located at the center and the corner, respectively.

Figure 3.33 shows typical force-deformation loops for link elements. The short link undergoes shear yielding and has a force-displacement loop shown in Figure 3.33(a). The force-displacement loop for long links due to flexural yielding is shown in Figure 3.33(b). An intermediate link yields due to the combined action of shear yielding along the length and flexural yielding near ends. It has a force-displacement characteristic that is shown in Figure 3.33(c).

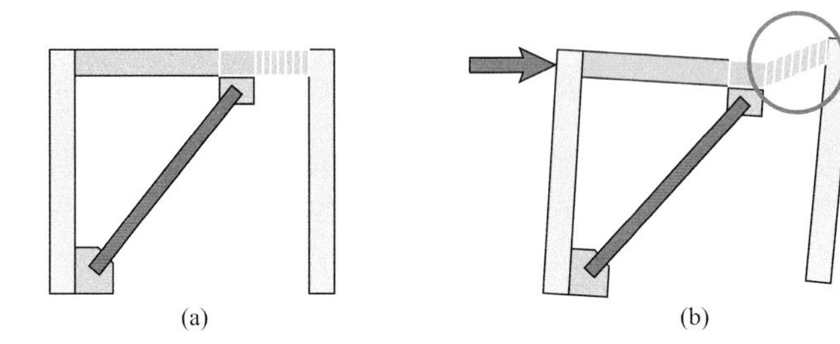

FIGURE 3.30 Eccentrically located link element in a single-story frame: (a) link element at beam column joint and (b) deformed link element.

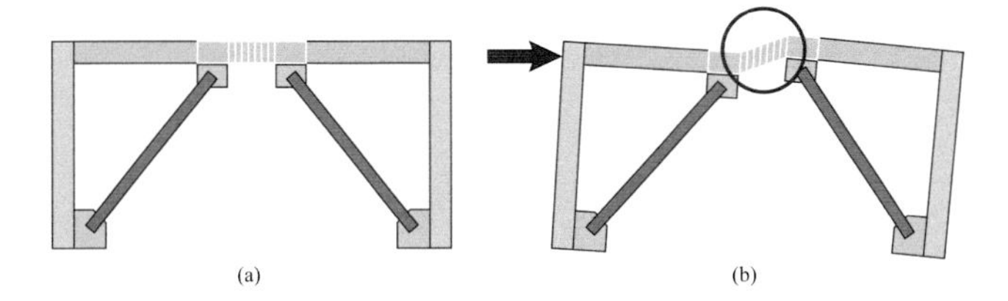

FIGURE 3.31 Centrally located link element in a single-story frame: (a) link element at center and (b) deformed link element.

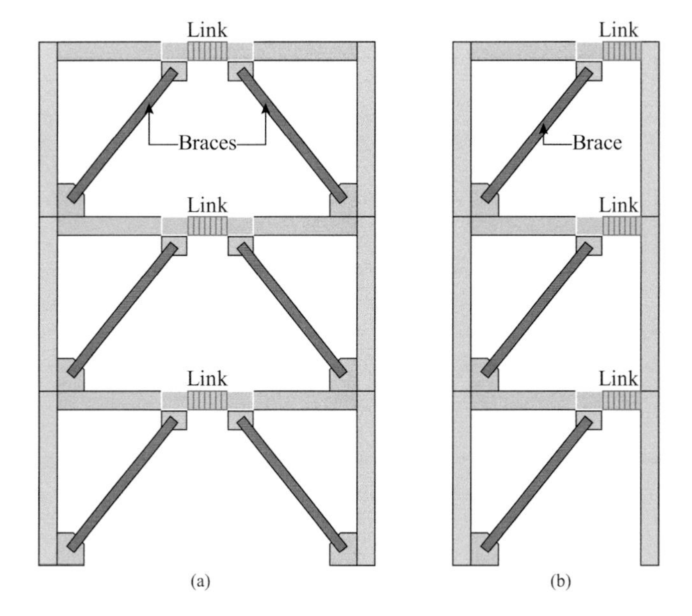

FIGURE 3.32 Link element in a three-story frame: (a) link element at center and (b) link element at beam column joint.

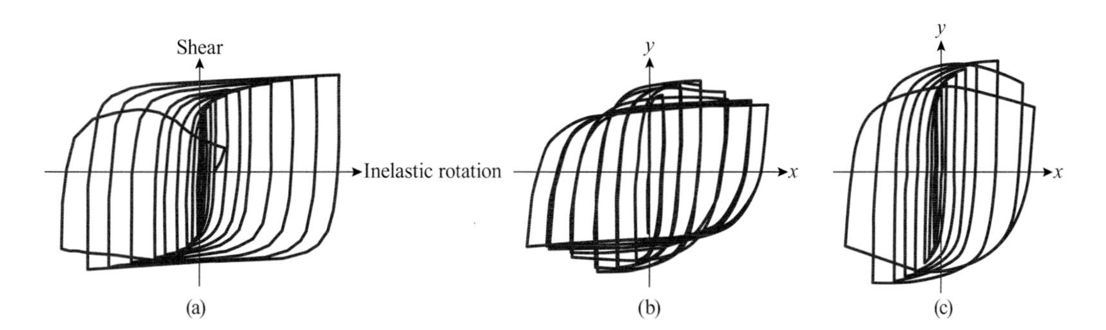

FIGURE 3.33 Yielding patterns of links: (a) shear yielding, (b) flexural yielding, (c) intermediate yielding.

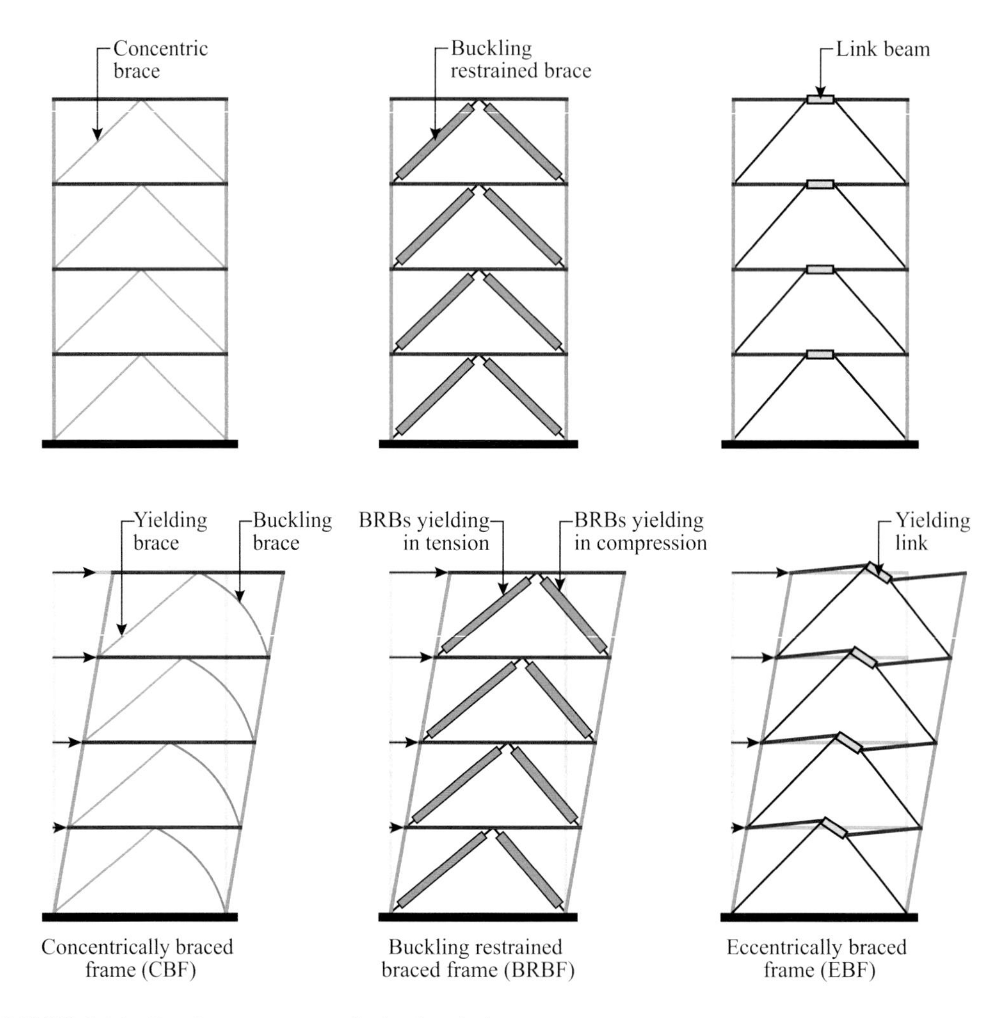

FIGURE 3.34 Bracing type energy dissipation devices.

Attaching vertical links to the story beam is an alternative proposal to overcome the limitations of horizontal links. The vertical links act as fuses and are similar to the horizontal links, which provide the required ductility and energy dissipation capacity, and withstand most of the shear during the dynamic excitation by deforming plastically. There is no rotational restraint at the lower end of the vertical link. The floor beam does not suffer significant damage because the internal energy dissipation and inelastic deformation are accumulated in the link. Hence, the replacement of the vertical links is efficient and easier than horizontal links after dynamic excitation.

Figure 3.34 shows the bracing type energy dissipation devices, viz., CBF, buckling-restrained brace, and link beam with their deformation patterns under lateral loading.

Seismic Design Considerations for EBF

According to the capacity design concept, the computation of the link strength should be based on the expected yield strength of the steel and considerations of strain-hardening and overstrength due to the composite action of the slab. Flexural hinges always form at two ends of the link when the moment reaches the plastic moment. A shear hinge is said to form when the shear force reaches the plastic shear capacity. The presence of an axial force in a link reduces the flexural, shear, and inelastic deformation capacities. Composite action due to the presence of slabs can significantly increase the link shear capacity during the first few cycles of large inelastic deformations. However, the composite action deteriorates rapidly in subsequent cycles due to local concrete floor damage at both ends of the link. For design purposes, it is conservative to ignore the contribution of composite action for calculating the link shear strength. However, the overstrength produced by the composite slab effect must be considered when estimating the maximum forces imposed by the link on other structural components. When detailing a link, full-depth web stiffeners must be placed symmetrically on both sides of the link web at the diagonal brace ends of the link. The link section must satisfy the same compactness requirement as the beam section for special MRFs. Further, the link must be stiffened to delay the onset of web buckling and to prevent flange local buckling. The stiffening requirement depends on the length of the link. Intermediate link web stiffeners must be provided for the full depth. Two-sided stiffeners are required at the end of the link at the intersection of the diagonal brace. For links of small depth, intermediate stiffeners placed on one side of the link web are adequate.

To ensure stable hysteresis, a link must be laterally braced at each end to avoid out-of-plane twisting. Lateral bracing also stabilizes the eccentric bracing and the beam segment outside the link. The concrete slab alone cannot be relied upon to provide lateral bracing. Therefore, both the top and bottom flanges of the link beam must be braced. At the connection between the diagonal brace and the beam, the intersection of the brace and beam centerlines shall be at the end of the link or within the length of the link. If the intersection point lies outside the link length, the eccentricity and the brace axial force produce additional moments in the beam and brace, which should be accounted for in the design. If the connection is designed as a pin, the gusset plate must be properly stiffened at the free edge to avoid local buckling. It is desirable to use a split V-braced EBF to avoid a moment connection between the link and column. Test results have shown that a fully restrained welded connection between a column and a link, particularly if the link is relatively long, is vulnerable to brittle fracture. When cover plates are used to reinforce a link-to-column connection, the link over the reinforced length must be designed so that no yielding occurs in this region. For the preferred EBF configuration, where the link is not adjacent to a column, a simple connection between the beam and column is considered adequate, if it provides some restraint against torsion in the beam. Although the link end moment is distributed between the brace and the beam outside of the link according to their relative stiffness, in preliminary design, it is conservative to assume that all the link end moment is resisted by the beam. The horizontal component of the brace produces a significant axial force in the beam, particularly if the angle between the diagonal brace and the beam is small. Therefore, the beam outside the link must be designed as a beam-column.

3.6 X-TYPE BRACE

A single-story X-type brace is made of two diagonals. It may sometimes be challenging to employ X-type braces practically because the doorways and other openings may conflict with the X braces. It is required to provide additional splicing for the diagonal members where they cross each other. The behavior of the single-story X brace and two-story X brace is like a truss. Therefore, it is essential to check the columns for axial compression, as they have to resist the gravity loads and the vertical component of the brace tension. The two-story X brace shown in Figure 3.18(e) requires special considerations of gravity loads and seismic forces compared to the single-story X brace. Particular attention should be given to columns as they have to resist the vertical component of the brace tensile yield force without buckling.

3.7 K-TYPE BRACE

K bracing allows for circulations through the middle of the bay, unlike X bracing that virtually eliminates the possibility of circulation. K-type braces shown in Figure 3.18(d) are not recommended as compared to SCBFs. Many studies have reported that K braces should not be used in areas of moderate-to-high seismicity. The braces' unequal buckling and tension-yielding strengths create an unbalanced horizontal force at the mid-height of the columns, jeopardizing the ability of the column to carry gravity loads if a plastic hinge forms at its mid-height.

3.8 V BRACE AND INVERTED V BRACE (CHEVRON BRACE)

High elastic stiffness and strength are the characteristics of the chevron braces. Chevron braces are most effective in the architectural functionality, as they allow the positioning of windows and doorways at bay of the braces. In chevron braced frames, compression and tension are exerted on each side of braces. Both braces distribute the lateral load equally in the elastic range, prior to buckling. The compression braces buckle and the tension braces sustain tensile forces. Large bending moments are created at the intersection of beam and braces. The unbalanced distributed force has major influence of the tension and compression braces that can cause large deflections. Since the chevron bracing has a non-ductile behavior, using a deep beam or heavy beam which is able to provide adequate strength, the undesirable deterioration of frame can be prevented. The strong beam mechanism is highly ductile and can dissipate considerable energy; thus, delaying the failure of braces in fracture. Superior hysteretic response can be achieved from the strong beam mechanism, wherein the ductile braces distribute damages over the height of the building. The most efficient architectural bracing configuration is the V type, shown in Figure 3.18(b) or inverted V type, shown in Figure 3.18(a). In the V bracing, the beams are designed for the unbalanced loading; that occurs when the compression brace buckles and the tension brace pulls the beam down. This potential failure mode is largely present in V bracing as compared to in other brace configurations. When V braces are used, the ends of the SCBF beams are generally assumed as pinned-pinned. However, this is a very conservative assumption that may lead to large beam size because most of the beam bending moment results from the unbalanced load. Gusset plate provided at the beam ends provides partial or full fixity, thus reducing the moment.

3.9 OUTRIGGER BRACED SYSTEM

Outriggers have been historically used in the ships to resist wind loading, and the same concept has been implemented in high-rise structures as a lateral load resisting system. The outriggers are used to connect exterior columns at the outboard of the building to the lateral load resisting core, which can be either a shear wall or braced frame. This approach mobilizes the axial strength and stiffness of exterior columns to provide resistance to the overturning moment caused by lateral forces.

3.10 TOGGLE BRACE

Stiff structural systems respond to dynamic excitation with small drifts and small inter-story velocities. Further, implementing damping devices either as diagonal elements or as horizontal elements on top of chevron bracing leads to device displacements that are smaller than or equal to the drift. This necessitates damping devices that can offer significant forces for effective energy dissipation. Therefore, stiff structural systems may not be suitable as energy dissipation systems.

Toggle mechanism of energy dissipation results in device displacements that are larger than the structural drift. The amount of magnification can be in the range of 2–5, depending on the geometry of the toggle braces. As a result of the energy dissipation and device displacement magnification, the required control force is reduced by the same amount, leading to highly effective systems that are highly suitable for stiff structural systems. Figure 3.35 shows a toggle braced damper system installed in a frame. The damper is placed in a perpendicular position to the member DA connecting to pin A and the lower floor. The beam can also be connected to the upper floor.

Figure 3.36 shows the mechanism of the lower toggle brace damper system. Member DA rotates due to the movement of point C with respect to point D, creating story drift u. The displacement (u_d) of the damper related to the drift is the resulting distance between points B and A. The damping

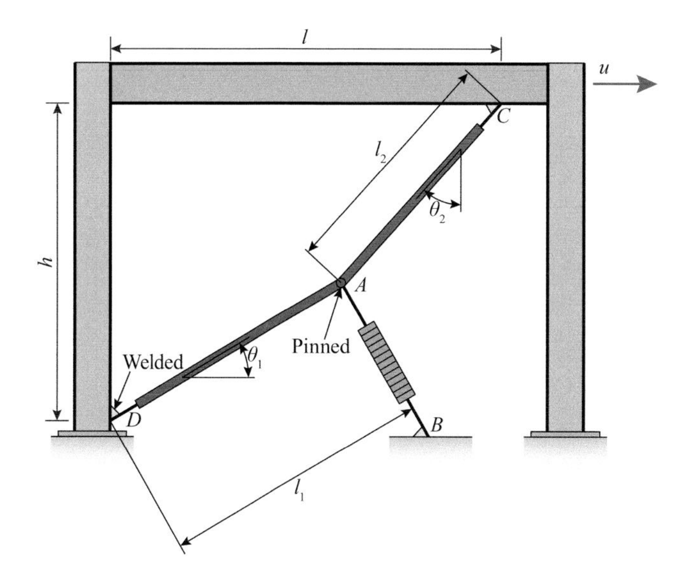

FIGURE 3.35 Toggle brace damper system in frame.

forces in the toggle brace damping system are relatively small. They are higher in the shallow truss configurations of the system and are delivered to the frame by compression or tension in the braces. The use of small sections and standard section details is possible, as there is no bending in the toggle braced damping system. In the toggle bracing configuration, shown in Figure 3.36 for drift toward the right, point A moves upwards.

The relative displacement (u_d) of the damper along its axis in the lower position (movement of point A with respect to the point of attachment) is

$$u_d = \pm l_1 \left[\sqrt{\left(1 + \frac{1}{cos^2\theta_1} - \frac{2(cos\theta_1 \pm \phi)}{cos\theta_1} \right)} - tan\theta_1 \right] \tag{3.8}$$

In Equation (3.8), + and – signs of ϕ indicate positive and negative rotations, respectively. A nonlinear complex relation between u_d and u is shown by Equation (3.10). Consider ϕ and u as relatively small angle and displacement. By comparing the dimensions, the following terms are obtained

$$\phi = f\left(\frac{u}{l_1}\right) \tag{3.9}$$

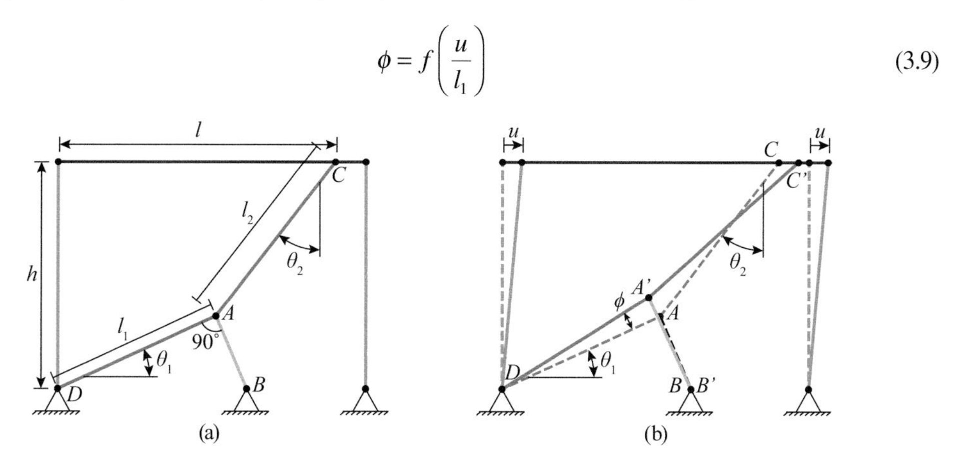

FIGURE 3.36 Mechanism of lower toggle brace damper system: (a) undeformed shape and (b) deformed position.

$$u_d = fu \tag{3.10}$$

$$f = \frac{sin\theta_2}{cos(\theta_1 + \theta_2)} \tag{3.11}$$

where u is the inter-story drift and f is the displacement magnification factor that depends on the inclination of the toggles. For the chevron brace configuration, $f = 1.0$, whereas for the diagonal configuration, $f = cos\theta$, where θ is the angle of inclination of the damper. The force (f_d) along the axis of the damper is related to the horizontal component of force F exerted by the damper on the frame.

$$F = f \times f_d \tag{3.12}$$

A high magnification factor can be achieved if the resulting system is very sensitive to small changes in the angles. However, magnification factor in the range of 2–3 is insensitive to small variations in the inclination of the toggles.

3.11 NEGATIVE STIFFNESS DAMPER

One of the strategies implemented in the seismic design of structures is to use members with reduced stiffness. This approach could sometimes be advantageous in reducing the seismic force imposed on the structure. However, it could lead to significant inelastic deformations of structural members. The negative stiffness damper (NSD) is a device that can help deliver the required characteristics of the system through the introduction of negative stiffness (Nagarajaiah et al. 2013; Pasala et al. 2013; Sarlis et al. 2013). The schematic diagram of a typical NSD, which depicts its key components and mechanism of operation, is shown in Figure 3.37, whereas Figure 3.38 shows the placement of NSD on a three-story structure. The force in the NSD (F_{NSD}) is obtained as the sum of the forces due to the horizontal and vertical springs, i.e., F_{HS} and F_{VS}, respectively.

$$F_{NSD} = F_{HS} + F_{VS} \tag{3.13}$$

The force-deformation characteristics of the NSD as well as the force-deformation behavior of an assembly of a structure and NSD are depicted in Figure 3.39. Further details of the computation of force in the NSD and its characteristics can be obtained in Nagarajaiah et al. (2013), Pasala et al. (2013), and Sarlis et al. (2013).

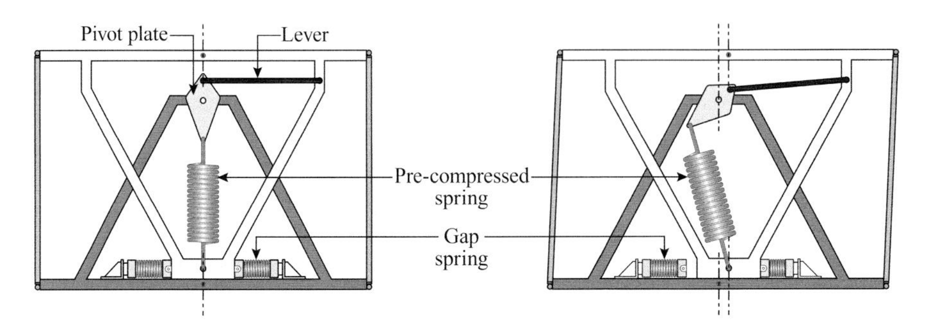

FIGURE 3.37 Schematic diagram of a typical negative stiffness damper (NSD).

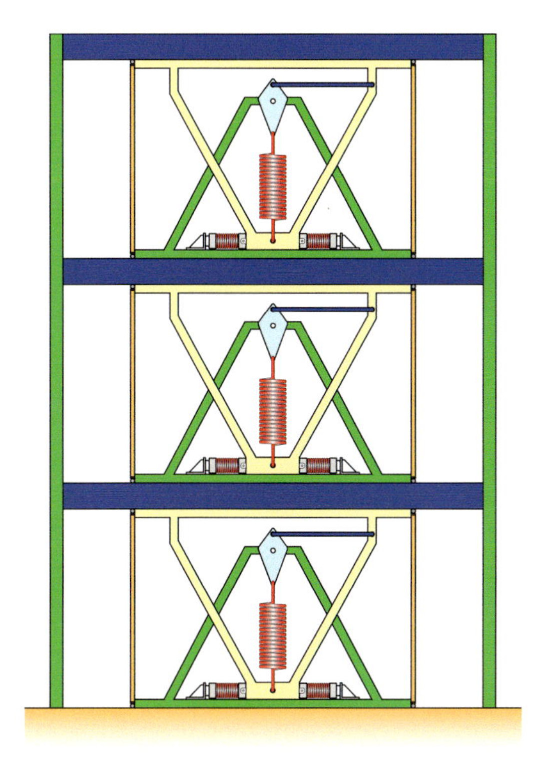

FIGURE 3.38 Negative stiffness damper (NSD) installed in a multistory building.

3.12 LEAD RUBBER SHEAR DAMPER

A lead rubber shear damper was first invented by Robinson and Tucker (1977). Several tests performed on a lead rubber shear damper demonstrate that the damper works well under successive major earthquakes and during many small movements caused by wind or thermal expansions. Good performance of the damper is because lead is capable of recovering most of its mechanical properties immediately after its deformation. Further, the lead is confined by the steel and rubber plates. The lead rubber shear damper has many interesting features that make it suitable for use in a base isolation system. Rubber bearing is capable of controlling the strain pattern of the lead, forcing it to deform into shear, and thus improving the damping characteristics of the device. The damper

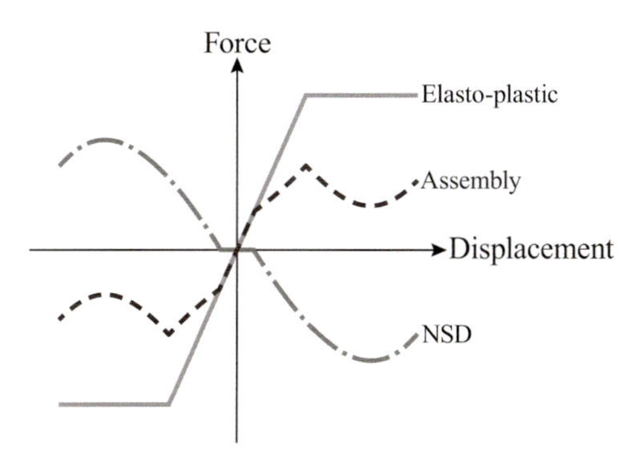

FIGURE 3.39 Force-deformation characteristics of a typical NSD and an assembly of structure and NSD.

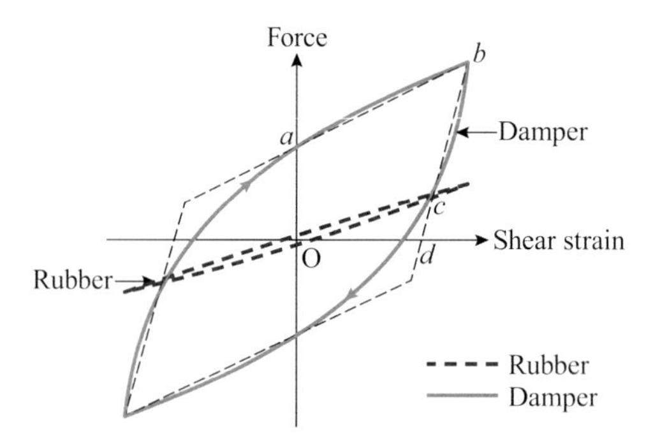

FIGURE 3.40 Force-displacement loop of lead rubber shear damper.

provides a horizontally flexible mount and an elastic centering force, both required for most of the base isolation systems. Hysteretic damping, good load bearing capacity, and centering force are the three important features that make a lead rubber shear bearing a convenient option to implement. Figure 3.40 depicts the resulting force-displacement loop obtained for the rubber alone and for the combined lead-rubber system. The performance of the hysteresis damper could be defined by the ratio of the actual hysteresis loop area of the device to the area of a circumscribing rectangle that has sided parallel to the coordinate axis Robinson and Tucker (1977). However, as seen in Figure 3.40, the hysteresis loop can be approximated by the circumscribing parallelogram passing through points a, b, c, and d, whose sides are parallel to the rubber loop axis.

3.13 PASSIVE TENDON SYSTEM

A new concept of passive control of bridges is based on a prestressed tendon system, which is typically suitable for bridges of 200- to 300-m span. The concept is based on cable-supported beam systems with additional prestressing control of the support cables to optimize the structural behavior of the system. The passive tendon system can effectively address the problem of excessive bridge deck deformation that happens due to the high ratio of moving loads to permanent loads. Two different prestressing systems are used for this purpose, as shown in Figure 3.41.

The first level prestressing (S_a at level a) is used to relieve the permanent loading of the structure. In addition to the bridge system with the first layer/set of prestressing cables, the effects of the moving loading are supported by a second system of prestressed cables that consists of a series of prestressed tendons, which are connected to the primary structure with different initial prestressing. Thus, the second level of prestressing (S_b at level b) mainly supports the moving loads and the additional displacements due to this loading case. Due to the additional choice provided by the initial prestressing of the second set of prestressing cables and their design parameters, they can be used as passive control devices. The advantages of cable-supported bridges for large-scale structures are effectively used by eliminating the disadvantages of large displacements due to the moving loading, which are typically associated with this type of structural system. The passive control system

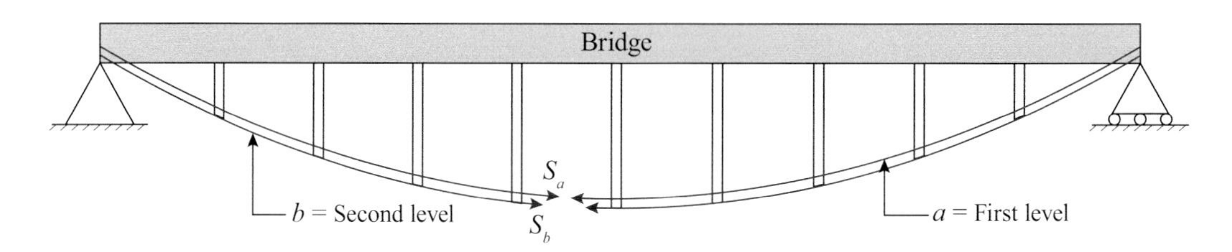

FIGURE 3.41 The two-level prestressing cables concept.

is a partial control system, whereas a part of the moving loading is relieved by the prestressed cables that act as the control devices. The whole composite structural system is composed of the beam elements, the reinforcing cable, and the vertical connecting rods. The prestressing cables (substructures) are divided into two levels, and this two-level prestressing system can considerably enhance the dynamic behavior of the bridge by using the prestressed tendons as passive control devices.

REFERENCES

Andrews, B. M., Song, J., and Fahnestock, L. A. (2009). Assessment of buckling-restrained braced frame reliability using an experimental limit-state model and stochastic dynamic analysis. *Earthquake Engineering and Engineering Vibration*, 8(3), 373–385.

Comerio, M., Tobriner, S., and Fehrenkamp, A. (2006). A guide to seismic safety on the UC Berkeley campus. Pacific Earthquake Engineering Research Center, California, United States. PEER 2006/01.

Nagarajaiah, S., Pasala, D. T. R., Reinhorn, A., Constantinou, M., Sirilis, A. A., and Taylor, D. (2013). Adaptive negative stiffness: A new structural modification approach for seismic protection. *Advanced Materials Research*, 639–640, 54–66.

Pasala, D. T. R., Sarlis, A. A., Nagarajaiah, S., Reinhorn, A. M., Constantinou, M. C., and Taylor, D. (2013). Adaptive negative stiffness: New structural modification approach for seismic protection. *Journal of Structural Engineering*, 139(7), 1112–1123.

Robinson, W. H., and Tucker, A. G. (1977). A lead-rubber shear damper. Bulletin of the New Zealand Society for Earthquake Engineering, 10(3), 151–153.

Sarlis, A. A., Pasala, D. T. R., Constantinou, M. C., Reinhorn, A. M., Nagarajaiah, S., and Taylor, D. P. (2013). Negative stiffness device for seismic protection of structures. *Journal of Structural Engineering*, 139(7), 1124–1133.

Takeuchi, T., and Wada, A. (Eds.). (2017). Buckling-restrained braces and applications. Tokyo, Japan. Japan Society of Seismic Isolation.

Wen, Y. K. (1976). Method for random vibration of hysteretic systems. *Journal of the Engineering Mechanics Division*, 102(2), 249–263.

4 Viscous and Viscoelastic Dampers

4.1 INTRODUCTION

Among various energy dissipation devices, fluid viscous dampers (FVDs) have been widely used in vibration control of several structural and mechanical systems, for dynamic response mitigation, where their performance has been very promising and satisfactory. The performance of structures can be greatly enhanced by using the right type and right configuration of FVDs. The dampers have been used in shock and vibration isolation of military and aerospace hardware, wind vibration control of missile launching platforms, and several other projects. Fluid dampers are now widely used in applications for seismic energy dissipation in buildings, bridges, water tanks, and other structures because of their characteristics, such as (i) linear viscous behavior, (ii) lower sensitivity to temperature changes, (iii) small size in comparison to stroke and output force, and (iv) their reliability and longevity.

Fluid dampers have the following intrinsic and noteworthy advantages, compared to hysteretic, viscoelastic, tuned mass, and elastoplastic energy dissipators.

i. They are self-reliant; no supplementary equipment or external power is required for their operation.
ii. The contemporary fluid dampers operate at significantly high fluid pressure level. Due to ease of installation, adaptability, good coordination with other structural members, compactness and diversity in their sizes, viscous dampers have found many applications in new structures and in retrofitting works.
iii. Fluid dampers have a low initial cost and almost negligible maintenance cost. They are economical and can reduce the overall cost of the project, when used at high damping ratios in the range of 15%–40%.
iv. Fluid dampers are especially attractive for enhancing the performance of structures because they reduce the deformation demand and also the force transferred to the structure. Because of their significant energy dissipation capability, fluid dampers can effectively reduce both internal shear forces and bending moments (usually governing limit state of collapse); and story drift and displacement (usually governing limit state of serviceability).

4.2 FLUID VISCOUS DAMPERS

In 1980, FVDs were effectively employed for applications in civil engineering projects for seismic energy dissipation. Based on the laws of fluid mechanics, various types of fluid viscous devices have been researched, tested, and brought in use. Viscous damping walls (VDWs) were proposed in Japan in 1986, in which three steel plates were used to sandwich the viscous fluid inside them. These walls proved their effectiveness in minimizing the earthquake-induced forces, and also in reducing base shear, story drift, and structural displacements, by shearing action of viscous fluid sandwiched between the plates. Other classes of FVDs are linear and nonlinear fluid devices, where velocity is linear and nonlinear function of the force of the damper, respectively. In order to mitigate the effect of temperature rise in piping system in nuclear power plants, a pipework (GERB) damper is used. In an orifice fluid damper, a cylindrical tube consisting of piston head with orifices is attached to the damper and is filled with highly viscous fluid. Orifice fluid dampers are used for effectively

DOI: 10.1201/9781315269269-4

reducing the earthquake effects on structures. The common mechanism behind energy dissipation in all fluid viscous devices is the smooth transfer of mechanical energy to the surrounding environment in the form of heat. FVDs can also be used in combination with seismic isolation systems for enhancing the energy dissipation capacity, and for controlling the excessive displacements at the isolator level, during earthquakes.

Characteristics of FVD

Figure 4.1(a) shows a longitudinal cross section of a typical FVD. It consists of a stainless-steel piston, with a bronze orifice head and a self-contained piston displacement accumulator. The damper is filled with a compressible, nonflammable, nontoxic, environmentally safe, and thermally stable viscous fluid, such as silicone oil. The piston head utilizes specially shaped passages, which alter the flow characteristics with fluid speed so that the force output is proportional to $|\dot{u}_d|^\alpha$, where $\dot{u}_d$ is the velocity of piston rod and α is an exponential term (velocity exponent). This behavior of FVD dominates for frequencies of motion below a pre-determined cutoff frequency that is related to the characteristics of the accumulator valves. Beyond this frequency, the fluid dampers exhibit strong stiffness, in addition to significant ability to dissipate energy. The existence of the cutoff frequency is desirable, since the lower modes of vibration are only damped, while the higher modes of vibration are damped and stiffened so that their contribution is completely suppressed.

The resistive output force (f_D) of the damper acts in a direction opposite to that of the input motion. Thus, energy dissipation is by heat transfer, i.e., the mechanical energy dissipated by the damper causes heating of the viscous fluid and mechanical parts, and this thermal energy is transferred to the environment. Figure 4.1(b) illustrates the working principle of FVD.

Operation of FVD

Pure viscous behavior of fluid dampers can be achieved by forcing fluid through orifices. When the piston moves in the viscous fluid medium, the damper offers an internal resisting force, proportional to the pressure differential across the piston head in the two chambers. The fluid flows from one chamber to the other, causing energy dissipation. During the motion of the piston head, the volume of compressible fluid is changed by the product of travel and piston rod area. When the damper is

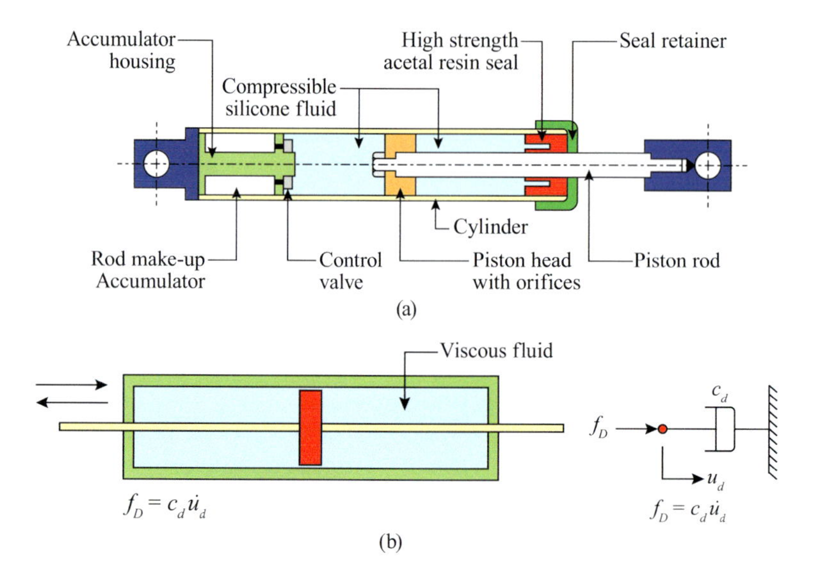

FIGURE 4.1 Fluid viscous damper: (a) schematic diagram and (b) free body diagram. (Taylor and Constantinou, 2000)

subjected to a compressive force, the fluid volume inside the cylinder is decreased as a result of the piston rod area movement. The decrease in fluid volume results in a restoring force. At higher frequency, FVDs exhibit strong stiffness. An accumulator works by collecting the volume of fluid that is displaced by the piston rod and storing it in the makeup area. As the rod retreats, a vacuum is created, which draws the fluid out.

Mathematical Modeling

FVD is a velocity-dependent device, in which, the total resisting force offered by the damper varies with respect to the translational velocity of fluid. The resisting force depends on the velocity of movement, the fluid viscosity, and the size of orifice of the piston. Figure 4.2 shows the force-velocity relationship of FVD for different values of the velocity exponent α, which decide the behavior of the damper, i.e., linear or nonlinear. The velocity exponent α depends on the viscosity properties of the fluid and the size of orifice of the piston.

In case of linear viscous damper $(\alpha = 1)$, the relationship between velocity of the damper and the damping force is linear, whereas the damper with $(\alpha < 1)$ exhibits nonlinear performance. It is perceived from Figure 4.2 that the peak damper force is limited when $(\alpha < 1)$. This property demonstrates the efficiency of nonlinear FVD to limit peak damper force in controlling earthquake-induced forces. This behavior of nonlinear FVD is contrary to that of the linear FVD and thus makes it attractive for control of dynamic loads, effectively. FVDs have the unique advantage of reducing the shearing and bending stresses at the same time, as the velocity-dependent maximum damping force is 90° out of phase with the maximum deflection of the structure, as seen from the idealized force-displacement relationship, shown in Figure 4.3. Typically, the two ends of a damper are attached to two different floor levels of a building. Due to this arrangement, the damper observes differential displacement, velocity, and acceleration, leading to the development of energy dissipating force.

Figures 4.3(a)–(c) show the elliptical force-displacement relationship of an FVD, linear force-displacement relationship of structure without damper, and the combined effect on the structure which is installed with FVD, respectively. As seen from the elliptical hysteresis loop in Figure 4.3(a), FVD allows significant amount of energy dissipation for a pure viscous linear behavior. The natural frequency of the structure remains unaltered, even after installation of FVD; thus, the structural design procedure with supplemental viscous dampers is simplified. As the damper develops restoring force, the hysteresis loop changes from viscous behavior (Figure 4.3(a)) to viscoelastic behavior shown in Figure 4.3(c). The maximum energy dissipation capacity of the damper, during the short time of the dynamic forces, is limited by the thermal capacity of lead and steel tube.

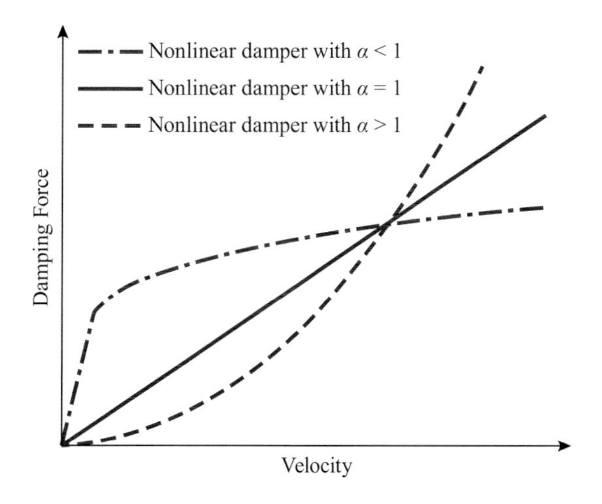

FIGURE 4.2 Force-velocity relationship of fluid viscous damper.

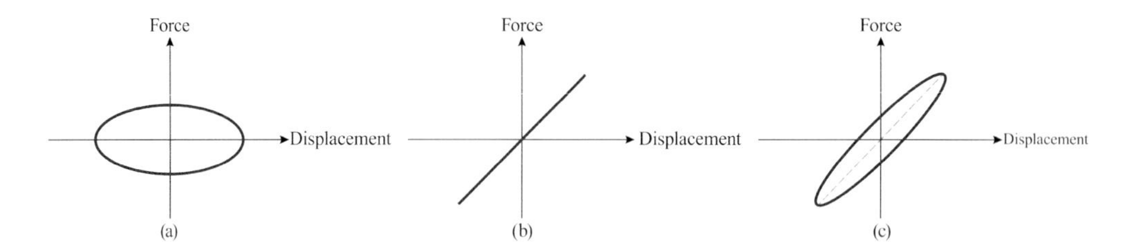

FIGURE 4.3 Force-displacement relationship: (a) viscous damper, (b) structure, and (c) combined response.

The resisting force in the damper is expressed as:

$$f_D = k_1 u_d + \left(\frac{k_2}{\varpi}\right)\dot{u}_d$$
$$f_D = k_1 u_d + c_d \dot{u}_d \tag{4.1}$$

where k_1 and k_2 are the storage stiffness and loss stiffness of the damper, respectively. The first term $(k_1 u_d)$ represents the force due to stiffness of the damper, which is in phase with the motion, and the second term $(c_d \dot{u}_d)$ represents the force due to viscosity of damper, which is 90° out of phase with the motion.

4.2.1 Linear Fluid Viscous Damper

Linear viscous damping is a property of the computational model and is not a property of a real structure. Viscous dampers are normally mounted in structural locations where some elastic recentering forces are provided by the structural frame itself. For linear dampers $(\alpha = 1)$, the damper force increases linearly with damper velocity.

4.2.2 Nonlinear Fluid Viscous Damper

Nonlinear FVDs have the apparent advantage of limiting the peak damper force at large velocities while still providing sufficient supplemental damping. Nonlinear damping system provides higher damping, compared to the linear damping system; therefore, such system is quite effective in absorbing and mitigating the high-velocity shocks. The resisting force of a nonlinear FVD is proportional to a fractional power law of the velocity, with velocity exponent α ranging between 0.1 and 1.0, i.e., the damping force is nonlinear function of the velocity of fluid. The supplemental damping ratio can be obtained by evaluating the equivalence between a nonlinear and a linear FVD. The criteria for evaluating the supplemental damping ratio for a nonlinear FVD presented in literature are expressed in terms of energy dissipated. One of the most common methods for defining equivalent viscous damping is to equate the energy dissipated in a vibration cycle of the actual nonlinear system to that of equivalent viscous system. The damping force induced by viscous damper, connected to the structure at its both ends, is expressed as in Equation (4.2).

$$f_D = c_d \left(\dot{u}_d\right)^{\alpha} sgn\left(\dot{u}_d\right) \tag{4.2}$$

where $\dot{u}_d$ and c_d are relative velocity between the two connected ends of the damper, and the damping constant, respectively.

Energy Dissipation in Linear FVD

The resisting output force of fluid damper acts in a direction opposite to that of the input motion, over a finite displacement, thus causing energy dissipation. At any instant of time, the resisting force varies with respect to the translational velocity of the damper. Consider a structural system undergoing steady-state vibrations under the action of force $F(t) = F_0 \ sin(\varpi t)$. The displacement of the damper end at any time instant is given by:

$$u_d(t) = u_0 sin(\varpi t) \tag{4.3}$$

where u_0 is the amplitude of the displacement and ϖ is the forcing frequency. The energy dissipated by viscous damping in one cycle of harmonic vibration is:

$$E_D = \int f_D \, du = \int_0^{\bar{T}} \left(c_d \dot{u}_d\right) \dot{u}_d dt = \int_0^{\bar{T}} c_d \dot{u}_d^2 dt \tag{4.4}$$

where $\bar{T} = \frac{2\pi}{\varpi}$ is the time period of the forcing function.

$$E_D = c_d \int_0^{\bar{T}} \left[\varpi u_0 cos(\varpi t)\right]^2 dt = \pi c_d \varpi u_0^2 = 2\pi \xi_d \left(\frac{\varpi}{\omega}\right) k u_0^2 \tag{4.5}$$

Damping ratio of the damper is written as $\xi_d = \frac{c_d}{c_{cr}}$.
Therefore, the energy dissipated (work done) is:

$$\begin{aligned} W_D &= \pi c_d u_0^2 \varpi = \pi \xi_d c_{cr} u_0^2 \varpi \\ &= 2\pi \xi_d \left(\sqrt{km}\right) u_0^2 \varpi \\ &= 2\pi \xi_d k u_0^2 \left(\frac{\varpi}{\omega}\right) \\ &= 4\pi \xi_d W_s \left(\frac{\varpi}{\omega}\right) \end{aligned} \tag{4.6}$$

where c_{cr}, k, m, and ω are the critical damping coefficient, stiffness, mass, and natural frequency of the system, respectively. The elastic strain energy of the system $\left(W_s\right)$ is given by:

$$W_s = \left(\frac{k u_0^2}{2}\right) \tag{4.7}$$

Thus, the damping ratio attributed to the viscous damper is:

$$\xi_d = \frac{W_D \omega}{4\pi W_s \varpi} \tag{4.8}$$

The dissipated energy W_D and the stored strain energy W_s in FVD are shown in Figure 4.4.
For the condition that $\omega = \varpi$, Equation (4.8) reduces to:

$$\xi_d = \frac{W_D}{4\pi W_s} \tag{4.9}$$

The total effective damping ratio of the system is:

$$\left(\xi_{eff} = \xi + \xi_d = \frac{c}{c_{cr}} + \frac{W_D}{4\pi W_s}\right) \tag{4.10}$$

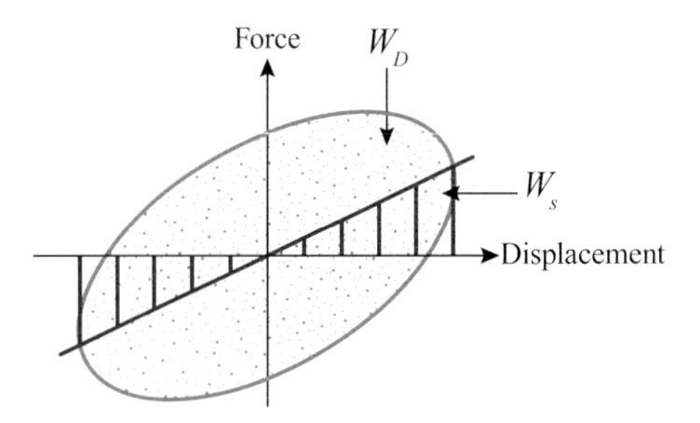

FIGURE 4.4 Dissipated energy and stored strain energy in fluid viscous damper.

where ξ is the inherent damping ratio (without dampers) of the system and ξ_d is the damping ratio of the FVD. The energy dissipated is proportional to the square of the amplitude of motion, as shown in Figure 4.5. For any given amount of damping and amplitude, energy dissipated linearly increases with the excitation frequency ϖ.

The external harmonic force $F(t)$ inputs energy to the system. For each cycle of vibration, the input energy is:

$$E_I = \int F(t)du = \int_0^{\bar{T}} F(t)\dot{u}_d dt$$

$$E_I = \int_0^{\bar{T}} \left[F_0 \, sin(\varpi t)\right]\left[\varpi u_0 \, cos(\varpi t)\right]dt = \pi F_0 u_0 \tag{4.11}$$

Equation (4.11) for (E_I) indicates that in steady-state vibration, the energy input to the system due to the applied force $F(t)$ is dissipated in viscous damping. Over each cycle of harmonic vibration, there is no change in the potential energy, i.e., the strain energy of the spring, and kinetic energy. The input energy (E_I) varies linearly and the dissipated energy (E_D) varies quadratically with the displacement amplitude, as seen in Figure 4.5.

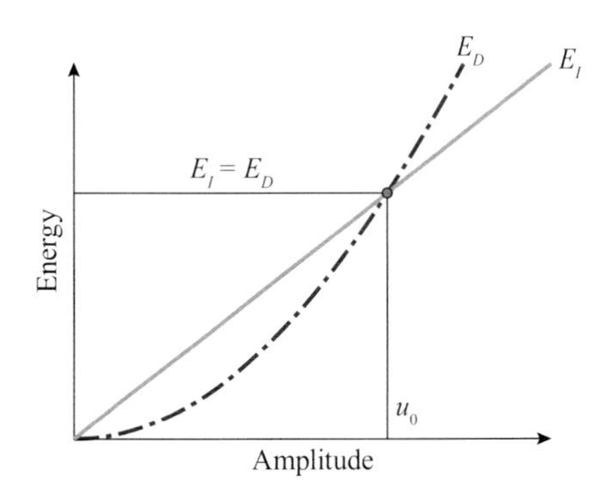

FIGURE 4.5 Input energy (E_I) and energy dissipated (E_D) in viscous damping.

$$\begin{aligned} f_D &= c_d \dot{u}_d(t) \\ &= c_d \varpi u_0 \ cos(\varpi t) \\ &= c_d \varpi \sqrt{u_0^2 - u_0^2 \sin^2(\varpi t)} \\ &= c_d \varpi \sqrt{u_0^2 - \left[u_d(t)\right]^2} \end{aligned} \tag{4.12}$$

Hence, we can write:

$$\left(\frac{u_d}{u_0}\right)^2 + \left(\frac{f_D}{c_d \varpi u_0}\right)^2 = 1 \tag{4.13}$$

Equation (4.13) is the equation of ellipse shown in Figure 4.6. The $(f_D - u)$ curve is known as hysteresis loop. The area enclosed by the elliptical hysteresis loop gives the energy dissipated by damping and is given by Equation (4.14). It is seen that the area enclosed by the elliptical hysteresis loop is proportional to the excitation frequency ϖ.

$$\pi(u_0)(c_d \varpi u_0) = \pi c_d \varpi u_0^2 \tag{4.14}$$

The total resisting force offered by the damper is the sum of the elastic force (f_S) and the damping force (f_D).

$$\begin{aligned} (f_S) + (f_D) &= k u_d(t) + c_d \dot{u}_d(t) \\ &= k u_d(t) + c_d \varpi \sqrt{u_0^2 - u_d^2} \end{aligned} \tag{4.15}$$

The graph of total resisting force $\left[(f_S) + (f_D)\right]$ against the displacement is the rotated position of ellipse, which represents the combined hysteresis loop of spring and viscous damper in parallel. The ellipse rotates due to the $k u_d$ term. The hysteresis loop associated with viscous damping, shown in Figure 4.6(a), is related to the dynamic nature of loading.

Energy Dissipation in Nonlinear FVD

When a structure installed with nonlinear FVD is subjected to sinusoidal excitation, the velocity is:

$$\dot{u}_d(t) = \varpi u_0 \ cos(\varpi t) \tag{4.16}$$

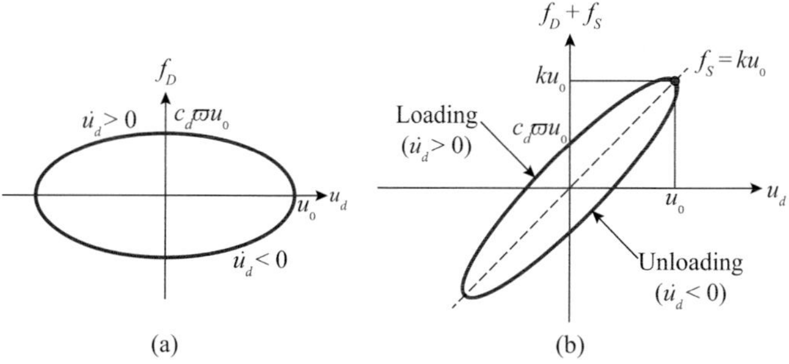

FIGURE 4.6 Hysteresis loops of (a) viscous damper and (b) spring and viscous damper in parallel.

The work done (energy dissipated) by the damper in one cycle of sinusoidal loading is given as:

$$W_D = \int f_D \, du \tag{4.17}$$

Knowing that $f_D(t) = c_d\left(\dot{u}_d\right)^{\alpha}$ and substituting Equation (4.16) by Equation (4.17), the work done by the nonlinear FVD is:

$$\begin{aligned} W_D &= \int_0^T f_D \dot{u}_d dt = \int_0^T c_d\left(\dot{u}_d\right)^{\alpha} \dot{u}_d dt \\ &= \int_0^{\bar{T}} \left| c_d\left(\dot{u}_d\right)^{1+\alpha} \right| dt = c_d\left(\varpi u_0\right)^{1+\alpha} \int_0^{\bar{T}} \left| sin(\varpi t)^{1+\alpha} \right| dt \end{aligned} \tag{4.18}$$

$$\begin{aligned} \varpi t &= 2\theta \\ t &= \frac{2}{\varpi}\theta \\ dt &= \frac{2}{\varpi} d\theta \end{aligned} \tag{4.19}$$

Using Equation (4.17), Equation (4.18) can be rewritten as:

$$\begin{aligned} W_D &= c_d\left(\varpi u_0\right)^{1+\alpha} \frac{2}{\varpi} \int_0^{\bar{T}} \left| sin^{1+\alpha}(2\theta) \right| d\theta = 2^{2+\alpha} c_d \varpi^{\alpha} u_0^{1+\alpha} \int_0^{\bar{T}} sin^{1+\alpha}(\theta) cos^{1+\alpha}(\theta) d\theta \\ &= 2^{2+\alpha} c_d \varpi^{\alpha} u_0^{1+\alpha} \frac{\Gamma^2\left(1+\dfrac{\alpha}{2}\right)}{\Gamma(2+\alpha)} \end{aligned} \tag{4.20}$$

where $\Gamma(\cdot)$ is the gamma function. Thus, the equivalent damping ratio of the system contributed by nonlinear damper can be expressed by equating the work done by nonlinear viscous damper to that of the equivalent viscous damper.

The equivalent (added) damping is calculated using Equation (4.9) and Equation (4.20).

$$\begin{aligned} 4\pi\xi_d W_s &= 2^{2+\alpha} c_d \omega^{\alpha} u_0^{1+\alpha} \frac{\Gamma^2\left(1+\dfrac{\alpha}{2}\right)}{\Gamma(1+\alpha)} \\ \xi_d &= \frac{2^{2+\alpha} c_d \omega^{\alpha} u_0^{1+\alpha} \dfrac{\Gamma^2\left(1+\dfrac{\alpha}{2}\right)}{\Gamma(1+\alpha)}}{4\pi W_s} \end{aligned} \tag{4.21}$$

4.2.3 GERB Fluid Viscous Damper

The pipework damper shown in Figure 4.7 is a unique FVD, which is first manufactured in 1970 by GERB in Germany. Pipework damper is used to reduce the dynamic response in piping system in nuclear power plants, where the piping system in operation may get excited by

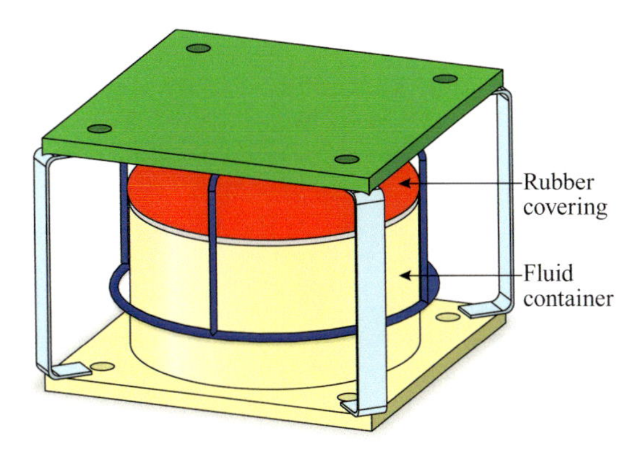

FIGURE 4.7 GERB fluid viscous damper.

undesirable dynamic loads caused by earthquake and vibrations caused by regular operations. The working principle of pipework damper is similar to that of the FVD, where the to and fro movement of piston through highly viscous fluid causes dissipation of vibration energy in the form of heat. GERB viscous damper reacts in no-time delay and requires almost no maintenance.

Assembly of GERB pipework damper includes outer box, highly viscous fluid, piston, protective sleeve, and a simple indicator of piston position. The viscous fluid should be nontoxic, resistant to moisture, vapor, and steam, should not react with oxygen, and should not corrode the material with which it is in contact. There are no adjusting components such as seals and guides. The movement of the piston is multidirectional; the volume and shape of the damper can be varied as per the application requirements. The pipework damper reacts to loads in any direction, i.e., it can be operated in all translational and rotational degrees of freedom. GERB dampers are available for specified operating temperatures and they can remain operative even if there are fluctuations in the temperature of the contained viscous fluid. It is recommended that the temperature of the fluid should never drop by more than 20% below that of the operating temperature. Higher damping is observed at lower temperature, whereas damping efficiency is considerably reduced at higher temperature. As the temperature of viscous fluid returns to the operating temperature, the GERB damper regains its original characteristics. Commonly, GERB dampers are manufactured to operate at 20°C. If they are not able to operate at the specified temperature, then temperature-dependent viscous fluids are used. However, such types of viscous fluids are costly and they reduce the damping efficiency.

Installation of GERB Damper

It is recommended to install pipework viscous damper in vertical position at the location of highest deflection, so as to achieve the highest efficiency of damping. There is a provision for placement of piston with a certain eccentricity inside the damper to take care of any unilateral heat expansion. However, at the initial installation stage, the piston is positioned centrally.

Mathematical Model

Generally, piping systems are supported at various locations along their length and, hence, need to be modeled by straight or curved pipe finite elements. Boundary conditions are applied to describe the support conditions. The matrix form of the governing differential equations of motion for the finite element model of piping system is written as:

$$[M]\{\ddot{u}\}+[C]\{\dot{u}\}+[K]\{u\}=\{F\}-\{Q\} \tag{4.22}$$

where $[M]$,$[C]$, and $[K]$ are mass, internal pipe damping, and stiffness matrices, respectively; $\{\ddot{u}\}$, $\{\dot{u}\}$, and $\{u\}$ are the vectors of nodal acceleration, velocity, and displacement, respectively; $\{F\}$ is the external nodal force vector; and $\{Q\}$ is the vector of nonlinear internal forces. When the piping system is subjected to seismic excitation, the external nodal force vector becomes:

$$\{F\} = -[M]\{\ddot{u}_g\} \tag{4.23}$$

where $\{\ddot{u}_g\}$ is the vector of absolute ground acceleration.

Equation (4.22) includes damping due to internal piping and does not include damping due to viscous damper. Considering damping due to the viscous damper:

$$\{Q\} = [C]_d\{\dot{u}\} + [K]_d\{u\} \tag{4.24}$$

where $[C]_d$ and $[K]_d$ are damping and pseudo-stiffness matrices of viscous damper, respectively.

Substitution of Equation (4.24) in Equation (4.22) gives:

$$[M]\{\ddot{u}\} + [C]_T\{\dot{u}\} + [K]_T\{u\} = \{F\} \tag{4.25}$$

where $[C]_T = [C] + [C]_d$ and $[K]_T = [K] + [K]_d$ are equivalent damping and stiffness matrices of the piping system, installed with GERB viscous damper. Equation (4.25) is the governing differential equation of pipework GERB viscous damper.

4.2.4 Viscous Damping Wall

A viscous damping wall (VDW) is a passive structural control system. Like other passive devices, it does not require any external source of power supply for seismic energy dissipation. It utilizes the motion of the structure for effectively controlling its dynamic response. Broadly, it is one of the types of FVD, where viscous fluid plays an important role in dissipating seismic energy. VDWs are found to be very effective and cost-efficient energy dissipaters. Lots of research works have been done on VDW in the last three decades. The concept of VDW was first proposed by a Japanese researcher Miyazaki et al. (1986). A large number of numerical and experimental studies have proved the effectiveness of VDWs for alleviating the dynamic response of structures. Dynamic structural response measurements of scaled-down and full-scale models of multistory reinforced concrete (RC) and steel buildings, installed with VDW, have established their effectiveness in reducing base shear and story drift, as well as providing additional damping and stiffness to the structure. VDW also proved effective in controlling wind-induced vibrations. Sumitomo construction company first manufactured VDWs in 1992. Furthermore, parameter optimization techniques for VDW are proposed recently in order to reduce the overall cost of the project, including material and installation costs.

Characteristics of VDW

Miyazaki et al. (1986) injected highly viscous fluid in between steel plate walls, to make them work as a passive control device, which on its own can dissipate seismic energy. He developed a simple device of three plates, where the viscous fluid is sandwiched between them. For sandwiching the high viscosity fluid, a tiny gap of 4–10 mm is maintained between the plates. The schematic diagram of a VDW shown in Figure 4.8 consists of one inner plate and two outer plates. The inner plate is connected to the bottom of upper floor and the two outer plates stand on the lower floor of the same story.

The mechanism of VDW is based on Newton's Viscosity Law, which states that shear stress is directly proportional to the velocity gradient or rate of change of shear strain. The sandwiched highly viscous fluid within the gaps maintained between the plates is capable of moving the inner plate, when relative movement occurs between the two floors.

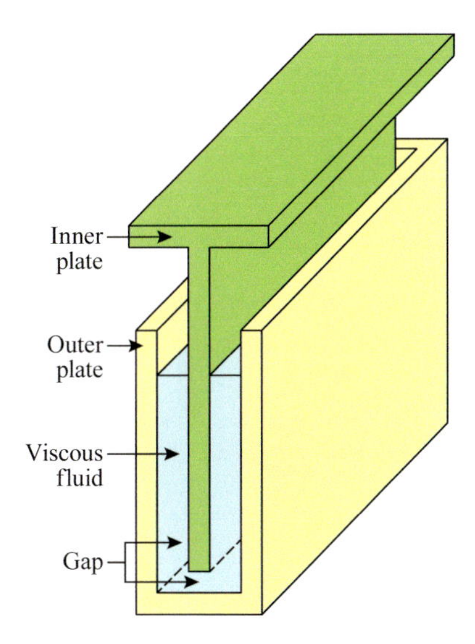

FIGURE 4.8 Schematic diagram of viscous damping wall.

The three important factors that govern the viscous restoring force (Q_w) and energy-absorbing capacity (E_W) are viscosity of fluid, gap between the plates, and surface area of wall plates. By altering the values of these three parameters, the magnitude of viscous restoring force (Q_w) and energy-absorbing capacity (E_W) of wall can be easily adjusted.

Materials and Their Properties

The two outer plates and one inner plate, which comprise the unit of VDW, are commonly made of steel. The desired gap clearance between the plates is maintained by using spacers. The outer surface of the wall is covered with RC or fireproof material so as to make it resistant to fire and impacts. The external cover also provides thermal insulation in normal functioning conditions. Calcium silicate insulation boards are regularly used to keep the temperature of fluid below 200°C. The sandwiched viscous fluid within the gap of the plates has viscosity in the range of 3,000–100,000 P. This viscosity is known as standard viscosity, which corresponds to the standard temperature of 30°C. For building constructions in Japan, polybutene was widely used as viscous fluid where VDWs were installed on a large scale. The viscosity of polybutene is in the range 60,000–90,000 P. Researchers have also used silicone fluid as viscous fluid, which has a viscosity of 12,000 centistokes at 25°C.

Operation of Dampers

Significant dynamic inertial forces are induced in a building under earthquake excitation. Due to this, each story experiences horizontal displacement between its upper and lower floor. The difference of horizontal displacements between upper and lower floor, i.e., the inter-story drift, causes movement of inner plate relative to the movement of the outer plates. This plate action of VDW is opposed by the highly viscous fluid sandwiched between the plates, in the form of a viscous damping force (Q_C) and elastic restoring force (Q_K). This reduces the overall dynamic response of building. As large contact area of wall is available for viscous fluid to act over, a high energy-absorbing (E_W) mechanism is created. The diagrammatic representation of operation of VDW is shown in Figure 4.9.

Design Formula of VDWs

The viscous restoring force (Q_w) generated in a VDW unit is given by Kelvin-Voigt model. The model is named after British engineer Lord Kelvin and German physicist Woldemar Voigt. It is also termed Voigt model. In this model, the purely viscous damper and purely elastic spring are

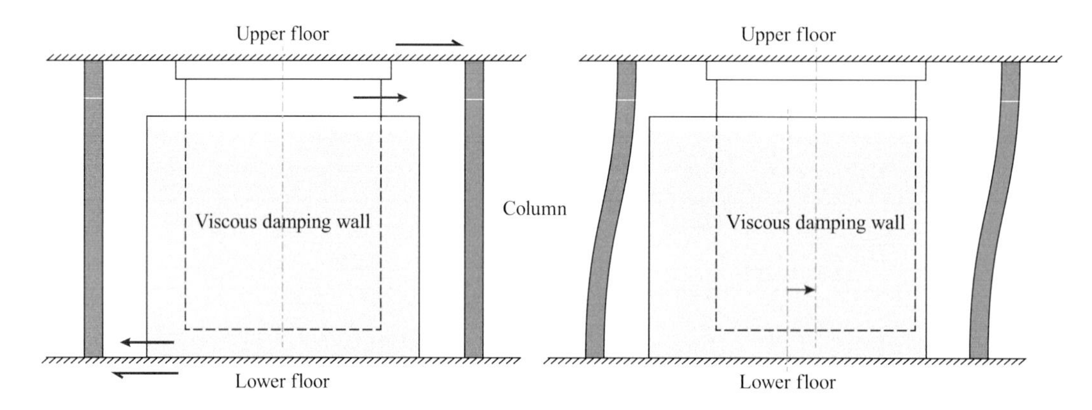

FIGURE 4.9 Undeformed and deformed shapes of viscous damping wall.

connected parallel to each other as shown in Figure 4.10(a). In Maxwell model, viscous damper and spring are connected in series as shown in Figure 4.10(b).

The viscous restoring force (Q_w) is a generalized sum of viscous damping force (Q_C) and elastic restoring force (Q_K).

$$Q_w = Q_C + Q_K \tag{4.26}$$

As represented in Figure 4.10(c), according to Newton's Viscosity Law, the viscous damping force (Q_C) can be designated as $\bar{\mu}\bar{A}\left(\frac{dv}{dy}\right)$, where $\bar{\mu}$ is viscosity of fluid, $\bar{A}$ is the area of wall in contact with fluid, dv is the relative velocity, and dy is the gap maintained between the inner and outer plates. The term $\left(\frac{dv}{dy}\right)$ is known as rate of change of shear strain or velocity gradient.

Since the fluid used in VDW is not a perfectly Newtonian fluid, the term $\left(\frac{dv}{dy}\right)$ is modified by a velocity exponent factor α, obtained experimentally. Thus, the viscous damping force is expressed as:

$$Q_C = \bar{\mu}\bar{A}\left(\frac{dv}{dy}\right)^{\alpha} \tag{4.27}$$

The spring force (elastic restoring force) depends upon the relative displacement between inner and outer plates. It is expressed as:

$$Q_K = \bar{\mu}\bar{A}\left(\frac{\delta^{\beta}}{dy^2}\right) \tag{4.28}$$

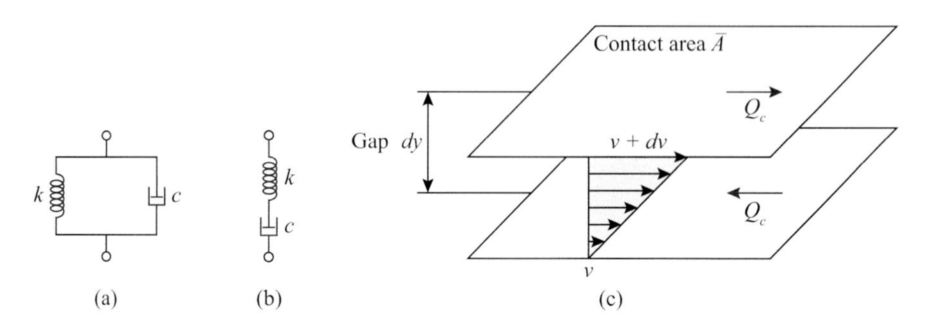

FIGURE 4.10 Modeling and mechanism of viscous damping wall: (a) Kelvin-Voigt model, (b) Maxwell model, and (c) viscous damping mechanism.

where δ is the relative displacement between inner and outer plates of VDW. The exponent factor β is obtained experimentally.

Mathematical Model of VDW

It has been demonstrated experimentally that VDW exhibits viscoelastic fluid behavior; i.e., it possesses property of elasticity, in addition to viscosity. Considering linear viscous damper as velocity-dependent, the damping force (f_D) is:

$$f_D(t) = c\dot{u}(t) \tag{4.29}$$

where c is the damping constant of viscous damper and $\dot{u}(t)$ represents the relative velocity between the two floors, at time t. Considering sinusoidal response of VDW, the relative displacement between the two stories is given as:

$$u(t) = u_0 sin(\varpi t) \tag{4.30}$$

$$\dot{u}(t) = \varpi u_0 cos(\varpi t) \tag{4.31}$$

The damping force (f_D) is

$$f_D(t) = c\varpi u_0 cos(\varpi t) \tag{4.32}$$

The force-displacement-velocity relationship is given as:

$$F_D(t) = Ku(t) + C\dot{u}(t) \tag{4.33}$$

where $F_D(t)$ is the total restoring force offered by the VDW, K is the storage stiffness, and C is the damping coefficient. This model is known as Kelvin-Voigt model, in which the total stress is the addition of stress in each element, whereas, strain is the same for both elements. The damping coefficient C and storage stiffness K are expressed by Equations (4.34) and (4.35), respectively.

$$C = \left(\frac{W_d}{\pi \varpi u_0^2} \right) \tag{4.34}$$

$$K = \frac{F_0}{u_0} \sqrt{\left\{ 1 - \left(\frac{C\varpi u_0}{F_0} \right)^2 \right\}} \tag{4.35}$$

In Equation (4.34), W_d is the energy dissipated, i.e., the area enclosed by the force-displacement loop. Damping matrix $\left[C_W\right]$ and stiffness matrix $\left[K_W\right]$ are added to the dynamic equation of equilibrium, given by Equation (4.36). The effect of non-Newtonian fluid is considered by parameters α and β.

$$[M]\{\ddot{u}\} + [C]\{\dot{u}\} + \left[C_W\right]\{\dot{u}^\alpha\} + [K]\{u\} + \left[K_W\right]\{u^\beta\} = -[M]\left\{\ddot{u}_g\right\} \tag{4.36}$$

In Equation (4.36), $[M]$, $[C]$, and $[K]$ are mass, damping, and stiffness matrices, respectively; and $\left\{\ddot{u}_g\right\}$, $\{\ddot{u}\}$, $\{\dot{u}\}$, and $\{u\}$ are the absolute ground acceleration, relative acceleration, relative velocity, and relative displacement vectors, respectively.

Merits of VDW

The merits of VDWs are given as follows.

1. VDWs are inexpensive, reliable, and do not require any digitally managed computerized system for their operation.
2. VDWs can sustain a temperature rise up to 200°C. Their outer surfaces are protected with fireproof material. Therefore, the temperature changes in the structure do not affect the functioning of VDW. The fireproofing material also serves as a thermal insulator under normal working conditions.
3. VDWs substantially reduce the dynamic response of structures subjected to earthquake excitations and wind loading, such that the structure always remains elastic. Hence, the possibility of formation of plastic hinges in the structure is reduced substantially.
4. VDWs constructed with two inner plates offer twice the damping force with a very small increase in wall thickness.
5. VDWs are almost maintenance-free because they do not require any high-maintenance components, such as sealing system.
6. VDWs provide the designer with a greater architectural flexibility. They can be manufactured so as to get accommodated easily, as structural walls of building.
7. Because of their compact rectangular shape, they can be installed easily, compared to the diagonal bracing systems or dampers.
8. VDWs reduce the seismically induced inertial forces and inter-story drift in buildings substantially.

Applications of VDWs

Since the inception of VDWs in Japan, a lot of research has been done and is still ongoing so that their benefits can be exploited for designing cost-effective and robust seismic-resistant structures. Some of the notable worldwide applications of VDWs are mentioned subsequently.

1. VDWs were installed in the SUT building shown in Photo 4.1, at the center of Shizuoka City, 150 km west of Tokyo, Japan. The 14-story building is 78.6 m tall. It is designed for the musical hall of the capacity of 400, a high-quality video theater, restaurants, and shops. The building consists of four VDWs of height 5 m, in each direction, at each story. Japanese engineers Miyazaki et al. (1986) were able to increase the damping of building by 20%, due to installation of the VDWs.
2. In 1995, VDWs were installed in the 31-story NTT building, shown in Photo 4.2, located in Tokyo, Japan.
3. VDWs were installed in 31-story Omiya Postal Building, in Japan (1996).
4. A combination of VDWs and viscous damping outrigger was used in the Kunming Dianchi Project, China, in a 42-story building. The building consists of 42 VDWs from 4th floor to 24th floor, in X-direction, and 28 VDWs from 8th floor to 21st floor, in Y-direction. Installation of VDWs, along with viscous damping outrigger, reduced the base shear substantially.
5. The 33-storied China Union Pay Operation Centre located in Shanghai is installed with 60 VDWs, from 2nd floor to 16th floor; 30 wall units are arranged in cross-type manner, in X-direction, and 30 VDWs are arranged in continuous type, in Y-direction. The VDWs reduced the base shear considerably.
6. In the United States, the first project employing VDWs was completed in 2012. The installation of 113 walls was completed by Dynamic Isolation System (DIS), United States, in Gery Van-Ness Medical Centre, located in San Francisco, California, shown in Photo 4.3.

PHOTO 4.1 SUT building, Shizuoka, Japan.

PHOTO 4.2 NTT building, Tokyo, Japan.

PHOTO 4.3 Gery Van-Ness Medical Centre, United States.

4.3 VISCOELASTIC DAMPER

A purely elastic material regains its original shape and size on the removal of the stress acting over it. Thus, all the energy stored in the elastic body during loading is recovered rapidly on the removal of the externally applied load. Most of the elastic materials clearly show linear elastic behavior and are well represented by Hooke's Law. According to Hooke's Law, stress (σ) is directly proportional to strain (ε) and the ratio of stress to strain is known as modulus of elasticity (E). Hence, the stress and strain curves for elastic materials always move in phase, as shown in Figure 4.11. For a member of cross-sectional area (A) and thickness (T), subjected to cyclic loading $F(t)$, with corresponding displacement response $u(t)$, the cyclic stress in material is given by input load per unit cross-sectional area. The cyclic strain in the material is defined as displacement divided by thickness of the member.

The material that exhibits property exactly opposite to the elastic material, i.e., the material that does not obey Hooke's Law, is known as a purely viscous material. Viscous material does

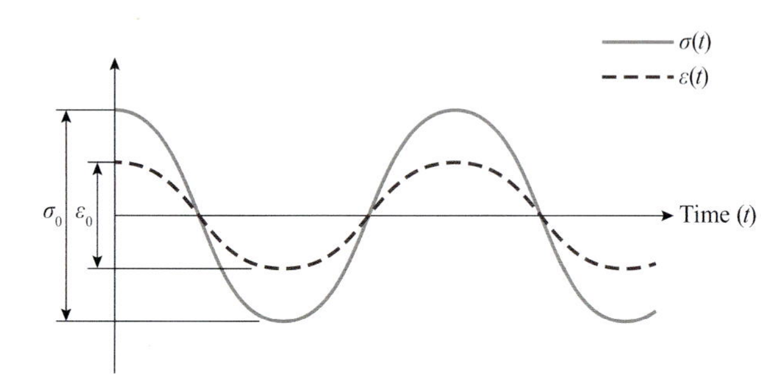

FIGURE 4.11 Stress and strain curves for elastic material.

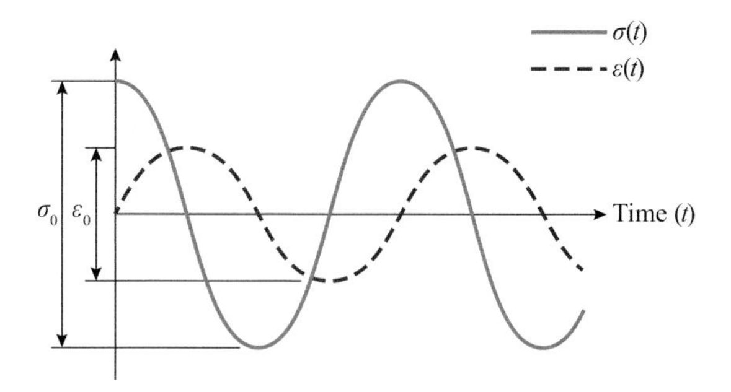

FIGURE 4.12 Stress and strain curves for viscous material.

not recover any energy lost during loading, after removing the load acting over it. Once the load is removed, all the energy is lost or absorbed by material as pure damping. In this case, stress is proportional to the rate of change of strain. The ratio of stress rate to strain rate is known as viscosity (μ). For viscous material, stress lags behind strain by a 90° phase lag, as shown in Figure 4.12.

The material that does not fall under the category of either elastic material or viscous material is known as viscoelastic material. Such material features both properties; elasticity and viscosity. Upon removal of the applied load, some amount of energy gets recovered, while the remaining energy is dissipated in the form of heat. The cyclic stress (σ) is out of phase with the cyclic strain (ε) by an angle (ϕ) at loading frequency (ϖ) as shown in Figure 4.13.

Dynamic modulus, also referred to as complex modulus, is used to represent modulus of viscoelastic material. It is the ratio of cyclic stress to cyclic strain under vibratory conditions. Dynamic modulus is calculated from the experimental data obtained from either free or forced vibration tests, in shear and compression. The cyclic strain and cyclic stress in a viscoelastic material are represented by Equations (4.37) and (4.38), respectively.

$$\varepsilon = \varepsilon_0 sin(\varpi t) \tag{4.37}$$

$$\sigma = \sigma_0 sin(\varpi t + \phi) \tag{4.38}$$

Upon removal of load, the storage modulus (E_1) is represented as:

$$E_1 = \left(\frac{\sigma_0}{\varepsilon_0}\right) cos\phi \tag{4.39}$$

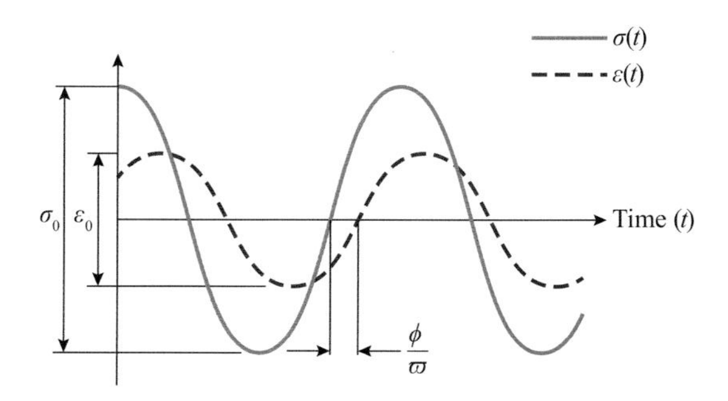

FIGURE 4.13 Stress and strain curves for viscoelastic material.

After recovery, some amount of energy is dissipated in the form of heat, representing viscous portion and is known as loss modulus (E_2), represented by Equation (4.40).

$$E_2 = \left(\frac{\sigma_0}{\varepsilon_0}\right) sin\phi \tag{4.40}$$

Likewise, the shear storage modulus (G_1) and shear loss modulus (G_2) can also be defined. Complex variables (E^*) and (G^*) are used to express the complex modulus, as shown in Equations (4.41) and (4.42), respectively.

$$E^* = E_1 + iE_2 \tag{4.41}$$

$$G^* = G_1 + iG_2 \tag{4.42}$$

The properties of viscoelastic material are influenced by several parameters such as temperature, excitation frequency, dynamic strain rate, creep and relaxation, aging, and supplementary irreversible effects. Variations in the behavior of viscoelastic material can be well illustrated by Figure 4.14. In glassy region, the storage modulus (E_1) is at its highest value for the material, whereas the loss modulus (E_2) is at its lowest value. In this region, the material possesses glass-like structure with high stiffness and low damping. Further, a transition takes place in the behavior of material; it starts getting transmitted from glassy to rubbery region. Hence, this region is known as transition region, in which the viscoelastic material goes through a rapid rate of change in energy storage, eventually possessing higher damping characteristics. In the rubbery region, the material reaches to its lowest stiffness and damping too is lower, but at a reasonable level. In this region, the storage modulus (E_1) and the loss modulus (E_2) vary with the respect to temperature and frequency; hence, the material is best suited for providing damping and for use in isolation devices.

Damping of viscoelastic material is represented by hysteresis model shown in Figure 4.15(a). Kelvin-Voigt model shown in Figure 4.15(b) is used for representing damping of elastic and viscous material. In hysteresis model, damping is directly proportional to strain, unlike Kelvin-Voigt model, wherein damping is proportional to rate of strain. In hysteresis model, damping is achieved by eliminating viscous dashpot and representing energy dissipation in the system by a dynamic (complex) spring element (K^*), thus resulting in a complex elastic modulus (E^*) and complex shear modulus (G^*).

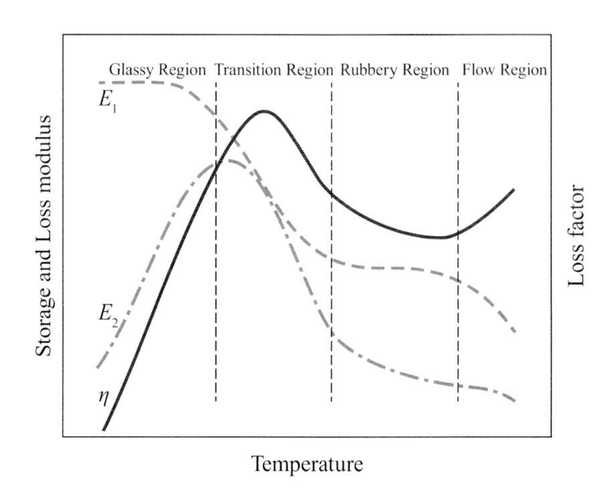

FIGURE 4.14 Variations in behavior of viscoelastic material.

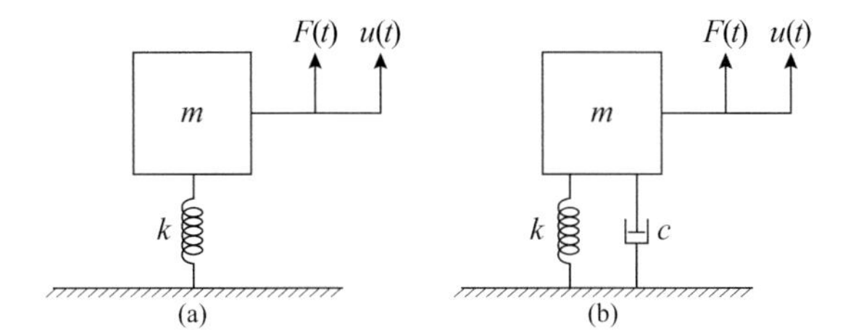

FIGURE 4.15 Models representing damping of viscoelastic material: (a) hysteresis model and (b) Kelvin-Voigt model.

Practically, the concept of using properties of elastic material and viscous material together is widely used for stiffness enhancement and energy dissipation. Furthermore, the energy is dissipated either by using solid or fluid action of damper. Hence, viscoelastic dampers are further classified as viscoelastic solid damper and viscoelastic fluid damper.

4.3.1 Viscoelastic Solid Damper

Characteristics of Viscoelastic Solid Damper

Viscoelastic solid dampers generally employ copolymers or glassy substances that dissipate energy, when subjected to shear deformation. Viscoelastic solid damper consists of layers of viscoelastic material, bonded with steel plates, as shown in Figure 4.16. The two outer steel flanges are connected to center steel plate with solid elastomeric pads of viscoelastic material. This is done in order to obtain large tensile and compressive strains in viscoelastic material.

The behavior of viscoelastic solid damper can be modeled using Kelvin-Voigt model of viscoelasticity, represented by Equation (4.43).

$$F(t) = Ku(t) + C\dot{u}(t) \tag{4.43}$$

where K is the storage stiffness of damper and C is the damping coefficient, which is equal to the ratio of loss stiffness (K_1) to the frequency of motion (ϖ), $u(t)$ is the relative displacement, $\dot{u}(t)$ is the relative velocity, and $F(t)$ is the force in damper. In Kelvin-Voigt model, the dashpot or purely viscous damper C and purely elastic spring K are connected parallel to each other, as shown in Figure 4.15(b), wherein the restoring force is directly proportional to the displacement, and the damping force is directly proportional to the velocity. Thus, viscoelastic solid damper possesses the ability not only to dissipate energy but also to store it.

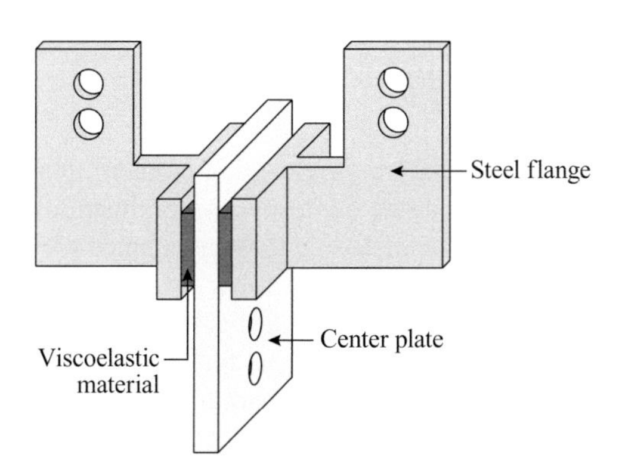

FIGURE 4.16 Typical configuration of viscoelastic solid damper.

The dynamic response of a viscoelastic solid damper depends upon parameters such as frequency of motion, the level of strain, and the ambient temperature. Temperature has an inverse effect on energy dissipation; as temperature tends to increase, the energy dissipated per cycle tends to decrease. A lot of research has been conducted in the past for employing high damping rubber as a viscoelastic material, because of its properties depending on both temperature and time. Under constant stress, rubber shows an increase in deformation with time and under the condition of constant strain, there is a decrease in stress with time. Rubber possesses the property of both creep and relaxation, which results in increased energy dissipation.

Working Principle of Viscoelastic Solid Damper

Diagonal bracings are employed to attach the steel flange plates of damper to the structure. During earthquake excitations or wind vibrations, one end of the damper gets displaced with respect to the other. This results in shearing of the elastomeric pads of viscoelastic material, causing development of heat, which is dissipated to the environment. By virtue of this shearing activity of pads between the two steel flanges, viscoelastic solid possesses both viscosity and elasticity properties. Thus, a viscoelastic solid damper is displacement-dependent, as well as velocity-dependent.

Mathematical Model of Viscoelastic Damper

Usually, mechanical behavior of viscoelastic material is represented in the form of shear stresses and strains, instead of forces and displacements. Thus, the mechanical properties of importance are storage modulus and loss modulus of the viscoelastic material; rather than the storage stiffness and loss stiffness of the viscous damper. The relationship between shear stress (τ), initial shear strain (γ), and the final shear strain $(\dot{\gamma})$ under harmonic excitations of frequency (ϖ) can be expressed as:

$$\tau(t) = G_1(\varpi)\gamma(t) + \frac{G_2(\varpi)}{\varpi}\dot{\gamma}(t) \tag{4.44}$$

where (G_1) and (G_2) are the shear storage modulus and shear loss modulus, respectively. The ratio of the shear loss modulus and the shear storage modulus is defined as loss factor (η), expressed as:

$$\eta(\varpi) = \frac{G_2(\varpi)}{G_1(\varpi)} \tag{4.45}$$

In case of general excitations for polymer materials, Boltzmann's superposition principle can be implemented. The principle states that the response of the material to a given load is independent of its response, subjected to any other load acting on it. Thus, in a linear viscoelastic material, accumulated viscoelastic creep strain is the superimposed sum of individual creep increments. The constitutive relation for polymer material can be expressed as:

$$\tau(t) = \int_0^t G(\tau)\dot{\gamma}(t-\tau)\,d\tau + G(t)\gamma \tag{4.46}$$

In Equation (4.46), $G(t)$ represents the stress relaxation modulus defined as the ratio of stress to strain at constant deformation, which can be determined experimentally. Various expressions can be assumed for stress relaxation modulus, including those associated with the classical Kelvin-Voigt and Maxwell models.

Overall force-deformation model for the viscoelastic solid damper can be easily established after successful construction of the consecutive model for the response of the damper material in shear. For a typical viscoelastic damper possessing shear area (A_S) and displacement (δ), the force-displacement relationship for response under harmonic loading can be expressed as:

$$F(t) = K_1 u(t) + C_1 \dot{u}(t) \tag{4.47}$$

where

$$K_1 = \frac{A_S G_1(\varpi)}{\delta} \quad \text{and} \quad C_1 = \frac{A_S G_2(\varpi)}{\varpi\delta}$$

Merits of Viscoelastic Solid Damper

The main advantages of viscoelastic solid dampers are given as follows.

1. A viscoelastic solid damper is effective in reducing wind vibrations and is also capable of providing the structure with increased damping for frequent low-level ground shaking as well as for severe earthquake ground motions.
2. Viscoelastic solid dampers are inexpensive as compared to viscoelastic fluid dampers.
3. Viscoelastic behavior of damper is linear and it gets easily activated at very low displacements, thus reducing the displacement demand on the structure.
4. A viscoelastic solid damper is capable of dissipating energy in early stages of cracking of concrete elements, thus preventing further damage.

Disadvantages of Viscoelastic Solid Damper

The primary disadvantages of viscoelastic solid dampers are as follows:

1. Viscoelastic solid damper has a very limited deformation capacity.
2. Properties of damper are frequency and temperature dependence. As ambient temperature tends to increase, the energy dissipated per cycle tends to decrease.
3. There is a possibility of debonding and tearing of viscoelastic material.

Worldwide Applications of Viscoelastic Solid Damper

Initially, viscoelastic solid dampers were used for mitigation of wind-induced vibrations. As the research progressed, the solid dampers proved their effectiveness in reduction of earthquake-induced vibrations. Some of the notable applications of viscoelastic solid dampers in the world are presented subsequently.

1. The first application of viscoelastic solid damper was in 1961, in the twin towers of the World Trade Centre in New York (NY), United States, as an integral part of the structural system as shown in Photo 4.4. There are 10,000 viscoelastic solid dampers installed in each tower, methodically disturbed from the 10th to the 110th floor.
2. In 1982, the 76th floor of Columbia Sea First Building in Seattle, shown in Photo 4.5, was incorporated with 260 viscoelastic solid dampers, in order to control the wind-induced vibrations. The installation of solid dampers was found to increase the damping ratio of the building in the fundamental mode from 0.8% to 6.4% for frequent storms, and from 0.8% to 3.2% at design wind speed.
3. In 1988, viscoelastic solid dampers were used to control wind vibrations in the Two Union Square Building in Seattle, shown in Photo 4.6. A total of 16 dampers were installed on four columns in each floor.
4. The first application of viscoelastic solid dampers for mitigating earthquake-induced vibrations was in 1993, for the 13-story Santa Clara building in San Jose, California (CA), United States, shown in Photo 4.7. The building was constructed in 1976 and the seismic upgradation plan was developed for installing viscoelastic solid dampers. Two dampers on each face of the building, for each floor, were installed, which provided substantial reduction to building response under all levels of ground vibrations.

PHOTO 4.4 World Trade Centre, New York (NY), United States.

PHOTO 4.5 Columbia Sea First Building, Seattle.

PHOTO 4.6 Two Union Square Building, Seattle.

PHOTO 4.7 Santa Clara building in San Jose, California (CA), United States.

4.3.2 Viscoelastic Fluid Damper

A viscoelastic solid damper utilizes the action of solid viscoelastic material in order to improve the performance of structures subjected to wind-induced vibrations, as well as seismic shaking. However, viscoelastic fluids can also be effectively be used for achieving the desired viscoelastic damping characteristics. This section presents the application of viscoelastic fluid dampers for structural vibration control purpose.

Characteristics of Viscoelastic Fluid Dampers

In viscoelastic fluid dampers, viscoelastic fluid is used instead of the solid viscoelastic material. Several types of viscous dampers employed with fluid applications have been developed. Cylindrical plot fluid damper and VDW are the noteworthy examples. These devices cause dissipation of energy through the deformation of a viscous fluid, residing in an open container. However, in order to increase the efficiency of energy dissipation, fluids with larger viscosities, which exhibit both frequency- and temperature-dependent behaviors, are recommended.

The cylindrical plot fluid damper shown in Figure 4.17 consists of classical dashpot, wherein mechanical energy gets converted to heat energy, causing energy dissipation. The piston provided in the damper deforms a highly viscous fluid, such as silicone gel (viscosity in the range 60,000–90,000 P).

Mathematical Model

Although construction of viscoelastic fluid damper is different from viscoelastic solid damper, the analytical models appropriate for the overall force-displacement response of both dampers are similar. Both solid and fluid devices are frequency- and temperature-dependent. In the past, various constitutive models were proposed for viscoelastic fluid dampers. One of the approaches is the fractional derivative technique applied to classical Maxwell model, to represent the behavior of fluid dampers mathematically (Makris et al. 1993).

Fractional derivative of Maxwell force-displacement model is expressed as:

$$F(t)+\lambda^{v}\frac{d^{v}F(t)}{dt^{v}}=C_0\frac{du(t)}{dt} \tag{4.48}$$

where $F(t)$ is the force applied to the piston; $u(t)$ is the resulting piston displacement; and C_0, λ, and v represent the damping coefficient, relaxation time, and order of fractional derivative, respectively. The relaxation time for the damper is defined as $\lambda=\left(\frac{C_0}{K_0}\right)$, where K_0 is the storage stiffness at a definite frequency. Considering a complex derivative, Maxwell model is used to characterize a highly viscous fluid over a broad range of frequency and temperature. Assuming a very small incompressible deformation, at some reference temperature T_0, the above proposed model can be represented in terms of shear stress (τ) and shear strain (γ), as shown in Equation (4.49).

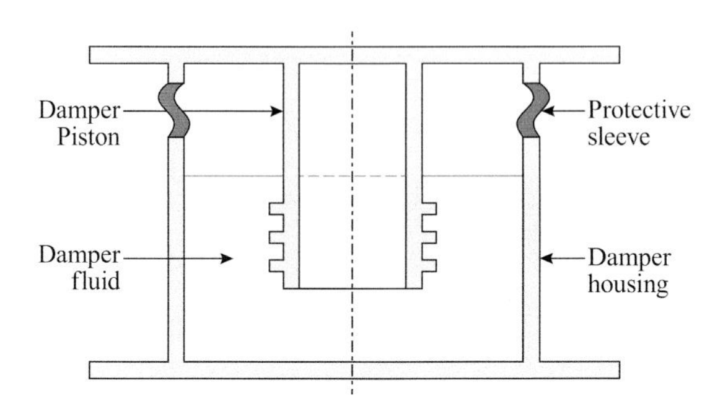

FIGURE 4.17 Cylindrical plot fluid damper.

$$\tau + \left[\lambda(T_0)\right]^{v} \frac{d^{v}\tau}{dt^{v}} = \mu(T_0)\frac{d\gamma}{dt} \tag{4.49}$$

where $\lambda = (\lambda_1 + i\lambda_2), v = (v_1 + iv_2)$, and $\mu = (\mu_1 + i\mu_2)$ are complex-valued material parameters and μ is fraction of temperature. The term $\left(\frac{d^{v}\tau}{dt^{v}}\right)$ denotes a generalized derivative of order v with respect to time t. Equation (4.49) reduces to the classical Maxwell model for $v = 1$. The values of the shear storage modulus and the shear loss modulus analogous to the above model can be obtained by performing a Fourier transform. Orifice fluid damper has a very less frequency dependency. Therefore, Equation (4.49) can be reduced to Equation (4.50):

$$F(t) + \lambda \frac{dF(t)}{dt^{v}} = C_0 \frac{du(t)}{dt} \tag{4.50}$$

For a very small relaxation time λ, Equation (4.50) can be further reduced to Equation (4.51), which simplifies the structural analysis with high degree of accuracy.

$$F(t) = C_0 \frac{du(t)}{dt} \tag{4.51}$$

Merits of Viscoelastic Fluid Damper

The benefits of using viscoelastic fluid dampers are given as follows:

1. A viscoelastic fluid damper can improve the seismic performance of a structure by considerably reducing displacements and damage.
2. As compared to solid damper, fluid damper is capable of providing longer strokes.
3. Mechanical properties of fluid dampers are less sensitive to the variation of ambient temperature, in comparison to solid dampers. Thus, unlike solid dampers, properties of fluid damper are less dependent on temperature and frequency of excitation.
4. The damper gets easily activated at very low displacements also and is capable of providing nominal restoring force for controlling very low displacements.
5. In military applications, the effectiveness of viscoelastic fluid damper is well established.

Demerits of Viscoelastic Fluid Damper

The only demerit of viscoelastic fluid dampers is the susceptibility of the acetyl resin seal to damage. There have been some reported problems of seal leakage, which has negatively affected the effectiveness of the dampers.

Worldwide Applications of Viscoelastic Fluid Damper

The early applications of viscoelastic fluid dampers were in the aerospace and military. Recently, they are employed in civil engineering structural applications, mostly, in combination with seismic isolation systems. Some of the notable applications of viscoelastic fluid dampers in the world are mentioned subsequently.

1. A total of 62 fluid dampers were incorporated for seismic response mitigation of the Pacific Bell North Area Operations Center, Sacramento, California, United States. The center, shown in Photo 4.8, was completed in 1995. Each damper can offer a restoring force of 130 kN and a stroke of ±50 mm. For the structure to remain operational in seismic events, the dampers in chevron configuration were installed throughout the structure.
2. In 1995, viscoelastic fluid dampers were used for five buildings of the new San Bernardino Medical Center, California, United States, shown in Photo 4.9. These buildings are located

PHOTO 4.8 Pacific Bell North Area Operations Center, Sacramento, California (CA), United States.

PHOTO 4.9 San Bernardino Medical Center, California (CA), United States.

PHOTO 4.10 Woodland Hotel, California (CA), United States.

very close to the two major active fault lines. Total 18 dampers were incorporated into base isolation systems, each with an output force of 1,400 kN, and a stroke of ±600 mm.

3. The four-story Woodland Hotel, shown in Photo 4.10, a historical masonry structure in Woodland, California, United States, was employed with viscoelastic fluid dampers. The dampers installed in the hotel offered a total resisting force of 16,450 kN. The construction of hotel was completed in 1996.
4. In 1996, another project was undertaken with the installation of 40 orifice fluid dampers in the 35-story building located in downtown Boston, MA, United States. Each damper has a capacity of 670 kN with a stroke of ±25 mm.

4.4 OIL DAMPER

Oil dampers make use of the resistance generated by passing of hydraulic fluid through an opening provided in the closed cylinder. All types of oil dampers work on the principle of converting kinetic energy generated by the movement of oil inside the cylinder, into heat energy. Assembly of oil damper consists of a closed cylindrical container with sealing on both sides so as to avoid leakage of oil. Piston rod is provided coaxially within the damper, which can move to and fro along the cylinder. The effectiveness of the damping realized by oil damper is directly proportional to the velocity of oil and is independent of the temperature variation. The most useful feature of oil dampers is that they are low-maintenance devices and allow a smooth flow of oil along the damper length. Broadly, there are two types of oil dampers, i.e., linear type and bilinear type.

Linear-type oil dampers are equipped with a sole pressure regulating valve. The diagrammatic representation of linear-type oil damper is shown in Figure 4.18.

In bilinear-type oil damper, there is an extra relief valve provided along with the pressure regulating valve as shown in Figure 4.19. In this type, damping increases proportionately, until certain speed is reached and thereafter it increases at a lower rate. The damping-velocity relationship of bilinear-type oil damper is as shown in Figure 4.20.

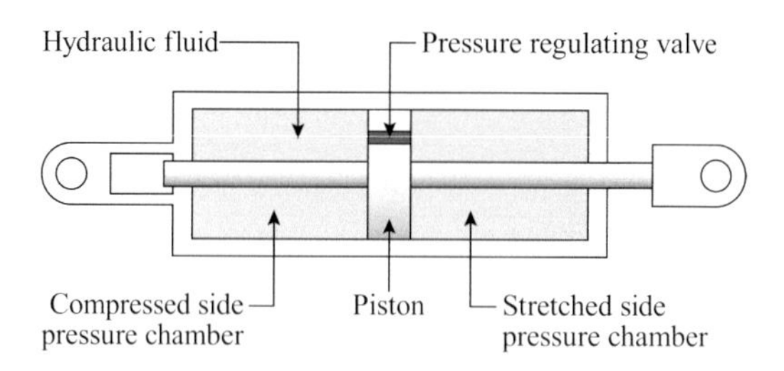

FIGURE 4.18 Structure of linear-type oil damper.

4.5 ORIFICE FLUID DAMPER

Orifice fluid damper consists of a tight cylinder with a piston rod, which is allowed to move backward and forward through the cylinder. Generally, the stainless-steel piston rod is provided with a bronze head. The cylinder has two chambers for storage of highly viscous fluid (silicone) and a self-contained piston displacement accumulator. The fluid is placed in a closed container, where it is allowed to deform and is forced to pass through small orifices, as shown in Figure 4.21. With this arrangement, it is possible to achieve a higher level of energy dissipation. High-strength resin seal is placed at one end of the cylinder so as to prevent any leakage of viscous fluid through cylinder and thus the liquid remains within the closed container for the design period of the damper.

Operation of Orifice Fluid Damper

The volume of fluid inside the cylinder is reduced, as the damper is subjected to compressive force. The amount of fluid volume displaced by the forward movement of piston rod is collected by the accumulator and is deposited in the makeup area. The backward movement of the piston rod then creates a vacuum, drawing the fluid out.

4.6 LEAD EXTRUSION DAMPER

Extrusion is a process in which metal is pushed through a die opening placed in a closed container to mold it in a desired shape. When a polycrystalline metal deforms, its grains are dislocated. Elongation of such grains with large gaps is seen in metal deformation. Further, such metals try to regain their original configurations by the three inter-linked processes, i.e., recovery, recrystallization, and grain growth. Recrystallization is the process of replacing the displaced grains with a fresh set of defect-free grains, until growth of new grains takes place. The temperature at which half of the total recrystallization of grains happens for a period of one hour, is known as recrystallization temperature. Recrystallization temperature differs for various metals; for example, lead has recrystallization temperature well below 20°C. As the recrystallization temperature of lead is very less, compared to other metals, any deformation of lead is simultaneous with the processes of recovery, recrystallization, and grain growth. This property of lead makes it suitable than all other metals as an energy absorber.

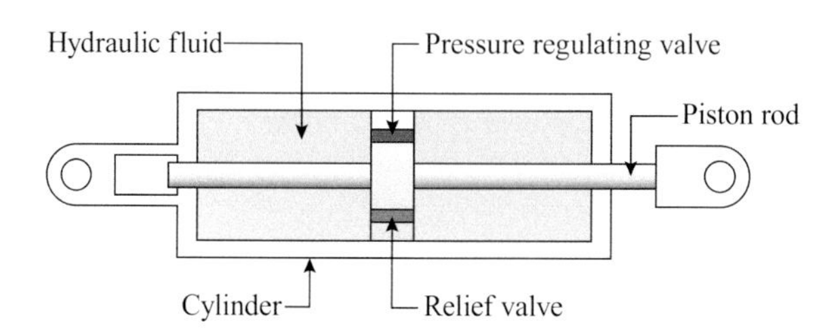

FIGURE 4.19 Structure of bilinear-type oil damper.

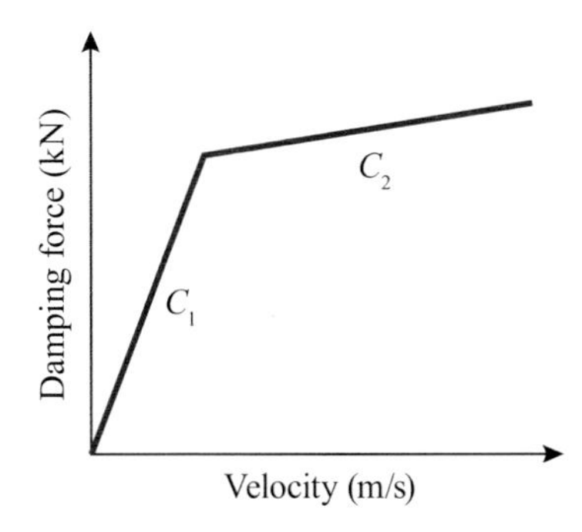

FIGURE 4.20 Damping force-velocity relationship.

Description and Working of Damper

Lead extrusion damper consists of a thick-walled tube with two pistons on either side, which are connected by means of tie-rod, and the space between them is filled with lead. A thin layer of lubricant is kept in position by hydraulic seals that are used to decouple lead from the tube walls. The hydraulic seals are placed near the piston. Both ends of the damper are connected to two different ends of the structure. As the structure vibrates, the two ends of damper cause the piston to move forward and backward along the tube portion. This movement of piston forces the lead to be extruded back and forth through the orifice provided in the damper. Furthermore, the extrusion of lead mitigates the buildup of destructive oscillations in the structure (Robinson and Greenbank 1976). After extrusion of lead, its recrystallization starts immediately and its mechanical properties are recovered before the next lead extrusion.

Robinson was the first to introduce lead extrusion dampers for mitigating the earthquake-induced vibrations of base-isolated structures in New Zealand (Skinner et al. 1991). Robinson and Greenbank (1976) introduced two types of devices as shown in Figure 4.22. The first device shown in Figure 4.22(a) consists of two pistons embedded in a thin-walled tube coaxially, and the second device shown in Figure 4.22(b) has a bulge at the central portion of the shaft and has bearings instead of hydraulic seals, to hold the thick layer of lead in position.

Mathematical Model of Lead Extrusion Damper

Lead extrusion damper, attached to the two different ends of the structure, has resulting motion defined by:

$$u(t) = u_0 \, sin(\varpi t) \tag{4.52}$$

$$\dot{u}(t) = \varpi u_0 \, cos(\varpi t) \tag{4.53}$$

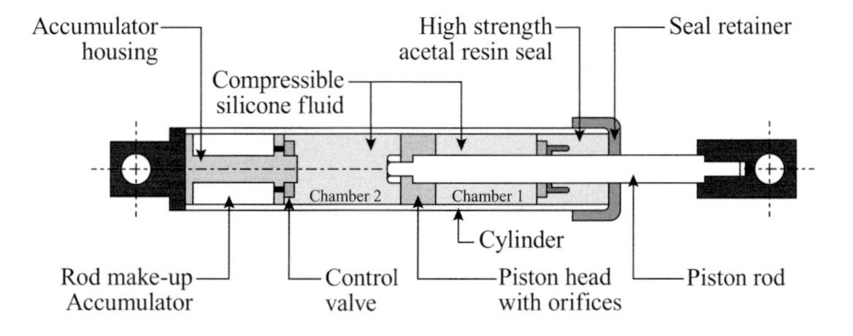

FIGURE 4.21 Orifice fluid damper.

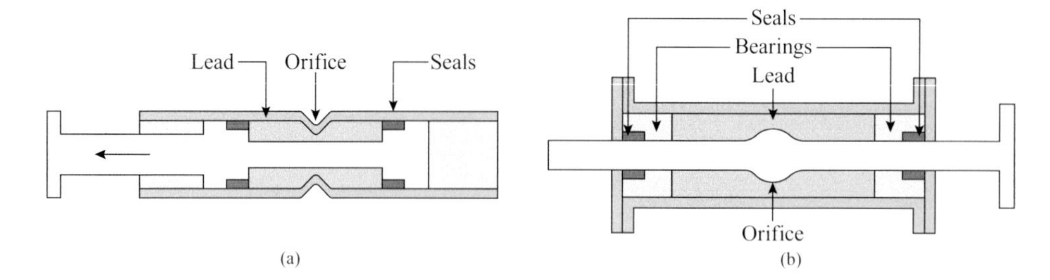

FIGURE 4.22 Longitudinal section of lead extrusion damper: (a) tube-type lead extrusion damper and (b) bulged shaft-type lead extrusion damper.

where u_0 and ϖ are the amplitude of displacement and frequency of motion, respectively. For a steady-state function, the force required to continue the motion is given by:

$$\begin{aligned} F(t) &= F_0 \ sin(\varpi t + \phi) \\ &= F_0 \left(sin\varpi t \ cos\phi + cos\varpi t \ sin\phi \right) \end{aligned} \tag{4.54}$$

where F_0 is the amplitude of force and ϕ is the phase angle.

The work done (energy dissipated) by the damper in one cycle of motion is given as:

$$W_d = \int_0^{\frac{2\pi}{\varpi}} F(t)dt \tag{4.55}$$

Denoting K_1 and K_2 as the storage stiffness and loss stiffness:

$$K_1 = \left(\frac{F_0}{u_0} \right) cos\phi \text{ and } K_2 = \left(\frac{F_0}{u_0} \right) sin\phi \tag{4.56}$$

Introducing K_1 and K_2, Equation (4.54) can be written as:

$$F(t) = K_1 u_0 sin\varpi t + K_2 u_0 cos\varpi t \tag{4.57}$$

Equation (4.57) can also be written in the form:

$$F(t) = K_1 u + \left(\frac{K_2}{\varpi} \right) \dot{u} = K_1 u + C\dot{u} \tag{4.58}$$

where the first term and second term indicate the restoring force due to stiffness and damping of the damper, respectively. From experimentally measured values of W_d, F_0, and u_0, the mechanical properties of LED can be determined using Equations (4.56) and (4.58).

EXAMPLES

Example 4.1 Application of viscous damper to adjacent buildings

Two fixed-base building frames, shown in Figure 4.23, have masses, m_1 = 10,000 kg, m_2 = 15,000 kg, and the total lateral stiffnesses of the columns, k_1 = 2,470 kN/m, k_2 = 1,470 kN/m. Furthermore, for both the buildings, the damping ratio of the superstructure (ξ_s) is 0.02, and the base excitation is given as $\ddot{u}_g = 3sin(5\pi t)$. Answer the following questions by performing time

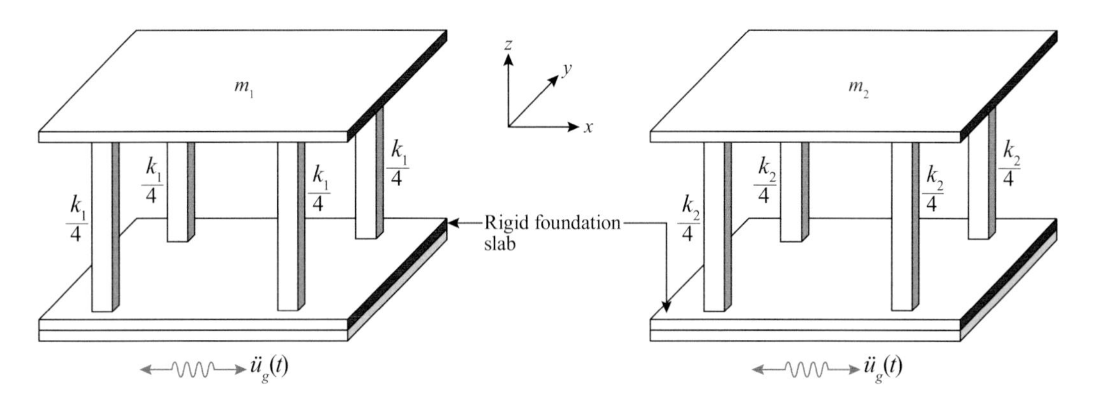

FIGURE 4.23 Fixed-base frames, Building 1 and Building 2, placed adjacent to each other.

history analysis under the base excitation $(\ddot{u}_g)$ acting for a total duration such that at least two full cycles of damper hysteresis loops would be completed [use a suitable time step (Δt) for the time history analysis].

a. Perform a time history analysis and determine the peak values of the top floor displacement and absolute acceleration response quantities of the uncontrolled building frames modeling them as single degree of freedom (SDOF) systems.
b. Considering that the adjacent buildings are connected (coupled) by a viscous damper as shown in Figure 4.24, construct the mass, the stiffness, and the damping matrices of the multi-degree of freedom (MDOF) system. Subsequently, evaluate the peak values of the floor displacement and absolute acceleration of the coupled buildings. Select the properties of the passive control device appropriately; e.g., a damping constant (c_d) of 5 kNs/m for viscous damper. Compare the peak values of the two response quantities of the coupled building frames with that of the individual uncontrolled building frames.

SOLUTION

First, the data given regarding the two structures is provided.

- **Given data**

Structure 1	**Structure 2**	
Mass	$m_1 = 10{,}000$ kg	$m_2 = 15{,}000$ kg
Stiffness	$k_1 = 2{,}470$ kN/m	$k_2 = 1{,}470$ kN/m
Damping ratio	$\xi_s = 0.02$	$\xi_s = 0.02$

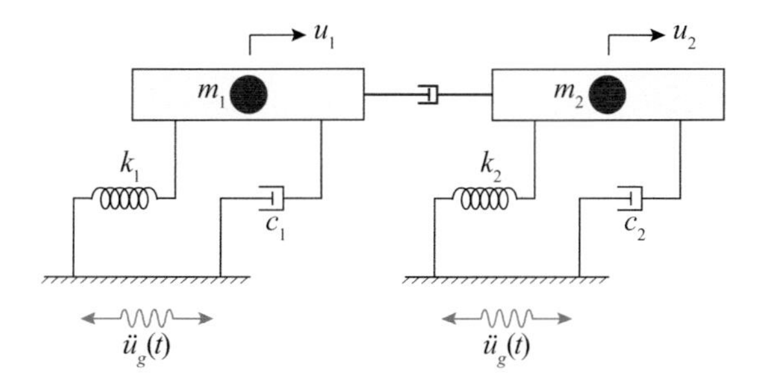

FIGURE 4.24 Adjacent building frames coupled by viscous damper.

- **Time period of the structures**
  The time period of an SDOF structure can be calculated as:

$$T = 2\pi\sqrt{\frac{m}{k}} \tag{4.1}$$

  where m is the mass of the structure and k is the stiffness of the structure. Therefore, the time periods of the two structures are evaluated using Equation (4.1) and presented subsequently.
  Time period $T_1 = 0.4$ s $T_2 = 0.635$ s
- **Angular frequency of the structures**
  The angular frequency of an SDOF structure can be calculated as:

$$\omega = \frac{2\pi}{T} \tag{4.2}$$

  Therefore, the angular frequencies of the two structures are evaluated using Equation (4.2) and presented subsequently.
  Angular frequency $\omega_1 = 15.72$ rad/s $\omega_2 = 9.899$ rad/s
- **Damped angular frequency**
  The damped frequency of an SDOF structure can be calculated as:

$$\omega_D = \omega\sqrt{1-\xi_s^2} \tag{4.3}$$

  Therefore, the damped frequencies of the two structures are evaluated using Equation (4.3) and presented subsequently.
  Damped frequency $\omega_{D1} = 15.71$ rad/s $\omega_{D2} = 9.897$ rad/s
- **Damping constant**
  The damping constant of an SDOF structure can be calculated as:

$$c = 2\xi_s m\omega \tag{4.4}$$

  Therefore, the damping constants of the two structures are evaluated using Equation (4.4) and presented subsequently.
  Damping constant $c_1 = 6{,}286.5$ N s/m, $c_2 = 5{,}939.7$ N s/m

a. **Time history analysis of the uncontrolled structures**
   The time history response quantities of the two uncontrolled SDOF structures under sinusoidal/harmonic base excitation are evaluated analytically.

- **Base excitation**
  The sinusoidal base excitation is given as:

$$\ddot{u}_g = 3sin(5\pi t) \tag{4.5}$$

  The amplitude of the sinusoidal base excitation is 3 m/s^2, and the excitation frequency is given as:

$$\varpi = 5\pi \text{ rad/s.}$$

- **Equation of motion**
  The equation of motion for an SDOF structure can be written as:

$$m\ddot{u} + c\dot{u} + ku = -m\ddot{u}_g \tag{4.6}$$

$$m\ddot{u} + c\dot{u} + ku = -3m\ sin(5\pi t) \tag{4.7}$$

where $u, \dot{u}$ and, $\ddot{u}$ are the displacement, velocity, and acceleration of the structure relative to the ground, respectively.

The analytical solution of the equation of motion under the sinusoidal/harmonic excitation can be obtained as:

$$u = e^{-\xi_s \omega t}\left[A\cos(\omega_D t) + B\sin(\omega_D t)\right] + \left[C\sin(\varpi t) + D\cos(\varpi t)\right] \tag{4.8}$$

where

$$C = \frac{P_0}{k}\left\{\frac{1-\left(\frac{\varpi}{\omega}\right)^2}{\left[1-\left(\frac{\varpi}{\omega}\right)^2\right]^2 + \left[2\xi_s\left(\frac{\varpi}{\omega}\right)\right]^2}\right\} \tag{4.9}$$

$$D = \frac{P_0}{k}\left\{\frac{-2\xi\left(\frac{\varpi}{\omega}\right)}{\left[1-\left(\frac{\varpi}{\omega}\right)^2\right]^2 + \left[2\xi_s\left(\frac{\varpi}{\omega}\right)\right]^2}\right\} \tag{4.10}$$

$$P_0 = -3m \tag{4.11}$$

The calculated values of C and D for the two structures are:

Structure 1	**Structure 2**
$C_1 = 0.008$	$C_2 = 0.0201$
$D_1 = 0.3036$	$D_2 = 0.00084197$

Additionally, A and B can be obtained by applying the initial conditions to the analytical equation given as a solution for the equation of motion.

Both structures are assumed to be at rest at the beginning of the excitation, and the initial conditions of both structures at $t = 0$ s are considered $u_0 = 0$ and $\dot{u}_0 = 0$. Equation (4.8) can be differentiated with respect to time to obtain the velocity, which is given as:

$$\begin{aligned}\dot{u} = {} & e^{-\xi_s \omega t}\left[B\omega_D\cos(\omega_D t) - A\omega_D\sin(\omega_D t)\right] + C\varpi\cos(\varpi t) - D\varpi\sin(\varpi t) \\ & -\xi_s \omega e^{-\xi_s \omega t}\left[A\cos(\omega_D t) + B\sin(\omega_D t)\right]\end{aligned} \tag{4.12}$$

By applying the initial conditions in Equations (4.8) and (4.9), values of A and B for the two structures are obtained as:

Structure 1	**Structure 2**
$A = 0.3036$	$A = -0.000841975$
$B = 0.0019$	$B = -0.0320$

Now, the time histories of the displacement response for both structures can be calculated using Equation (8) by taking appropriate time step, i.e., $\Delta t = 0.01$ s.

Furthermore, the acceleration response can be obtained by differentiating Equation (12) with respect to time, which gives the following equation.

$$\begin{aligned}\ddot{u} = {} & (\xi_s \omega)^2 e^{-\xi_s \omega t}\left[A\cos(\omega_D t) + B\sin(\omega_D t)\right] - D\varpi^2\cos(\varpi t) - C\varpi^2\sin(\varpi t) \\ & - e^{-\xi_s \omega t}\left[A\omega_D^2\cos(\omega_D t) + B\omega_D^2\sin(\omega_D t)\right] - 2\xi_s \omega e^{-\xi_s \omega t}\left[B\omega_D\cos(\omega_D t) + A\omega_D\sin(\omega_D t)\right]\end{aligned} \tag{4.13}$$

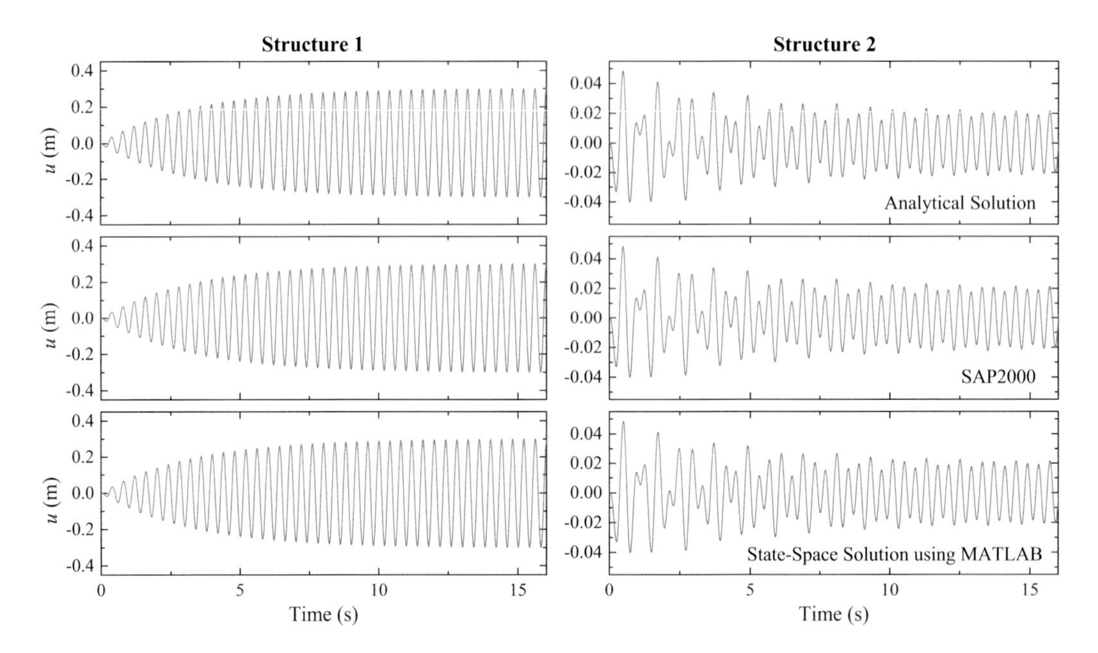

FIGURE 4.25 Displacement time history plots of the uncontrolled SDOF structures.

The absolute acceleration of the structures can be evaluated by adding the ground acceleration and the relative acceleration at each time instant. Accordingly, the calculated response quantities of Structure 1 and Structure 2 (uncontrolled) for some of the time steps are given in Table Ex.4.1.1. The response of the structures for the required duration can be evaluated by implementing the same procedure iteratively.

The plot of the time histories of the response quantities for both structures is given in Figures 4.25 and 4.26. In addition, the response quantities are evaluated using two additional approached: (i) SAP2000 commercial software and (ii) State-Space method

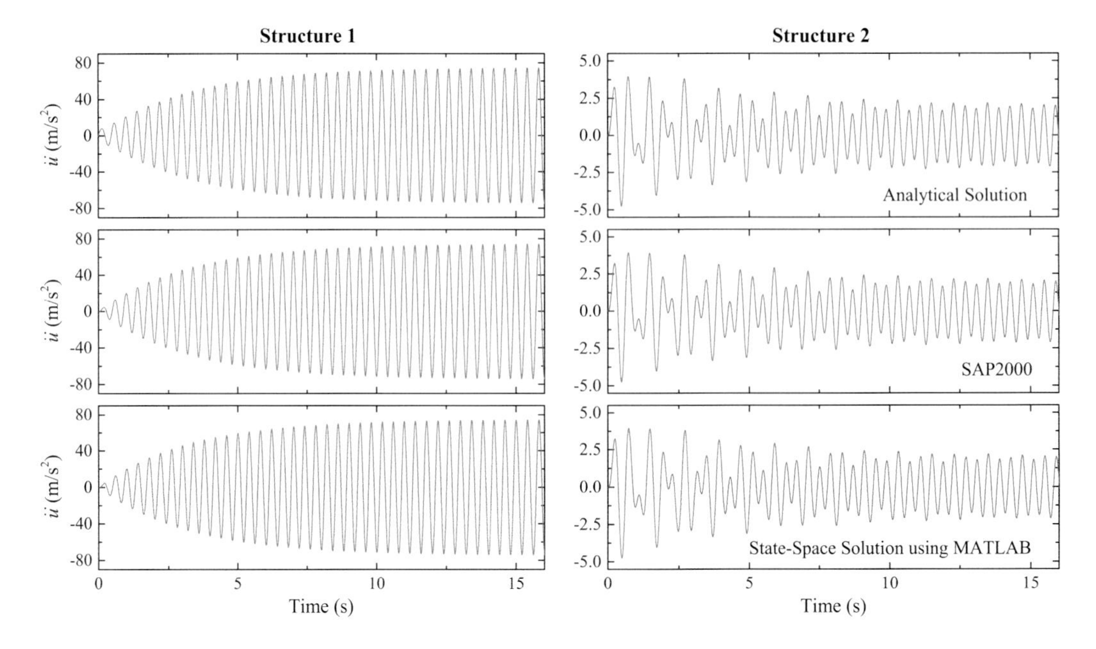

FIGURE 4.26 Acceleration time history plots of the uncontrolled SDOF structures.

TABLE EX.4.1.1
The Displacement and Absolute Acceleration Response History Data for the Uncontrolled SDOF Structures

		Analytical Solution				SAP2000 Solution				MATLAB Solution – State-Space Method			
		Structure 1		Structure 2		Structure 1		Structure 2		Structure 1		Structure 2	
Time (s)	$\ddot{u}_g$ (m/s²)	u (m)	$\ddot{u}_{Abs}$(m/s²)	u (m)	$\ddot{u}_{Abs}$(m/s²)	u (m)	$\ddot{u}_{Abs}$(m/s²)	u (m)	$\ddot{u}_{Abs}$(m/s²)	u (m)	$\ddot{u}_{Abs}$(m/s²)	u (m)	$\ddot{u}_{Abs}$(m/s²)
0	0.0000	0.00000	0.00000	0.00000	0.00000	0.00000	0.00000	0.00000	0.00000	0.00000	0.00000	0.00000	0.00000
0.01	0.4693	–0.00001	0.93927	–0.00001	0.00232	–0.00001	0.00339	–0.00001	0.00169	0.00000	0.00000	0.00000	0.00000
0.02	0.9271	–0.00006	1.86375	–0.00006	0.01105	–0.00006	0.02107	–0.00006	0.00976	–0.00001	0.00356	–0.00001	0.00177
0.03	1.3620	–0.00021	2.76218	–0.00021	0.03044	–0.00021	0.06359	–0.00021	0.02848	–0.00006	0.02162	–0.00006	0.01000
0.04	1.7634	–0.00048	3.62312	–0.00049	0.06440	–0.00048	0.14022	–0.00049	0.06175	–0.00021	0.06474	–0.00021	0.02897
0.05	2.1213	–0.00092	4.43527	–0.00094	0.11630	–0.00091	0.25844	–0.00093	0.11295	–0.00049	0.14214	–0.00049	0.06258
0.06	2.4271	–0.00154	5.18745	–0.00158	0.18885	–0.00153	0.42355	–0.00158	0.18479	–0.00092	0.26128	–0.00094	0.11417
0.07	2.6730	–0.00236	5.86869	–0.00246	0.28404	–0.00235	0.63831	–0.00245	0.27924	–0.00155	0.42740	–0.00160	0.18646
0.08	2.8532	–0.00338	6.46834	–0.00357	0.40299	–0.00337	0.90278	–0.00356	0.39744	–0.00237	0.64323	–0.00247	0.28140
0.09	2.9631	–0.00459	6.97617	–0.00492	0.54592	–0.00459	1.21412	–0.00491	0.53962	–0.00339	0.90874	–0.00358	0.40012
0.1	3.0000	–0.00598	7.38252	–0.00652	0.71215	–0.00597	1.56663	–0.00651	0.70509	–0.00461	1.22104	–0.00494	0.54281
0.11	2.9631	–0.00752	7.67849	–0.00834	0.90002	–0.00750	1.95184	–0.00833	0.89220	–0.00600	1.57438	–0.00654	0.70878
0.12	2.8532	–0.00915	7.85612	–0.01038	1.10696	–0.00913	2.35871	–0.01036	1.09840	–0.00753	1.96023	–0.00837	0.89636
0.13	2.6730	–0.01083	7.90865	–0.01258	1.32951	–0.01081	2.77399	–0.01256	1.32023	–0.00916	2.36749	–0.01040	1.10296
0.14	2.4271	–0.01250	7.83073	–0.01492	1.56341	–0.01247	3.18258	–0.01489	1.55346	–0.01084	2.78286	–0.01260	1.32514
0.15	2.1213	–0.01409	7.61866	–0.01735	1.80369	–0.01406	3.56808	–0.01731	1.79311	–0.01250	3.19121	–0.01494	1.55861
0.16	1.7634	–0.01553	7.27066	–0.01981	2.04477	–0.01551	3.91333	–0.01977	2.03363	–0.01409	3.57610	–0.01736	1.79839
0.17	1.3620	–0.01677	6.78703	–0.02224	2.28060	–0.01673	4.20106	–0.02219	2.26899	–0.01553	3.92037	–0.01982	2.03893
0.18	0.9271	–0.01771	6.17044	–0.02458	2.50484	–0.01768	4.41448	–0.02453	2.49285	–0.01675	4.20674	–0.02224	2.27417
0.19	0.4693	–0.01831	5.42602	–0.02676	2.71099	–0.01828	4.53802	–0.02671	2.69872	–0.01769	4.41844	–0.02457	2.49777
0.2	0.0000	–0.01850	4.56153	–0.02872	2.89258	–0.01847	4.55789	–0.02866	2.88016	–0.01828	4.53991	–0.02675	2.70325
0.21	–0.4693	–0.01823	3.58743	–0.03039	3.04332	–0.01820	4.46274	–0.03033	3.03088	–0.01846	4.55741	–0.02870	2.88415
0.22	–0.9271	–0.01747	2.51687	–0.03171	3.15731	–0.01745	4.24418	–0.03165	3.14501	–0.01819	4.45965	–0.03036	3.03420
0.23	–1.3620	–0.01620	1.36564	–0.03262	3.22918	–0.01617	3.89726	–0.03256	3.21717	–0.01742	4.23831	–0.03167	3.14751
0.24	–1.7634	–0.01440	0.15204	–0.03308	3.25428	–0.01438	3.42085	–0.03301	3.24272	–0.01614	3.88851	–0.03257	3.21875

(Continued)

TABLE EX.4.1.1 (*Continued*)

The Displacement and Absolute Acceleration Response History Data for the Uncontrolled SDOF Structures

		Analytical Solution				SAP2000 Solution				MATLAB Solution – State-Space Method			
		Structure 1		Structure 2		Structure 1		Structure 2		Structure 1		Structure 2	
Time (s)	$\ddot{u}_g$ (m/s^2)	u (m)	$\ddot{u}_{Abs}$(m/s^2)	u (m)	$\ddot{u}_{Abs}$(m/s^2)	u (m)	$\ddot{u}_{Abs}$(m/s^2)	u (m)	$\ddot{u}_{Abs}$(m/s^2)	u (m)	$\ddot{u}_{Abs}$(m/s^2)	u (m)	$\ddot{u}_{Abs}$(m/s^2)
0.25	−2.1213	−0.01208	−1.10332	−0.03303	3.22877	−0.01206	2.81787	−0.03297	3.21783	−0.01433	3.40920	−0.03301	3.24326
–	–	–	–	–	–	–	–	–	–	–	–	–	–
0.3	−3.0000	0.00582	−7.16572	−0.02452	2.28151	0.00580	−1.70130	−0.02448	2.27609	0.00170	−0.67732	−0.02723	2.56788
0.4	0.0000	0.03587	−8.84558	0.02292	−2.45083	0.03581	−8.83810	0.02286	−2.43912	0.03544	−8.80337	0.01782	−1.95376
0.5	3.0000	−0.00569	6.96638	0.04811	−4.68350	−0.00565	1.82980	0.04801	−4.66676	0.00102	0.17096	0.04855	−4.75116
0.6	0.0000	−0.05219	12.86880	0.00805	−0.55897	−0.05209	12.85754	0.00806	−0.55850	−0.05156	12.80757	0.01371	−1.11642
0.7	−3.0000	0.00558	−6.78319	−0.03745	3.75287	0.00552	−1.95238	−0.03735	3.73754	−0.00355	0.30057	−0.03511	3.54530
0.8	0.0000	0.06751	−16.64707	−0.02634	2.45030	0.06739	−16.63210	−0.02629	2.43942	0.06670	−16.56827	−0.02954	2.76959
0.9	3.0000	−0.00549	6.61491	0.00752	−0.82277	−0.00540	2.06928	0.00747	−0.81947	0.00592	−0.73963	0.00515	−0.60701
1	0.0000	−0.08189	20.19530	0.01172	−1.11462	−0.08175	20.17670	0.01165	−1.10586	−0.08092	20.10026	0.01252	−1.19616
1.1	−3.0000	0.00542	−6.46041	0.00550	−0.55600	0.00529	−2.18074	0.00549	−0.54969	−0.00813	1.14843	0.00533	−0.52377
1.2	0.0000	0.09540	−23.52749	0.01636	−1.65314	0.09524	−23.50534	0.01637	−1.64993	0.09428	−23.41745	0.01512	−1.53321
1.3	3.0000	−0.00537	6.31864	0.01220	−1.09763	−0.00520	2.28699	0.01222	−1.10104	0.01019	−1.52902	0.01435	−1.32690
1.4	0.0000	−0.10809	26.65679	−0.02430	2.53369	−0.10791	26.63119	−0.02424	2.52165	−0.10682	26.53291	−0.02051	2.16478
1.5	−3.0000	0.00534	−6.18861	−0.03776	3.63477	0.00512	−2.38825	−0.03772	3.62617	−0.01211	1.88333	−0.03907	3.78547
1.6	0.0000	0.12001	−29.59555	0.00442	−0.64940	0.11980	−29.56660	0.00435	−0.64109	0.11860	−29.45889	−0.00091	−0.12574
1.7	3.0000	−0.00532	6.06943	0.04093	−4.04539	−0.00505	2.48473	0.04081	−4.02876	0.01390	−2.21315	0.03976	−3.95402
1.8	0.0000	−0.13120	32.35537	0.01809	−1.60368	−0.13097	32.32317	0.01810	−1.59990	−0.12967	32.20693	0.02225	−2.01554
1.9	−3.0000	0.00531	−5.96025	−0.01974	2.01514	0.00499	−2.57664	−0.01962	2.00318	−0.01558	2.52014	−0.01747	1.81142
2	0.0000	0.14170	−34.94715	−0.01694	1.58418	0.14146	−34.91179	−0.01686	1.57325	0.14006	−34.78783	−0.01877	1.76766

using MATLAB®. The comparison of the results obtained using different methods is also presented in Table Ex.4.1.1 and Figures 4.25 and 4.26.

b. **Time history analysis of the structures connected with viscous damper**
The viscous damper is considered to connect the two adjacent SDOF structures, Structure 1 and Structure 2. The damper is placed between the floors of the structures. The vibration response of the structures connected using the damper is evaluated subsequently.

When the two structures are connected using viscous damper, the mass and the stiffness matrices of the connected system become diagonal matrices because there is no mass and stiffness coupling in the system. The mass and the stiffness matrices of the connected system, respectively, are given as:

$$\begin{bmatrix} m_1 & 0 \\ 0 & m_2 \end{bmatrix} = \begin{bmatrix} 10,000 & 0 \\ 0 & 15,000 \end{bmatrix} \text{kg} \tag{4.14}$$

and

$$\begin{bmatrix} k_1 & 0 \\ 0 & k_2 \end{bmatrix} = \begin{bmatrix} 2,470 & 0 \\ 0 & 1,470 \end{bmatrix} \text{kN/m} \tag{4.15}$$

The damping force in the viscous damper is obtained as a function of the relative velocity of the two structures, and therefore, there is a coupling of damping in the system of structures connected using the viscous damper.

The damping matrix of the combined system, which incorporates the damping constant of the viscous damper, can be given as:

$$\begin{bmatrix} c_1 + c_d & -c_d \\ -c_d & c_2 + c_d \end{bmatrix} = \begin{bmatrix} 6.2865+5 & -5 \\ -5 & 5.9397+5 \end{bmatrix} \text{kNs}^2/\text{m} = \begin{bmatrix} 11.2865 & -5 \\ -5 & 10.9397 \end{bmatrix} \text{kNs/m} \tag{4.16}$$

where c_d is the damping constant of the viscous damper.

Considering the damping matrix given in Equation (4.16), the matrix form of the equations of motion for the system of structures connected by the viscous damper can be expressed as:

$$\begin{bmatrix} m_1 & 0 \\ 0 & m_2 \end{bmatrix} \begin{Bmatrix} \ddot{u}_1 \\ \ddot{u}_2 \end{Bmatrix} + \begin{bmatrix} c_1 + c_d & -c_d \\ -c_d & c_2 + c_d \end{bmatrix} \begin{Bmatrix} \dot{u}_1 \\ \dot{u}_2 \end{Bmatrix} + \begin{bmatrix} k_1 & 0 \\ 0 & k_2 \end{bmatrix} \begin{Bmatrix} u_1 \\ u_2 \end{Bmatrix} = -\begin{bmatrix} m_1 & 0 \\ 0 & m_2 \end{bmatrix} \begin{Bmatrix} 1 \\ 1 \end{Bmatrix} \ddot{u}_g \tag{4.17}$$

$$\begin{bmatrix} m_1 & 0 \\ 0 & m_2 \end{bmatrix} \begin{Bmatrix} \ddot{u}_1 \\ \ddot{u}_2 \end{Bmatrix} + \begin{bmatrix} c_1 + c_d & -c_d \\ -c_d & c_2 + c_d \end{bmatrix} \begin{Bmatrix} \dot{u}_1 \\ \dot{u}_2 \end{Bmatrix} + \begin{bmatrix} k_1 & 0 \\ 0 & k_2 \end{bmatrix} \begin{Bmatrix} u_1 \\ u_2 \end{Bmatrix} = -\begin{bmatrix} m_1 & 0 \\ 0 & m_2 \end{bmatrix} \begin{Bmatrix} 1 \\ 1 \end{Bmatrix} 3sin(5\pi t) \tag{4.18}$$

Alternatively, the damping matrix can be constructed without incorporating the damping constant of the viscous damper as:

$$\begin{bmatrix} c_1 & 0 \\ 0 & c_2 \end{bmatrix} = \begin{bmatrix} 6.2865 & 0 \\ 0 & 5.9397 \end{bmatrix} \text{kNs/m} \tag{4.19}$$

When the damping matrix given in Equation (4.19) is used, the equations of motion for the system of structures connected by the viscous damper can be expressed in matrix form as:

$$\begin{bmatrix} m_1 & 0 \\ 0 & m_2 \end{bmatrix} \begin{Bmatrix} \ddot{u}_1 \\ \ddot{u}_2 \end{Bmatrix} + \begin{bmatrix} c_1 & 0 \\ 0 & c_2 \end{bmatrix} \begin{Bmatrix} \dot{u}_1 \\ \dot{u}_2 \end{Bmatrix} + \begin{bmatrix} k_1 & 0 \\ 0 & k_2 \end{bmatrix} \begin{Bmatrix} u_1 \\ u_2 \end{Bmatrix} + \begin{Bmatrix} -1 \\ 1 \end{Bmatrix} F_d = -\begin{bmatrix} m_1 & 0 \\ 0 & m_2 \end{bmatrix} \begin{Bmatrix} 1 \\ 1 \end{Bmatrix} 3sin(5\pi t) \tag{4.20}$$

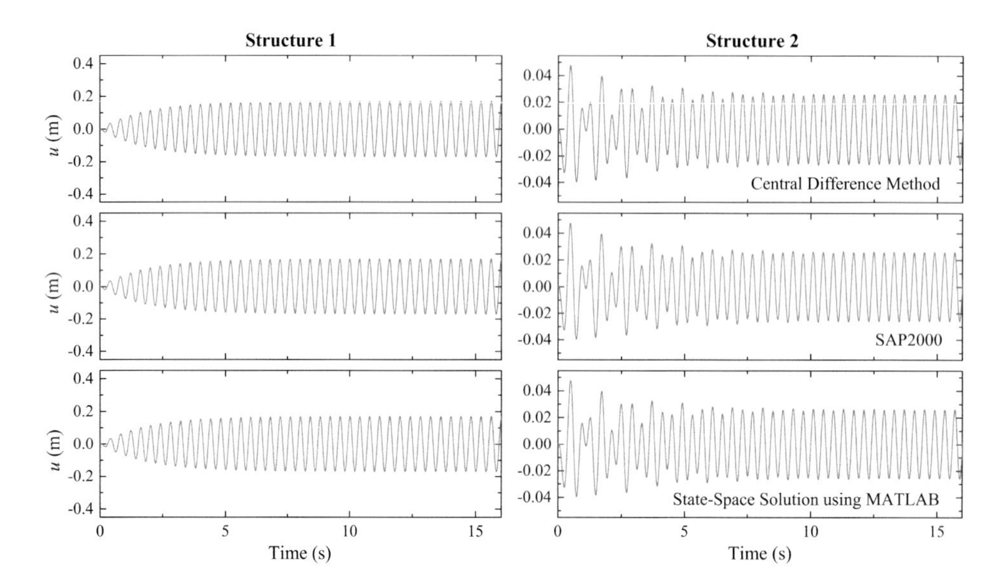

FIGURE 4.27 Displacement time history plots of the structures connected with viscous damper.

where $F_d = c_d(\dot{u}_2 - \dot{u}_1)$ is the force in the viscous damper; the damping constant of the viscous damper is taken as $c_d = 5$ kNs/m in this problem.

The equations of motion can be solved numerically to obtain the time history response quantities of the connected structures. There are various methods that can be implemented to calculate the response of the structures numerically. Some of the common methods include the Central Difference method, Newmark's β method, Wilson-θ method, Hilber-Hughes-Taylor (HHT) method, state-space method, and modal time history analysis. For this problem, the Central Difference method is used.

The plot of the time histories of the response quantities for the structures connected with viscous damper is given in Figures 4.27 and 4.28. Furthermore, the comparison of the peak values of the response quantities obtained using different methods is also presented in Table Ex.4.1.2.

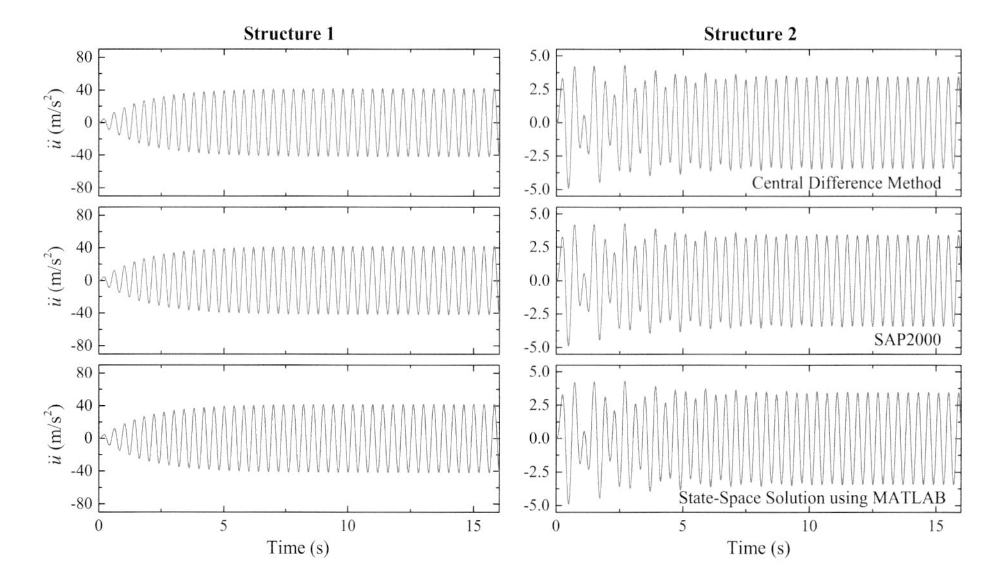

FIGURE 4.28 Acceleration time history plots of the structures connected with viscous damper.

TABLE EX.4.1.2
The Displacement and Absolute Acceleration Response History Data for the Structures Connected Using Viscous Damper

		Solution Using Central Difference Method				SAP2000 Solution				MATLAB Solution – State-Space Method			
		Structure 1		Structure 2		Structure 1		Structure 2		Structure 1		Structure 2	
Time (s)	$\ddot{u}_g$ (m/s²)	u (m)	$\ddot{u}_{Abs}$ (m/s²)	u (m)	$\ddot{u}_{Abs}$ (m/s²)	u (m)	$\ddot{u}_{Abs}$ (m/s²)	u (m)	$\ddot{u}_{Abs}$ (m/s²)	u (m)	$\ddot{u}_{Abs}$ (m/s²)	u (m)	$\ddot{u}_{Abs}$ (m/s²)
0	0.0000	0.00000	0.00000	0.00000	0.00000	0.00000	0.00000	0.00000	0.00000	0.00000	0.00000	0.00000	0.00000
0.01	0.4693	–0.00001	–0.04216	–0.00001	–0.04435	–0.00001	0.00339	–0.00001	0.00169	–0.00001	0.00355	–0.00001	0.00177
0.02	0.9271	–0.00007	–0.02068	–0.00007	–0.03358	–0.00006	0.02104	–0.00006	0.00978	–0.00006	0.02159	–0.00006	0.01002
0.03	1.3620	–0.00023	0.02772	–0.00023	–0.01057	–0.00021	0.06345	–0.00021	0.02857	–0.00021	0.06460	–0.00021	0.02907
0.04	1.7634	–0.00052	0.11201	–0.00052	0.02854	–0.00048	0.13982	–0.00049	0.06202	–0.00049	0.14173	–0.00049	0.06285
0.05	2.1213	–0.00097	0.23929	–0.00099	0.08703	–0.00091	0.25751	–0.00093	0.11358	–0.00092	0.26034	–0.00094	0.11481
0.06	2.4271	–0.00161	0.41440	–0.00166	0.16753	–0.00153	0.42173	–0.00158	0.18602	–0.00155	0.42556	–0.00160	0.18771
0.07	2.6730	–0.00245	0.63961	–0.00256	0.27187	–0.00235	0.63513	–0.00245	0.28142	–0.00237	0.64001	–0.00247	0.28361
0.08	2.8532	–0.00349	0.91444	–0.00369	0.40101	–0.00337	0.89767	–0.00356	0.40098	–0.00339	0.90358	–0.00358	0.40369
0.09	2.9631	–0.00473	1.23551	–0.00507	0.55500	–0.00459	1.20645	–0.00491	0.54498	–0.00461	1.21332	–0.00494	0.54823
0.1	3.0000	–0.00614	1.59664	–0.00669	0.73289	–0.00598	1.55575	–0.00650	0.71280	–0.00601	1.56343	–0.00654	0.71655
0.11	2.9631	–0.00768	1.98889	–0.00853	0.93280	–0.00751	1.93711	–0.00832	0.90279	–0.00754	1.94541	–0.00836	0.90702
0.12	2.8532	–0.00933	2.40086	–0.01058	1.15186	–0.00914	2.33956	–0.01035	1.11239	–0.00917	2.34823	–0.01039	1.11704
0.13	2.6730	–0.01102	2.81900	–0.01280	1.38631	–0.01083	2.74997	–0.01254	1.33812	–0.01086	2.75871	–0.01259	1.34311
0.14	2.4271	–0.01269	3.22806	–0.01514	1.63156	–0.01250	3.15345	–0.01487	1.57563	–0.01253	3.16193	–0.01492	1.58088
0.15	2.1213	–0.01428	3.61159	–0.01757	1.88229	–0.01410	3.53384	–0.01729	1.81984	–0.01413	3.54168	–0.01733	1.82522
0.16	1.7634	–0.01573	3.95252	–0.02001	2.13255	–0.01556	3.87430	–0.01973	2.06502	–0.01558	3.88112	–0.01978	2.07042
0.17	1.3620	–0.01695	4.23380	–0.02242	2.37595	–0.01681	4.15786	–0.02214	2.30495	–0.01683	4.16331	–0.02219	2.31023
0.18	0.9271	–0.01789	4.43897	–0.02473	2.60579	–0.01778	4.36814	–0.02446	2.53306	–0.01779	4.37186	–0.02450	2.53807
0.19	0.4693	–0.01848	4.55291	–0.02687	2.81522	–0.01840	4.48996	–0.02661	2.74261	–0.01841	4.49161	–0.02665	2.74720
0.2	0.0000	–0.01866	4.56239	–0.02878	2.99747	–0.01863	4.50992	–0.02854	2.92685	–0.01862	4.50923	–0.02858	2.93088
0.21	–0.4693	–0.01838	4.45665	–0.03038	3.14598	–0.01840	4.41705	–0.03018	3.07924	–0.01838	4.41379	–0.03021	3.08256
0.22	–0.9271	–0.01761	4.22798	–0.03163	3.25462	–0.01768	4.20330	–0.03146	3.19362	–0.01766	4.19731	–0.03148	3.19606
0.23	–1.3620	–0.01633	3.87210	–0.03246	3.31786	–0.01646	3.86401	–0.03233	3.26430	–0.01642	3.85521	–0.03234	3.26578
0.24	–1.7634	–0.01453	3.38854	–0.03282	3.33094	–0.01471	3.39819	–0.03274	3.28646	–0.01466	3.38661	–0.03274	3.28682

(Continued)

TABLE EX.4.1.2 (*Continued*)

The Displacement and Absolute Acceleration Response History Data for the Structures Connected Using Viscous Damper

		Solution Using Central Difference Method				SAP2000 Solution				MATLAB Solution – State-Space Method			
		Structure 1		Structure 2		Structure 1		Structure 2		Structure 1		Structure 2	
Time (s)	$\ddot{u}_g$ (m/s²)	*u* (m)	$\ddot{u}_{Abs}$(m/s²)	*u* (m)	$\ddot{u}_{Abs}$(m/s²)	*u* (m)	$\ddot{u}_{Abs}$(m/s²)	*u* (m)	$\ddot{u}_{Abs}$(m/s²)	*u* (m)	$\ddot{u}_{Abs}$(m/s²)	*u* (m)	$\ddot{u}_{Abs}$(m/s²)
0.25	−2.1213	−0.01221	2.78084	−0.03267	3.29001	−0.01245	2.80899	−0.03265	3.25595	−0.01239	2.79463	−0.03263	3.25517
–	–	–	–	–	–	–	–	–	–	–	–	–	–
0.3	−3.0000	0.00563	−1.71161	−0.02359	2.22021	0.00518	−1.60095	−0.02387	2.25090	0.00528	−1.62442	−0.02380	2.24363
0.4	0.0000	0.03584	−8.65388	0.02412	−2.77182	0.03577	−8.59555	0.02356	−2.66712	0.03576	−8.59436	0.02367	−2.67944
0.5	3.0000	−0.00406	1.69953	0.04683	−4.72016	−0.00335	1.52471	0.04686	−4.72707	−0.00351	1.56423	0.04683	−4.72474
0.6	0.0000	−0.05052	12.24963	0.00423	0.05657	−0.05040	12.18515	0.00484	−0.05556	−0.05039	12.18376	0.00469	−0.04003
0.7	−3.0000	0.00232	−1.64516	−0.03844	4.16502	0.00138	−1.41225	−0.03819	4.14348	0.00161	−1.46757	−0.03824	4.14928
0.8	0.0000	0.06269	−15.36802	−0.02120	1.75475	0.06254	−15.29850	−0.02156	1.84376	0.06253	−15.29672	−0.02147	1.83391
0.9	3.0000	−0.00146	1.65561	0.01270	−1.72730	−0.00032	1.36977	0.01245	−1.70536	−0.00060	1.44056	0.01253	−1.71263
1	0.0000	−0.07344	18.13573	0.00928	−0.74675	−0.07327	18.06312	0.00941	−0.81337	−0.07326	18.06023	0.00940	−0.81095
1.1	−3.0000	0.08377	−20.67594	0.01417	−1.56381	0.08359	−20.60118	0.01398	−1.49124	0.00046	−1.49018	−0.00142	0.52748
1.2	0.0000	−0.00140	1.81131	0.01572	−1.84731	0.00013	1.42984	0.01594	−1.87074	0.08357	−20.59688	0.01399	−1.49417
1.3	3.0000	−0.09364	23.01085	−0.02116	2.40058	−0.09344	22.93363	−0.02071	2.30108	−0.00028	1.53131	0.01586	−1.86318
1.4	0.0000	0.00063	−1.80413	−0.03646	3.96027	−0.00108	−1.37979	−0.03658	3.97552	−0.09341	22.92811	−0.02080	2.31267
1.5	−3.0000	0.10229	−25.08668	0.00605	−1.02209	0.10207	−25.00626	0.00544	−0.90551	−0.00061	−1.49622	−0.03652	3.96938
1.6	0.0000	0.00057	1.74186	0.03975	−4.46327	0.00244	1.27886	0.03962	−4.45401	0.10204	−25.00005	0.00560	−0.92424
1.7	3.0000	−0.10936	26.87693	0.01120	−0.72040	−0.10911	26.79354	0.01169	−0.82719	0.00191	1.40942	0.03965	−4.45653
1.8	0.0000	−0.00142	−1.70234	−0.02457	3.08479	−0.00342	−1.20459	−0.02433	3.06401	−0.10908	26.78674	0.01155	−0.81083
1.9	−3.0000	0.11531	−28.43740	−0.01091	0.81768	0.11505	−28.35228	−0.01117	0.90288	−0.00283	−1.34838	−0.02442	3.07204
2	0.0000	0.00000	0.00000	0.00000	0.00000	0.00000	0.00000	0.00000	0.00000	0.11502	−28.34445	−0.01113	0.89491

TABLE EX.4.1.3
Comparison of the Peak Response Quantities of the Unconnected and Connected Structures Using Different Response Analysis Methods

		Structure 1		Structure 2	
Structure	Analysis Method	Displacement (m)	Acceleration (m/s²)	Displacement (m)	Acceleration (m/s²)
Uncontrolled/ unconnected	Analytical	0.30359	74.90839	0.04866	4.77627
	SAP2000	0.30319	74.80871	0.04855	4.75863
	State-Space	0.30250	74.78020	0.04860	4.76620
Viscous damper connected	Central Difference	0.17034	42.02523	0.04788	4.93030
	SAP2000	0.16985	41.90785	0.04770	4.90250
	State-Space	0.17000	41.88990	0.04780	4.90450

TABLE EX.4.1.4
Percentage Reduction in the Peak Response Quantities of the Connected Structures as Compared to the Unconnected Response Quantities

		Structure 1		Structure 2	
Structure	Analysis Method	Displacement Reduction (%)	Acceleration Reduction (%)	Displacement Reduction (%)	Acceleration Reduction (%)
Viscous damper connected	Central Difference	43.89	43.90	1.60	−3.22
	SAP2000	43.98	43.98	1.75	−3.02
	State-Space	43.80	43.98	1.65	−2.90

- **Peak values of the response quantities**
  The peak values of the floor displacements and absolute accelerations are obtained by taking the absolute maximum of the response quantities. The comparison of the peak values of the displacements and the absolute accelerations for the unconnected and connected structures is given in Table Ex.4.1.3. Furthermore, the percentage reduction in the peak response quantities of the connected structures relative to the unconnected response quantities is presented in Table Ex.4.1.4.
  Response time histories of the displacement and force of the viscous damper together with the response of the connected structures are presented in Figure 4.29.
- **Force-deformation curves of the dampers**
  The force-deformation curve of the viscous damper is given in Figure 4.30.

Example 4.2 Building frame controlled using viscous and viscoelastic damper

The building frame shown in Figure 4.31 is fitted with passive dampers in all floors in the form of X bracings. Find the top floor acceleration, velocity, and displacement of the frame under the 1940 Imperial Valley earthquake recorded at El Centro. Consider two types of dampers having the following properties: (i) viscous damper with 10% damping and (ii) viscoelastic damper having stiffness $k_d = 15$ M N/m and 5% damping (ξ_d). Compare the response quantities of the controlled building frame with that of the uncontrolled building. The damping ratio of the superstructure is given as 0.02.

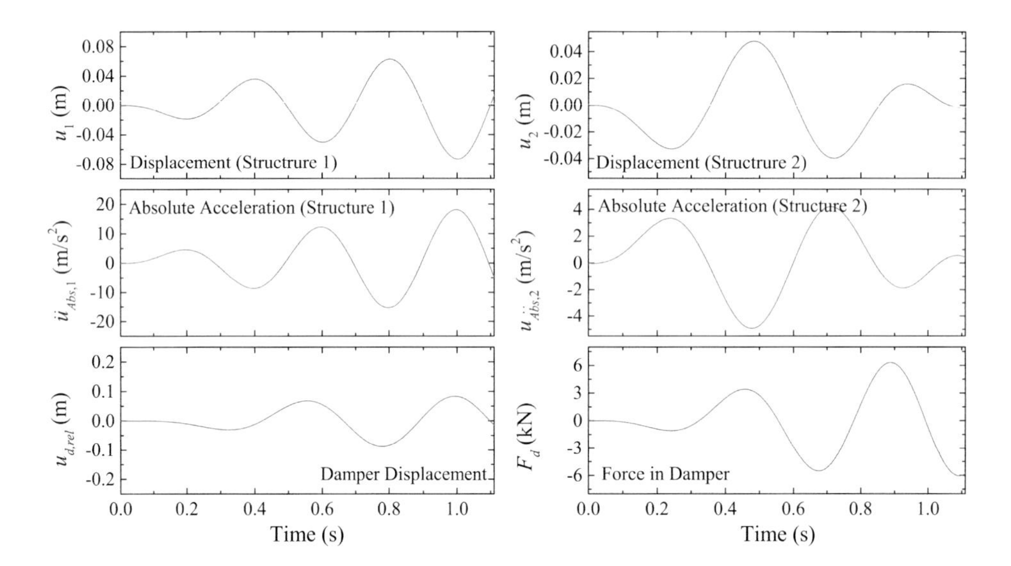

FIGURE 4.29 Response time histories of the structures connected with viscous damper.

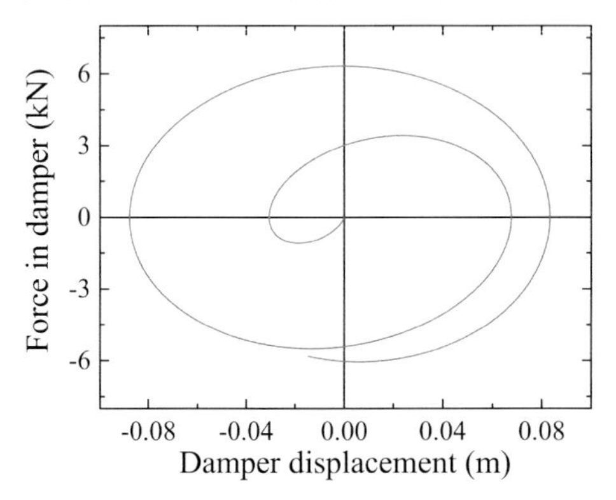

FIGURE 4.30 Force-deformation curve of viscous dampers.

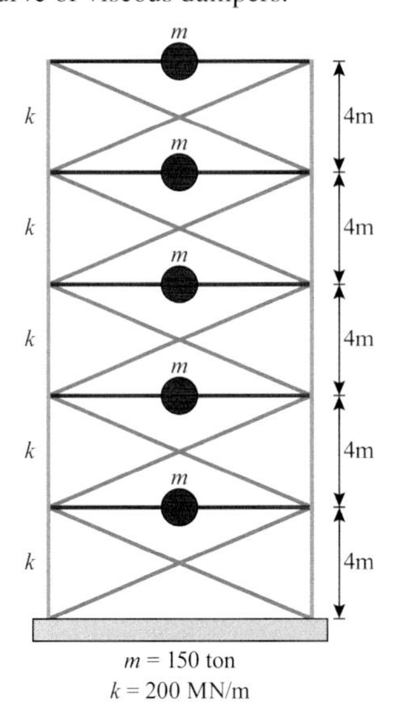

FIGURE 4.31 Five-story building with X-bracing configuration of dampers.

TABLE EX.4.2
Response of the Five-Story Building Frame Equipped with Viscous and Viscoelastic Dampers

Response Quantity	Uncontrolled	Viscous Damper – Controlled	Viscoelastic Damper – Controlled
Displacement (mm)	109.000	106.100	105.400
Velocity (m/s)	1.172	1.130	1.205
Acceleration (m/s^2)	11.800	11.610	11.930

SOLUTION

The response quantities of the uncontrolled structure and the structure controlled using the viscous and viscoelastic dampers are obtained by performing response history analysis. The time history analysis of the uncontrolled and controlled structures under the 1940 Imperial Valley earthquake is performed, and the results are presented in Table Ex.4.2.

REFERENCES

Makris, N., Constantinou, M. C., and Dargush, G. F. (1993). Analytical model of viscoelastic fluid dampers. *Journal of Structural Engineering*, 119(11), 3310–3325.

Miyazaki, M., Arima, F., Kidata, Y., and Hristov, I. (1986). Earthquake response control design of buildings using viscous damping walls. In *Proceedings of 1st East Asia Pacific Conference*, Bangkok, 3, 1882–1891.

Robinson, W., and Greenbank, L. (1976). An extrusion energy absorber suitable for the protection of structures during an earthquake. *Earthquake Engineering and Structural Dynamics*, 4(3), 251–259.

Skinner, R. I., Robinson, W. H., and McVerry, G. H. (1991). Seismic isolation in New Zealand. *Nuclear Engineering and Design*, 127(3), 281–289.

Taylor, D. P., and Constantinou, M. C. (2000). *Fluid dampers for applications of seismic energy dissipation and seismic isolation*. Taylor Devices, Inc., North Tonawanda, NY.

5 Tuned Dampers

5.1 TUNED MASS DAMPER

Structures could be subjected to severe vibrations due to dynamic excitations, such as earthquake and wind. When the vibration amplitudes are large, there will always be safety issues to the occupants. Further, the occurrence of excessive vibration also affects the comfort of the occupants. One of the most widely used passive response control devices is the tuned mass damper (TMD), which uses the concept of tuning the frequency of the control device to the frequency of the structure. Passive TMDs are typically reliable, efficient, and have low maintenance cost. TMDs are generally installed at the rooftops of buildings to control the response of buildings produced due to wind or an earthquake. TMDs may be installed in other structures also, e.g., flexible bridges, such as, suspension/cable-stayed bridges to control the wind-induced vibration. A TMD consists of a mass, a spring, and a damper, which is attached to one side of the building. Figure 5.1(a) shows a building without any control device, whereas a schematic representation of a typical multistory structure fitted with a single TMD (S-TMD) on the top floor is shown in Figure 5.1(b). The most important feature of the TMD is the tuning of frequency. Usually, the frequency of the TMD is tuned to be equal to the fundamental frequency of the structure.

Principle of TMD

A TMD is a simple mass attached to the primary structure in such a way that the vibration response is countered. Harmonic vibrations make violent motion, and TMDs stabilize against them. This suppression of the vibration of the primary structure is generally achieved by tuning the frequency of the TMD to that of the fundamental frequency of the structure. During excitation, a properly designed TMD tends to move in the opposite direction to that of the structure. This, coupled with the energy dissipation due to damping of the TMD, suppresses the vibration of the structure. The vibration of a system can be reduced by a tuned damper with a comparatively lightweight component so that the worst case vibrations are less intense. Mass dampers are frequently implemented with a frictional or hydraulic component that turns mechanical kinetic energy into heat. Application of a dynamic absorber or a TMD to a building structure is more complex as building structures are large and heavy. The TMD responds to structural vibrations, and part of the energy is transferred to the vibration energy of the TMD. The TMD damping dissipates its vibration energy, and as a result, the vibration energy of the structure is absorbed by TMD damping. On the basis of the mechanism of the undamped vibration absorber, the energy-absorbing capacity of the TMD is related to (a) the tuning frequency ratio of the TMD, which is influenced by the mass ratio of the TMD, the stiffness ratio of the TMD and (b) the damping ratio of the TMD. To control the response quantities in two directions, the TMDs may be placed in two directions on the top of a building. Figure 5.2 shows the mathematical model of a bidirectional TMD, controlling the response of the structure in both horizontal directions.

A TMD in its simplest form consists of an auxiliary mass (m_d), spring of stiffness (k_d), and dashpot (c_d). It is attached to the main mass (m_s), with spring stiffness (k_s). The TMD can be modeled as just another additional degree of freedom to the structure. The concept of a TMD originated from the attempt made by Frahm (1911). The author tried to use a vibration absorber to control rolling motion in ships. TMDs are most effective when the first mode contribution to the response is dominant. This is generally the case for tall and slender structural systems. The optimum absorber can reduce the peak response for input frequencies close to the natural frequency of the main system.

DOI: 10.1201/9781315269269-5

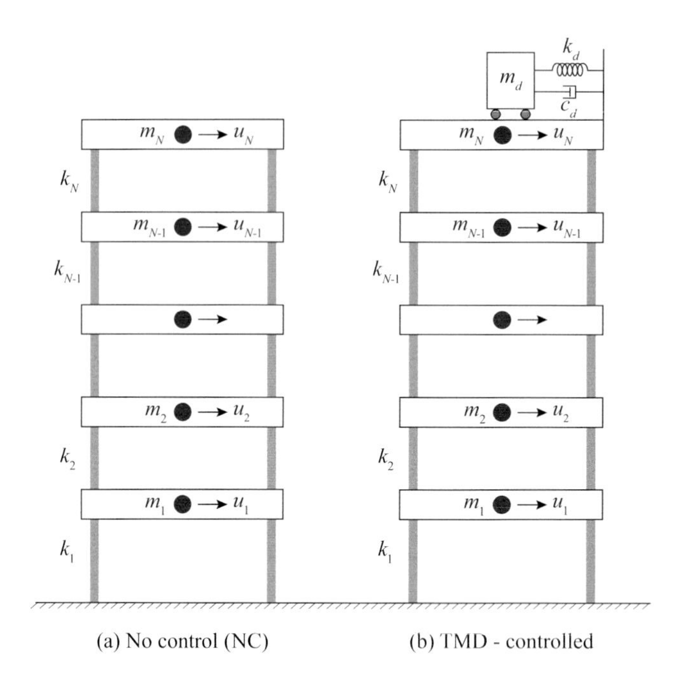

FIGURE 5.1 Comparison of MDOF systems (a) with no control and (b) installed with tuned mass damper.

Uncertainties in the dynamic properties of the main structure, as well as those in the characteristics of the excitation, may affect the performance of TMD. Therefore, the choice of the design parameters of the TMD must be specified carefully.

5.1.1 Simple Pendulum

A simple pendulum is a simple mechanism consisting of a mass hanging at the free end of a massless string or a rigid rod, pivoted at point A, as shown in Figure 5.3. The mass is in the form of a small metal sphere that is allowed to swing back and forth. Figure 5.4 shows a mass supported by a platform, which is suspended by means of two ropes. The nature of motion is periodic when a simple pendulum is allowed to swing back and forth or when it is set in motion. The time taken by a pendulum to complete one cycle oscillation is defined as the time period (T). Along with period, T, another quantity that is used to describe periodic motion is known as the cyclic frequency of

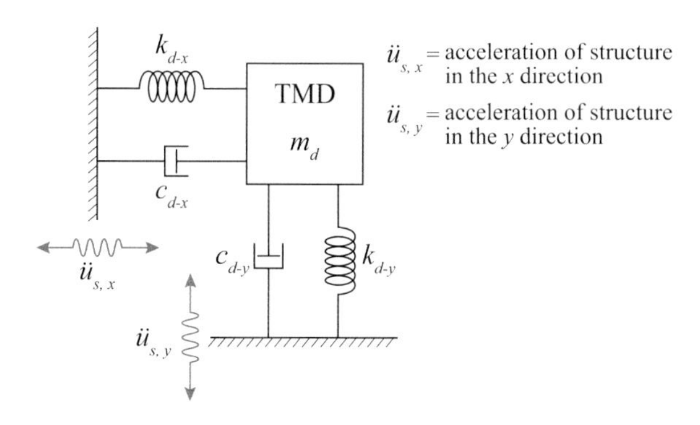

FIGURE 5.2 Bidirectional response control by TMD.

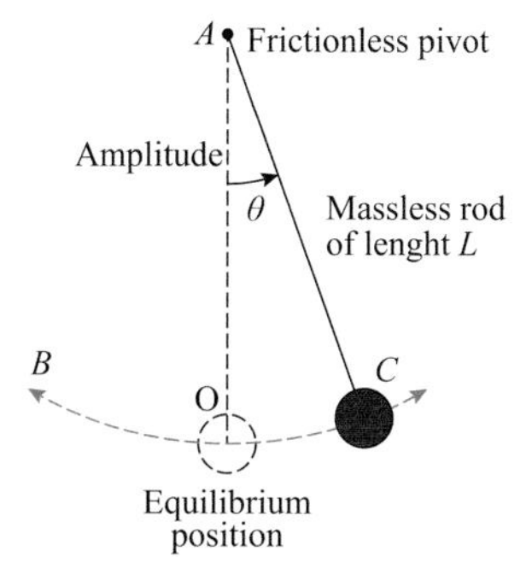

FIGURE 5.3 Simple pendulum system with mass suspended at pivot A.

oscillation, f, defined as the number of oscillations that can occur per unit time period, T. Thus, a simple relationship can be established between f and T as follows.

$$f = \frac{1}{T} \text{ and } T = \frac{1}{f} \tag{5.1}$$

$$f = \frac{\omega}{2\pi} = \frac{1}{2\pi}\sqrt{\frac{\pi g}{L} tanh\left(\frac{\pi h}{L}\right)} \tag{5.2}$$

$$\xi = \frac{2\omega\mu\left(1 + \frac{2h}{B} + S\right)L}{\left(\frac{4\pi h}{\sqrt{gh}}\right)} \tag{5.3}$$

The frequency of oscillation is inversely proportional to the time period and vice versa. The amplitude of oscillation A is defined as the maximum distance covered by the mass from its equilibrium position (point O) as shown in Figure 5.3.

Mathematical Model of Simple Pendulum

When a pendulum is displaced from its equilibrium position, point O, it is set into motion and reaches point B. A restoring force is developed, which pulls back the pendulum to the equilibrium position, point O. Further, it is displaced in the opposite direction to point C, and again attains its equilibrium position O, thus completing one oscillation. Thus, the motion of the pendulum is a simple harmonic motion. The time period of pendulum can be calculated using Equation (5.4), as:

$$T = \frac{2\pi}{\omega} = 2\pi\sqrt{\frac{m}{k}} \tag{5.4}$$

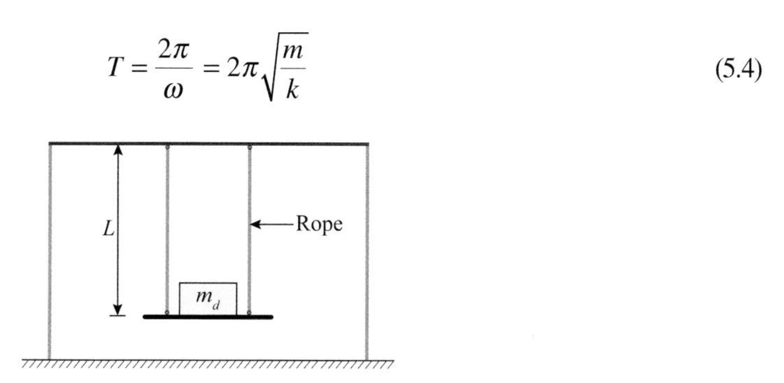

FIGURE 5.4 Simple pendulum system with mass supported by platform.

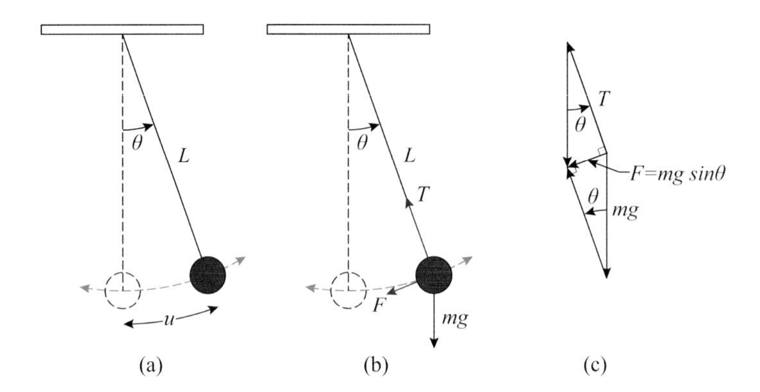

FIGURE 5.5 Restoring forces on a simple pendulum: (a) displaced position, (b) forces acting, and (c) equilibrium of forces.

where ω is the angular frequency of simple harmonic motion, m is the mass of pendulum, and k is the stiffness. For the simple pendulum, the stiffness is obtained as a function of gravitational force on the mass (mg) and the length (L). Consider that the mass m, suspended by the string of length, L, gets displaced from its equilibrium position, by an angle θ. The displacement along the arc through which the mass moves is u, as shown in Figure 5.5(a).

As shown in Figure 5.5(b), the gravitational force is resolved into two components, one along the radial direction, away from the point of suspension, and the other along the arc direction along which the mass moves; providing the restoring force, F, expressed as:

$$F = mg\ sin\theta \tag{5.5}$$

where θ is the angle measured in radian, by which the pendulum is displaced, g is the acceleration due to gravity, and the negative sign indicates that the restoring force acts opposite to the displacement. For very small amplitudes, θ is quite small; hence, $sin\theta$ can be approximated as θ, so that Equation (5.5) can is expressed as:

$$F = mg\ \theta \tag{5.6}$$

From Figure 5.5(a), $\theta = u/L$, where u is the arc length, and L is the radius of the circle through which the mass moves. Thus, the restoring force can be expressed as:

$$F = mg\left(\frac{u}{L}\right) \tag{5.7}$$

It can be seen from Equation (5.7) that the restoring force of pendulum is directly proportional to the displacement u. Therefore, the stiffness k can be written as:

$$k = \frac{mg}{L} \tag{5.8}$$

Therefore, the time period of pendulum is expressed as:

$$T = 2\pi\sqrt{\frac{m}{\left(\frac{mg}{L}\right)}} = 2\pi\sqrt{\frac{L}{g}} \tag{5.9}$$

Thus, the time period of a simple pendulum for very small amplitudes is directly proportional to its length, L, and inversely proportional to the acceleration due to gravity g.

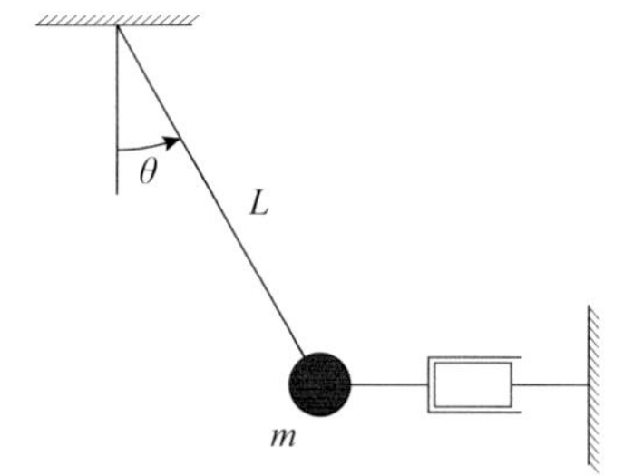

FIGURE 5.6 Pendulum tuned mass damper (PTMD).

One limitation shown by simple pendulum damper is that the time period depends on the length of pendulum L. Thus, the required pendulum length, L, for large time period structures may be very large, which could prove challenging for practical implementation.

5.1.2 Pendulum with Damper

The performance of TMDs in delivering satisfactory response reduction is dependent on the characteristics of the excitation, i.e., wind-induced vibrations, earthquake excitations. Simple pendulums require a large area for installation and, hence, the overall assembly may become massive and costly. Further, uncertainties from ground motions, the structural nonlinearities, and the effect of soil structure interaction have to be taken into consideration during modeling. For such cases, these limitations of the pendulum TMD (PTMD) can be alleviated by modifying the TMD mechanism by providing damper to simple pendulum, as shown in Figure 5.6.

Such suspended mass system is known as PTMD. PTMDs replace the spring and damper with a pendulum; commonly modeled as a simple pendulum. The assembly of suspended mass system and damper counters the vibration of the primary structure and dissipates the energy induced by the excitation. Therefore, it suppresses the vibration and reduces the overall damage of the structure, economically.

Mathematical Model

As shown in Figure 5.6, the damper is restricted to move only in the horizontal direction. D'Alembert's principle of dynamic equilibrium is used for writing the equation of motion for the PTMD model. A free body diagram of the PTMD model and various forces acting on the PTMD model are shown in Figure 5.7.

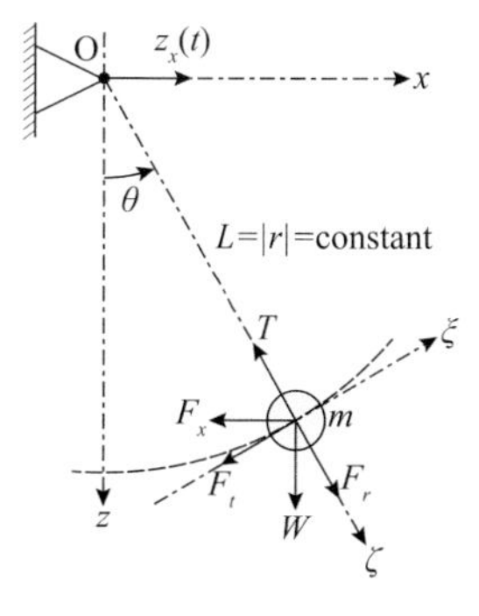

FIGURE 5.7 Free body diagram of the PTMD model.

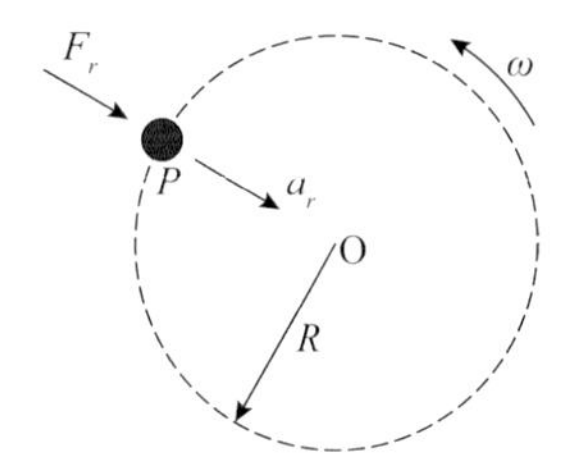

FIGURE 5.8 Centrifugal force.

i. Centrifugal force

Due to the presence of angular velocity, there exists a fictitious centrifugal force and such force is observed from a no-inertial reference frame. For example, consider Figure 5.8, in which a particle P is firmly sitting on a turntable, which is rotating at an angular velocity ω about point O.

The radius of the turntable measured from point O to the location of particle is R. The centripetal acceleration (a_r) of the particle with respect to ground, pointing toward the center of rotation O is:

$$a_r = \omega^2 R \tag{5.10}$$

Due to friction between the rotating particle P and the turntable, which holds the particle in place, the friction force F_r comes into picture. The friction force acts in the direction of centripetal acceleration a_r and is determined by Newton's second law:

$$F_r = ma_r = m(\omega^2 R) \tag{5.11}$$

where m is the mass of the particle. The angular velocity ω is represented by $\dot{\theta}$ and the radius R is represented by L. Then, the centrifugal force can be expressed as:

$$F_r = ma_r = m(\omega^2 R) = m(\dot{\theta}^2 L) \tag{5.12}$$

ii. Inertia force

There are two types of centrifugal forces acting on the pendulum, one due to the trajectory of motion, and the other due to horizontal movement of suspension point of the pendulum.

- Inertia force due to trajectory of motion

  The trajectory of motion is defined as the path traveled by moving object with mass through space as shown by the dotted line in Figure 5.9. The inertia force acting tangentially along this path is known as Euler force. In Figure 5.9, a particle P is firmly sitting on a turntable, rotating in anticlockwise direction, at an angular acceleration α about point O.

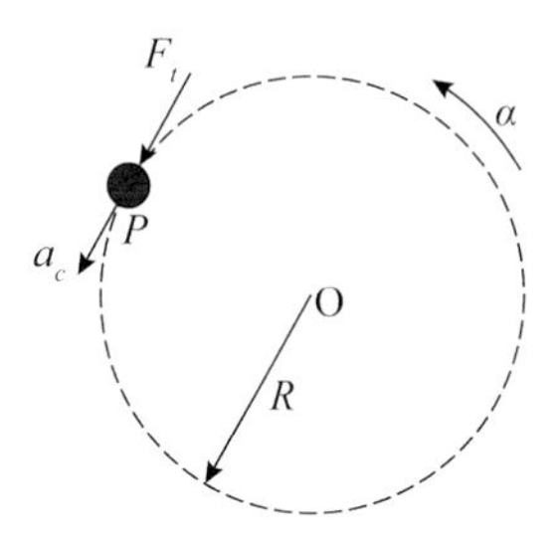

FIGURE 5.9 Inertia force.

The tangential acceleration or circumferential acceleration a_C of the particle can be expressed as:

$$a_C = \alpha R \tag{5.13}$$

The particle tries to slide along the circumferential direction and the turntable prevents it from sliding. This gives rise to the tangential force F_t; which is determined by applying Newton's second law in the direction of circumferential acceleration a_C. The angular acceleration α is represented by $\ddot{\theta}$ and the radius R is represented by L. Thus,

$$F_t = ma_C = m(\alpha R) \tag{5.14}$$

Using Equation (5.14) and Figure 5.9, the centrifugal force can be expressed as:

$$F_t = ma_C = m(\alpha R) = m\left(L\ddot{\theta}\right) \tag{5.15}$$

- Inertia force due to horizontal movement of pendulum's suspension point
  Inertia force due to the horizontal movement of pendulum's suspension point is represented by F_x, and can be expressed as:

$$F_x = m\ddot{Z}_x. \tag{5.16}$$

iii. Gravitational force
The gravitational force depicts attraction between any two masses or any two bodies. It is denoted by $W = mg$, as shown in Figure 5.7.

iv. Tension
There exists a reaction force represented by T, which is the tension in the massless rod as shown in Figure 5.7. Using D'Alembert's principle in ξ and ζ directions as shown in Figure 5.7, following Equations (5.17) and (5.18) can be derived.

$$\sum \xi = -F_t - F_x cos\theta - W sin\theta = 0 \tag{5.17}$$

$$\sum \zeta = -T + F_r - F_x sin\theta + W cos\theta = 0 \tag{5.18}$$

Thus, Equation (5.19) can be written as:

$$\ddot{\theta} + \frac{g}{L} sin\theta + \frac{\ddot{Z}_x}{L} cos\theta = 0 \tag{5.19}$$

Equation (5.19) is a nonlinear equation that describes vibration of a pendulum with mobile suspension point. For very small displacement of the pendulum, Equation (5.19) can be written, by neglecting high-order differential components.

$$m\ddot{\theta} + \frac{mg}{L} sin\theta = -\frac{\ddot{Z}_x}{L} cos\theta \tag{5.20}$$

Equation (5.20) is the basic equation for PTMD system.

5.1.3 Inverted Pendulum with Damper and Spring

Inverted pendulum is a passive damper that can transfer or absorb the input energy effectively and minimize the dynamic effect of vibrations.

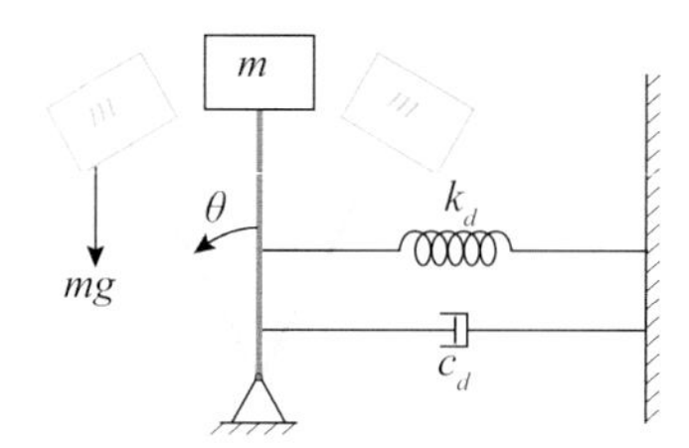

FIGURE 5.10 Inverted pendulum with damper and spring.

Mathematical Model

Figure 5.10 shows the mathematical model of inverted pendulum of mass m located at the top of massless rod of length l. The components of gravitational force are a force parallel to the rod $(mgcos\theta)$ and the force perpendicular to the rod $(mgsin\theta)$. The force parallel to the rod is resisted by tension in the rod. The force perpendicular to the rod is given as:

$$F_p = -mg\, sin\theta \tag{5.21}$$

For very small value of $\theta, sin\theta \approx \theta$, and hence, Equation (5.21) becomes:

$$F_p = -mg\theta \tag{5.22}$$

A spring attached to the rod tries to keep the rod in upright position. The force from the spring is given by:

$$F_s = -k_d x = -k_d L\theta \tag{5.23}$$

The equation of motion is:

$$\begin{aligned} m\ddot{u} &= F_p + F_s + F_d \\ mL\ddot{\theta} &= mg\theta - k_d L\theta - c_d L\dot{\theta} \end{aligned} \tag{5.24}$$

Dividing both sides of Equation (5.24) by mL gives:

$$\ddot{\theta} = \frac{mg}{mL}\theta - \frac{k_d L}{mL}\theta - \frac{c_d L}{mL}\dot{\theta} = \left(\frac{g}{L} - \frac{k_d}{m}\right)\theta - \frac{c_d}{m}\dot{\theta} = \omega^2\theta - \frac{c_d}{m}\dot{\theta} \tag{5.25}$$

5.1.4 Two-Mass Damper

Two-mass damper is a system that refers to the application of a TMD as an extra mass attached to the primary mass in order to reduce the dynamic response of the structure as shown in Figure 5.11. The concept of providing an extra mass is to tune the frequency of damper to the frequency of a particular structure. Hence, when the frequency of the structure is excited, the damper mass will oscillate out of phase with the structural motion. This concept was first put forth by Frahm (1911) to suppress the rolling motion of ships. Later, researchers started applying the concept of TMD in civil engineering structures.

i. Mathematical model of two-mass system with undamped structure and undamped TMD
   The mathematical model of the undamped two-mass damper, the single degree of freedom (SDOF)-TMD system, is presented in Figure 5.12. The undamped SDOF system consists

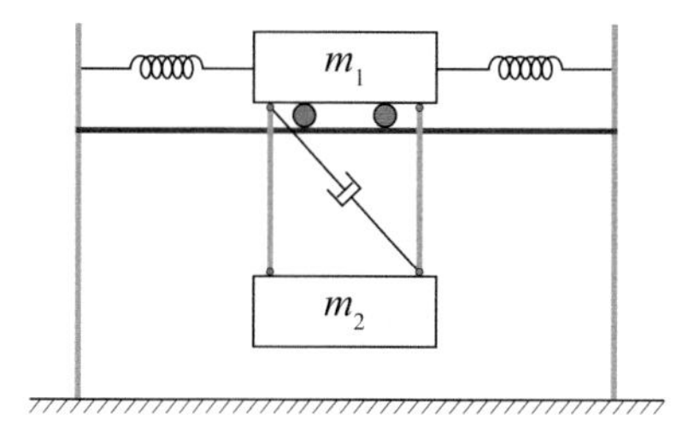

FIGURE 5.11 Two-mass damper.

of primary mass m and stiffness k. A TMD with secondary mass m_d and stiffness k_d is attached to the primary mass. The SDOF-undamped TMD system is subjected to both; the external force P acting on the primary mass, and ground acceleration $\ddot{u}_g$.

The equation of motion for undamped SDOF system without TMD is given as:

$$m\ddot{u} + ku = Psin\varpi t \tag{5.26}$$

Using Equation (5.26), the equation of motion for primary mass m and primary stiffness k, shown in Figure 5.12, can be expressed as:

$$m\ddot{u} + ku - k_d u_d = P - m\ddot{u}_g \tag{5.27}$$

where u and u_d are the displacements of primary mass and the secondary mass, respectively. Similarly, the equation of motion for the secondary mass m_d and secondary stiffness k_d, shown in Figure 5.12 can be expressed as:

$$m_d\left(\ddot{u}_d + \ddot{u}\right) + k_d u_d = -m_d\ddot{u}_g \tag{5.28}$$

The external force acting on the primary system is assumed to be sinusoidal in nature with frequency of excitation ϖ. Hence,

$$P = P_0 sin(\varpi t),\ u = u_0 sin(\varpi t),\ u_d = u_{d0} sin(\varpi t),\ \text{and}\ \ddot{u}_g = \ddot{u}_{g0} sin(\varpi t) \tag{5.29}$$

By substituting the terms from Equation (5.29), Equations (5.27) and (5.28), respectively, can be written as:

$$\left(k - m\varpi^2\right)u_0 - k_d u_{d0} = P_0 - m\ddot{u}_{g0} \tag{5.30}$$

$$\left(k_d - m_d\varpi^2\right)u_{d0} - m_d\varpi^2 u_0 = -m_d\ddot{u}_{g0} \tag{5.31}$$

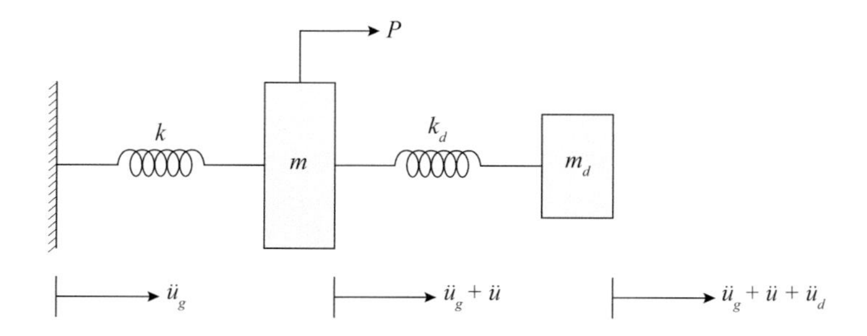

FIGURE 5.12 Undamped SDOF system coupled with TMD.

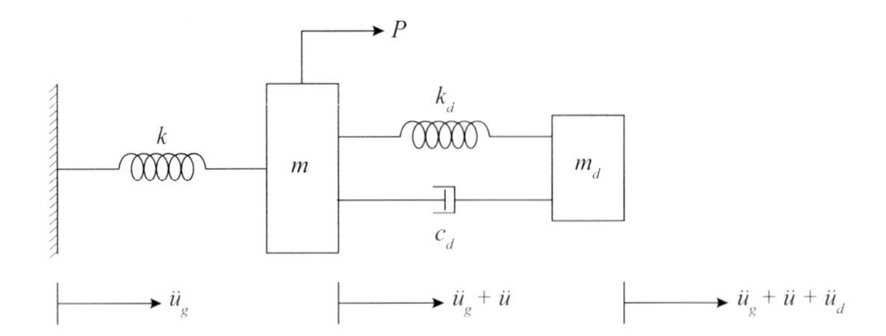

FIGURE 5.13 Undamped SDOF system coupled with damped TMD system.

Summation of Equations (5.30) and (5.31) leads to Equation (5.32).

$$m_d\left(1+\frac{1}{\bar{m}}\right)\left(\ddot{u}+\ddot{u}_g\right)-m_d\varpi^2u_d+ku=P \tag{5.32}$$

where $\bar{m}$ is the mass ratio, defined as the ratio of secondary mass m_d to the primary mass m; $\left(\bar{m}=m_d/m\right)$. Equations (5.30), (5.31), and (5.32) are the governing equations of motion for SDOF-undamped TMD system.

ii. Mathematical model of two-mass system undamped structure and damped TMD
Figure 5.13 shows an undamped SDOF system with primary mass m and primary stiffness k. A TMD with secondary mass m_d and stiffness k_d is attached to the primary mass along with the secondary viscous damper c_d, as shown. The SDOF undamped system coupled with the damped TMD system is subjected to both the external force p acting on the primary mass and ground acceleration $\ddot{u}_g$.

The equation of motion for primary mass m and primary stiffness k shown in Figure 5.13 can be expressed as follows.

$$m\ddot{u}+ku-k_du_d-c_d\dot{u}_d=P-m\ddot{u}_g \tag{5.33}$$

where u and u_d are the displacements of primary and secondary mass respectively, $\dot{u}_d$ is the velocity of secondary damper, $\ddot{u}_g$ is the ground acceleration to which the primary mass is subjected to, P is the external force applied to the primary mass, and c_d is the viscous damper attached to the secondary mass. The equation of motion for secondary mass m_d, secondary stiffness k_d, and secondary viscous damper c_d can be expressed as follows.

$$m_d\ddot{u}_d+c_d\dot{u}_d+k_du_d+m_d\ddot{u}=-m_d\ddot{u}_g \tag{5.34}$$

Due to the inclusion of damping term in Equations (5.33) and (5.34), a phase shift is produced between the periodic excitations and the response. Thus, the solution is initially expressed in terms of complex quantities. Hence, load and excitations can be expressed as:

$$P=P_0e^{i\varpi t}\text{ and }\ddot{u}_g=\ddot{u}_{g0}e^{i\varpi t} \tag{5.35}$$

The response is expressed in terms of complex quantities as follows.

$$u=\bar{u}e^{i\varpi t}\text{ and }u_d=\bar{u}_de^{i\varpi t} \tag{5.36}$$

The real and imaginary parts of the excitation correspond to cosine and sinusoidal inputs, respectively. Therefore, the corresponding solutions are either given by the real part of u and u_d for the cosine inputs or by the imaginary part of u and u_d for the sinusoidal input. Substituting terms from Equations (5.35) and (5.36), Equations (5.33) and (5.34) can be written as follows.

$$\left(-k_d - ic_d\varpi\right)\bar{u}_d + \left(k - m\varpi^2\right)\bar{u} = P_0 - m\ddot{u}_{g0} \tag{5.37}$$

$$\left(k_d + ic_d\varpi - m_d\varpi^2\right)\bar{u}_d - m_d\varpi^2\bar{u} = -m_d\ddot{u}_{g0} \tag{5.38}$$

Summation of Equations (5.37) and (5.38) leads to:

$$-m_d\varpi^2\bar{u}_d + \left[k - m_d\varpi^2\left(1 + \frac{1}{\bar{m}}\right)\right]\bar{u} = P_0 - m_d\left(1 + \frac{1}{\bar{m}}\right)\ddot{u}_{g0} \tag{5.39}$$

where $\bar{m}$ is the mass ratio of secondary mass m_d to the primary mass $m\left(\bar{m} = m_d/m\right)$. Thus, Equations (5.37), (5.38), and (5.39) are the governing equations for SDOF undamped system coupled with damped TMD system.

iii. Mathematical model of two-mass system with damped structure and damped TMD
Figure 5.14 shows a damped SDOF system of primary mass m, primary stiffness k, and primary viscous damper c. All real systems show some amount of damping, and analyzing the effect of damping in the real system on the optimal tuning of the absorber is an important design consideration. A TMD with secondary mass m_d, secondary stiffness k_d, and secondary viscous damper c_d is attached to the primary mass as shown in Figure 5.14. The SDOF damped system coupled with the damped TMD system is subjected to both the external force p acting on the primary mass and ground acceleration $\ddot{u}_g$.

The equation of motion for primary mass m, primary stiffness k, and primary viscous damper c, shown in Figure 5.14, can be expressed as follows.

$$m\ddot{u} + c\dot{u} + ku - k_d u_d - c_d\dot{u}_d = P - m\ddot{u}_g \tag{5.40}$$

where u and u_d are the displacements of primary and secondary mass respectively, $\dot{u}_d$ and $\dot{u}$ are the velocity of secondary damper and primary damper respectively, $\ddot{u}$ is the acceleration of primary mass, $\ddot{u}_g$ is the absolute ground acceleration, P is the external force applied to the primary mass, and c_d and c are the damping coefficients of the viscous damper attached to the secondary mass and primary mass, respectively. Similarly, equation of motion for secondary mass m_d, secondary stiffness k_d, and secondary viscous damper c_d as shown in Figure 5.14 can be expressed as:

$$m_d\ddot{u}_d + c_d\dot{u}_d + k_d u_d + m_d\ddot{u} = -m_d\ddot{u}_g \tag{5.41}$$

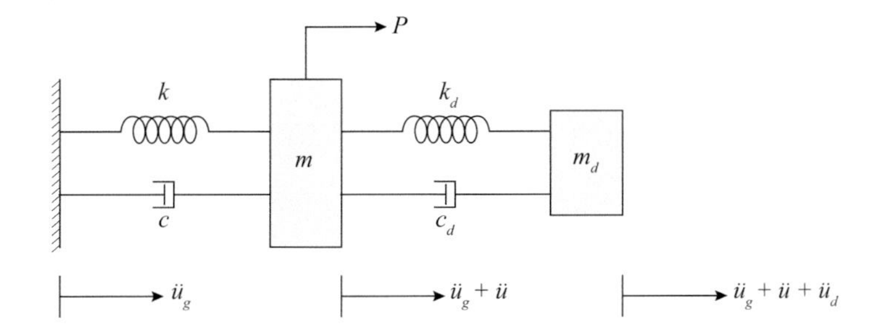

FIGURE 5.14 Damped SDOF system coupled with damped TMD system.

Analogous to the undamped case, the load and response terms are as per Equations (5.35) and (5.36). By substituting them, Equations (5.40) and (5.41) can be rewritten as follows.

$$\left(-k_d - ic_d\varpi\right)\bar{u}_d + \left(-m\varpi^2 + k + ic\varpi\right)\bar{u} = P_0 - m\ddot{u}_{g0} \tag{5.42}$$

$$\left(k_d + ic_d\varpi - m_d\varpi^2\right)\bar{u}_d - m_d\varpi^2\bar{u} = -m_d\ddot{u}_{g0} \tag{5.43}$$

Summation of Equations (5.42) and (5.43) leads to Equation (5.44).

$$-m_d\varpi^2\bar{u}_d + \left[k + ic\varpi - \varpi^2 m_d\left(1 + \frac{1}{\bar{m}}\right)\right]\bar{u} = P_0 - \ddot{u}_{g0}m_d\left(1 + \frac{1}{\bar{m}}\right) \tag{5.44}$$

iv. Mathematical model of a multi-degree of freedom (MDOF) system equipped with TMD

The concept of an SDOF system coupled with TMD is extended to MDOF system. The assembly shown in Figure 5.15 is called as MDOF system coupled with TMD.

Equation of motion for an uncontrolled MDOF system is given as:

$$[M]\{\ddot{u}\} + [C]\{\dot{u}\} + [K]\{u\} = -[M]\{r\}\ddot{u}_g \tag{5.45}$$

where $[M]$, $[C]$, and $[K]$, respectively, are the mass, damping, and stiffness matrices of the structure; $\{u\}$ is the global displacement vector; and $\ddot{u}_g$ is the ground acceleration. Due to the addition of TMD in the system, as shown in Figure 5.15, Equation (5.45) is modified by adding the effect of force coming from the TMD to the equation of motion as follows.

$$[M]\{\ddot{u}\} + [C]\{\dot{u}\} + [K]\{u\} = -[M]\{r\}\ddot{u}_g + \{F_{TMD}\} \tag{5.46}$$

where $\{F_{TMD}\}$ is the vector of the resisting force provided to the structure by the attached TMD. The governing equations of motion for the system shown in Figure 5.15 are as follows.

$$m_1\ddot{u}_1 + c_1\dot{u}_1 + k_1u_1 - k_2\left(u_2 - u_1\right) - c_2\left(\dot{u}_2 - \dot{u}_1\right) = P_1 - m_1\ddot{u}_g \tag{5.47}$$

$$m_2\ddot{u}_2 + c_2\left(\dot{u}_2 - \dot{u}_1\right) + k_2\left(u_2 - u_1\right) - c_d\dot{u}_d - k_du_d = P_2 - m_2\ddot{u}_g \tag{5.48}$$

$$m_d\ddot{u}_d + c_d\dot{u}_d + k_du_d = -m_d\left(\ddot{u}_2 + \ddot{u}_g\right) \tag{5.49}$$

Combining Equations (5.47) and (5.48); and expressing them in a similar way, as that of SDOF case, reduces the problem to an equivalent SDOF system, for which formulation and

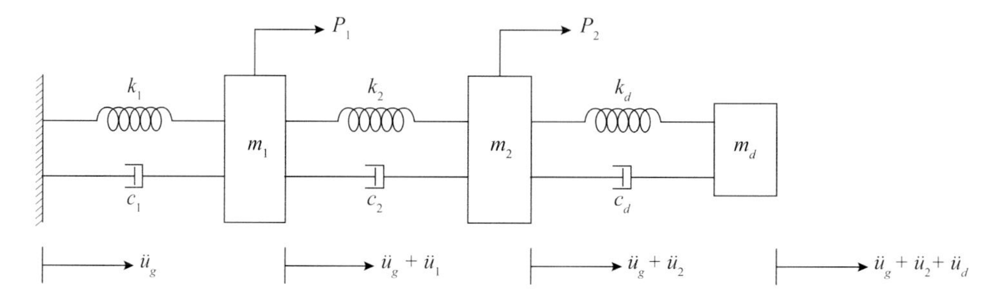

FIGURE 5.15 Damped MDOF system coupled with damped TMD system.

theory of SDOF system is applicable. It is followed by transforming the original matrix equation to scalar modal equations. Equations (5.47) and (5.48) are written in matrix form as follows.

$$[M]=\begin{bmatrix} m_1 & 0 \\ 0 & m_2 \end{bmatrix} \tag{5.50}$$

$$[K]=\begin{bmatrix} k_1+k_2 & -k_2 \\ -k_2 & k_2 \end{bmatrix} \tag{5.51}$$

$$[C]=\begin{bmatrix} c_1+c_2 & -c_2 \\ -c_2 & c_2 \end{bmatrix} \tag{5.52}$$

$$\{u\}=\begin{Bmatrix} u_1 \\ u_2 \end{Bmatrix} \tag{5.53}$$

$$-[M]\{r\}\ddot{u}_g+\{F_{TMD}\}=\begin{bmatrix} P_1-m_1\ddot{u}_g \\ P_2-m_2\ddot{u}_g \end{bmatrix}+\begin{bmatrix} 0 \\ c_d\dot{u}_d+k_du_d \end{bmatrix} \tag{5.54}$$

$$\begin{bmatrix} m_1 & 0 \\ 0 & m_2 \end{bmatrix}\begin{Bmatrix} \ddot{u}_1 \\ \ddot{u}_2 \end{Bmatrix}+\begin{bmatrix} c_1+c_2 & -c_2 \\ -c_2 & c_2 \end{bmatrix}\begin{Bmatrix} \dot{u}_1 \\ \dot{u}_2 \end{Bmatrix} +\begin{bmatrix} k_1+k_2 & -k_2 \\ -k_2 & k_2 \end{bmatrix}\begin{Bmatrix} u_1 \\ u_2 \end{Bmatrix}=\begin{Bmatrix} P_1-m_1\ddot{u}_g \\ P_2-m_2\ddot{u}_g \end{Bmatrix}+\begin{Bmatrix} 0 \\ c_d\dot{u}_d+k_du_d \end{Bmatrix} \tag{5.55}$$

Thus, Equation (5.46) leads to Equation (5.55), which is the governing equation of motion for MDOF system, installed with TMD. Equation (5.55) can be solved together with Equation (5.49) to obtain the response of the entire system.

5.1.5 Multistage Tuned Mass Damper

In multistage TMD (MSTMD) system, oscillators are placed at the different floors of the building. All the oscillators are tuned to the natural frequency of the structure so as to mitigate the dynamic response by decoupling the vibration modes of the structure. The idea behind using MSTMD system is that whenever any of the dampers becomes inoperative during strong earthquake shaking, all the remaining dampers will take up and work in combination to dissipate seismic energy. This leads to the steady dynamic performance as compared to the conventional S-TMD. Also, the mass values of dampers used in MSTMD system are very small, as compared to one large mass that is used in the conventional TMD.

Mathematical Model of MSTMD

Figure 5.16 shows a multistory uncontrolled structure, whereas Figure 5.17 depicts the MDOF structure equipped with an oscillator installed at each floor level. The oscillators attached to each floor have identical damping and frequency characteristics.

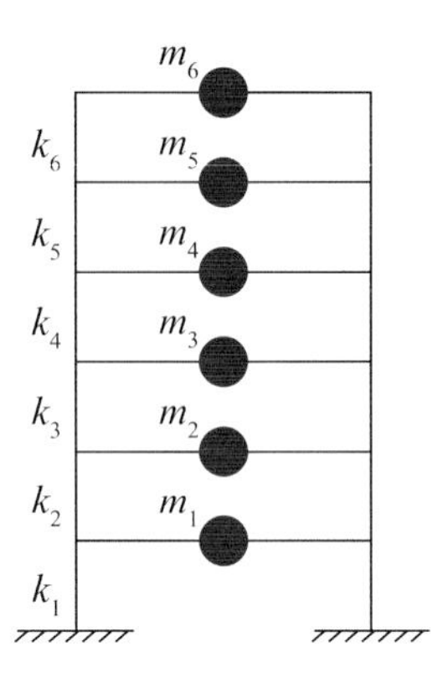

FIGURE 5.16 Uncontrolled MDOF structure.

The governing equation of motion of the structure shown in Figure 5.17 can be constructed based on the combined mass matrix, $[M]$; the combined damping matrix, $[C]$; and the combined stiffness matrix, $[K]$, which are of order $(s+d)\times(s+d)$; where s indicates the degree of freedom for the structure and d indicates the degree of freedom for the multistage oscillators. The mass matrix of the system is:

$$[M]=\begin{bmatrix} [M_s] & 0 \\ 0 & [M_d] \end{bmatrix} \tag{5.56}$$

where $[M_s]$ is the mass matrix the structure, and $[M_d]$ is the mass matrix of the multistage oscillators. The stiffness matrix and the damping matrix are expressed by Equations (5.57) and (5.58), respectively.

$$[K]=\begin{bmatrix} [K_{ss}] & [K_{sd}] \\ [K_{ds}] & [K_{dd}] \end{bmatrix} \tag{5.57}$$

In Equation (5.57), $[K_{ss}]$ is given as $[K_{ss}]=[K_{ss,0}]+[K_{dd}]$; $[K_{ss,0}]$ is the stiffness matrix of the structure without the TMDs; $[K_{dd}]$ are the stiffness matrices of the multistage oscillators, respectively; and $[K_{sd}]=-[K_{dd}]$ and $[K_{ds}]=-[K_{dd}]$ are the diagonal matrices.

$$[C]=\begin{bmatrix} [C_{ss}] & [C_{sd}] \\ [C_{ds}] & [C_{dd}] \end{bmatrix} \tag{5.58}$$

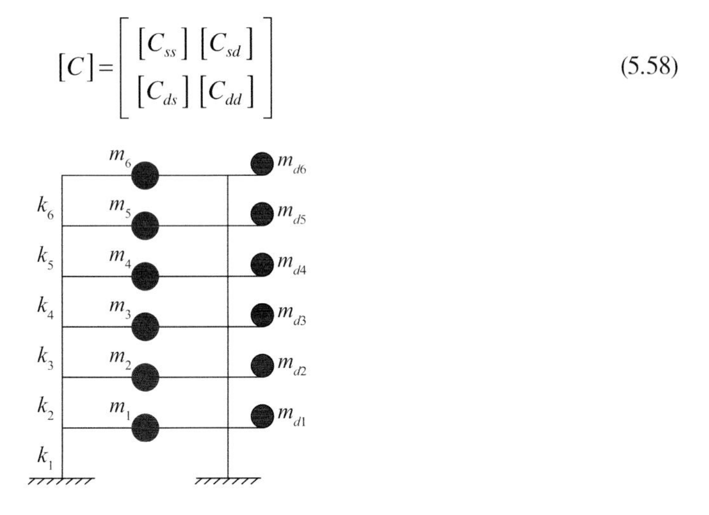

FIGURE 5.17 Structure equipped with multistage tuned mass damper system.

Similarly, $\left[C_{ss}\right],\left[C_{dd}\right],\left[C_{sd}\right]$, and $\left[C_{ds}\right]$ are damping matrices defining exactly the same structure as that of stiffness $\left[K_{ss}\right],\left[K_{dd}\right],\left[K_{sd}\right]$, and $\left[K_{ds}\right]$, respectively. Parameters from Equations (5.56), (5.57), and (5.58) are used to formulate the governing equation of motion of the structure installed with MSTMDs.

$$
\begin{aligned}
&\begin{bmatrix} [M_s] & 0 \\ 0 & [M_d] \end{bmatrix} \begin{Bmatrix} \{\ddot{u}_s\} \\ \{\ddot{u}_d\} \end{Bmatrix} + \begin{bmatrix} [C_{ss}] & [C_{sd}] \\ [C_{ds}] & [C_{dd}] \end{bmatrix} \begin{Bmatrix} \{\dot{u}_s\} \\ \{\dot{u}_d\} \end{Bmatrix} \\
&+ \begin{bmatrix} [K_{ss}] & [K_{sd}] \\ [K_{ds}] & [K_{dd}] \end{bmatrix} \begin{Bmatrix} \{u_s\} \\ \{u_d\} \end{Bmatrix} = -\begin{bmatrix} [M_s] & 0 \\ 0 & [M_d] \end{bmatrix} \ddot{u}_g
\end{aligned}
\tag{5.59}
$$

In Equation (5.59), $\{u_s\} = \{u_{1s}, u_{2s}, \ldots, u_{Ns}\}^T$ and $\{u_d\} = \{u_{1d}, u_{2d}, \ldots, u_{Nd}\}^T$ represent the displacement vectors of the structure and the multistage oscillators, respectively.

5.1.6 Mass on Rubber Bearing

A multistage rubber bearing consists of a large number of laminated rubber bearings (LRB), stacked over each other as shown in Figure 5.18. In order to achieve stability, LRB are connected to each other by means of stabilizers placed between them. Multistage rubber bearings supporting the large mass are capable of mitigating the dynamic response of structure by shearing deformation among the LRB and, thus, enabling the structure to be free from friction. The damper is recommended for installation in tall buildings, wherein, it is beneficial in reducing the wind- or earthquake-induced accelerations. The time period of the damper can be varied by changing the number of stages of LRB by reducing the mass and providing auxiliary coil springs. Furthermore, additional damping can be incorporated through oil dampers.

5.1.7 Multiple and Distributed Tuned Mass Dampers

An S-TMD installed in a structure has two main problems. A small offset in tuning of frequency may result in its reduced efficiency. Further, it requires a large floor space and a large mass. Because of various uncertainties inherent in the properties of the TMD and the structure, perfect tuning is very difficult to achieve. For earthquake application, investigators have shown that an S-TMD often is not effective in reducing seismic response, because earthquake-induced forces are typically impulsive in nature and reach their maximum values very fast during a short duration. A TMD subjected

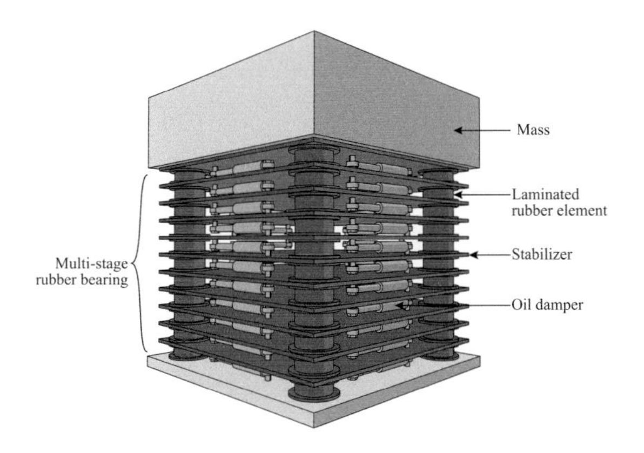

FIGURE 5.18 Schematic diagram of the mass on rubber bearing system. (Masaki et al. 2004)

to a dynamic load filtered by the building structure is usually not set into significant motion in such a short period. As a result, its energy-absorbing capability is not fully developed when it is needed the most. The heavier the TMD, the slower it reaches its full potential. Furthermore, earthquake ground motions include a wide spectrum of frequency components and often induce significant vibration in the fundamental and higher modes of a tall building. Therefore, an S-TMD may sometimes not be effective in reducing the total response quantities of the structures. Several researchers have proposed and successfully verified the concept of installing multiple TMDs (M-TMDs) with varying dynamic characteristics.

M-TMDs

The concept of M-TMD was explicitly used in mechanical engineering and allied fields till late seventies. Warburton and Ayorinde (1980) extended the concept of the M-TMDs, used in mechanical and automotive equipment, for civil engineering applications. M-TMDs can be appropriately used where the damping of the oscillator is limited to low values. Many experimental and numerical studies have confirmed the effectiveness of M-TMDs and their insensitivity to the offset in tuning frequency. M-TMD configuration is more effective in controlling the motion of the primary system. It offers the advantages of portability and ease of installation, because of the reduced size of an individual damper. This makes it attractive not only for new installation but also for temporary use during construction or for retrofitting existing structures. For the same mass ratio, the optimally designed M-TMD systems are more effective than the optimum S-TMD system. The optimum damping ratio for the M-TMD system is quite low, as compared to that of an S-TMD. Tuned to different frequencies of a structure, M-TMD systems can suppress the vibration of multiple modes. Figure 5.19 shows M-TMDs installed at the rooftop of a building.

Optimizing the parameters of the M-TMDs is important for improving their effectiveness. It is reported that when the M-TMDs are placed at top floor of the building, optimum tuning frequency decreases with the increase in the mass ratio and increases with the increase in the number of M-TMDs.

Distributed-M-TMDs (d-M-TMDs)

Researchers tried to distribute the TMDs vertically along the height of the structure. Such distribution of M-TMDs along the height of the structure is known as distributed-TMDs (d-TMDs).

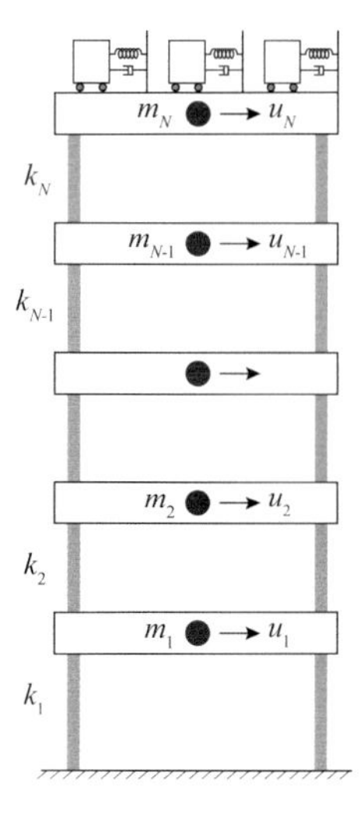

FIGURE 5.19 Idealized mathematical model of a multistory building equipped with M-TMD.

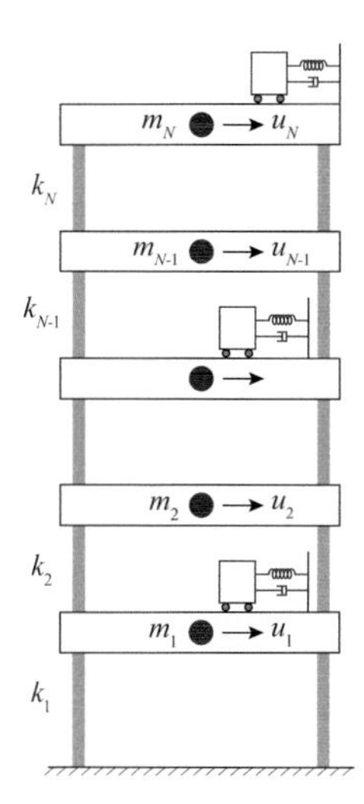

FIGURE 5.20 Idealized mathematical model of a multistory building equipped with d-M-TMD.

Figure 5.20 presents the idealized mathematical model of a multistory building equipped with d-M-TMD, whereas the idealized mathematical model of a chimney equipped with d-M-TMD is given in Figure 5.21.

M-TMDs are more effective in suppressing the accelerations at lower floors than at upper floors. The loss of effectiveness of the M-TMDs is minimal if they are distributed vertically based on the mode shapes of the main system. d-TMDs are placed according to mode shapes of the controlled and uncontrolled structure, and each one is tuned to the corresponding modal frequency. They are effective for multi-mode control of the low-rise buildings with closely spaced frequencies under earthquake ground excitations. Figure 5.22 shows application of d-TMD system to a continuous bridge. The d-TMDs are installed at the center of each span, for controlling mid-span deflection. Moreover, Figure 5.23 shows the application of d-TMD system to a steel truss.

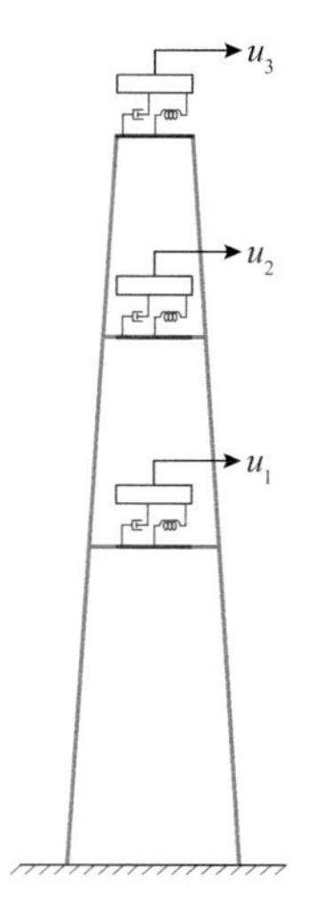

FIGURE 5.21 Idealized mathematical model of a chimney equipped with d-M-TMD.

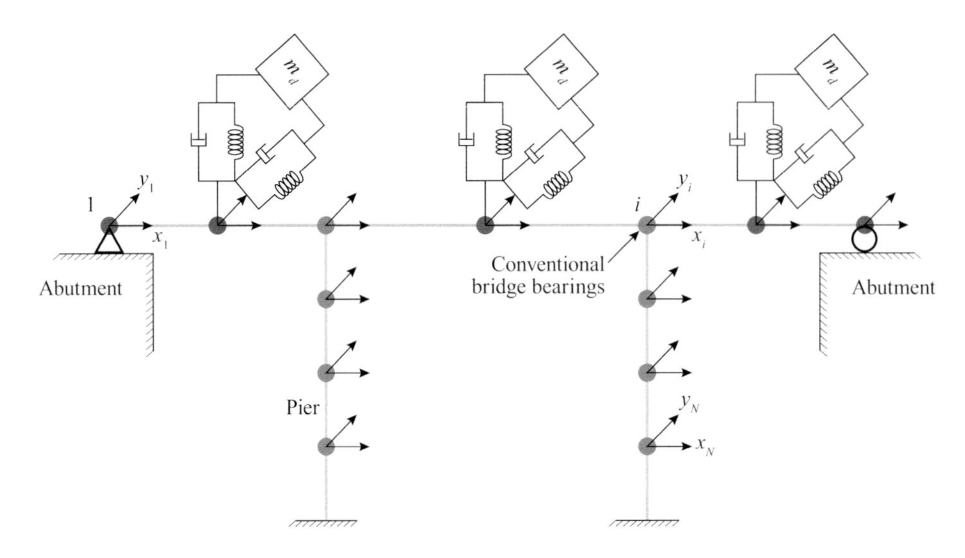

FIGURE 5.22 Application of distributed TMD system to bridge.

Mathematical Model

Figure 5.24(a) and (b) shows the idealized models of a building installed with TMD schemes, consisting of TMDs located at the rooftop, and at different floors. In general, consider that the N-degree of freedom building is installed with n number of TMDs; thereby, the total DOF of the controlled system becomes ($N + n$). The governing equations of motion for the structure installed with TMDs are obtained by considering the equilibrium of forces at the location of each DOF. The matrix form of the equations of motion under earthquake excitation is given as:

$$[M_s]\{\ddot{u}_s\}+[C_s]\{\dot{u}_s\}+[K_s]\{u_s\}=-[M_s]\{r\}\ddot{u}_g \tag{5.60}$$

where $[M_s]$, $[C_s]$, and $[K_s]$ are the mass, damping, and stiffness matrices of the building installed with TMDs, respectively. $\{u_s\}=\{U_1,U_2,\ldots,U_N,u_1,\ldots,u_n\}^T$, $\{\dot{u}_s\}$, and $\{\ddot{u}_s\}$ are the unknown relative floor displacement, velocity, and acceleration vectors, respectively. The earthquake ground acceleration is represented by $\ddot{u}_g$, whereas $\{r\}$ is the vector of influence coefficients.

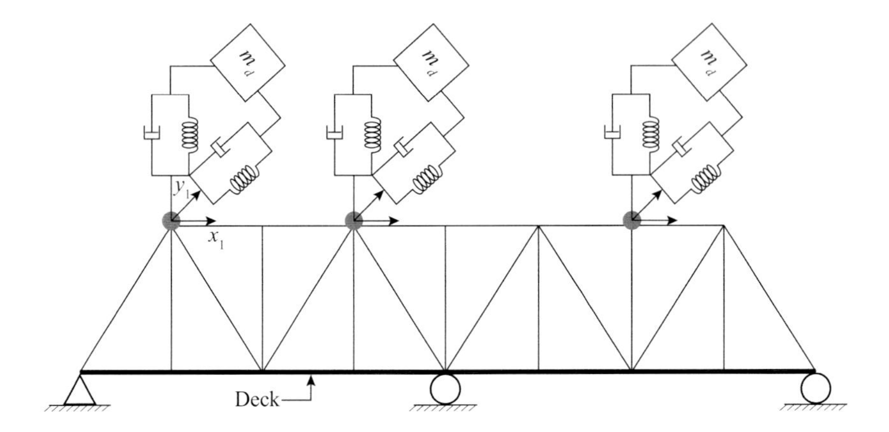

FIGURE 5.23 Application of distributed TMD system to steel truss.

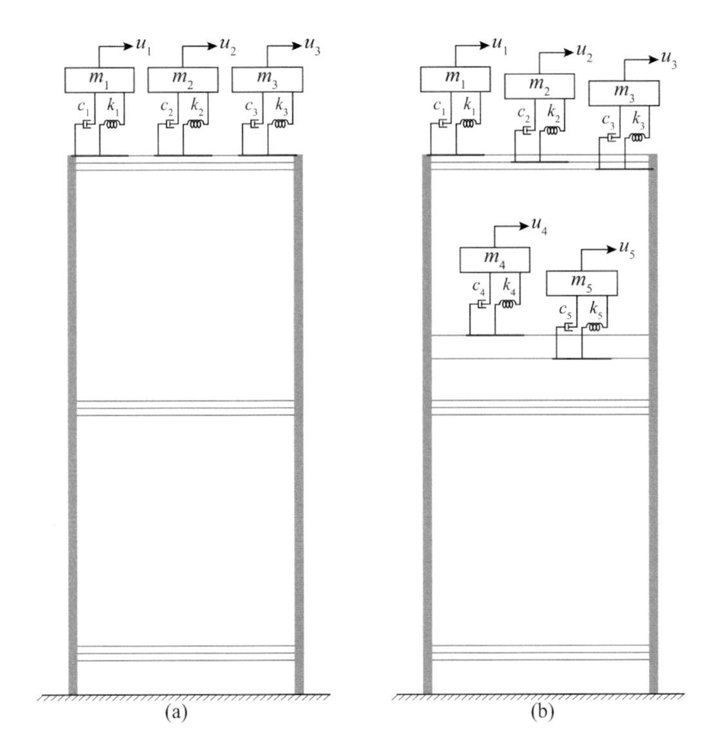

FIGURE 5.24 Configurations of TMDs in a multistory building: (a) multiple TMDs at top and (b) distributed TMDs.

The mass, damping, and stiffness matrices are defined as:

$$[M_s] = \begin{bmatrix} [M_N]_{N\times N} & [0]_{N\times n} \\ [0]_{n\times N} & [m_n]_{n\times n} \end{bmatrix} \tag{5.61}$$

$$[C_s] = \begin{bmatrix} [C_N]_{N\times N} + [C_n]_{N\times N} & -[C_n]_{N\times n} \\ -[C_n]_{n\times N} & [C_n]_{n\times n} \end{bmatrix} \tag{5.62}$$

$$[K_s] = \begin{bmatrix} [K_N]_{N\times N} + [K_n]_{N\times N} & -[K_n]_{N\times n} \\ -[K_n]_{n\times N} & [K_n]_{n\times n} \end{bmatrix} \tag{5.63}$$

where $[M_N]$, $[C_N]$, and $[K_N]$ are the mass, damping, and stiffness matrices of the building, respectively. The mass, damping, and stiffness matrices of the TMDs, respectively, are $[M_n]$, $[C_n]$, and $[K_n]$. The vectors of unknown relative displacement, velocity, and acceleration of the TMDs are $\{x_n\}$, $\{\dot{x}_n\}$, and $\{\ddot{x}_n\}$, respectively. Modal analysis is conducted to determine the natural frequencies, mode shapes and modal masses contribution of the uncontrolled and controlled building. The locations for installation of the TMDs are identified based on the mode shapes of the uncontrolled and controlled buildings. A TMD is placed where the mode shape amplitude of the building is largest/larger in the particular mode and is tuned to the corresponding modal frequency. Next, larger amplitude is chosen instead of the largest when already a TMD is installed previously on the particular floor for controlling the lower mode. Thereby, installation intricacies of the TMDs are eased as not more than one TMD is placed on one floor at a time.

Designing and Placement of TMDs

Frequency of each TMD is calculated using the following relationship based on the tuning frequency ratio of each mode (f_i).

$$f_i = \frac{\omega_i}{\Omega_i} \quad (i = 1 \text{ to } n) \tag{5.64}$$

where $\omega_i = \omega_1 \cdots \omega_5$ are the frequencies of the five TMDs, and $\Omega_i = \Omega_1 \cdots \Omega_5$ are the first five natural frequencies of the building. In cases of all M-TMDs placed at the top as shown in Figure 5.24(a) and d-MTMDs as shown in Figure 5.24(b); the TMDs are tuned with the modal frequencies of the uncontrolled building. The effectiveness of installing TMD in a structure depends on the mass ratio $\left(\mu = \frac{m_t}{M_t}\right)$ between the total mass of the TMDs, $m_t = \sum_{i=1}^{i=n} m_i$ and the total mass of the building, $M_t = \sum_{i=1}^{i=N} M_i$, M_i is the mass lumped at each floor. If all the masses of the TMDs are kept equal, i.e., $m_1 = m_2 = m_3 = \cdots m_n$, the mass (m_i) of TMD can be calculated by Equation (5.65).

$$m_i = \frac{m_t}{n} \quad (i = 1 \text{ to } n) \tag{5.65}$$

The stiffness is used for frequency tuning of each TMD. The stiffness values of the TMDs installed on different floors are determined using Equation (5.66).

$$k_i = m_i \omega_i^2 \quad (i = 1 \text{ to } n) \tag{5.66}$$

When the damping ratio of the TMDs is kept the same (i.e., $\xi_d = \xi_1 = \xi_2 = \cdots \xi_n$), the damping coefficients c_i are calculated using Equation (5.67).

$$c_i = 2\xi_d m_i \omega_i \quad (i = 1 \text{ to } n) \tag{5.67}$$

5.1.8 Tuned Mass Damper Inerter

The TMD inerter (TMDI) is a passive vibration response control device proposed for the mitigation of unwanted building vibration. The TMDI is composed of the traditional TMD configured together with an inerter device. The inerter, which is paired with the TMD, generates a control force proportional to the relative acceleration between its two ends. A single-story structure equipped with a typical TMDI is shown in Figure 5.25.

The control force generated by the TMDI is influenced by the inertance. The force in the inerter (f_I) can be obtained as $f_I = b_d \ddot{u}_I$, where $\ddot{u}_I$ is the relative acceleration between the two ends of the inerter and b_d is the constant of proportionality, which is given as a mass unit to characterize the behavior of the inerter.

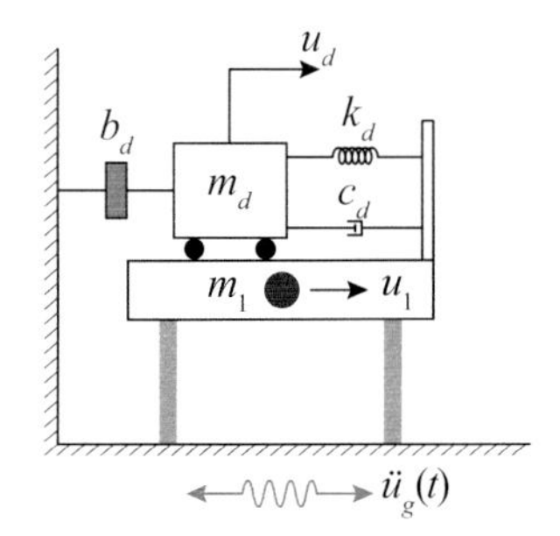

FIGURE 5.25 A single-story structure equipped with a tuned mass damper inerter.

5.1.9 Merits of Tuned Mass Dampers

TMDs have been widely implemented to different structures, worldwide, because of their several advantages. TMDs are inexpensive, provide large structural damping, and reduce the resonant peak of the amplitude of vibration. Pendulum-type TMD requires less space for installation as compared to the traditional TMD, and, hence, it does not affect the architectural design. MSTMDs or M-TMDs have a large deformation capability than that of S-TMD, with the same horizontal and vertical stiffness.

5.1.10 Demerits of Tuned Mass Dampers

TMDs are specialized control devices, the performances of which are sensitive to the dynamic properties of the structure. Therefore, they require personnel specialized knowledge for design and implementation. Furthermore, they occupy a relatively large space and create additional gravity load to the structure. Mistuning of TMDs could lead to reduced effectiveness.

5.1.11 Worldwide Applications of Tuned Mass Dampers

Originally, TMD was used to reduce the impact caused due to wind vibrations. Later on as the research progressed, TMDs have been effectively used for mitigation of earthquake-induced vibrations. Some of the notable applications of TMDs in the world are mentioned below.

- Center point tower in Sydney, Australia
  The TMD was first installed at the Center point Tower in Sydney, Australia. The height of the structure is 250.4 m as seen in Figure 5.26. The water tank located at the top of the tower, shown in Figure 5.27, is of capacity of 132.4 m^3. In addition to supply water to the

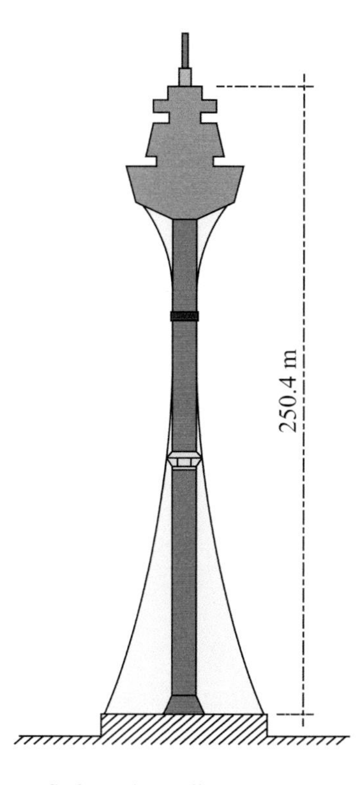

FIGURE 5.26 The center point tower, Sydney, Australia.

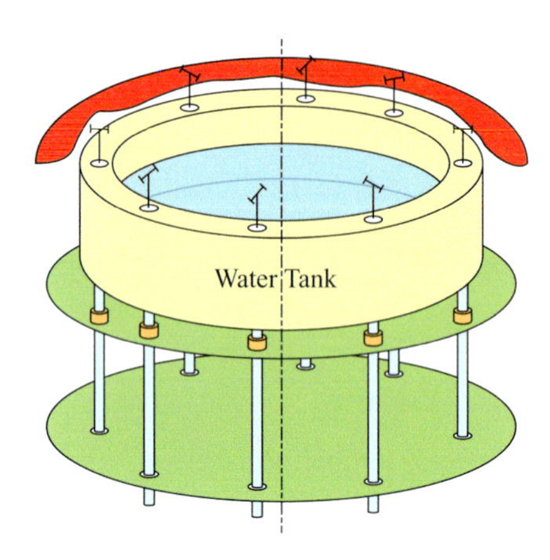

FIGURE 5.27 Water tank TMD at the center point tower, Sydney, Australia.

tower and fire protection supply, the tank, designed as TMD, controls the wind-induced vibrations. A 40-ton secondary mass was later installed in the tank, which increased the vibration control effect in first mode by 20%.

- Canadian National (CN) tower, Toronto, Canada
  The CN railway tower, shown in Photo 5.1, was built by the railway company. It is a free-standing prestressed concrete telecommunications tower in Toronto, Ontario, Canada. A PTMD was installed at the top of the 553.33-m tall CN Tower. The structure includes

PHOTO 5.1 CN Tower, Toronto, Canada.

PHOTO 5.2 Crystal Tower, Osaka, Japan.

102-m steel antenna mass installed at the top of tower. Two lead dampers are provided to prevent the antenna from deflecting excessively and to reduce the bending moment in the antenna under wind loads. At 488- and 503-m elevations, two doughnut-shaped steel rings, 0.35 m wide, 0.30 m deep, and 2.4 and 3 m in diameter respectively, are provided. Each ring contains 9 tons of lead. Sitting on three steel support structure, four universal joints pivoting in all directions connect the rings to the support beam. Four hydraulically activated fluid viscous dampers, provided between the mast and the center of universal joints, help in energy dissipation. The pendulum-type TMD is tuned to the second and fourth modes of vibrations, which are critical for the tower. In first and third modes, antenna response is not critical.

- Crystal Tower, Osaka, Japan
  Crystal Tower, Osaka, Japan, is a 39-story office building, constructed in 1990. The 157-m tall building, shown in Photo 5.2, consists of 37 floors above ground and 2 floors below ground. The tower is slender, with height-to-width ratio 6. The pendulum-type TMD weighing 540 tons is installed at the top of the tower. It reduces the lateral wind-induced motion by 50%. Crystal Tower is the first high-rise building with PTMD, in Japan. The ice thermal storage tanks for air conditioning are used as the moving mass of the damper. The TMD of Crystal Tower, therefore, needs no additional mass weight nor space for installing.
- Citigroup Center, United States
  Photo 5.3 shows the 221-m high Citigroup Center, located in United States. To help stabilize the building, a TMD is installed in the mechanical space at its top. This substantial piece of stabilizing concrete block weighs 400 tons. Figure 5.28 shows the details of the TMD.

PHOTO 5.3 Citigroup Center, United States.

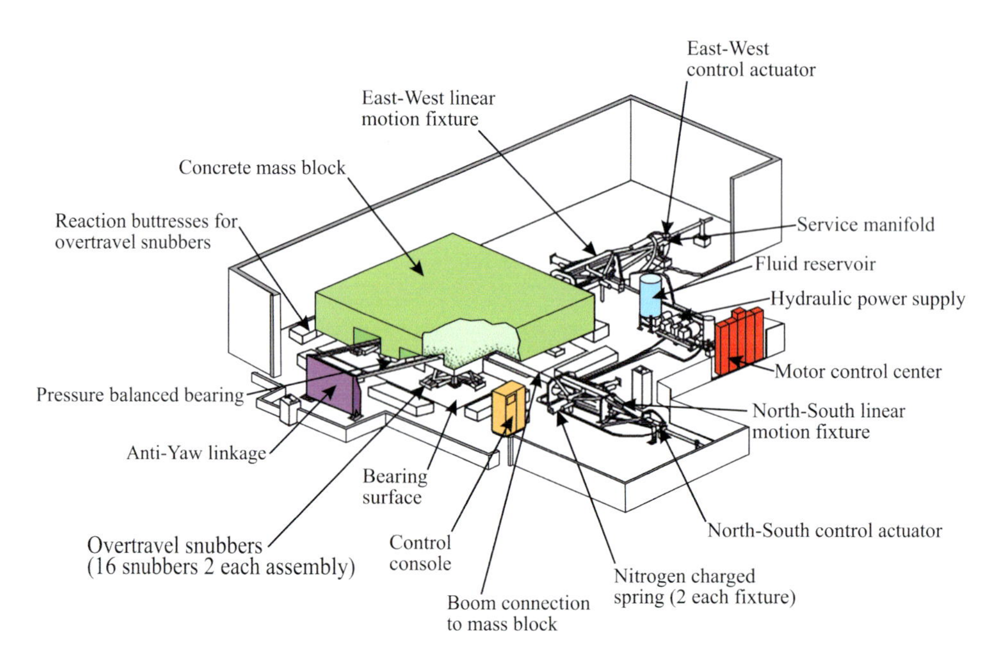

FIGURE 5.28 Details of the TMD at the Citicorp Center, United States.

PHOTO 5.4 John Hancock Tower, United States.

The damper is designed to counteract swaying motions due to the effect of wind on the building and reduces the building's movement due to wind by as much as 50%.

- John Hancock Tower, Boston, United States
The John Hancock, is a 62-story, 240-m tall skyscraper in Boston, shown in Photo 5.4. The occupants of the upper floor of the building suffered from motion sickness, when the building swayed in the wind. To reduce the movement, a dual TMD system was installed on the 58th floor. The TMD consists of 300-ton mass blocks as shown in Figure 5.29, placed at the opposite ends. The blocks move in-phase to provide lateral response control, and out-of-phase to provide control for torsion. Each weight is a box of steel, filled with square

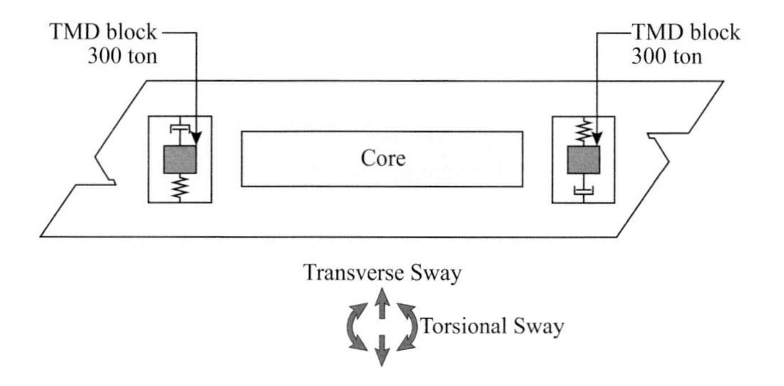

FIGURE 5.29 Dual TMD at John Hancock Tower, Boston, United States.

PHOTO 5.5 Chiba Port Tower, Tokyo Bay, Japan.

lead block of side 5.1 m, and height 0.9 m, resting on a steel plate. The plate is covered with lubricant, allowing the weight to slide freely; however, the weight is attached to the steel frame of the building by means of springs and shock absorbers, as depicted in Figure 5.29. When the tower sways, the weight tends to remain still, allowing the floor to slide underneath it. The dampers are free to move relative to the floor. Then, as the springs and shock absorbers start functioning, they begin to pull the building back, thus stabilizing the tower. Two weights are provided, instead of one, so that they can pull the building in opposite directions, when it tends to twist.

- Chiba Port Tower, Tokyo Bay, Japan
  The Chiba Port Tower, in Japan, shown in Photo 5.5, is a 125-m high steel structure, rhombus in plan, with a side length of 15 m. It is the first tower in Japan, installed with TMD. Figure 5.30 shows the particulars of TMD located at the 63rd floor. The mass can move in X or Y direction, but it is not allowed to rotate. The TMD consists of $9.1 \times 9.1 \times 2.6$ m concrete block of mass 360 tons, which provides 15% effective damping. It reduces the displacement and bending moment considerably.
- Taipei 101 tower, Taiwan
  Taipei tower located in Taiwan, shown in Photo 5.6 has 101 floors, and is 501-m tall above ground. The building is used for communications, conference center, library, observation decks, office space, restaurants, and retail shopping malls, to Taipei Financial Corporation. Construction of the building started in 1999 and was completed in 2004. The building's most noteworthy structural element is its main TMD, shown in Photo 5.7. The TMD can

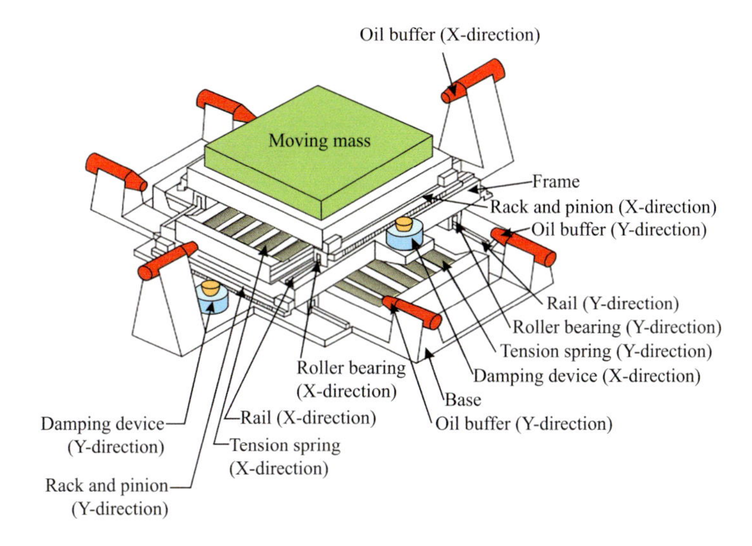

FIGURE 5.30 Schematic diagram of the TMD at the Chiba Port tower, Tokyo Bay, Japan.

PHOTO 5.6 Taipei 101 tower, Taiwan.

PHOTO 5.7 Close up view of the TMD used in Taipei 101 tower.

be viewed from the indoor public observatory. The primary damper is a steel mass of 726 tons, built up from stacked steel plates to form a sphere visible to the visitors. The mass of the damper is 0.24% of the total mass of the building, and is suspended between the 92nd and 87th floors at the building's center. The swing rate is set up by simple pendulum action as it hangs from the 92nd floor.

Under conditions of seismic stress or typhoon-force winds, the sway of the damper tends to counteract and thereby dampen any sway of the building. The inherent structural damping is supplemented by the massive TMD that uses building's motion to push and pull dashpots attached to it. The TMD is tuned to sway freely at about the same period or the same sway rate as that of the building. Fluid viscous dampers are bolted between the building frame and the TMD mass, as seen in Photo 5.7. The damping effect of the sealed dashpots varies with the square of the velocity of the mass. The regular wind-induced sway creates a relatively small resistance force that provides damping, while permitting the mass to swing. In the event of sudden shock, like earthquake, the dashpot resistance will rise intensely and create a lockdown effect that limits the motion of the mass.

- Statue of Unity, India

The Statue of Unity is the world's tallest statue today. The total height of the statue is 240 m, with a base of 58 m and the statue measuring 182 m, as shown in Photo 5.8. It is a colossal statue of Indian statesman and independence activist Sardar Vallabhbhai Patel (1875–1950) who was the first Home minister of India. It is located in Narmada District, in the state of Gujarat, on a river island facing the Sardar Sarovar dam on river Narmada, 100 km SE of the city of Vadodara.

The scale of the project is reflected in the staggering size and weight of its materials. The construction of the statue took 25 lakh cubic feet of concrete, 5,700 tons of steel structure, and 18,500 tons of rebars. There are approximately 12,000 bronze panels covering the structure, weighing around 1,700 tons. Visitors have approach up to the viewing gallery, located near the chest of the statue at a height of 150 m. The statue is designed and built to withstand winds of up to 180 kmph and earthquakes measuring 6.5 on the Richter scale, which are at a depth of 10 km and within a radius of 12 km of the statue. The unique

PHOTO 5.8 Statue of Unity, Gujarat, India.

feature about the statue is that it is aided by the use of two 250-ton TMDs ensuring maximum stability. The TMDs are located at the chest level of the statue, above the observation deck. This provides comfort to the visitors against high wind forces by controlling the sway in high winds.

5.2 TUNED LIQUID DAMPER

A TLD is a passive response control device that operates in a similar manner as the TMD, i.e., by countering the vibration of the primary system. In the case of TLD, the mass is replaced by a liquid, which is usually water, that serves as the mass in motion, and the restoring force is generated by gravity. The structural vibration shakes the TLD and induces the liquid movement inside the container. The turbulence of the liquid flow and the friction between the liquid flow and the container convert the dynamic energy of the fluid flow to heat, thus absorbing structural vibration energy. A TLD has the same elementary principle as a TMD, which is to absorb structural vibration energy. The difference is that all characteristics of auxiliary system mass, damping, and restoring mechanisms of TLD are provided by the liquid. In a typical TLD, a rigid tank containing shallow water is connected rigidly to the structure. The liquid in the tank counteracts the vibration of the structure. The fundamental linear sloshing frequency of the TLD can be tuned to the natural frequency of the structure. This causes large amount of sloshing and wave breaking of the liquid. Such sloshing and wave breaking that happens at the resonant frequencies of the combined TLD-structure system dissipates a significant amount of energy; which helps in reducing the vibration of the parent structure and improving the level of protection the TLD can deliver. Broadly, TLDs are classified into tuned sloshing damper (TSD) and tuned liquid column damper (TLCD). Further, TSDs are subdivided into more subgroups, such as shallow and deep TSD. In addition, TLCDs are further classified into more subgroups, such as liquid column vibration absorbers (LCVA), double TLCD (DTLCD), hybrid TLCD (HTLCD), and pressurized TLCD (PTLCD).

5.2.1 Sloped Bottom Tuned Liquid Damper

Since 1950s, TLDs were used for stabilizing marine vessels against rolling operation in anti-rolling tanks. As the research progressed, the same concept was applied to control wobbling motion of satellite in space in the 1960s. Bauer (1984) proposed the idea of applying TLDs in civil engineering structures with the use of rectangular containers. The performance of liquid tank, however, can be modeled in two approaches. In the first approach, potential flow theory is used to work out the dynamic equations of motion. In the second approach, TLD is modeled as TMD, where the properties of the liquid damper are converted to corresponding mass, stiffness, and damping ratio. Sloshing of water is the motion of free surface of liquid in the container, in which it is placed. For the flat (horizontal) bottom tank shown in Figures 5.31 and 5.32, Laplace equation, based on the linear water wave theory, is used for determining the natural frequency and damping ratio of the sloshing of liquid.

According to the Laplace's theory, gravity waves are propagated linearly on the surface of the homogenous liquid layer. Using Laplace equation, natural frequency (f) and damping ratio $\left(\bar{\xi}\right)$ are given as below.

$$\frac{d^2\phi}{dx^2} + \frac{d^2\phi}{dy^2} = 0 \tag{5.68}$$

$$f = \frac{\varpi}{2\pi} = \frac{1}{2\pi}\sqrt{\frac{\pi g}{L} tanh\left(\frac{\pi h}{L}\right)} \tag{5.69}$$

$$\bar{\xi} = \frac{2\varpi\mu\left(1 + \frac{2h}{B} + S\right)L}{\left(\frac{4\pi h}{\sqrt{gh}}\right)} \tag{5.70}$$

where h is the water depth, g is the gravitational acceleration, L is the length of the tank, B is the width of the tank, $\bar{\omega}$ is the angular frequency of the liquid sloshing, μ is the kinematical viscosity of the liquid, and S is the is the surface contamination factor.

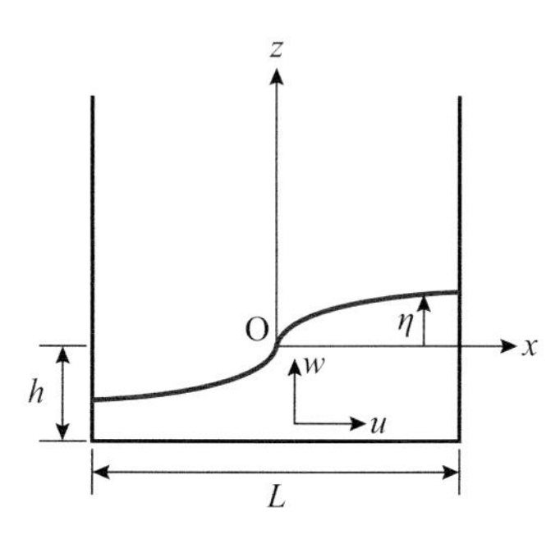

FIGURE 5.31 Sloshing of liquid in flat (horizontal) rectangular tank.

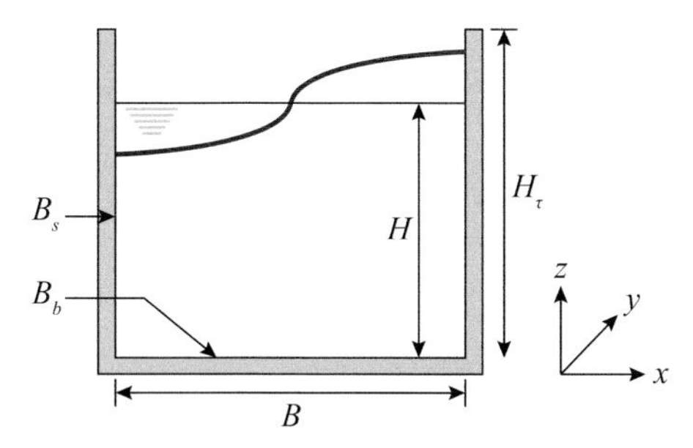

FIGURE 5.32 Tuned liquid damper.

Mathematical Model

The linear wave theory assumes that the liquid inside the tank is irrotational, inviscid, and incompressible; the pressure is constant everywhere on the free surface of the liquid; and tank walls are considered rigid. Therefore, the liquid depth h is motionless and, hence, constant. These assumptions put forward the presence of velocity potential. The velocity potential with the boundary conditions determines the water wave height η, which is a function of the damping ratio $\left(\xi\right)$ and the peak acceleration $\left(\ddot{u}_b\right)$ at the base, given by Equation (5.71).

$$\eta_{\max} = \frac{4L}{\pi^2 g}\left(\frac{\left(\dfrac{w^2}{\varpi^2}\right)}{\left\{1-\left(\dfrac{w^2}{\varpi^2}\right)^2\right\}^2+\left\{2\xi\left(\dfrac{w}{\varpi}\right)\right\}^2 L}+\frac{\pi^2}{8}\right)\ddot{u}_b \tag{5.71}$$

The governing differential equation in terms of pressure variable for the liquid is given as below.

$$\nabla^2 P = 0 \text{ on } V \tag{5.72}$$

where V is the volume of the liquid, $P = P(x,z,t)$ is the liquid dynamic pressure head and

$$\nabla^2 = \frac{d^2}{dx^2}+\frac{d^2}{dz^2} \tag{5.73}$$

Following boundary conditions considered in the mathematical formulations.

i. Pressure is constant everywhere on free surface of liquid.

$$\therefore \frac{d^2P}{dt^2}+g\frac{dP}{dh}=0 \text{ on } B_f \tag{5.74}$$

ii. On rigid solid boundary, the liquid depth h is motionless and constant.

$$\therefore \frac{dP}{dh}=0 \text{ on } B_b \tag{5.75}$$

Divergence theorem is applied to the governing differential equation in order to reduce volume term to surface area. Mathematically, divergence theorem is expressed as follows.

$$\iiint_E \text{div } F_0.dv = \iint_s F_0.ds \tag{5.76}$$

where E is a solid region enclosed by surface s, and F_0 is a continuously differentiable vector oriented outward. Using Equations (5.76), (5.72), and (5.73) can be reduced to Equation (5.77) by minimizing energy terms.

$$\int_V \left(\frac{dN_i}{dx} \sum_1^N \frac{dN_j}{dx} \bar{P}_j \right) dV + \int_V \left(\frac{dN_i}{dz} \sum_1^N \frac{dN_j}{dz} \bar{P}_j \right) dV = \int_B \left(N_i \sum_1^N \frac{dP}{dh} \right) ds \tag{5.77}$$

where N_i and N_j are the shape functions at x and y coordinate at a particular point, respectively. Using boundary conditions and making simplifications, Equation (5.77) can be rewritten as follows.

$$\left[M_f \right] \{ P \} + \left[K_f \right] \{ P \} = \{ F_p \} \tag{5.78}$$

where $\left[M_f \right]$ is the liquid-free surface matrix, $\left[K_f \right]$ is the liquid matrix, and $\{ F_p \}$ is the force vector. The dynamic equation of the structure-TLD relation model is as follows.

$$[M]\{\ddot{u}\} + [C]\{\dot{u}\} + [K]\{u\} = -[M]\{r\}\ddot{u}_g + \{F_{TLD}\} \tag{5.79}$$

where $[M]$ is the combined mass matrix, $[C]$ is the combined damping matrix, $[K]$ is the combined stiffness matrix, $\{F_{TLD}\}$ is the force vector of the structure-TLD model, respectively, and $\ddot{u}_g$ is the ground acceleration. The natural frequency is the function of the wet length of the sloped bottom tank. By coordinate transformation based on the basic shape function shown in Figure 5.33, the potential velocity term in the Laplace equation is changed from the sloped bottom to the horizontal bottom.

From Figure 5.33, it is clear that the flat length is equal to the projection of inclined length to the horizontal axis as, $L' = L/\cos\theta$. The coordinate transformations are explained as below.

$$X = \sum_{i=1}^{4} N_i(r,s) x_i \tag{5.80}$$

$$Y = \sum_{i=1}^{4} N_i(r,s) y_i \tag{5.81}$$

where x_i and y_i are the coordinates of four nodes of the left reference system of Figure 5.33. Hence, the four nodes have the shape functions as given below.

$$N_1 = \left(\frac{s}{2L'h} \right) (2r - L') \tag{5.82a}$$

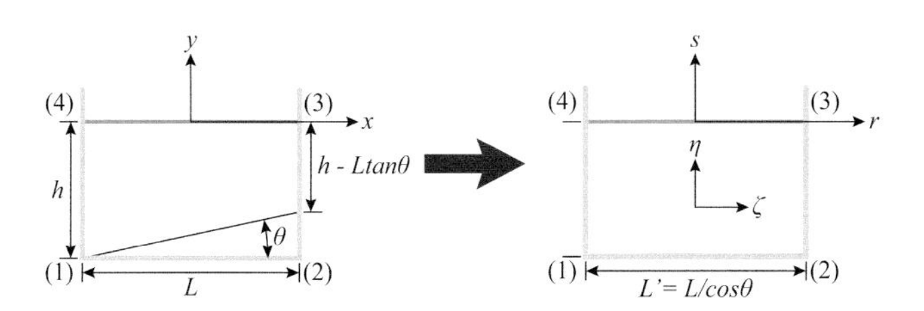

FIGURE 5.33 Coordinate transformation.

$$N_2 = \left(\frac{s}{2L'h}\right)(-L' - 2r) \quad (5.82b)$$

$$N_3 = \left(\frac{1}{2L'h}\right)(L' + 2r)(h + s) \quad (5.82c)$$

$$N_4 = \left(\frac{1}{2L'h}\right)(L' - 2r)(h + s) \quad (5.82d)$$

Differentiating the potential, ϕ, with respect to the coordinates r and s gives the following expression.

$$\frac{d^2\phi}{dx^2} + \frac{d^2\phi}{dy^2} = \frac{d^2\phi}{dr^2}\left[\left(\frac{dr}{dx}\right)^2 + \left(\frac{dr}{dy}\right)^2\right] + \frac{d^2\phi}{ds^2}\left[\left(\frac{ds}{dx}\right)^2 + \left(\frac{ds}{dy}\right)^2\right] + \cdots + \frac{d\phi}{dr}\left[\frac{d^2r}{dx^2} + \frac{d^2r}{dy^2}\right]$$
$$+ \frac{d\phi}{ds}\left[\frac{d^2s}{dx^2} + \frac{d^2s}{dy^2}\right] + \frac{d^2\phi}{dsdr}\left[2\frac{ds}{dx}\frac{dr}{dx} + 2\frac{dr}{dy}\frac{ds}{dy}\right] \quad (5.83)$$

The right-hand side expression of Equation (5.81) is found by differentiation of X and Y about x and y for each one by differentiating shape function as below.

$$\frac{dX}{dx} = \sum_{i=1}^{4} \frac{dN_i}{dx} x_i = \frac{dN_1}{dx} x_1 + \frac{dN_2}{dx} x_2 + \frac{dN_3}{dx} x_3 + \frac{dN_4}{dx} x_4 \quad (5.84a)$$

$$\frac{dX}{dy} = \sum_{i=1}^{4} \frac{dN_i}{dy} x_i = \frac{dN_1}{dy} x_1 + \frac{dN_2}{dy} x_2 + \frac{dN_3}{dy} x_3 + \frac{dN_4}{dy} x_4 \quad (5.84b)$$

$$\frac{dY}{dx} = \sum_{i=1}^{4} \frac{dN_i}{dx} y_i = \frac{dN_1}{dx} y_1 + \frac{dN_2}{dx} y_2 + \frac{dN_3}{dx} y_3 + \frac{dN_4}{dx} y_4 \quad (5.84c)$$

$$\frac{dY}{dy} = \sum_{i=1}^{4} \frac{dN_i}{dy} y_i = \frac{dN_1}{dy} y_1 + \frac{dN_2}{dy} y_2 + \frac{dN_3}{dy} y_3 + \frac{dN_4}{dy} y_4. \quad (5.84d)$$

Replacing the x and y coordinates of the four nodes and taking derivative of the shape functions results in the following relation.

$$\left(\frac{dr}{dx}\right)^2 + \left(\frac{dr}{dy}\right)^2 = \left(\frac{1}{cos\theta}\right)^2, \frac{d^2r}{dx^2} + \frac{d^2r}{dy^2} = 0$$
$$\frac{ds}{dy} = \frac{2h}{(2h - Ltg\theta - 2r\ sin\theta)}, \frac{d^2s}{dy^2} = 0;\ \text{and}\ \frac{ds}{dx} = \frac{d^2s}{dx^2} = 0$$

Hence, the right-hand side expression of Equation (5.83) is given as below.

$$\left(\frac{dr}{dx}\right)^2 + \left(\frac{dr}{dy}\right)^2 = \left(\frac{1}{cos\,\theta}\right)^2, \frac{d^2r}{dx^2} + \frac{d^2r}{dy^2} = 0 \quad (5.85a)$$

$$\left(\frac{dr}{dx}\right)^2 + \left(\frac{dr}{dy}\right)^2 = \left(\frac{2h}{2h - Ltg\theta - 2r\ sin\theta}\right)^2 \tag{5.85b}$$

$$\frac{d^2s}{dx^2} + \frac{d^2s}{dy^2} = 0 \tag{5.85c}$$

$$2\left(\frac{ds}{dx}\right)\left(\frac{dr}{dx}\right) + 2\left(\frac{dr}{dy}\right)\left(\frac{ds}{dy}\right) = 0 \tag{5.85d}$$

By substituting Equations (5.85a)–(5.85d) into Equation (5.83), Laplace equation in the (r,s) coordinate system is as given below.

$$\frac{d^2\phi}{dx^2} + \frac{d^2\phi}{dy^2} = \left(\frac{1}{cos^2\theta}\right)\frac{d^2\phi}{dr^2} + \left(\frac{2h}{2h - Ltg\theta - 2r\ sin\theta}\right)^2 \frac{d^2\phi}{ds^2} = 0 \tag{5.86}$$

For small value of θ, Equation (5.86) can be written as given below.

$$\frac{d^2\phi}{dx^2} + \frac{d^2\phi}{dy^2} = 0 \tag{5.87}$$

Equation (5.87) is the Laplace equation in the new reference system i.e., (r,s) coordinate system. Thus, the natural frequency f and damping ratio $\bar{\xi}$ of water sloshing in the sloped bottom tank is given below.

$$f = \frac{\varpi}{2\pi} = \frac{1}{2\pi}\sqrt{\frac{\pi g}{\left\{\frac{L}{cos\theta}\right\}} tanh\left(\frac{\pi h}{\left\{\frac{L}{cos\theta}\right\}}\right)} \tag{5.88}$$

$$\bar{\xi} = \frac{2\varpi\mu\left(1 + \frac{2h}{\left\{\frac{L}{cos\theta}\right\}} + s\right)L}{\left(\frac{4\pi h}{\sqrt{gh}}\right)} \tag{5.89}$$

Considering an SDOF structure with TLD at its top and ground motion applied to it as shown in Figure 5.34. The governing equation of motion of the structure is as given below.

$$m_s\ddot{u}_s + c_s\dot{u}_s + k_s u_s = -m_s\ddot{u}_g + F \tag{5.90}$$

Where $m_s, c_s,$ and k_s, respectively, are the mass, stiffness, and damping of the structure; $\ddot{u}_s, \dot{u}_s,$ and u, respectively, are the acceleration, velocity, and displacement of the structure with respect to ground; $\ddot{u}_g$ is the ground acceleration; and F represents the shear force at the base of the TLD because of sloshing of water. Normalizing about the structural mass, Equation (5.90) can be given as below.

$$\ddot{u}_s + 2\xi_s\omega_s\dot{u}_s + \omega_s^2 u_s = -\ddot{u}_g + \frac{F}{m_s} \tag{5.91}$$

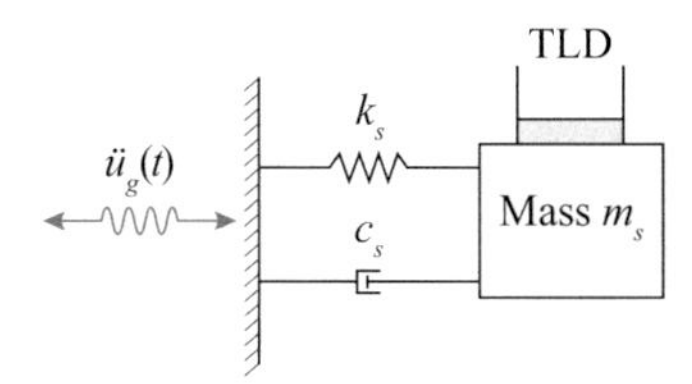

FIGURE 5.34 SDOF structure with rectangular TLD.

where $(\omega_s = 2\pi/T_s)$ and ξ_s are the natural frequency and damping ratio of the structure, respectively, and T_s is the natural time period.

TLDs can be installed on an MDOF system, for the TLD-MDOF structure, e.g., consider Figure 5.35(a) in which three TLDs are attached on the roof of a three-story shear building idealized as a three-DOF lumped mass-spring-dashpot system. In Figure 5.35(b), at all three floors, the primary structure is under external forces that are proportionate to the floor masses. The matrix form of the governing equations of motion for the MDOF structure installed with TLD can be given below.

$$[M_s]\{\ddot{u}_s\}+[C_s]\{\dot{u}_s\}+[K_s]\{u_s\}=-[M_s]\{r\}\ddot{u}_g \tag{5.92}$$

where $[M_s]$, $[C_s]$, and $[K_s]$ are the mass, damping, and stiffness matrices of the building installed with TLDs, respectively; whereas $\{u_s\}=\{u_1,u_2,\ldots,u_N,u_{d1},\ldots,u_{dn}\}^T$, $\{\dot{u}_s\}$, and $\{\ddot{u}_s\}$ are the unknown relative floor displacement, velocity, and acceleration vectors, respectively. Further, $\{r\} = \{1, 1, \ldots, 1\}^T$, and the matrices $[M_s]$, $[C_s]$, and $[K_s]$ are formulated as follows.

$$[M_s]=\begin{bmatrix} m_{s1} & 0 & 0 & 0 & 0 & 0 \\ 0 & m_{s2} & 0 & 0 & 0 & 0 \\ 0 & 0 & m_{s3} & 0 & 0 & 0 \\ 0 & 0 & 0 & m_{d1} & 0 & 0 \\ 0 & 0 & 0 & 0 & m_{d2} & 0 \\ 0 & 0 & 0 & 0 & 0 & m_{d3} \end{bmatrix} \tag{5.93}$$

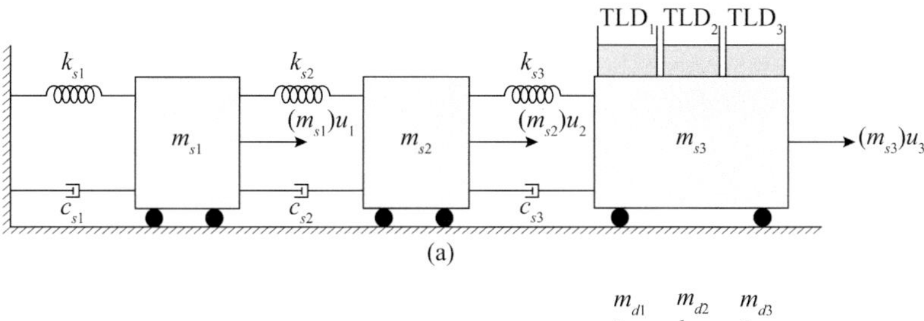

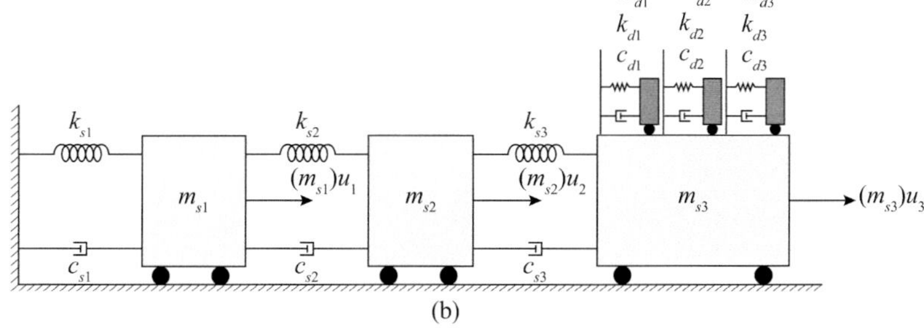

FIGURE 5.35 3-DOF structure with: (a) three TLDs and (b) three equivalent TMDs.

$$[K_s]=\begin{bmatrix} k_{s1}+k_{s2} & -k_{s2} & 0 & 0 & 0 & 0 \\ -k_{s2} & k_{s2}+k_{s3} & -k_{s3} & 0 & 0 & 0 \\ 0 & -k_{s3} & k_{s3}+k_{d1}+k_{d2}+k_{d3} & -k_{d1} & -k_{d2} & -k_{d3} \\ 0 & 0 & -k_{d1} & k_{d1} & 0 & 0 \\ 0 & 0 & -k_{d2} & 0 & k_{d2} & 0 \\ 0 & 0 & -k_{d3} & 0 & 0 & k_{d3} \end{bmatrix} \tag{5.94}$$

$$[C_s]=\begin{bmatrix} c_{s1}+c_{s2} & -c_{s2} & 0 & 0 & 0 & 0 \\ -c_{s2} & c_{s2}+c_{s3} & -c_{s3} & 0 & 0 & 0 \\ 0 & -c_{s3} & c_{s3}+c_{d1}+c_{d2}+c_{d3} & -c_{d1} & -c_{d2} & -c_{d3} \\ 0 & 0 & -c_{d1} & c_{d1} & 0 & 0 \\ 0 & 0 & -c_{d2} & 0 & c_{d2} & 0 \\ 0 & 0 & -c_{d3} & 0 & 0 & c_{d3} \end{bmatrix} \tag{5.95}$$

5.2.2 Tuned Liquid Column Damper

U-shaped TLCDs are one of the most popular TLDs. Other than providing liquid in water tank, it is possible to induce oscillations in a liquid column. The energy dissipation is controlled through orifices provided at the base of the tank as shown in Figure 5.36.

Mathematical Modeling of TLCD

When the structural system experiences vibration, the same horizontal translational motion is experienced by the liquid column tube shown in Figure 5.37. Also, a relative motion is experienced by the column with respect to the tube. With reference to Figure 5.36, consider ρ = the density of the liquid column, L = the overall length of the liquid column, B = the horizontal length of the liquid column, A_1 = the horizontal cross-sectional area of the liquid column, and A = vertical cross-sectional area of the liquid column. By applying continuity equation at any section, the liquid velocity and liquid acceleration are given by as follows.

$$A\dot{w} = A_1\dot{w}_1 \Rightarrow \dot{w}_1 = \alpha\dot{w} \tag{5.96}$$

$$A\ddot{w} = A_1\ddot{w}_1 \Rightarrow \ddot{w}_1 = \alpha\ddot{w} \tag{5.97}$$

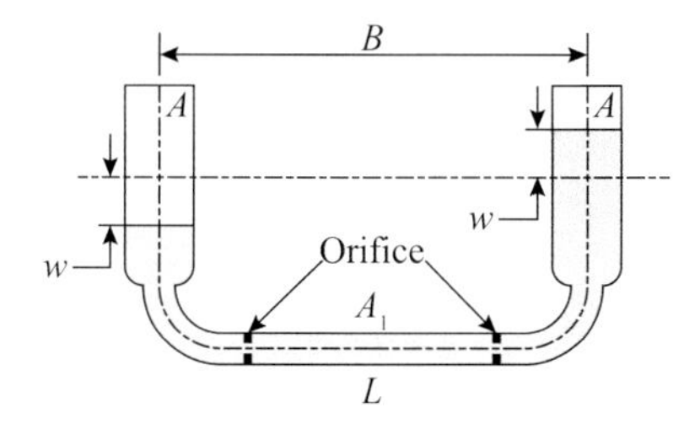

FIGURE 5.36 U-shaped tuned liquid column damper.

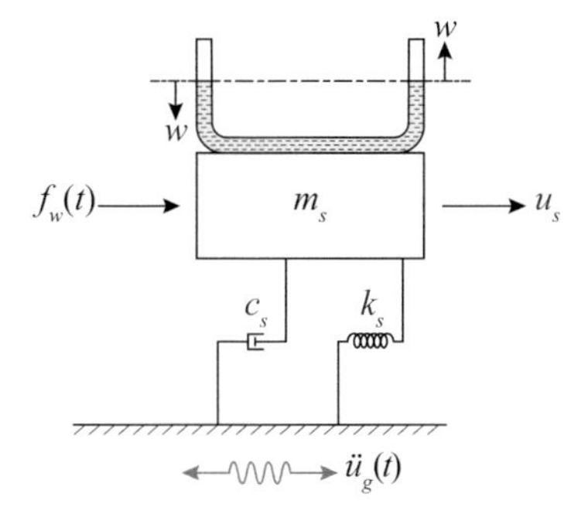

FIGURE 5.37 Simplified TLD-Structure system.

where $(\alpha = A/A_1)$ is the area ratio. Assuming that the liquid is incompressible, and the TLCD pipe is kept open, the total kinetic energy E_k and total potential energy E_p of the U-shaped tube shown in Figure 5.36 are as given below.

$$\begin{aligned}
E_k &= \frac{1}{2}mv^2 = \frac{1}{2}\rho A\left(\frac{L-B}{2} - w\right)\dot{w}^2 + \frac{1}{2}\rho A_1 B\dot{w}_1^2 + \frac{1}{2}\rho A\left(\frac{L-B}{2} + w\right)\dot{w}^2 \\
E_k &= \frac{1}{2}\rho A\dot{w}^2\left(\frac{L-B}{2} - w + \frac{L-B}{2} + w\right) + \frac{1}{2}\rho A_1 B\dot{w}_1^2 = \frac{1}{2}\rho A\dot{w}^2(L-B) + \frac{1}{2}\rho A_1 B\dot{w}_1^2 \\
E_k &= \frac{1}{2}\rho A\dot{w}^2(L-B) + \frac{1}{2}\rho A_1 B\left(\frac{A^2\dot{w}^2}{A_1^2}\right) = \frac{1}{2}\rho A\dot{w}^2\left[(L-B) + AB/A_1\right] \\
E_k &= \frac{1}{2}\rho A L_{eff}\dot{w}^2
\end{aligned} \tag{5.98}$$

$$L_{eff} = (L-B) + \alpha B \tag{5.99}$$

$$E_p = mgh = \rho(Aw)gw = \rho g A w^2 \tag{5.100}$$

where m is the mass of the liquid, v is the velocity of liquid, g is the acceleration due to gravity, and h is the height L_{eff} is the length of an equivalent uniform cross-sectional area of liquid column having area A, which has the same energy as that of TLCD. Inertial and damping components combinedly act as external force; where inertia force causing the relative motion is given as follows.

$$F_i = -\rho A_1 B\ddot{v}_x \tag{5.101}$$

where $\ddot{v}_x$ is the absolute horizontal acceleration of the structure. The damping force of liquid motion is given as:

$$F_d = -\frac{1}{2}\rho A\xi|\dot{w}|\dot{w} \tag{5.102}$$

where ξ is the damping coefficient, which is a function of area ratio and opening ratio of orifice. Using the Lagrange equation:

$$\frac{d}{dt}\left(\frac{\partial T}{\partial \dot{q}_i}\right) - \frac{\partial T}{\partial q_i} + \frac{\partial V}{\partial q_i} = Q_i \tag{5.103}$$

where T is the total kinetic energy, and V is the total potential energy for the system, respectively. Q_i is the nonconservative force in the direction of q_i.

$$\therefore T = \sum_{i=1}^{n} \frac{1}{2} m_i \dot{x}_i^2 \text{ and } V = \sum_{i=1}^{n} m_i g x_i .$$

Thus, the Lagrange equation on the liquid system is written as:

$$\frac{d}{dt}\left(\frac{\partial E_k}{\partial \dot{w}}\right) - \frac{\partial E_k}{\partial w} + \frac{\partial E_p}{\partial w} = F_i + F_d \tag{5.104}$$

Substituting the values from Equations (5.100) to (5.103) into Equation (5.104) gives Equation (5.105).

$$\frac{d}{dt}\left(\frac{\partial\left(\frac{1}{2}\rho A L_{eff}\dot{w}^2\right)}{\partial \dot{w}}\right) - \frac{\partial\left(\frac{1}{2}\rho A L_{eff}\dot{w}^2\right)}{\partial w} + \frac{\partial(\rho g A w)^2}{\partial w}$$
$$= -\rho A_1 B \ddot{v}_x - \frac{1}{2}\rho A \xi |\dot{w}| \dot{w}. \tag{5.105}$$

Solution of Equation (5.105) leads to:

$$\rho A L_{eff}\ddot{w} + \frac{\rho A}{2}\xi |\dot{w}| \dot{w} + 2\rho g A w = -\rho A B \ddot{v}_x \tag{5.106}$$

$$\because |w| \le \frac{L-B}{2}.$$

Thus, the natural frequency ω_L obtained for the above relative motion for the liquid is:

$$\omega_L = \sqrt{\frac{2g}{L_{eff}}}. \tag{5.107}$$

Consider an SDOF structure and TLCD system. As the structure experiences vibration, motion is induced in liquid column. As a result, motion of liquid exerts a reaction on the structure. The governing equation of motion of the structure is given as below.

$$m_s \ddot{v}_s + c_s \dot{v}_s + k_s v_s = P - \rho A B \ddot{w} - m_d \ddot{v}_s \tag{5.108}$$

where m_s, c_s, k_s and v_s are the mass, damping, stiffness, and translational displacement of the structure. P is the external reaction exerted by liquid motion on the structure. Mass of the damper m_d can be expressed as follows.

$$m_d = \rho A\left[\frac{B}{\alpha} + (L-B)\right] = \rho A L_{em} \tag{5.109}$$

where L_{em} is the length of an equivalent uniform cross-sectional area of liquid column having area A which has the same mass as that of TLCD.

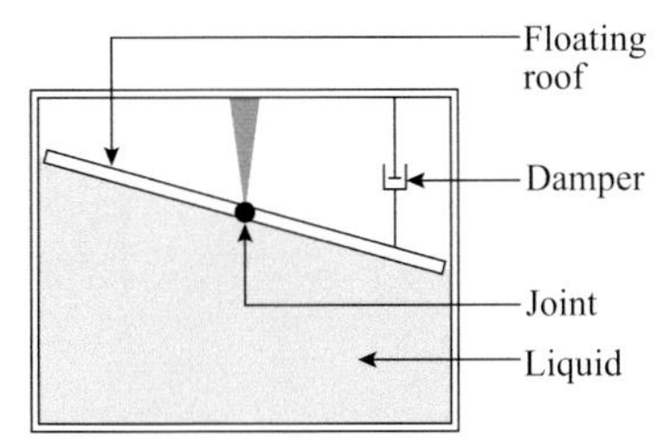

FIGURE 5.38 Rectangular TLD with floating roof.

5.2.3 Tuned Liquid Damper with Floats

Tuned liquid damper with floating roof (TLD-FR) generally consists of a traditional TLD incorporated with a floating roof. The float (floating material) is lightweight having very low density. The incorporation of floating roof reduces the additional self-weight and facilitates floating over the water surface. As the floating material is much stiffer than water, the goal of preventing wave breaking is well achieved by this modification to the TLD. The modification is capable of making the response linear even at large amplitudes. Another advantage of incorporating floating roof is that it makes possible to provide supplemental devices in order to carry out the dissipation of energy effectively. It is seen that TLD with floating roof possesses the characteristics of both TLD and TLCD systems. The arrangement of TLD-FR is shown in Figure 5.38.

Mathematical Model of Tuned Liquid Damper with Floats

Mathematical model for the TLD-FR can be developed by modeling the liquid and the floating roof separately, as shown in Figure 5.31, which shows sloshing of liquid in flat (horizontal) rectangular tank, and Figure 5.39, which shows the floating roof and combining them by the internal pressure generated at the liquid roof surface.

The liquid is modeled as described by Equation (5.83), which gives the sloshing of liquid, and Equation (5.84), which gives the dynamic equation of the structure-TLD relation model. Equation (5.83) is generated by considering the boundary condition that at the free surface of liquid, pressure is constant everywhere, i.e., pressure is zero, which is not applicable in the case of TLD-FR, wherein, the floating roof generates a nonzero pressure at the free surface of liquid. Hence, the pressure in liquid is introduced as follows.

$$M_f P + K_f P = F_p - \frac{P}{\rho} \tag{5.110}$$

The modeling of floating roof is done by considering it a flexible Euler-Bernoulli beam through a finite element methodology (FEM). It is considered a 2-D element comprising two nodes and four DOFs. Thus, the stiffness matrix of beam element is proportional to (EI/L^3); where E is the modulus of elasticity of beam material, I is the moment of inertia, and L is the length of the beam element. The elemental mass matrix is developed by employing a lumped-mass approach. A mesh is employed between liquid and floating roof in order to simplify the combined numerical models between liquid and floating roof shown in Figure 5.40.

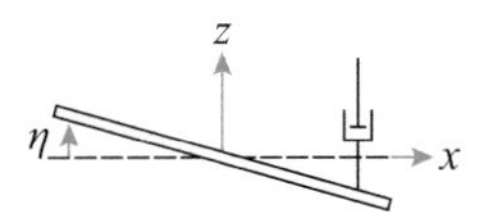

FIGURE 5.39 Modeling of floating roof.

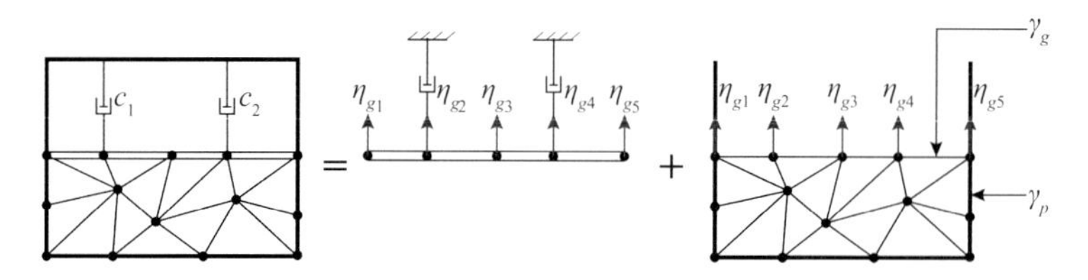

FIGURE 5.40 Meshing of liquid and floating roof for derivation of damping matrix.

Thus, assuming linear fluid viscous dampers governing equation of motion of the floating roof is as follows.

$$[M]\{\ddot{u}\}+[C]\{\dot{u}\}+[K]\{u\}=\{F\} \tag{5.111}$$

where $[M]$ is the mass matrix, $[C]$ is the damping matrix, $[K]$ is the stiffness matrix of floating roof, and $\{u\}$ is the vertical displacement of nodes of the floating roof. The mass matrix depends on the cross-section of the floating roof and stiffness matrix is directly proportional to EI and inversely proportional to the length L of the beam element. The damping matrix is formulated by considering external dampers connected to the roof, as shown in Figure 5.40. Considering R_d as the collocation matrix with the ith column relating DOF of the ith damper to the displacement vector $\{u\}$, the damping matrix is formulated as follows.

$$[C]=\{R_d\}^T[C_d]\{R_d\}=\{R_d\}^T\begin{bmatrix} C_{damper1} & 0 \\ 0 & C_{damperN} \end{bmatrix}\{R_d\} \tag{5.112}$$

where $[C_d]$ is the diagonal matrix comprising damping coefficients of each damper. Equations (5.111) and (5.112) constitute the final governing equations of motion. The TSDs primarily utilize circular containers for shallow configurations and rectangular ones for deep water TSDs, while the TLCDs rely on the traditional U-shaped vessel. Such applications work best for buildings with small vibrations and have been observed to reduce the structural response to 1/2 to 1/3 the original response in strong winds.

5.2.4 Merits of Tuned Liquid Damper

TLDs are effective against the strong motion of earthquakes and winds. The damper uses preexisting tanks for its damping function, thus results in use of less space. The system is inexpensive and easy to maintain. There is no mechanical friction in the system; hence, it is effective for even the slightest vibrations. TLDs are efficient for all amplitude levels, as it is easy to adjust the time period; the frequency can be controlled by adjusting height of liquid; and they are cheaper to incorporate in an existing structural design. The advantage of using TLCD system is that the containers are used for supplying building water, unlike a TMD where the dead weight is of no other use. TMDs require frictionless rubber bearings, special floor for installation, etc. for dissipating energy efficiently. TLDs on other hand are low-cost inertial devices having a superior performance as compared to TMDs. Failure of the system is virtually impossible.

5.2.5 Demerits of Tuned Liquid Damper

The main disadvantage of the majority of TLDs is that they provide only one main oscillation damped frequency, which is not adequate for the structure, as different vibration modes are required to be considered simultaneously. All the water mass does not participate in counteracting the structural motion. This results in extra cost in terms of added weight to the structure without the benefit of commensurate response control.

5.2.6 Worldwide Applications of Tuned Liquid Damper

TLD, which works on the similar principle as TMD, is widely employed for alleviating wind-induced vibrations of tall structures. Further, investigations have confirmed the effectiveness of TLDs in reducing the vibrations caused due to ground shaking. TLDs, encompassing both TSDs and TLCDs have become a popular form of inertial damping device since their first applications to ground structures in the 1980s. Low initial cost coupled with their low maintenance requirements has guaranteed their wide use. In particular, the TSDs are extremely practical, currently being proposed for existing water tanks on the building by configuring internal partitions into multiple dampers without adversely affecting the functional use of the water supply tanks. Considering only a small additional mass to the building, these systems can reduce acceleration response to 35–50% of the original response, depending on the amount of liquid mass. Currently, both deep and shallow water configurations of TSDs, which exploit the amplitude of fluid motion and wave-breaking patterns to provide additional damping, are in application worldwide. The shallow water configurations dissipate energy through the viscous action and wave breaking. The addition of floaters may also add to the dissipation of sloshing energy. Deep water TSDs require baffles or screens to increase the energy dissipation of the sloshing fluid. While the natural frequency of a TLD may be simply adjusted by the depth of water, and the dimension of the container, there are practical limitations on the water depth and on the frequency, which may be obtained by a given container design. Some of the notable applications of TLDs, installed in structures around the world are mentioned below.

- Hobart Tower, Tasmania, Australia
  Hobart Tower in Tasmania, Australia is 105 m tall. The tower was covered in a protective cylindrical shell for shielding the transmission antenna from the harsh conditions. However, the shell increased the wind-induced response of the tower. This demanded the installation of vibration control devices. The tower was then equipped with 80 TSDs.
- Gold Tower, Kagawa, Japan
  A total of 16 TSDs were installed in the top floor of the 158-m Gold Tower in Kagawa, Japan. The installation of 10 tons of TSDs was found to reduce the response to 35–50% of the original response. The tank, in the form of a cube, is filled with water and equipped with steel wire nets to dissipate the motion of the liquid. By adjusting these damping nets, the length of the tank, and the depth of water, the device is appropriately tuned.
- Nagasaki Airport Tower, Nagasaki, Japan
  TLD originally found its application in 1987, in the Nagasaki Airport Tower, Nagasaki, Japan. A multilayer configuration of 25 TSDs, weighing 950 kg was proposed for the 42-m tall tower. A total of 12 cylindrical, multilayered vessels of vinyl chloride of dimensions of 0.5-m height and 0.38-m diameter were installed on the air traffic control room floor. The remaining 13 devices were distributed on each stair landing. Each vessel was divided into seven 70-mm thick layers, each containing 48 mm of water of mass 38 kg. Tests conducted to calculate the frequency and damping ratio of the tower revealed that the displacement due to the across-wind component was more than the along-wind component. Further, it was noticed that beat phenomena was present, which was eliminated through the use of floating particles that helped to dampen the liquid motion in the containers. The installed TSDs have been found to improve the response of the tower considerably, even at higher wind velocities.

PHOTO 5.9 Yokohama Marine Tower, Japan.

- Yokohama Marine Tower, Yokohama, Japan
  In the same year, 1987, TLDs were installed in the Yokohama Marine Tower, Yokohama, Japan. The tower, shown in Photo 5.9, is a 106-m high lattice light house, with an observation deck at a height of 100 m.
- Shin Yokohama Prince Hotel (SYP), Yokohama, Japan
  A TSD configuration of multilayer stacks of nine circular fiber-reinforced plastic containers of 2-m diameter, each 0.22-m high was installed in 1991, in the 149.4-m tall skyscraper, Shin Yokohama Prince Hotel (SYP) in Yokohama, Japan, shown in Photo 5.10. Each layer of the TSD was equipped with 12 overhangs installed in a symmetric radial pattern to prevent the swirling motion of the liquid and to get adequate additional damping. From observations of the performance of this installation, the hotel has been shown to successfully meet minimum perception levels of RMS acceleration, with RMS response reductions of 30–50% in 20 m/s winds.
- Tokyo Airport Tower, Tokyo, Japan
  In 1993, TLD system was installed in the 77.6-m tall Tokyo Airport Tower, Tokyo, Japan. The tower consists of 1,400 vessels of water, and floating materials. Each of the 1,400 vessels is a shallow circular cylinder of diameter 0.6 m and height 0.125 m as shown in Figure 5.41. These vessels are stacked in layers and have injection taps and handles to serve as projections and four conical dents on the upside and base. These projections and dents provide additional stiffness for stacking the polyethylene vessels. The sloshing frequency of TLD is optimized at 0.743 Hz and the total mass of the system is about 3.5% of the first generalized mode of the tower. The TLD system reduced the dynamic response of the tower structure to 60% during high-intensity winds, as compared to the tower without TLD.

PHOTO 5.10 Shin Yokohama Prince Hotel, Yokohama, Japan.

- Shanghai World Financial Center
  Shanghai World Financial Center shown in Photo 5.11 is a supertall skyscraper, equipped with eight TSD units at its 91st floor. Each tank is 5.5 m in diameter, separated into six layers. The installation of the 800-ton TSD system (1% mass ratio) is anticipated to successfully reduce the story drift and peak and RMS accelerations to acceptable limits. The tower's trapezoid aperture is made up of structural steel and RC.

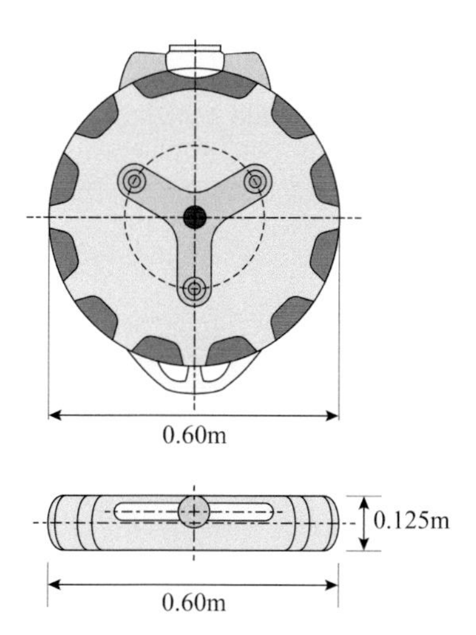

FIGURE 5.41 TLD vessel in Tokyo Airport Tower, Japan.

PHOTO 5.11 Shanghai World Financial Center, china.

EXAMPLES

Example 5.1

An SDOF structure, which has a mass (m_1) of 126,700.00 kg and a lateral stiffness (k_1) of 20,000 kN/m, and damping ratio (ξ_s) of 2%, is subjected to the NS component of the 1995 Kobe earthquake ground motion. Determine the floor acceleration and floor displacement response quantities of: (a) the uncontrolled SDOF structure and (b) the structure equipped with TMD. The damping ratio (ξ_d), mass ratio (μ_d), and tuning frequency ratio (f) of the TMD are given as 0.1, 0.02, and 1, respectively. The idealized model of the TMD-controlled SDOF structure is shown in Figure 5.42, whereas the time history of the 1995 Kobe earthquake ground motion is presented in Figure 5.43.

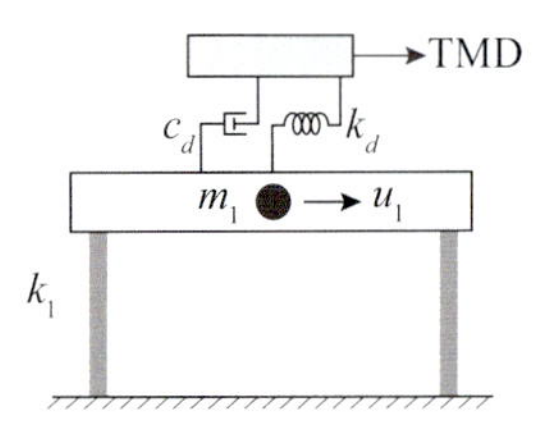

FIGURE 5.42 Idealized model of single degree of freedom (SDOF) structure equipped with tuned mass damper (TMD).

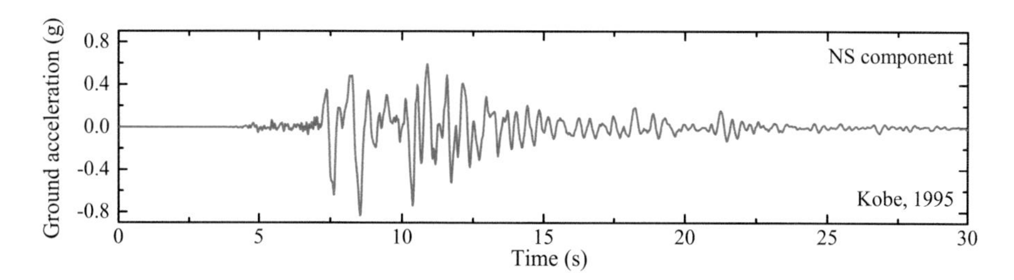

FIGURE 5.43 Time history of the 1995 Kobe earthquake ground motion used in this example.

SOLUTION

The time period of the SDOF structure is computed as:

$$T = 2\pi\sqrt{\frac{m_1}{k_1}} = 0.5 \text{ s}$$

Angular frequency of structure:

$$\omega = \frac{2\pi}{T} = 12.5661 \text{ rad/s}$$

Damping constant of the structure (c_1):

$$c_1 = 2\xi_s m_1 \omega = 63,663.2 \text{ N-s/m}$$

Mass of TMD (m_d):

$$\mu = \frac{m_d}{m_1}$$

$$m_d = \mu m_1 = 2,533.13 \text{ kg}$$

Angular frequency of TMD (ω_d):

$$f = \frac{\omega_d}{\omega}$$

$$\omega_d = f\omega = 1 \times 12.5661 \text{ rad/s} = 12.5661 \text{ rad/s}$$

The stiffness of the TMD (k_d):

$$k_d = m_d \omega_d^2 = 400 \text{ kN/m}$$

Damping constant of the TMD (c_d):

$$c_d = 2\xi_d m_d \omega_d = 6,366.32 \text{ N-s/m}$$

The matrix form of the governing equations of motion of the TMD-controlled structure is given as follows.

$$\begin{bmatrix} m_1 & 0 \\ 0 & m_d \end{bmatrix} \begin{Bmatrix} \ddot{u}_1 \\ \ddot{u}_d \end{Bmatrix} + \begin{bmatrix} c_1 + c_d & -c_d \\ -c_d & c_d \end{bmatrix} \begin{Bmatrix} \dot{u}_1 \\ \dot{u}_d \end{Bmatrix} + \begin{bmatrix} k_1 + k_d & -k_d \\ -k_d & k_d \end{bmatrix} \begin{Bmatrix} u_1 \\ u_d \end{Bmatrix} = -\begin{bmatrix} m_1 & 0 \\ 0 & m_d \end{bmatrix} \begin{Bmatrix} 1 \\ 1 \end{Bmatrix} \ddot{u}_g$$

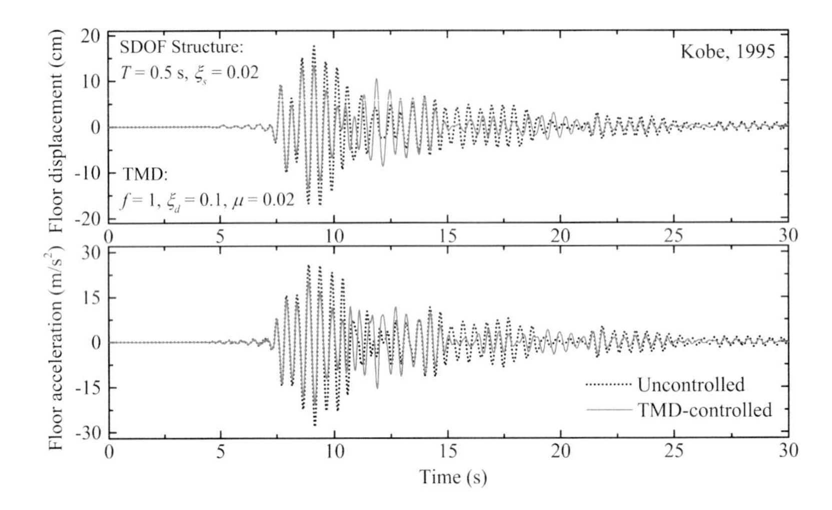

FIGURE 5.44 Time histories of floor displacement and floor acceleration of an SDOF structure with and without TMD under the 1995 Kobe earthquake ground motion.

The equations of motion are solved using a MATLAB® program considering the 1995 Kobe ground motion data. The comparison of the time histories of the floor displacement and the floor acceleration response quantities for the uncontrolled and TMD-controlled structures is presented in Figure 5.44. Furthermore, the comparison of the peak values of the response quantities is presented in Table 5.1. The comparisons show that the response quantities of the TMD-controlled structure are less than that of the uncontrolled structure.

Example 5.2

A four-story building (MDOF structure) shown in Figure 5.45 is subjected to the NS component of the 1995 Kobe earthquake ground motion (Figure 5.43). All the floors of the building have the same weight of 100 kN, whereas all the stories have lateral stiffness of 20,850 kN/m. Furthermore, the damping ratio of the structure (ξ_s) is given as 2%. Determine the top floor acceleration and top floor displacement response quantities of: (a) the uncontrolled building frame and (b) the TMD-controlled building under the earthquake excitation. The damping ratio (ξ_d), mass ratio (μ_d), and tuning frequency ratio (f) of the TMD are given as 0.1, 0.01, and 1, respectively.

SOLUTION

The properties of the TMD are obtained in a similar way as presented in Example 5.1, and the response quantities of the four-story building are obtained using a MATLAB program considering the 1995 Kobe ground motion data. The comparison of the time histories of the top floor displacement and the top floor acceleration response quantities for the uncontrolled and TMD-controlled building frames is presented in Figure 5.46. Furthermore, the comparison of the peak values of the response quantities is presented in Table 5.2. The comparisons show that the response quantities of the TMD-controlled four-story building are less than that of the uncontrolled building.

TABLE 5.1
Comparison of the Response Quantities of the Single-Story Structure with and without TMD

	Value of Peak Response and Percentage Reduction		
Response	**Uncontrolled**	**TMD-Controlled**	**Percentage Reduction (%)**
Displacement (cm)	17.56	13.59	22.60
Floor Acceleration (m/s²)	28.09	20.79	26.00

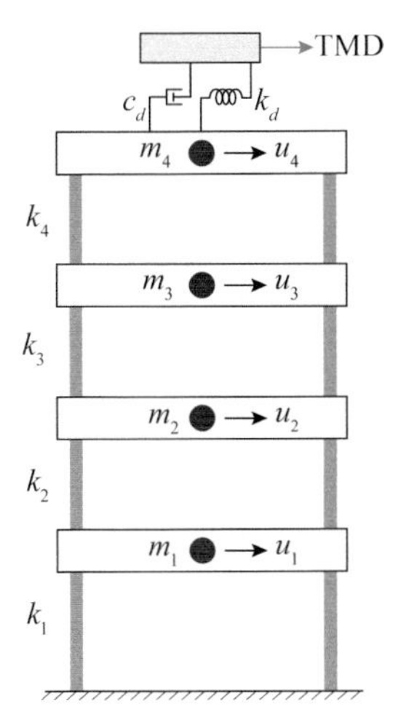

FIGURE 5.45 Idealized model of the four-story building equipped with tuned mass damper.

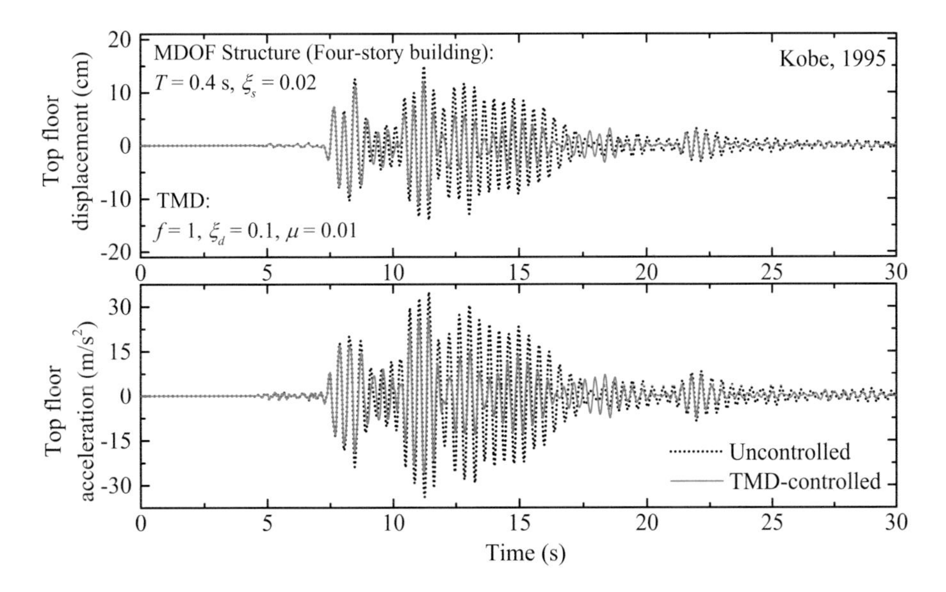

FIGURE 5.46 Time histories of floor displacement and floor acceleration of a four-story building with and without TMD under the 1995 Kobe earthquake ground motion.

TABLE 5.2
Comparison of the Response Quantities of the Four-Story Structure with and without TMD

	Value of Peak Response and Percentage Reduction		
Response	**Uncontrolled**	**TMD-Controlled**	**Percentage Reduction (%)**
Displacement (cm)	14.95	12.00	19.71
Floor acceleration (m/s^2)	34.89	26.34	24.49

MATLAB PROGRAM

MATLAB PROGRAM 5.1: MATLAB CODE FOR TIME HISTORY ANALYSIS OF SDOF AND MDOF STRUCTURES WITH AND WITHOUT TUNED MASS DAMPER (TMD)

Sr. No. **Program Code**

```
clc; clear; close all;
display ('Time History Analysis of')
display ('SDOF and MDOF Structure/
Building')
display ('With and Without Tuned Mass
Damper (TMD)')
display ('Using the State-Space
Method')
display ('_________________________')
%%
tic; % Start timing the execution

% "tsub" : Used to implement further
refined smaller time step for better
% accuracy especially for nonlinear
behaviour.
tsub=2; % Number of subdivision for the
time step

% Reading Input Data From Files
Input=textread('SuperstructureData.
txt'); % Superstructure Data
% Input=textread('SuperstructureDataS
DOF.txt');        % Superstructure Data
% InputTMD=textread('TMD_Data.txt');
% TMD Data
InputGM=textread(' GroundMotionData.
txt'); % Import Ground Motion Data

TMDfr=1;          % TMD frequency ratio
                    (wtmd/ws1)
TMDmr=0.01;       % TMD mass ratio (mtmd/
                    mtot)
zid=0.1;          % TMD damping ratio

%   Ground Motion
dt=InputGM(1);
nt=InputGM(2);
Uddg=zeros(nt,1);
for i=1:nt
    Uddg(i,1)=InputGM(2+i)*9.81;
end

%   Calculation of ground motion data
at further refineed time step
Uddgsub=zeros(nt*tsub,1);
for i=1:nt
    for j=1:tsub
    if i==1
    Uddgsub((i-1)*tsub+j)=(Uddg(i))/tsub;
    else
```

Comments

Variables:
Defined in the program wherever they are used.
Comments:
Input Files 1:
SuperstructureData.txt
In this input file, the following data are required to be organized in a single column adding one data per row.

1. Number of degrees of freedom,
2. Damping ratio of structure,
3. Weights of each floor in kN (One value per row), and
4. Lateral stiffnesses of each story in kN/m (One value per row).

For example, the following data was organized in the input text file "SuperstructureData.txt" for the problem presented in Example 5.2.

```
4
0.02
100
100
100
100
20850
20850
20850
20850
```

```
    Uddgsub((i-1)*tsub+j)=Uddgsub((i-
1)*tsub+j-1)+(Uddg(i)-Uddg(i-1))/tsub;
    end
  end
end
len=nt*tsub;
dt=dt/tsub;

%% Structure / Mass, Stiffness and
Damping Matrices
n=Input(1);              % n      = number
of stories
zi=Input(2);             % zi     =
superstructure damping ratio
% Prelocating the matrices for proper/
correct results as well as speed
ms=zeros(n,n);
ki=zeros(n,1);
ks=zeros(n,n);
m=zeros(n+1,n+1);
m1=0;
for i=1:n
    ms(i,i)=Input(2+i)/9.81*1000;
    ki(i)=Input(2+n+i)*1000;
    m1=m1+ms(i,i);
    m(i,i)=ms(i,i);
end
mtot=m1;    % mtot = total mass of
structure
for i=1:n-1
    ks(i,i)=ki(i)+ki(i+1);
end
ks(n,n)=ki(n);
for i=1:n-1
    ks(i,i+1)=-ki(i+1);
    ks(i+1,i)=-ki(i+1);
end

% Determination of natural frequencies
% Solve the eigen value problem
A=ms\ks;
[V,D]=eig(A);
T=zeros(n,1);
for i=1:n
    T(i)=1/(1/2/3.141592654*(D(i,i))
^0.5);
end
T;
% Fundamental frequency of building
fr(1)=abs(max(T));
fr(2)=abs(min(T));
T1=max(fr);
ws1=2*pi/T1;
ModeS=V;
```

Input Files 2:
GroundMotionData.txt
In this input file, the following data are required to be organized in a single column adding one data per row.

1. The time step of the data,
2. The total number of data points of the ground motion data, and
3. The values of the ground motion acceleration in g unit arranged one value per row.

```
% Calculate damping matrix
if n==1
cs(1,1)=2*zi*mtot*ws1;
else
AA1=0.5*[1/(D(1,1)^0.5) (D(1,1)^0.5);1/
(D(2,2)^0.5) (D(2,2)^0.5)];
AA2=[zi;zi];
AA3=AA1\AA2;
ao=AA3(1);
a1=AA3(2);
cs=ao*ms+a1*ks;
end

%%  TMD Properties
    wtmd=ws1*TMDfr;
    md=TMDmr*mtot;
    kd=md*(wtmd^2);
    cd=2*zid*md*wtmd;
%%  Mass, Stiffness and Damping
Matrices for the Structure With the TMD
m=zeros(n+1,n+1);
m(1:n,1:n)=ms;  m(n+1,n+1)=md;
k=zeros(n+1,n+1);
k(1:n,1:n)=ks;
k(n+1,n+1)=kd;
k(n,n)=ks(n,n)+kd;
k(n,n+1)=0-kd;
k(n+1,n)=0-kd;
    % Damping Matrix for Cd=Cdmax Case
c=zeros(n+1,n+1);
c(1:n,1:n)=cs;
c(n+1,n+1)=cd;
c(n,n)=cs(n,n)+cd;
c(n,n+1)=0-cd;
c(n+1,n)=0-cd;
    % Damping Matrix for Cd=Cdmin Case
% csa=zeros(n+1,n+1);
% csa(1:n,1:n)=cs;
% csa(n+1,n+1)=cd*0;
% csa(n,n)=cs(n,n)+cd*0;
% csa(n,n+1)=0-cd*0;
% csa(n+1,n)=0-cd*0;

m;
k;
c;
% csa;

%%  State Matrices for the Structure
With the TMD
AssAll=zeros(2*(n+1),2*(n+1));
Ass1all=zeros(n+1,n+1);
Ass2all=zeros(n+1,n+1);
Ass3all=zeros(n+1,n+1);
Ass4all=zeros(n+1,n+1);
```

```
BssAll=zeros(2*(n+1),1);
CssAll=zeros(3*(n+1),2*(n+1));
Css1all=zeros(n+1,n+1);
Css2all=zeros(n+1,n+1);
CssUall=zeros(n+1,n+1);
DssAll=zeros(3*(n+1),1);
RssAll=zeros(n+1,1);
for i=1:n+1
    Ass2all(i,i)=1;
    BssAll((n+1)+i,1)=-1;
    DssAll(2*(n+1)+i,1)=-1;
    CssUall(i,i)=1;
    RssAll(i,1)=1;
end
Ass3all=0-m\k;
Ass4all=0-m\c;
for i=1:(n+1)
    for j=1:(n+1)
         AssAll(i,(n+1)+j)=Ass2all(i,j);
         AssAll((n+1)+i,j)=Ass3all(i,j);
         AssAll((n+1)+i,(n+1)+j)=Ass4all
         (i,j);
    end
end
CssAll((n+1)+1:3*(n+1),1:2*(n+1))=Ass
All;
CssAll(1:(n+1),1:(n+1))=CssUall;
AssAll;
BssAll;
CssAll;
DssAll;

%%  State Matrices?
    %   Uncontrolled
Ass=zeros(2*n,2*n);
Ass1=zeros(n,n);    Ass2=zeros(n,n);
Ass3=zeros(n,n);    Ass4=zeros(n,n);
Bss=zeros(2*n,n);
Bss1=zeros(n,n);    Bss2=zeros(n,n);
BssU=zeros(n,n);
Css=zeros(3*n,2*n);
Css1=zeros(n,n);    Css2=zeros(n,n);
CssU=zeros(n,n);
Dss=zeros(3*n,n);
Rss=zeros(n,1);
for i=1:n
    Ass2(i,i)=1;
    BssU(i,i)=1;
    CssU(i,i)=1;
    Rss(i,1)=1;
end
Ass3=0-ms\ks;
Ass4=0-ms\cs;
Bss2=ms\BssU;
Css1=Ass3;  Css2=Ass4;
```

```
for i=1:n
    for j=1:n
        Ass(i,n+j)=Ass2(i,j);
        Ass(n+i,j)=Ass3(i,j);
        Ass(n+i,n+j)=Ass4(i,j);
        Bss(n+i,j)=Bss2(i,j);
        Css(i,j)=CssU(i,j);
        Css(n+i,n+j)=CssU(i,j);
        Css(2*n+i,j)=Css1(i,j);
        Css(2*n+i,n+j)=Css2(i,j);
        Dss(n+i,j)=Bss(i,j);
Dss(2*n+i,j)=Bss(n+i,j);

    end
end
Ass;
Bss;
Css;
%% For Structure With Passive TMD
    % Main Structure
Assm=Ass;
Bssm=zeros(2*n,2);
Cssm=Css;
Dssm=zeros(3*n,2);
Rssm1=zeros(n,1);
Rssm2=zeros(n,1);
for i=1:n
    Rssm1(i,1)=1;
end
Rssm2(n,1)=1;
Bssm1=-Rssm1;
Bssm2=ms\Rssm2;
for i=1:n
    Bssm(n+i,1)=Bssm1(i,1);
    Bssm(n+i,2)=Bssm2(i,1);
    Dssm(2*n+i,1)=Bssm1(i,1);
    Dssm(2*n+i,2)=Bssm2(i,1);
end

Rrg=zeros(2,1);     Rrf=zeros(2,1);
Rrg(1,1)=1;         Rrf(2,1)=1;

%% Uncontrolled Structure
Zo=zeros(2*n,1);
Yunc=zeros(len,3*n);
Exp1=exp(1);
eAssmdt=real(Exp1^(Assm*dt));
Time=zeros(len,1);
for i=1:len
    Fssm1=Rrg*Uddgsub(i);
    Fssm2=Rrf*0;
    FssM=Fssm1+Fssm2;
    FssMb=Bssm*FssM;
    Z=eAssmdt*Zo+eAssmdt*dt*FssMb;
```

```
        Yi=Cssm*Z+Dssm*FssM;
        Zo=Z;
        % Mechanical Energy
        Xsa(1:n,1)=Yi(1:n,1);
        Xdsa(1:n,1)=Yi(n+1:2*n,1);

        Yunc(i,:)=Yi;
        Time(i)=0+dt*(i-1);
    end

    %% Passive Tuned Mass Damper
    TMDf=0;
    Zo=zeros(2*n,1);
    ZoAll=zeros(2*(n+1),1);
    Yptmd=zeros(len,3*n);
    YptmdAll=zeros(len,3*(n+1));
    Exp1=exp(1);
    % eAssmdt=real(Exp1^(Assm*dt));
    eAssmdtAll=
    real(Exp1^(AssAll*dt));
    for i=1:len
        FssMall=Uddgsub(i);
        FssMball=BssAll*FssMall;
        Zall=eAssmdtAll*ZoAll+eAssmdtAll*dt
          *FssMball;
        YiAll=CssAll*Zall+DssAll*FssMall;
        ZoAll=Zall;
        YptmdAll(i,:)=YiAll;
        % Mechanical Energy
        Xsa(1:n,1)=YiAll(1:n,1);
        Xdsa(1:n,1)=YiAll(n+1:2*n,1);

        Yptmd(i,:)=Yi;
        Zo=Z;
    end

    THOutput(:,1)=Time(:);
    THOutput(:,2)=Yunc(:,n)*100;
    THOutput(:,3)=YptmdAll(:,n)*100;
    THOutput(:,4)=0;
    THOutput(:,5)=Time(:);
    THOutput(:,6)=Yunc(:,3*n);
    THOutput(:,7)=YptmdAll(:,3*(n+1)-1);
    THOutput(:,8)=0;
    THOutput(:,9)=Time(:);
    THOutput(:,10)=Yunc(:,2*n);
    THOutput(:,11)=YptmdAll(:,2*(n+1)-1);
    THOutput(:,12)=0;

    PeakOutputs=zeros(2,2);
        TopFlResp=Yunc(:,3*n);
        MaxUncAc=max(TopFlResp);
        MinUncAc=min(TopFlResp);
```

```
PeakOutputs(1,1)=max(abs(MaxUncAc),abs
(MinUncAc));
    TopFlResp=YptmdAll(:,3*(n+1)-1);
    MaxUncAc=max(TopFlResp);
    MinUncAc=min(TopFlResp);
PeakOutputs(1,2)=max(abs(MaxUncAc),abs
(MinUncAc));
    TopFlResp=Yunc(:,n)*100;
    MaxUncAc=max(TopFlResp);
    MinUncAc=min(TopFlResp);
PeakOutputs(2,1)=max(abs(MaxUncAc),abs
(MinUncAc));
    TopFlResp=YptmdAll(:,n)*100;
    MaxUncAc=max(TopFlResp);
    MinUncAc=min(TopFlResp);
PeakOutputs(2,2)=max(abs(MaxUncAc),abs
(MinUncAc));

% Write time history response outputs
to notepad file
filename=strcat('01_TimeHistoryOutput_
TMD',   '.txt');
dlmwrite(filename,THOutput,'\t')
%% Plots
subplot(2, 1, 1);
    plot (Time,Yunc(:,n)); axis([0 25
-inf inf]); xlabel('Time (s)');
ylabel('Top Fl. Disp. (m)');
    hold on
    plot (Time,YptmdAll(:,n)); axis([0
25 -inf inf]);  legend('Uncontrolled','
TMD-Controlled')
subplot(2, 1, 2);
    plot (Time,Yunc(:,3*n)); axis([0 25
-inf inf]);             xlabel('Time
(s)');     ylabel('Top Fl. Acc. (m/
s2)')
    hold on
    plot (Time,YptmdAll(:,3*(n+1)-1));
axis([0 25 -inf inf]);
legend('Uncontrolled','TMD-Controlled')
PeakOutputs
toc;    % End of timing the execution
```

REFERENCES

Bauer, H. F. (1984). Oscillations of immiscible liquids in a rectangular container: a new damper for excited structures. *Journal of Sound and Vibration*, 93(1), 117–133.

Frahm, H. (1911). Devices for damping vibration of bodies. United States Patent. Patent No.: 989958.

Masaki, N., Suizu, Y., Kamada, T., and Fujita, T. (2004). Development and application of tuned/hybrid mass dampers using multi-stage rubber bearings for vibration control of structures. In Proceedings of 13th World Conference on Earthquake Engineering, Vancouver, BC, Canada.

Warburton, G. B., and Ayorinde, E. O. (1980). Optimum absorber parameters for simple systems. *Earthquake Engineering and Structural Dynamics*, 8(3), 197–217.

6 Friction Dampers

6.1 INTRODUCTION

Different types of friction-based devices are used in vibration response control of structures under dynamic excitations, such as earthquakes. Friction dampers are effective, reliable, and cost-effective energy dissipation devices. The mitigation of the seismic response of structures is attained by friction-based energy dissipation that occurs when two solid plates slide over each other, as shown in Figure 6.1. The resisting force is developed due to the relative motion of two solid plates. Both plates are subjected to time-varying force $F(t)$ at their ends and static normal force N perpendicular to the contact point of the plates. In order to maintain equilibrium, the friction force (f_N) developed at the contact point of two plates opposing the time-varying force is expressed by Equation (6.1).

$$f_N = \mu N \tag{6.1}$$

where μ is the coefficient of friction, which is dependent on the type and present condition of the constituent materials of the two surfaces in contact. In Equation (6.1), the relative velocity between the plates sliding over each other does not influence the magnitude of friction force.

Effectiveness of Friction Dampers in Introducing Damping

The effectiveness of friction dampers in introducing damping is a function of several variables, including the number of devices installed, their location in the structure, and their physical parameters. The effectiveness of a friction damper depends on its important parameter, i.e., the slip load, which influences the amount of energy dissipated by the damper. The area under the hysteresis loop should be the largest for maximum energy dissipation. Figure 6.2 presents the hysteresis loops of friction damper for different slip load conditions. If the slip load is very high, as shown in Figure 6.2(a), the dampers will not slip even if a very high force is imparted to the structure. In this case, the area under the hysteresis loop is very small. Hence, the energy dissipation is small, and the dampers are less effective. Similarly, the area under the hysteresis loop is again very small for a very low slip load, as seen in Figure 6.2(b). Therefore, less energy is dissipated, and the dampers are less effective. In this case, the friction dampers will slip immediately when a very low force acts on them without delivering any response reduction. For maximum response reduction, an optimum slip load has to be achieved, at which the area under the hysteresis loop is maximum, resulting in maximum energy dissipation. Comparison of energy dissipated by friction dampers against slip load in the three cases is shown in graphical form in Figure 6.2.

Mayes and Mowbray (1974) and Keightley (1977) used frictional devices for mitigating the seismic response of structures. Further, Pall et al. (1980) and Pall and Marsh (1982) developed passive friction dampers for seismic response control. Notable progress has been made during the past few years, and a number of novel friction damping devices have been developed.

6.2 X-BRACED PALL FRICTION DAMPER

Pall and Marsh (1982) installed friction devices in a moment-resisting frame to enhance its performance during earthquakes. Pall damper consists of steel plates arranged in series, bolted together by high-strength bolts. During a major earthquake, the dampers slip at a predetermined optimum load before yielding occurs in other structural members and dissipate a significant portion of the seismic energy. By properly selecting the slip load, it is possible to reduce the response of the structure

DOI: 10.1201/9781315269269-6

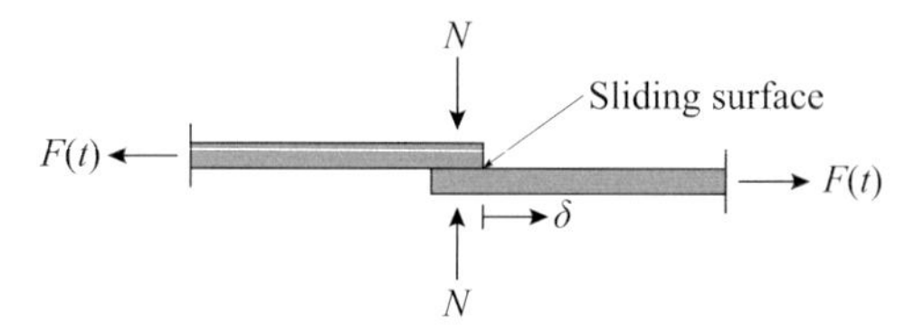

FIGURE 6.1 Friction between two sliding plates in contact.

significantly. The slippage is without any stick-slip phenomenon. The dampers are designed not to slip during service load and windstorms. Photo 6.1(a) shows a Pall friction damper installed in a structure, connected with four links, and Photo 6.1(b) shows a close-up view of the connection of the damper to the joint.

Figure 6.3(a) shows the connection details of the device to the links and the axial forces in bracing when the earthquake-induced lateral force acts towards the right. Further, Figure 6.3(b) shows the cross-section and the side view of the damper.

A friction-damped structure is an engineered structure wherein the forces exerted by the damper are predetermined. The possibility of determining the control force during design is a great technical and economic advantage of friction dampers over fluid viscous dampers. For fluid viscous damper, the forces could be very large for strong excitations that result in large relative velocities, e.g., maximum credible earthquakes (MCE). The shape of the hysteresis loop of friction dampers is rectangular, whereas the force-deformation loop is elliptical for fluid viscous dampers. For a given maximum force, the hysteresis loop area (energy dissipation or damping) of a fluid viscous damper is about 70% of that for friction damper. As friction mechanism is more effective for dissipating energy than any other method involving the yielding of steel, and very few Pall damping devices are sufficient to achieve substantial energy dissipation. This leads to significant savings in the cost of dampers, bracing, connections, columns, and foundations. While supplemental damping is beneficial in reducing the earthquake effects and amplitudes of vibration, added rigidity could also be beneficial. Pall friction dampers provide both added damping and rigidity (due to initial slip load) and serve as a suitable alternative to fluid viscous dampers (which primarily provide damping)

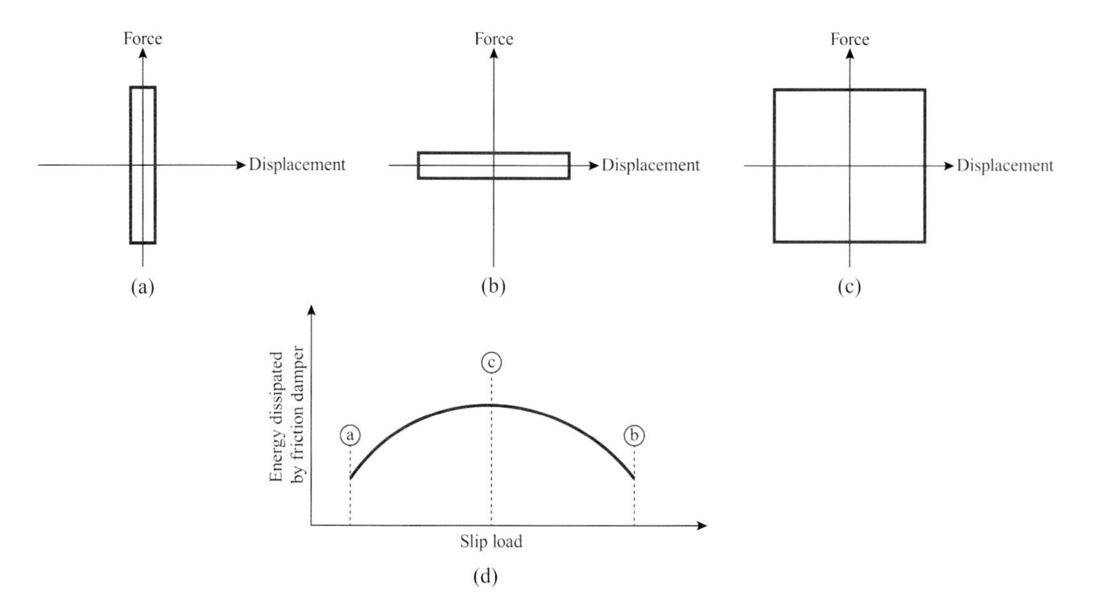

FIGURE 6.2 Comparison of the energy dissipation capacities of hysteresis loops of friction damper with: (a) very high slip load, (b) very low slip load, (c) optimum slip load, and (d) energy-slip load relationship.

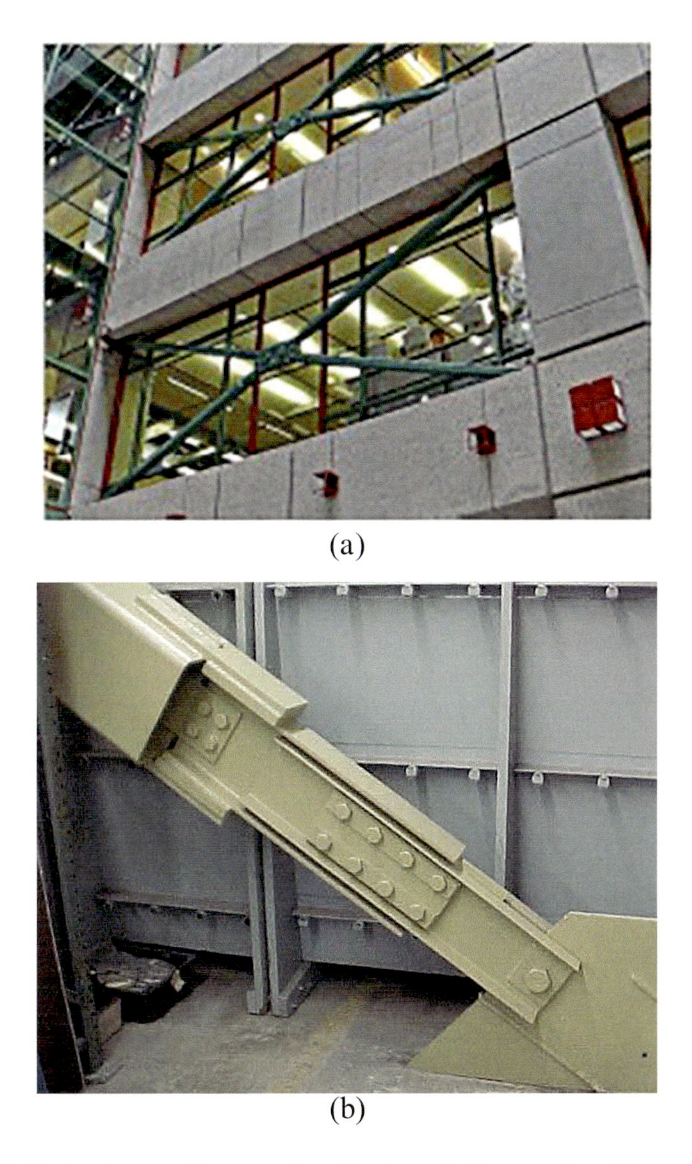

(a)

(b)

PHOTO 6.1 Pall friction damper: (a) structural application and (b) connection of damper to joint structure (Heysami 2015). (Courtesy of Tripat Pall.)

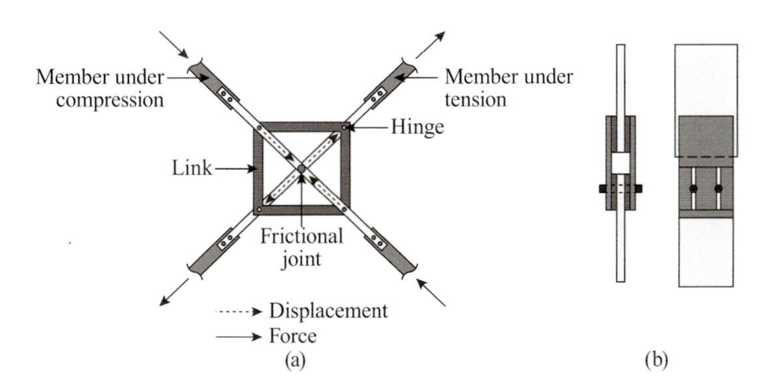

FIGURE 6.3 Pall friction damper: (a) connection details and (b) cross section and side view.

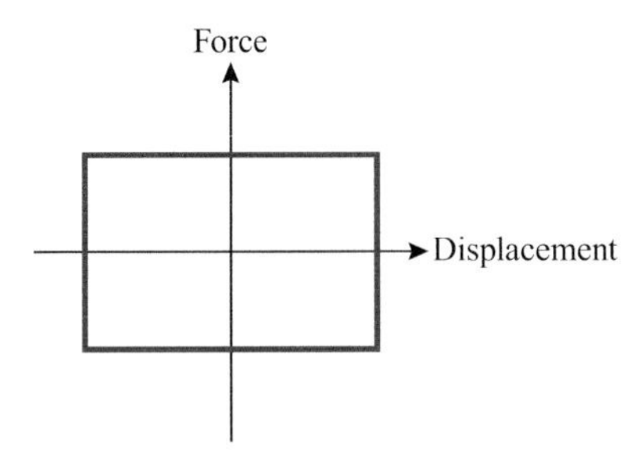

FIGURE 6.4 Idealized hysteresis loop of Pall friction damper.

for structures. Cyclic dynamic tests on Pall friction dampers have confirmed their reliable performance and large rectangular and stable hysteretic loops, as shown in Figure 6.4.

Friction dampers, such as Pall dampers, are displacement-dependent dampers, and their control effectiveness is not influenced by the frequency of the vibration. Therefore, they are ideal for the suppression of vibrations induced by dynamic excitations, such as earthquakes. Pall friction dampers can be incorporated in long slender tension-only cross bracing, single diagonal tension-compression bracing, and chevron bracing, as shown in Figures 6.5–6.7, respectively.

The X-braced Pall friction dampers are designed to work in tension only. However, the unique configuration of the mechanism ensures the slip to take place simultaneously in both tension and compression braces. The occurrence of tension in one of the braces forces the damper to slip, during which the mechanism of the damper shortens the other (compression) brace, thus preventing buckling. One brace is in tension and the other in compression, as shown in Figure 6.8. Further, Photo 6.2 shows a friction damper installed in bracing.

Due to lateral load, compression brace shortens, causing the slippage of braking pads before the brace in tension yields. This leads to the activation of friction devices of both braces connected to the four links of the device. For preventing slippage under wind loads, large numbers of devices are required, as Pall dampers require small activation force. Figure 6.9 portrays a typical hysteresis loop of Pall friction damper under earthquake.

Advantages of Pall Friction Damper

Pall dampers have gained popularity among friction damping devices because of their simple design, low construction cost, reliability, and repeatable performance. The dampers provide a constant control force under all types of earthquakes. Hence, their hysteresis loop is large rectangular, indicating a large amount of seismic energy dissipation. Their performance is independent of

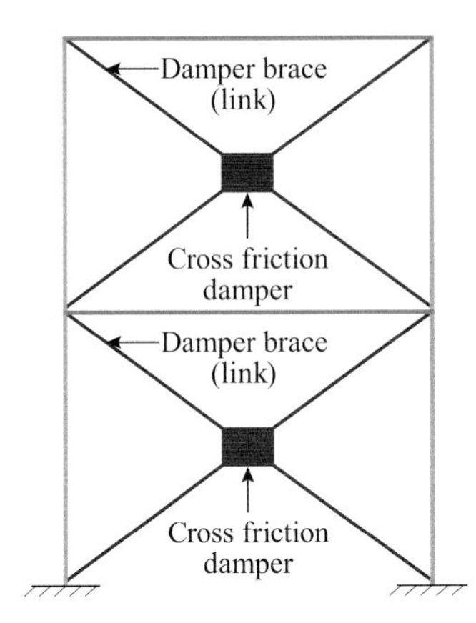

FIGURE 6.5 Long slender tension-only cross bracing employed with Pall friction damper.

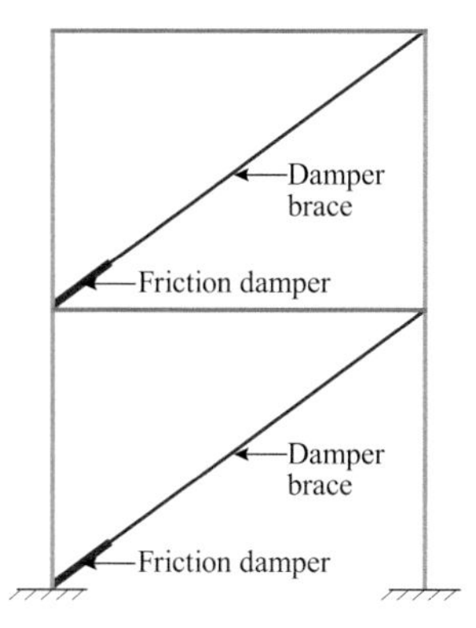

FIGURE 6.6 Single diagonal tension-compression bracing employed with Pall friction damper.

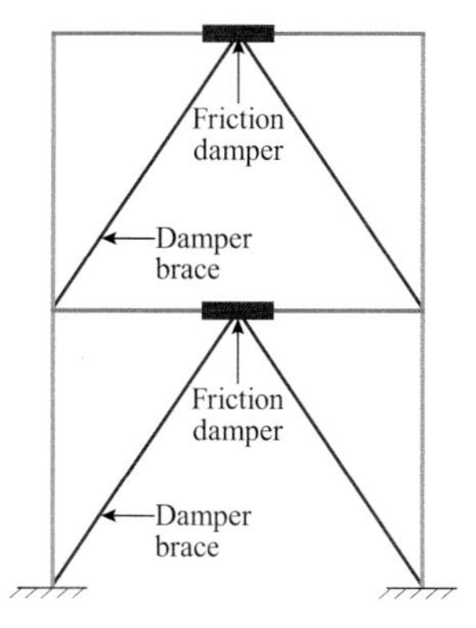

FIGURE 6.7 Chevron bracing employed with Pall friction damper.

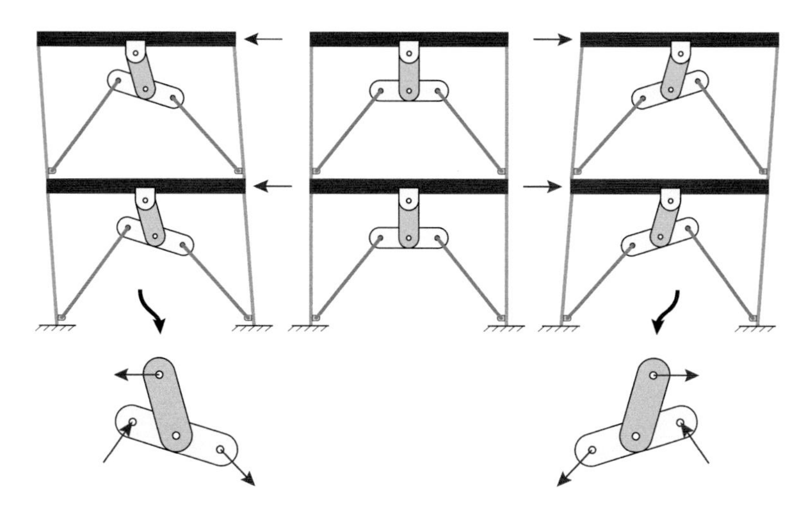

FIGURE 6.8 Mechanism of friction damper.

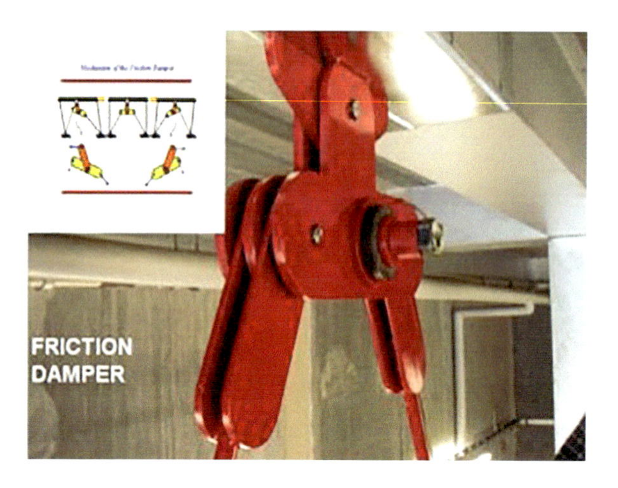

PHOTO 6.2 Friction damper installed in bracing. (Heysami 2015.)

velocity and temperature. Pall friction dampers have been found to perform satisfactorily during design-basis earthquakes (DBE) and maximum credible earthquakes (MCE). The dampers are not active during service loads and wind. They do not need frequent maintenance during the life span of the structure. Further, they are not sacrificial elements, and they typically do not need repair/replacement after an earthquake. Post-earthquake, the structure returns to its near-original alignment due to restoring action of an elastic structure. There is no component with fluid, and therefore, no risk of leakage, unlike fluid viscous dampers, where there is fear of fluid leakage. Pall dampers are compact and narrow enough to be hidden in the partitions. They can be installed along with bracing, including tension cross-bracing and expansion joints. Furthermore, Pall friction dampers can be tailored to demonstrate greater adaptability to the site conditions. For instance, they can be designed to accommodate foundation settlements.

Worldwide Applications of X-Braced Pall Friction Damper

Due to high reliability and best performance, Pall friction dampers have found applications in various structures worldwide since 1982. They are used in new construction as well as in the seismic retrofitting of existing structures. Some of the notable examples of applications of Pall friction dampers are specified subsequently, where the dampers have proved to be effective, economical, and reliable for seismic protection.

- **Boeing Factory, United States**
  The Mammoth Boeing Plant was constructed from 1968 to 1991. The plant is constructed from a steel frame with height and span of 37 m and 107 m, respectively, whereas it covers about 98 acres of land. An inside view of Boeing Factory is as shown in Photo 6.3.

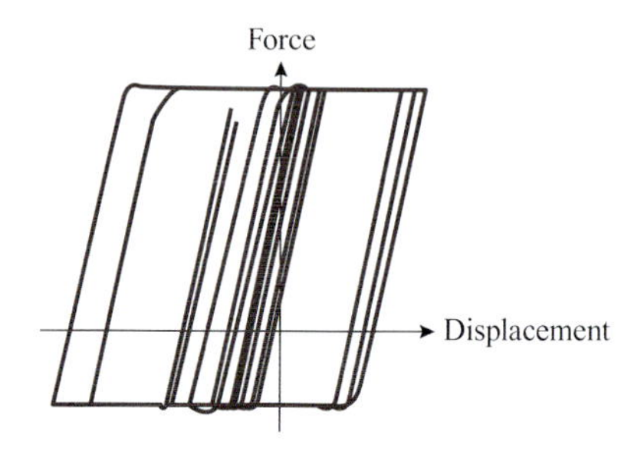

FIGURE 6.9 Hysteresis loop of Pall friction damper under earthquake.

PHOTO 6.3 Inside view of Boeing Factory, United States. (Courtesy of Tripat Pall.)

In 1998, the plant was retrofitted for seismic protection by installing Pall friction dampers in different existing bracing. Adaptation to the site conditions is an important factor for the design of different types of dampers.

Pall dampers were also used for seismic retrofit of Boeing Development Center, Cafeteria, and Auditorium buildings. Photo 6.4 shows the Boeing Development Center, United States, which contains three 2-storied Boeing development buildings and a 4-storied Boeing Cafeteria, Auditorium, and fitness center building. These steel frame buildings were built in 1980. Some of these buildings were damaged during the February 28, 2001, in Nisqually earthquake of magnitude 6.8. The epicenter was about 32 km from the building site. Due to soil liquefaction, the differential settlements in pile foundations were 0.1–0.125 m. This resulted in substantial nonstructural and structural damage to the Cafeteria and Auditorium building. The structural damage was primarily to rigid steel bracing. Several bracings in the Cafeteria and Auditorium building buckled or severely cracked.

PHOTO 6.4 Boeing Development Center, United States. (Courtesy of Tripat Pall.)

PHOTO 6.5 Moscone West Convention Centre, San Francisco, United States.

Friction dampers provide damping and stiffness and can be easily modified to suit site conditions. Further, they can be designed to accommodate any future foundation settlements. Hence, a total of 350 Pall friction dampers were used in the existing steel bracing. The seismic retrofit was completed in 2002. Savings are estimated to be more than 60%, compared to conventional retrofit.

- **Moscone West Convention Centre, San Francisco, United States**
  The Moscone West Convention Centre, located in San Francisco, United States, that is shown in Photo 6.5 is a 34 m tall four-storied steel frame building. The construction of the center started in 2001 and was completed in 2003. With the installation of Pall friction dampers, there was a substantial reduction in the cost of the project. In subsequent earthquakes, it was observed that the story drift recorded without dampers was reduced to half, with the installation of Pall friction dampers. Further, the dampers were capable of reducing the earthquake energy transmitted to the structure substantially, which minimizes the damage in the building.
- **Ambulatory Care Center, Sharp Memorial Hospital, San Diego, California, United States**
  The Ambulatory Care Center complex, shown in Photo 6.6, consists of two 4-story and 5-story buildings connected at two places by a pedestrian bridge and an elevator lobby.
  The steel-frame design was used from the consideration of cost-effectiveness, design flexibility, and speedy construction. Moment-resisting frames, in combination with Pall friction dampers in steel bracing, were selected to resist lateral seismic action. Photo 6.7 shows Pall friction dampers installed in chevron brace configuration. By the installation of Pall friction dampers, the story drift was reduced considerably. Despite the added cost of dampers, there was substantial saving in the overall construction cost.
- **Seismic retrofit of 3-million gallon reservoir, Sacramento, California, United States**
  The 3-million-gallon steel water tower shown in Photo 6.8 was built in 1956 in California, United States. The supporting structure of the 36 m tall tower consists of 27 steel columns with two levels of tension cross bracing. Installing Pall friction dampers in tension cross

PHOTO 6.6 Ambulatory Care Center, Sharp Memorial Hospital, San Diego, California, United States.

bracing was chosen as a retrofitting option. Due to the high damping provided by the dampers, it was not necessary to strengthen the columns and foundations. The seismic retrofit was completed in 1999. The option of installing Pall friction dampers saved more than 60%, compared to conventional.

- **Concordia University Library Building, Montreal, Canada**
  The 10-storied McConnell library building shown in Photo 6.9 is a masterpiece in innovative structural design. The building was constructed in 1991, and Pall friction dampers are installed at the junction of steel cross bracing in rigid RC frames. The use of steel bracing eliminated the need for RC shear walls. The supplemental damping provided by friction dampers eliminated the need for dependency on the ductility of structural members. Since the bracing were mostly not continuous one over the other, it provided greater flexibility in

PHOTO 6.7 Pall friction dampers in chevron brace configuration. (Courtesy of Tripat Pall.)

PHOTO 6.8 Three-million-gallon reservoir, Sacramento, California, United States. (Courtesy of Tripat Pall.)

PHOTO 6.9 Concordia University Library Building, Montreal, Canada. (Courtesy of Tripat Pall.)

space planning. The innovative structural system provided an economical design solution to safeguard the building from earthquake damage.

- **Quebec Provincial Police Headquarters, Montreal, Canada**
  The Quebec Provincial Police Headquarters, in Montreal, Canada, shown in Photo 6.10(a), is a 16-storied office building, with two levels of basement, that was built in 1964. Steel moment frames and some braced bays provided lateral resistance to the existing structure. A change of occupancy was planned in 1997 to house the provincial police headquarters. During the structural evaluation, it was found that the existing structure was not capable of resisting seismic action, and the story drifts were excessively high, especially at the lower soft stories. The structure was seismically upgraded in 1999, using Pall friction dampers, incorporated in existing and new bracing, as shown in Photo 6.10(b). Considering the significant amount of energy dissipation by the dampers, the story drifts and forces acting on the structure were found to reduce considerably. The innovative structural scheme offered savings of more than 50% over the conventional retrofit scheme.

(a)

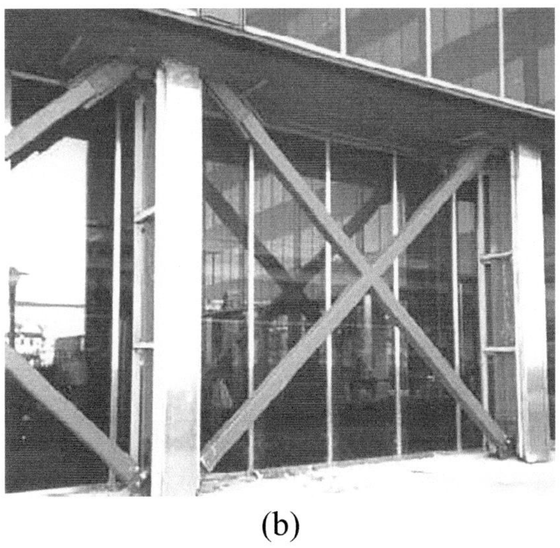

(b)

PHOTO 6.10 Quebec Provincial Police Headquarters, Montreal, Canada: (a) view of Police Headquarters and (b) Pall friction dampers in bracing. (Courtesy of Tripat Pall.)

6.3 SUMITOMO FRICTION DAMPER

Sumitomo friction dampers, manufactured by Sumitomo Metal Industries of Japan, were initially used for railway applications. Later, the dampers found applications in structural engineering for enhancing the seismic performance of structures. Figure 6.10 shows the assembly of a typical Sumitomo friction damper.

The Sumitomo friction damper consists of friction pads (generally copper pads) infused with graphite which slide along the cylindrical steel casing of the device. Graphite lubricates the contact surface to provide a consistent coefficient of friction between the copper pads and helps in the smooth functioning of the device. A pre-compressed internal spring provided in the device exerts force on the friction pads. The exerted force is converted into normal force through the action of a series of wedges which are always under compression due to the Belleville washer springs. The steel casing material is selected so that it does not suffer any additional corrosion when in contact with copper. Experimental studies performed by researchers reported that the chevron bracing configuration is a suitable arrangement for the installation of the Sumitomo damper. Figure 6.11 shows the placement of the Sumitomo friction damper with a chevron bracing configuration.

6.4 CONSTANT AND PROPORTIONAL FRICTION DAMPER

The capability of friction damper to reduce seismic response of structures largely depends on the hysteresis property, which in turn is influenced by the slip force of the damper. The constant friction damper (CFD) exhibits a rectangular hysteresis loop, wherein the slip force, i.e., the friction force, is constant and is developed by a constant normal force acting on the slip surfaces, which is generated by using either prestressed bolt mechanism or prestressed longitudinal spiral spring-wedge mechanism. In the proportional friction damper (PFD), the slip force is directly proportional to the displacement of the damper. The hysteresis loop of the PFD has a flag-shaped or triangular-shaped appearance. In PFD, a displacement-proportional normal force is used to produce displacement-proportional damping force. A longitudinal spiral spring wedge mechanism or a ring spring-wedge mechanism is used to achieve the proportional normal force. The PFD possesses a self-centering feature. For a given slip force and displacement amplitude, CFDs are usually capable of dissipating more energy per cycle than the PFDs.

Mathematical Model of CFD and PFD

Figure 6.12(a) depicts the mathematical model, and Figure 6.12(b) shows the hysteresis loop of the bilinear constant friction damper system.

Let u_0 be the initial displacement of the damper mechanism. The resisting force f_d provided by a PFD can be expressed as:

$$f_d = \begin{cases} (1+\eta')k(u+u_0) = k'_d(u+u_0) & \text{loading} \\ (1+\eta'')\ k_d(u+u_0) = k''_d(u+u_0) & \text{unloading} \end{cases} \tag{6.2}$$

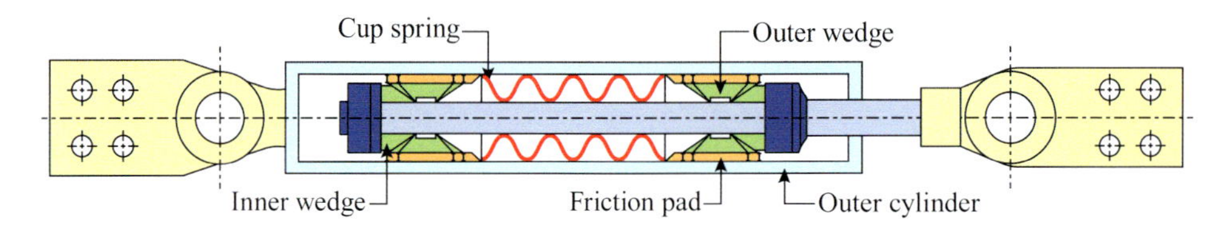

FIGURE 6.10 Assembly of Sumitomo friction damper.

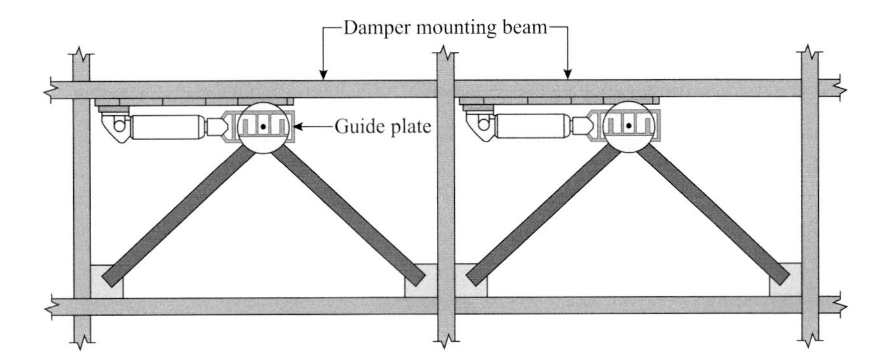

FIGURE 6.11 Installation of Sumitomo friction damper.

where u is the system displacement relative to the base; η' and η'' are the ratio of the slip force to the spring force for loading and unloading conditions, respectively; and k_d is the spring stiffness. Figure 6.13(a) shows the triangular hysteresis loop of a PFD, when $u_0 = 0$, and flag-shaped hysteresis loop when $u_0 > 0$.

The PFD can be connected with a bracing member to form a PFD unit. The bracing stiffness k_b considerably increases the effectiveness of PFD on seismic response reduction. The mathematical model of the PFD connected to a bracing member is shown in Figure 6.13(b). The displacement response of a PFD is considerably increased when connected with bracing, as mirrored in the hysteresis loop shown in Figure 6.13(b). Figure 6.13(c) shows the mathematical model and hysteresis loop of PFD, installed in a single-storied frame, excited by earthquake ground motion, where k_f is the frame stiffness. The total resisting force is the summation of the PFD unit force and the force contribution of the frame due to its elastic stiffness. The arrows in the hysteresis loops in Figure 6.13 show the loading and unloading paths. In a flag-shaped PFD system, slippage of damper starts at the slip displacement u_s and stops at the displacement smaller than u_s. However, PFD with triangular-shaped hysteresis loop behaves pseudo-elastically; its hysteresis characteristics do not change with respect to the displacement.

6.5 SLOTTED BOLTED FRICTION DAMPER

Slotted bolted friction damper is a modified version of bolted connections where the major limitation of Pall friction damper of inelastic member buckling is eliminated. This is possible by providing frictional resistance through slotted bolted connections. Generally, slotted bolted connections are concentric bracing member connections, where slip occurs at a designated friction resistance in long slotted bolt holes, as shown in Figure 6.14. Belleville spring washer assemblies are provided in slotted bolted connection so that the required level of bolt tension for the design normal force

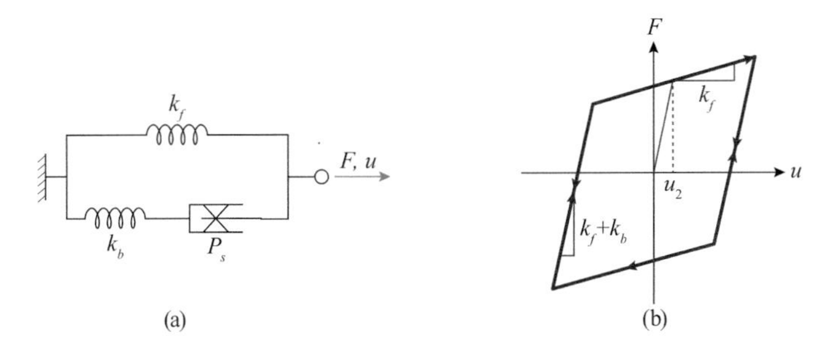

FIGURE 6.12 Bilinear constant friction damper system: (a) mathematical model and (b) hysteresis loop.

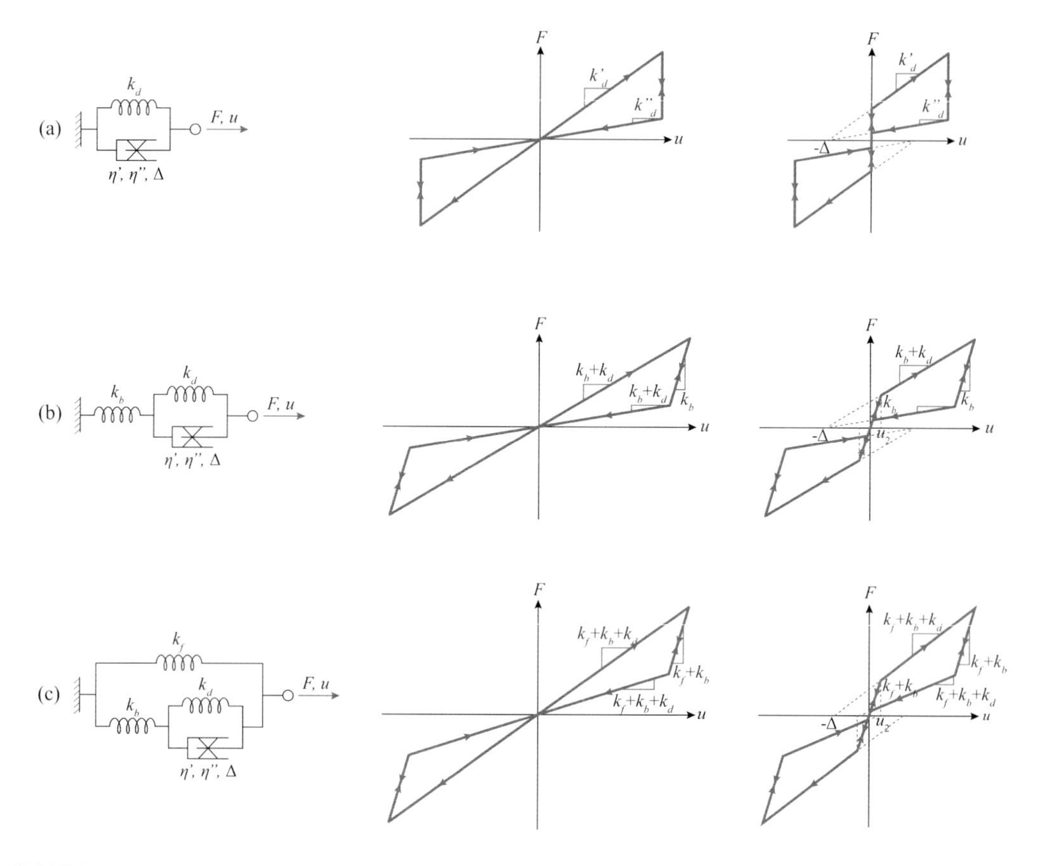

FIGURE 6.13 Mathematical model and hysteresis loops of proportional friction damper system: (a) PFD, (b) PFD unit, and (c) PFD installed in SDOF system.

is achieved. The number of washers utilized depends upon the specified bolt tension, and they are oriented in the concave side up or down position.

The washers are initially domed-shaped and later tend to become flat under bolt tension loads. A representative slotted bolted connection assembly, shown in Figure 6.15, consists of a gusset plate sandwiched between two channel sections placed back-to-back along with the cover plates, bolted with the help of a Belleville spring (Solon) washer.

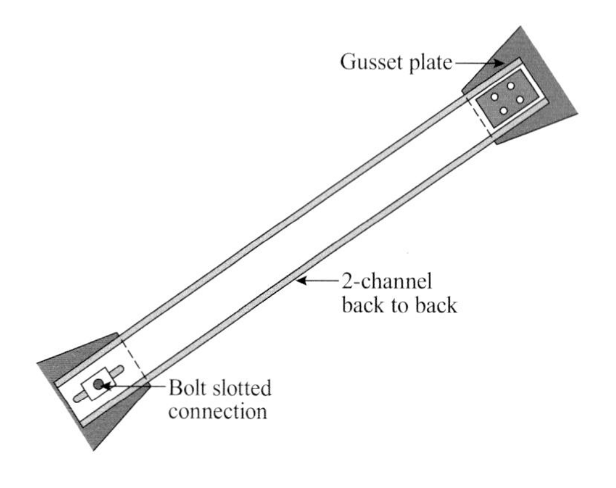

FIGURE 6.14 Configuration of slotted bolted connection.

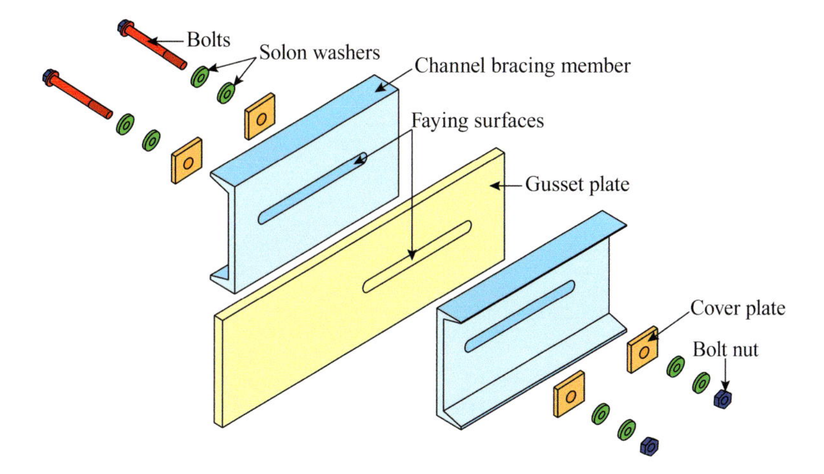

FIGURE 6.15 Assembly of slotted bolted connection.

The bolts are positioned at the center of the gusset plate, holding the channel slots in the same longitudinal position as shown in Figure 6.16. As the specimen is loaded, the gusset plate slips on one side, as shown in Figure 6.17; the state I, and thereafter the cover plates slide opposite to the gusset plate; the state II, shown in Figure 6.18.

Figure 6.19(a) portrays the hysteresis loop of slotted bolted friction damper for state I, whereas the hysteresis loop for state II is given in Figure 6.19(b).

Mathematical Model of Slotted Bolted Friction Damper

Free body diagram of state I, i.e., when the gusset plate slips on one side, is shown in Figure 6.17(b). After loading the specimen, the force F_1 is developed as a result of the friction between the gusset plate and channel plates on both sides and is given by Equation (6.3).

$$F_1 = 2\mu N \tag{6.3}$$

Similarly, the free body diagram of state II, i.e., when the channel plates slide opposite to the gusset plate to maintain equilibrium, is shown in Figure 6.18(b). The force F_2, exerted by the channel plate, is given as:

$$\begin{aligned} &\frac{F_2}{2} - F_1 = 0 \\ &F_2 = 2\mu N \times 2 = 4\mu N \end{aligned} \tag{6.4}$$

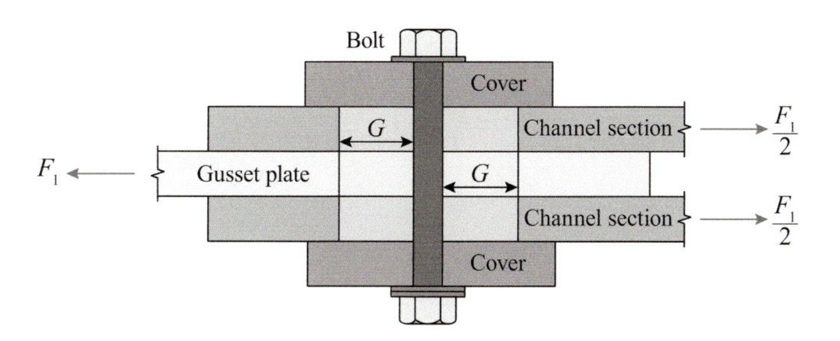

FIGURE 6.16 Initial assembly of slotted bolted connection.

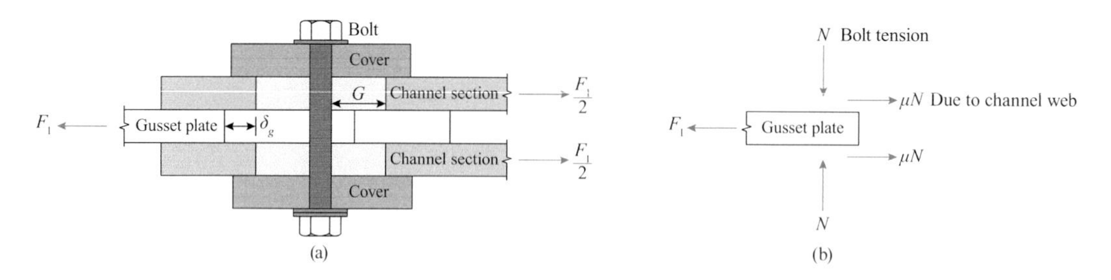

FIGURE 6.17 Slotted bolted connection – state I (gusset plate slip range $0 \le \delta_g \le G, \delta = \delta_g$): (a) assembly and (b) free body diagram.

In Equation (6.4), μ is the coefficient of friction between the sliding surfaces, and N is the initial tension in the bolts. The subscripts 1 and 2, respectively, correspond to state I and state II of the sliding condition of the connection.

6.6 ENERGY DISSIPATING RESTRAINT

Energy dissipating restraint (EDR) is a friction-based energy dissipating device that has a very good self-centering characteristic. The design of the EDR is similar to that of the Sumitomo friction damper, and it consists of a steel cylinder, steel compression wedges, bronze friction wedges, a steel shaft, steel spring, and internal stops. The internal stops are provided on both sides within the cylinder to regulate the tension and compression gaps appropriately. The relative motion of the shaft with respect to the cylinder is resisted by the frictional force developed due to the connection between the bronze friction wedges and the internal steel cylinder, as shown in Figure 6.20. During operation, EDR transforms the axial spring force into the normal force acting on the cylinder by the combined effect of steel compression wedges and bronze friction. The normal force is directly proportional to the total axial spring force. It is possible to change the length of the internal spring by providing variable slip force, which is dependent on the normal force and the coefficient of the friction between the bronze friction wedges and the steel cylindrical walls. The slip force is expressed as:

$$P_{sl} = \mu \alpha P_{sp} \tag{6.5}$$

$$P_{sp} = k_{sp} \Delta_{sp} \tag{6.6}$$

where P_{sl} is the slip force in the EDR device, μ is the coefficient of friction between bronze wedges and steel casing, α is the constant relating P_{sp} to the normal force on the cylinder wall, P_{sp} is the force in the spring, k_{sp} is the spring constant, and Δ_{sp} is the deformation of the spring.

Figures 6.21 and 6.22, respectively, show the double flag-shaped hysteresis loop and the triangular lobed hysteresis loop of EDR.

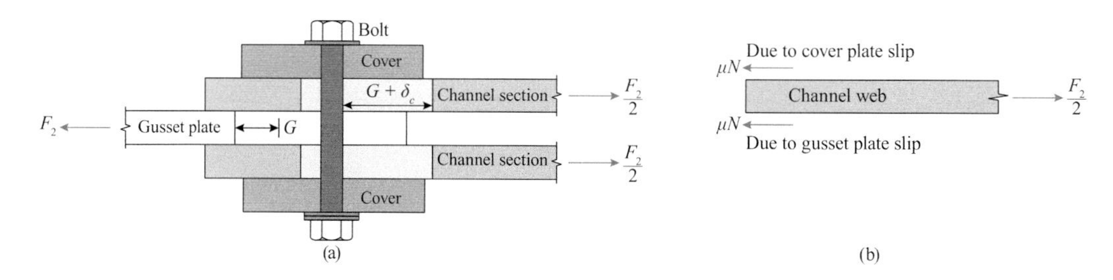

FIGURE 6.18 Slotted bolted connection – state II (gusset plate slip range $0 \le \delta_c \le G, \delta = G + \delta_c$): (a) assembly and (b) free body diagram.

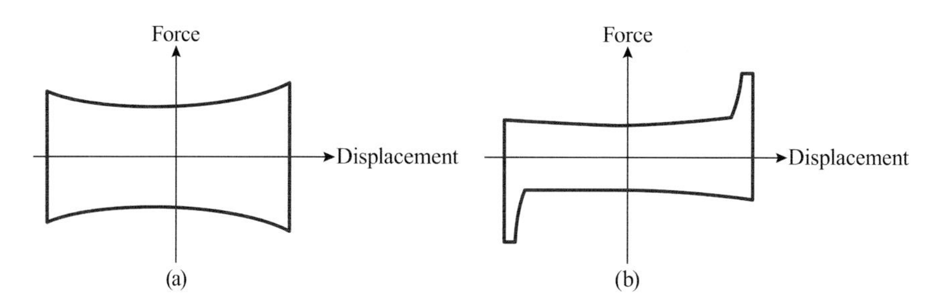

FIGURE 6.19 Typical force deformation diagrams for slotted bolted connections: (a) state I and (b) state II.

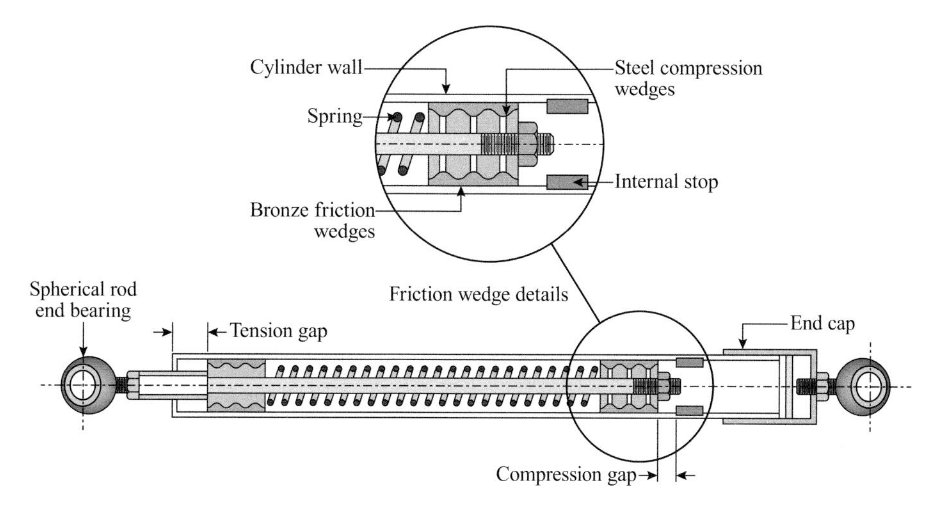

FIGURE 6.20 Energy dissipating restraint.

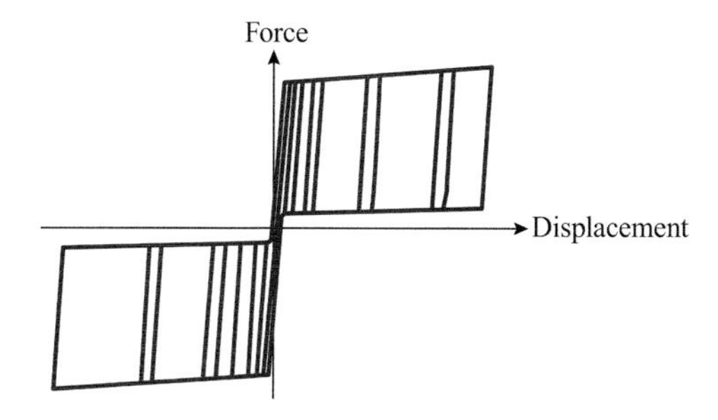

FIGURE 6.21 Double flag-shaped hysteresis loop of EDR.

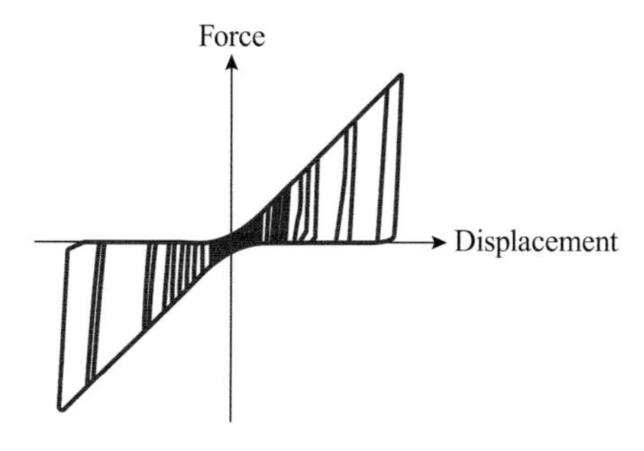

FIGURE 6.22 Triangular lobed hysteresis loop of EDR.

By changing the preload of the spring, and the lengths of tension and compression gaps, three different configurations of EDR are possible, viz., (i) zero gap and zero preload, (ii) very large gap and nonzero preload, and (iii) zero initial gap and nonzero preload.

6.7 PSEUDO-VISCOUS FRICTIONAL DAMPER

Pseudo-viscous frictional damper (PVFD) is an energy dissipator developed by combining the advantages of viscous energy dissipator and Pall friction damper. Viscous energy dissipators are capable of controlling the acceleration and base shear response without adding any axial pressure to the column. Pall friction dampers perform consistently and reduce the acceleration response even though the compression brace buckles. The construction of PVFD is similar to that of the Pall friction damper. It consists of crisscross plates connected by bolts with four links, as shown in Figure 6.23.

Two convexities are provided: one on each face of the PVFD, and a brass convexity on the inner face of the horizontal link. In Pall friction dampers, the brass convexity is provided with the flat faces of the crisscross plate and the horizontal links. Initial slip force in the PVFD is produced by the compressive force between the brass convex pad plate and the crisscross plate. The compressive force is a result of clip force in the slip bolt. The slip bolt is passed through the horizontal link over the brass convexity, brass pad plate, and crisscross plate. The initial slip force is overcome by the PVFD, and it gets deformed as the seismic load reaches its extreme limit. Due to deformation, the brass convexity slides down the convexity of the crisscross plate, and thus, the initial rectangular shape of the PVFD changes to rhomboidal. The change in shape shortens the distance between the convexities and thus decreases the clip force in the slip bolt. Thus, at maximum displacement, the restoring force of the PVFD reaches the minimum, and at zero displacement, it reaches maximum. Identical behavior is shown by fluid viscous dampers; hence, the name PVFD. Figure 6.24 shows the hysteresis loop of PVFD and the brace to which it is attached.

Advantages of PVFD

PVFDs are easy to manufacture and can be easily installed. The friction coefficient of PVFD is unaffected by temperature variations. The dampers are very effective in controlling seismic vibrations. In addition to the base shear, the displacement and acceleration response quantities are also efficiently controlled.

Limitations of PVFD

PVFDs impart additional stiffness to the structure, as they provide additional damping to the structure. This imposes a limit on the acceleration control as compared to other dampers.

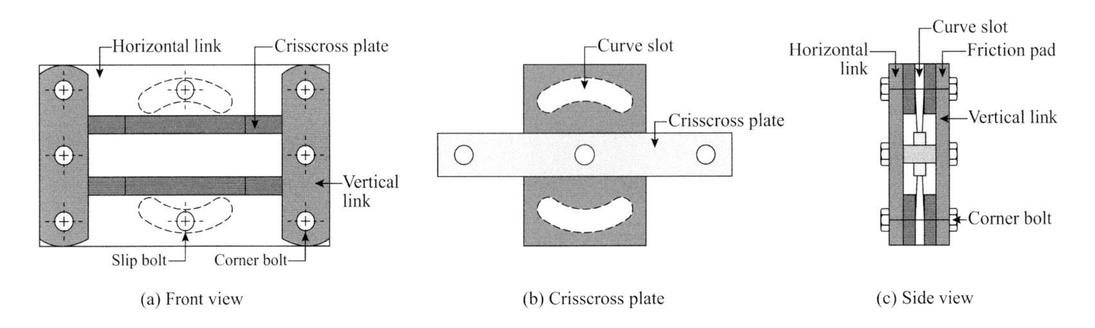

FIGURE 6.23 Pseudo-viscous frictional damper.

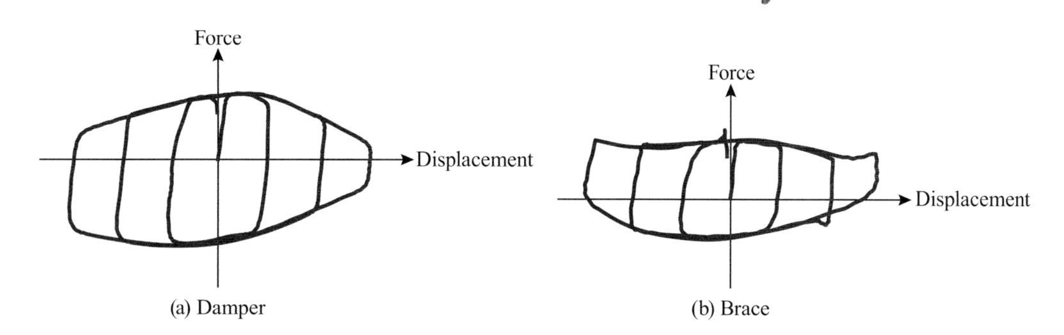

FIGURE 6.24 Hysteresis loop of pseudo-viscous frictional damper.

6.8 DOUBLE-FACE FRICTIONAL DAMPER

A conventional friction damper consists of a frictional sliding interface and a clamping mechanism that produce normal contact force on the inner face. With the view to enhance the seismic performance of conventional friction damper, Patil and Jangid (2009) proposed a modified frictional device named the double-face frictional damper. The conventional friction damper is modified by providing an additional plate between the two existing plates, as shown in Figure 6.25. This makes an additional interface available for the damper to resist the dynamic forces. It is reported that compared to the conventional friction damper, the double-face friction damper gives improved performance against displacement and acceleration; the accelerations are controlled more effectively.

6.9 MULTIDIRECTIONAL TORSIONAL DAMPER

The torsion-based friction damper is a mechanical device designed to make use of the torsional yielding of steel cylinders to dissipate the kinetic energy associated with large differential movements at its two anchoring ends. The anchoring ends are two end points of the damper, which are attached to two locations in a structure, e.g., for application in the bridge, as shown in Figure 6.26, one end of the damper is anchored to the bridge pier, and the other end to the bridge deck.

The bidirectional differential displacement at the two anchorage ends is converted to twisting in cylindrical energy dissipators. The multidirectional torsional damper consists of two major parts: the base device, attached to the substructure, and the rail system, attached to the superstructure.

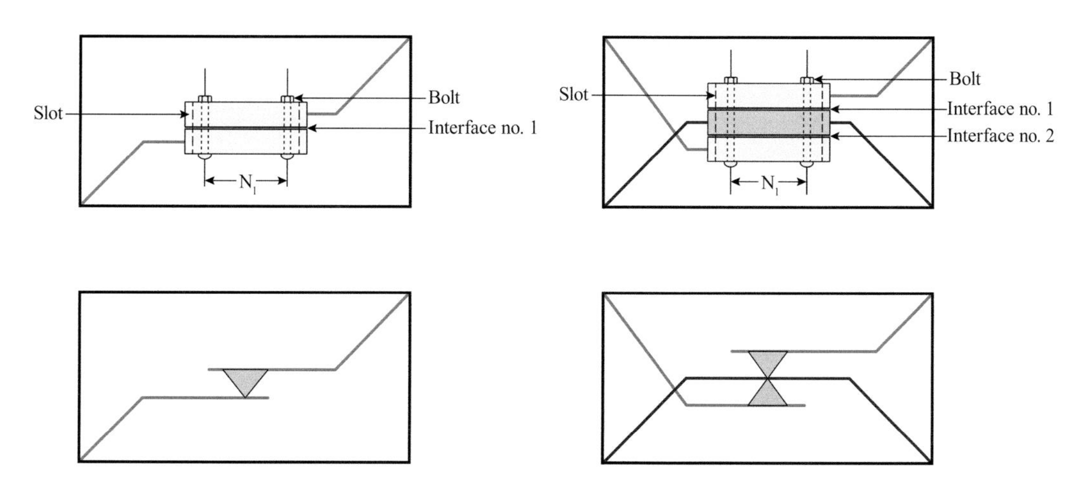

FIGURE 6.25 Schematic and mathematical models of conventional and double friction dampers.

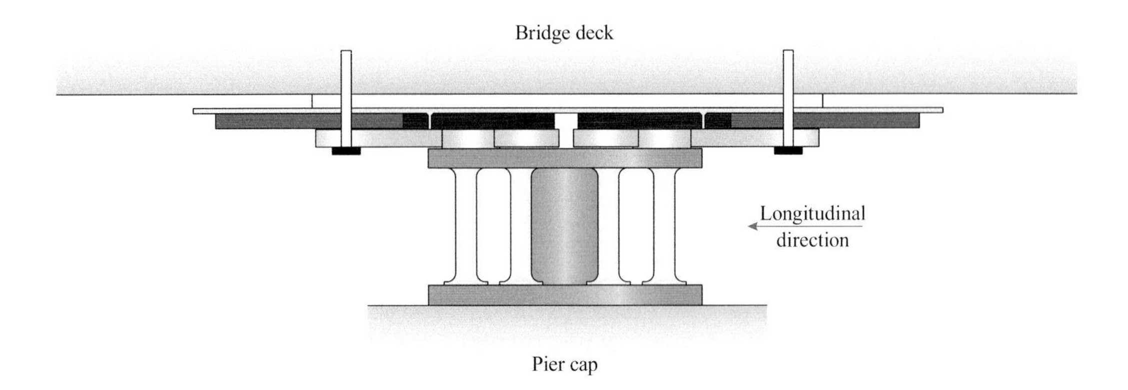

FIGURE 6.26 Installation of torsion-based friction damper in a bridge.

Figure 6.27 shows the base device and the rail system, whereas Figure 6.28 shows different parts of a multidirectional torsional damper.

The top view, showing the base device and the overlaid sketch of the rail system, is presented in Figure 6.29. Figure 6.30 shows the isometric view of section A-A marked in Figure 6.29.

Energy dissipators for the multidirectional torsional damper are eight identical cylinders, with enlarged top and bottom as shown in Figure 6.28, referred to as yielding cores or simply cores. The eight cylinders forming a core are attached to a base plate at the bottom and to a diaphragm plate at the top through sliding bearings, which allow for low-friction twisting of the cores. Each of the eight cylinders has a torsion arm attached through a plug connection. At the other end of the arms, a slider block is mounted through cylindrical shafts. The slider block made of solid plastic or steel houses the sliding pads. It is in contact with the guiding rails that are welded to the large top plate. The combination of the slider block and guiding rails creates a roller-hinge type connection.

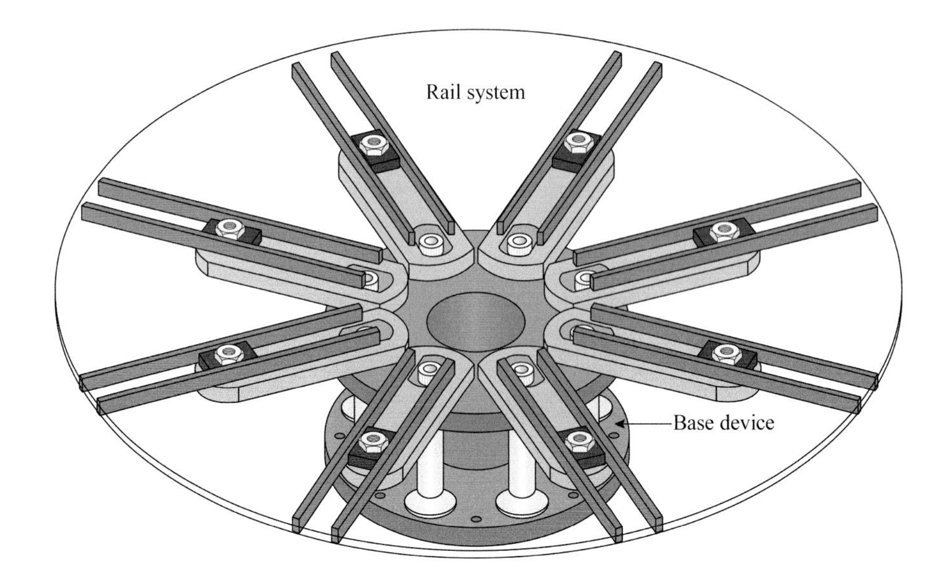

FIGURE 6.27 Major parts of torsion-based friction damper: base device and rail system.

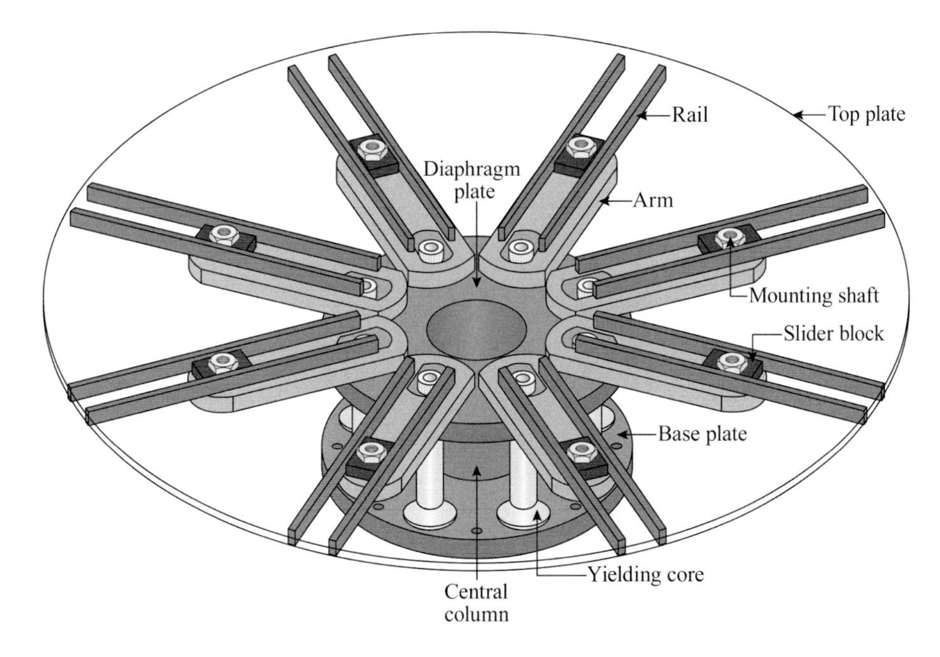

FIGURE 6.28 Different parts of torsion-based friction damper.

The concept behind this arrangement is to convert the translational motion at the end points of the arms to a twisting movement in cores, thus protecting the cores from bending. The central column of the device prevents bending of the cores.

The energy dissipation unit of the multidirectional torsional damper, shown in Figure 6.31, consists of core-arm-slider parts made of eight typical pairs. For each unit, a pair of guiding rails are present above and parallel to the arm in an undeformed configuration. Figure 6.32 shows the schematic top-view at undisplaced and displaced positions.

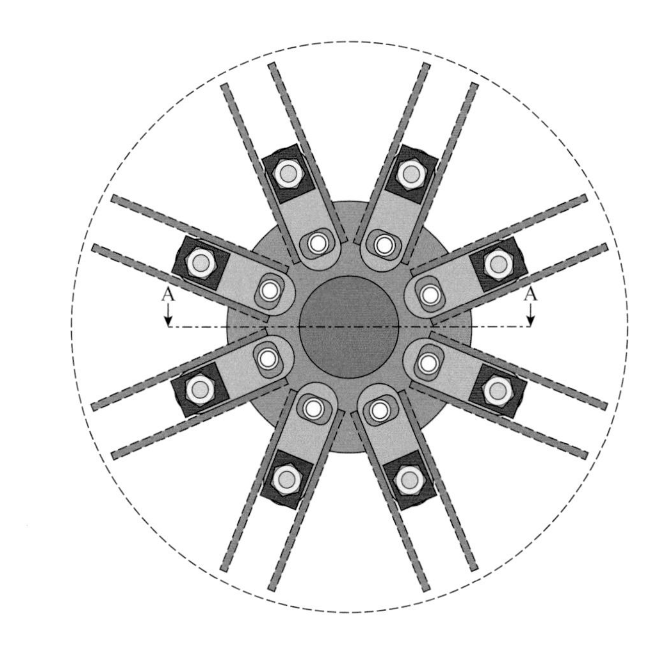

FIGURE 6.29 Top view showing the base device and the overlaid sketch of the rail system.

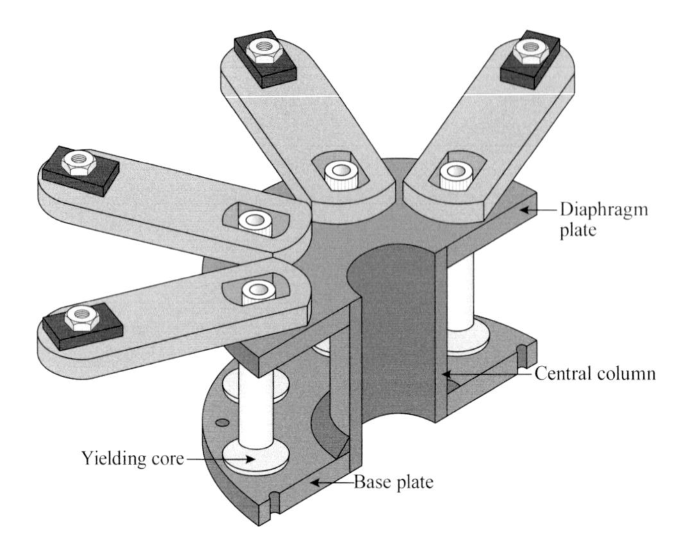

FIGURE 6.30 Isometric view (section A-A marked in Figure 6.29).

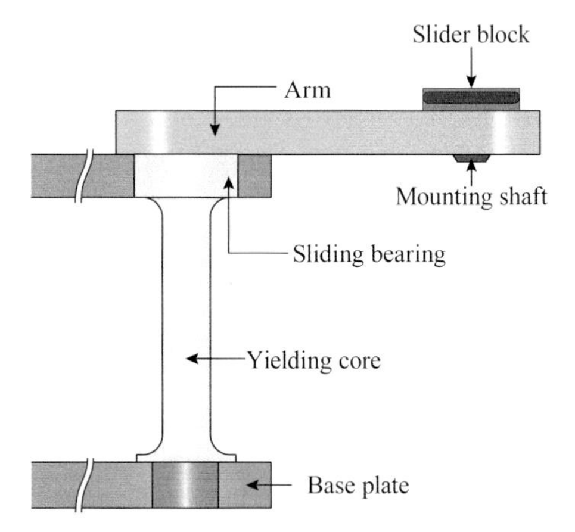

FIGURE 6.31 Energy dissipation unit of multidirectional torsional damper.

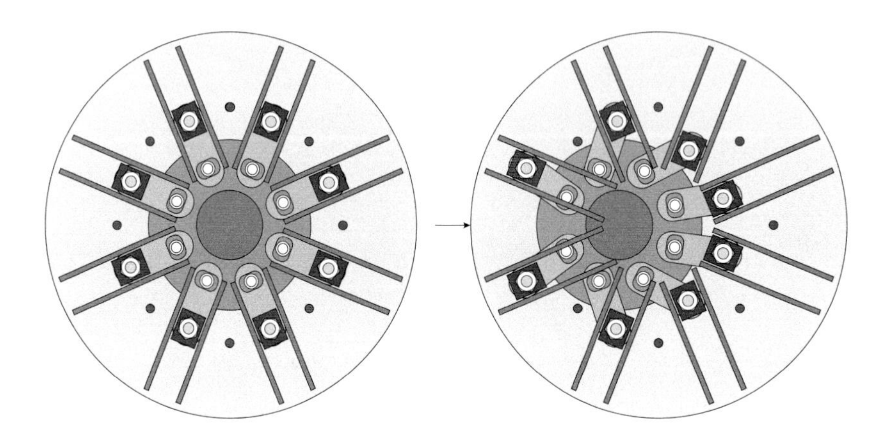

FIGURE 6.32 Schematic top-view at undisplaced and displaced positions.

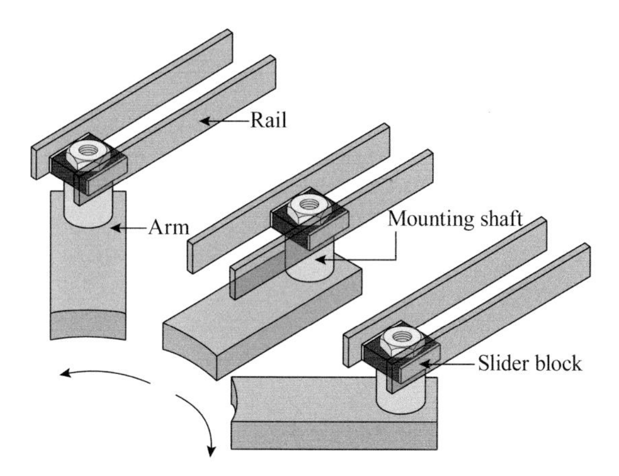

FIGURE 6.33 Mechanism of sliding and rotation of the slider block.

The combination of eight pairs of rails and the top plate is termed the rail system. There is no attachment between the slider block and the rails or the top plate and the contact between them is of frictional type. Figure 6.33 shows the mechanism of sliding and rotation of the slider block inside the rail and about the mounting shaft.

The connection between the end points of the arms, and rails is required due to the multidirectional movements of the superstructure, and it makes it impossible to directly connect the arms to the top plate or to the superstructure. The arm ends are not supposed to be restrained from moving with the superstructure in the axial direction of the arms. Thus, the rail system does not rotate, and the rails always stay parallel to the initial direction of their corresponding arms. Therefore, having the end points of the arms guided through a rail will free the arms from the unwanted displacement component of the superstructure, i.e., the parallel-to-arm component.

EXAMPLES

Example 6.1 Application of Friction Dampers to a Five-Storied Building

Here, the building frame given in Example 4.1 (see Figure 4.31) is considered to be equipped with friction dampers in all floors in the form of X bracings. Find the top floor acceleration, velocity, and displacement of the building frame under the 1940 Imperial Valley earthquake recorded at El Centro. Consider the friction damper with slip load = 0.1 times of the story weight. Furthermore, compare the response quantities of the friction damper-controlled building frame with that of the uncontrolled building, viscous damper-controlled building, and viscoelastic damper-controlled building, which are obtained in Example 4.1. The damping ratio of the superstructure is given as 0.02.

SOLUTION

The response quantities of the uncontrolled structure and the structure controlled using the friction damper are obtained by performing response history analysis. The time history analysis of the uncontrolled and controlled structures under the 1940 Imperial valley earthquake is performed, and the results are presented in Table Ex.6.1.1.

Example 6.2 Application of Friction Damper to Adjacent Buildings

Two fixed-base building frames, shown in Figure 6.34, have masses, m_1 = 10,000 kg, m_2 = 15,000 kg, and the total lateral stiffnesses of the columns, k_1 = 2,470 kN/m, k_2 = 1,470 kN/m. Furthermore, for both the buildings, the damping ratio of the superstructure (ξ_s) is 0.02, and the

TABLE EX.6.1.1
Response of the Five-Storied Building Frame Equipped with Friction Damper

Response Quantity	Uncontrolled	Viscous Damper-Controlled	Viscoelastic Damper-Controlled	Friction Damper-Controlled
Displacement (mm)	109.000	106.100	105.400	90.790
Velocity (m/s)	1.172	1.130	1.205	0.998
Acceleration (m/s^2)	11.800	11.610	11.930	9.872

base excitation is given as $\ddot{u}_g = 3sin(5\pi t)$. Answer the following questions by performing time history analysis under the base excitation $(\ddot{u}_g)$ acting for a total duration such that at least two full cycles of damper hysteresis loops would be completed (use a suitable time step (Δt) for the time history analysis).

a. Perform a time history analysis and determine the peak values of the top floor displacement and absolute acceleration response quantities of the fixed-base (uncontrolled) building frames modeling them as single degree of freedom (SDOF) systems.
b. Considering that the adjacent buildings are connected (coupled) by a friction damper, construct the mass, the stiffness, and the damping matrices of the multi-degree of freedom (MDOF) system. Subsequently, evaluate the peak values of the floor displacement and absolute acceleration of the coupled buildings. Select the properties of the passive control device appropriately, e.g., friction damper friction coefficient (μ_d) of 0.05 and clamping force (N) of 1% of the weight of an adjacent structure. Compare the peak values of the two response quantities of the coupled building frames with that of the individual uncontrolled building frames.

SOLUTION

a. **Time history analysis of the uncontrolled structures**
The basic structural properties and the uncontrolled response quantities of the structures are obtained as given in Example 4.1. Accordingly, the calculated response quantities of Structure 1 and Structure 2 (uncontrolled) for some of the time steps are given in Table Ex.6.2.1. The plots of the time histories of the response quantities for both structures are given in Figures 4.25 and 4.26. In addition, the response quantities are evaluated using two additional approached: (a) SAP2000 commercial software and (b) state-space method using MATLAB®. The comparison of the results obtained using different methods are also presented in Table Ex.6.2.1, Figures 4.25 and 4.26.

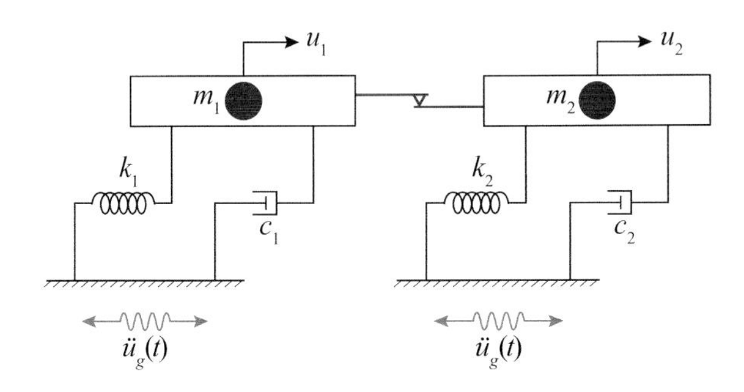

FIGURE 6.34 Adjacent building frames coupled by friction damper.

TABLE EX.6.2.1
The Displacement and Absolute Acceleration Response History Data for the Uncontrolled SDOF Structures

		Analytical Solution				SAP2000 Solution				MATLAB Solution – State-Space Method			
		Structure 1		Structure 2		Structure 1		Structure 2		Structure 1		Structure 2	
Time (s)	$\ddot{u}_g$ (m/s²)	u (m)	$\ddot{u}_{Abs}$(m/s²)	u (m)	$\ddot{u}_{Abs}$(m/s²)	u (m)	$\ddot{u}_{Abs}$(m/s²)	u (m)	$\ddot{u}_{Abs}$(m/s²)	u (m)	$\ddot{u}_{Abs}$(m/s²)	u (m)	$\ddot{u}_{Abs}$(m/s²)
0	0.0000	0.00000	0.00000	0.00000	0.00000	0.00000	0.00000	0.00000	0.00000	0.00000	0.00000	0.00000	0.00000
0.01	0.4693	–0.00001	0.93927	–0.00001	0.00232	–0.00001	0.00339	–0.00001	0.00169	0.00000	0.00000	0.00000	0.00000
0.02	0.9271	–0.00006	1.86375	–0.00006	0.01105	–0.00006	0.02107	–0.00006	0.00976	–0.00001	0.00356	–0.00001	0.00177
0.03	1.3620	–0.00021	2.76218	–0.00021	0.03044	–0.00021	0.06359	–0.00021	0.02848	–0.00006	0.02162	–0.00006	0.01000
0.04	1.7634	–0.00048	3.62312	–0.00049	0.06440	–0.00048	0.14022	–0.00049	0.06175	–0.00021	0.06474	–0.00021	0.02897
0.05	2.1213	–0.00092	4.43527	–0.00094	0.11630	–0.00091	0.25844	–0.00093	0.11295	–0.00049	0.14214	–0.00049	0.06258
0.06	2.4271	–0.00154	5.18745	–0.00158	0.18885	–0.00153	0.42355	–0.00158	0.18479	–0.00092	0.26128	–0.00094	0.11417
0.07	2.6730	–0.00236	5.86869	–0.00246	0.28404	–0.00235	0.63831	–0.00245	0.27924	–0.00155	0.42740	–0.00160	0.18646
0.08	2.8532	–0.00338	6.46834	–0.00357	0.40299	–0.00337	0.90278	–0.00356	0.39744	–0.00237	0.64323	–0.00247	0.28140
0.09	2.9631	–0.00459	6.97617	–0.00492	0.54592	–0.00459	1.21412	–0.00491	0.53962	–0.00339	0.90874	–0.00358	0.40012
0.1	3.0000	–0.00598	7.38252	–0.00652	0.71215	–0.00597	1.56663	–0.00651	0.70509	–0.00461	1.22104	–0.00494	0.54281
0.11	2.9631	–0.00752	7.67849	–0.00834	0.90002	–0.00750	1.95184	–0.00833	0.89220	–0.00600	1.57438	–0.00654	0.70878
0.12	2.8532	–0.00915	7.85612	–0.01038	1.10696	–0.00913	2.35871	–0.01036	1.09840	–0.00753	1.96023	–0.00837	0.89636
0.13	2.6730	–0.01083	7.90865	–0.01258	1.32951	–0.01081	2.77399	–0.01256	1.32023	–0.00916	2.36749	–0.01040	1.10296
0.14	2.4271	–0.01250	7.83073	–0.01492	1.56341	–0.01247	3.18258	–0.01489	1.55346	–0.01084	2.78286	–0.01260	1.32514
0.15	2.1213	–0.01409	7.61866	–0.01735	1.80369	–0.01406	3.56808	–0.01731	1.79311	–0.01250	3.19121	–0.01494	1.55861
0.16	1.7634	–0.01553	7.27066	–0.01981	2.04477	–0.01551	3.91333	–0.01977	2.03363	–0.01409	3.57610	–0.01736	1.79839
0.17	1.3620	–0.01677	6.78703	–0.02224	2.28060	–0.01673	4.20106	–0.02219	2.26899	–0.01553	3.92037	–0.01982	2.03893
0.18	0.9271	–0.01771	6.17044	–0.02458	2.50484	–0.01768	4.41448	–0.02453	2.49285	–0.01675	4.20674	–0.02224	2.27417
0.19	0.4693	–0.01831	5.42602	–0.02676	2.71099	–0.01828	4.53802	–0.02671	2.69872	–0.01769	4.41844	–0.02457	2.49777
0.2	0.0000	–0.01850	4.56153	–0.02872	2.89258	–0.01847	4.55789	–0.02866	2.88016	–0.01828	4.53991	–0.02675	2.70325
0.21	–0.4693	–0.01823	3.58743	–0.03039	3.04332	–0.01820	4.46274	–0.03033	3.03088	–0.01846	4.55741	–0.02870	2.88415
0.22	–0.9271	–0.01747	2.51687	–0.03171	3.15731	–0.01745	4.24418	–0.03165	3.14501	–0.01819	4.45965	–0.03036	3.03420
0.23	–1.3620	–0.01620	1.36564	–0.03262	3.22918	–0.01617	3.89726	–0.03256	3.21717	–0.01742	4.23831	–0.03167	3.14751
0.24	–1.7634	–0.01440	0.15204	–0.03308	3.25428	–0.01438	3.42085	–0.03301	3.24272	–0.01614	3.88851	–0.03257	3.21875

(Continued)

TABLE EX.6.2.1 (*Continued*)
The Displacement and Absolute Acceleration Response History Data for the Uncontrolled SDOF Structures

		Analytical Solution				SAP2000 Solution				MATLAB Solution – State-Space Method			
		Structure 1		Structure 2		Structure 1		Structure 2		Structure 1		Structure 2	
Time (s)	$\ddot{u}_g$ (m/s²)	u (m)	$\ddot{u}_{Abs}$(m/s²)	u (m)	$\ddot{u}_{Abs}$(m/s²)	u (m)	$\ddot{u}_{Abs}$(m/s²)	u (m)	$\ddot{u}_{Abs}$(m/s²)	u (m)	$\ddot{u}_{Abs}$(m/s²)	u (m)	$\ddot{u}_{Abs}$(m/s²)
0.25	−2.1213	−0.01208	−1.10332	−0.03303	3.22877	−0.01206	2.81787	−0.03297	3.21783	−0.01433	3.40920	−0.03301	3.24326
–	–	–	–	–	–	–	–	–	–	–	–	–	–
0.3	−3.0000	0.00582	−7.16572	−0.02452	2.28151	0.00580	−1.70130	−0.02448	2.27609	0.00170	−0.67732	−0.02723	2.56788
0.4	0.0000	0.03587	−8.84558	0.02292	−2.45083	0.03581	−8.83810	0.02286	−2.43912	0.03544	−8.80337	0.01782	−1.95376
0.5	3.0000	−0.00569	6.96638	0.04811	−4.68350	−0.00565	1.82980	0.04801	−4.66676	0.00102	0.17096	0.04855	−4.75116
0.6	0.0000	−0.05219	12.86880	0.00805	−0.55897	−0.05209	12.85754	0.00806	−0.55850	−0.05156	12.80757	0.01371	−1.11642
0.7	−3.0000	0.00558	−6.78319	−0.03745	3.75287	0.00552	−1.95238	−0.03735	3.73754	−0.00355	0.30057	−0.03511	3.54530
0.8	0.0000	0.06751	−16.64707	−0.02634	2.45030	0.06739	−16.63210	−0.02629	2.43942	0.06670	−16.56827	−0.02954	2.76959
0.9	3.0000	−0.00549	6.61491	0.00752	−0.82277	−0.00540	2.06928	0.00747	−0.81947	0.00592	−0.73963	0.00515	−0.60701
1	0.0000	−0.08189	20.19530	0.01172	−1.11462	−0.08175	20.17670	0.01165	−1.10586	−0.08092	20.10026	0.01252	−1.19616
1.1	−3.0000	0.00542	−6.46041	0.00550	−0.55600	0.00529	−2.18074	0.00549	−0.54969	−0.00813	1.14843	0.00533	−0.52377
1.2	0.0000	0.09540	−23.52749	0.01636	−1.65314	0.09524	−23.50534	0.01637	−1.64993	0.09428	−23.41745	0.01512	−1.53321
1.3	3.0000	−0.00537	6.31864	0.01220	−1.09763	−0.00520	2.28699	0.01222	−1.10104	0.01019	−1.52902	0.01435	−1.32690
1.4	0.0000	−0.10809	26.65679	−0.02430	2.53369	−0.10791	26.63119	−0.02424	2.52165	−0.10682	26.53291	−0.02051	2.16478
1.5	−3.0000	0.00534	−6.18861	−0.03776	3.63477	0.00512	−2.38825	−0.03772	3.62617	−0.01211	1.88333	−0.03907	3.78547
1.6	0.0000	0.12001	−29.59555	0.00442	−0.64940	0.11980	−29.56660	0.00435	−0.64109	0.11860	−29.45889	−0.00091	−0.12574
1.7	3.0000	−0.00532	6.06943	0.04093	−4.04539	−0.00505	2.48473	0.04081	−4.02876	0.01390	−2.21315	0.03976	−3.95402
1.8	0.0000	−0.13120	32.35537	0.01809	−1.60368	−0.13097	32.32317	0.01810	−1.59990	−0.12967	32.20693	0.02225	−2.01554
1.9	−3.0000	0.00531	−5.96025	−0.01974	2.01514	0.00499	−2.57664	−0.01962	2.00318	−0.01558	2.52014	−0.01747	1.81142
2	0.0000	0.14170	−34.94715	−0.01694	1.58418	0.14146	−34.91179	−0.01686	1.57325	0.14006	−34.78783	−0.01877	1.76766

b. **Time history analysis of the structures connected with friction damper**
The friction damper is considered to connect the two adjacent SDOF structures, Structure 1 and Structure 2. The damper is placed between the floors of the structures. The vibration response of the structures connected using the damper is evaluated subsequently.
The mass, stiffness, and damping matrices of the system of structures connected using friction damper, respectively, are given as:

$$\begin{bmatrix} m_1 & 0 \\ 0 & m_2 \end{bmatrix} = \begin{bmatrix} 10{,}000 & 0 \\ 0 & 15{,}000 \end{bmatrix} \text{kg} \tag{6.7}$$

$$\begin{bmatrix} k_1 & 0 \\ 0 & k_2 \end{bmatrix} = \begin{bmatrix} 2{,}470 & 0 \\ 0 & 1{,}470 \end{bmatrix} \text{kN/m}, \tag{6.8}$$

and

$$\begin{bmatrix} c_1 & 0 \\ 0 & c_2 \end{bmatrix} = \begin{bmatrix} 6.2865 & 0 \\ 0 & 5.9397 \end{bmatrix} \text{kNs/m} \tag{6.9}$$

The equations of motion for the structures connected by the friction damper can be expressed in matrix form as:

$$\begin{bmatrix} m_1 & 0 \\ 0 & m_2 \end{bmatrix} \begin{Bmatrix} \ddot{u}_1 \\ \ddot{u}_2 \end{Bmatrix} + \begin{bmatrix} c_1 & 0 \\ 0 & c_2 \end{bmatrix} \begin{Bmatrix} \dot{u}_1 \\ \dot{u}_2 \end{Bmatrix} + \begin{bmatrix} k_1 & 0 \\ 0 & k_2 \end{bmatrix} \begin{Bmatrix} u_1 \\ u_2 \end{Bmatrix}$$
$$+ \begin{Bmatrix} -1 \\ 1 \end{Bmatrix} F_d = -\begin{bmatrix} m_1 & 0 \\ 0 & m_2 \end{bmatrix} \begin{Bmatrix} 1 \\ 1 \end{Bmatrix} 3sin(5\pi t) \tag{6.10}$$

where $F_d = \mu_d N sgn(\dot{u}_2 - \dot{u}_1)$ is the force in the friction damper.
The friction coefficient of the friction damper is taken as: $\mu_d = 0.05$.
Three different cases/values of the clamping force (N) are considered. The values of N considered for each case are given as:

$$\text{Case 1: } N = 0.01m_1 \times 9.81 \text{ m/s}^2 = 981 \text{ N}$$

$$\text{Case 2: } N = 0.01m_2 \times 9.81 \text{ m/s}^2 = 1{,}471.5 \text{ N}$$

$$\text{Case 3: } N = 0.5(m_1 + m_2) \times 9.81 \text{ m/s}^2 = 122{,}625 \text{ N}$$

The response of the structures connected with the friction damper is evaluated using the central difference method described earlier. Furthermore, the results of the central difference method are also compared with that of the SAP2000 simulation and the state-space method. The plot of the time histories of the response quantities for the structures connected with friction damper (Case 3) are given in Figures 6.35 and 6.36. Furthermore, the comparison of the peak values of the response quantities obtained using different methods is also presented in Table Ex.6.2.2.

- **Peak values of the response quantities**
The peak values of the floor displacements and absolute accelerations are obtained by taking the absolute maximum of the response quantities. The comparison of the peak values of the displacements and the absolute accelerations for the unconnected and connected structures are given in Table Ex.6.2.3. Furthermore, the percentage reduction

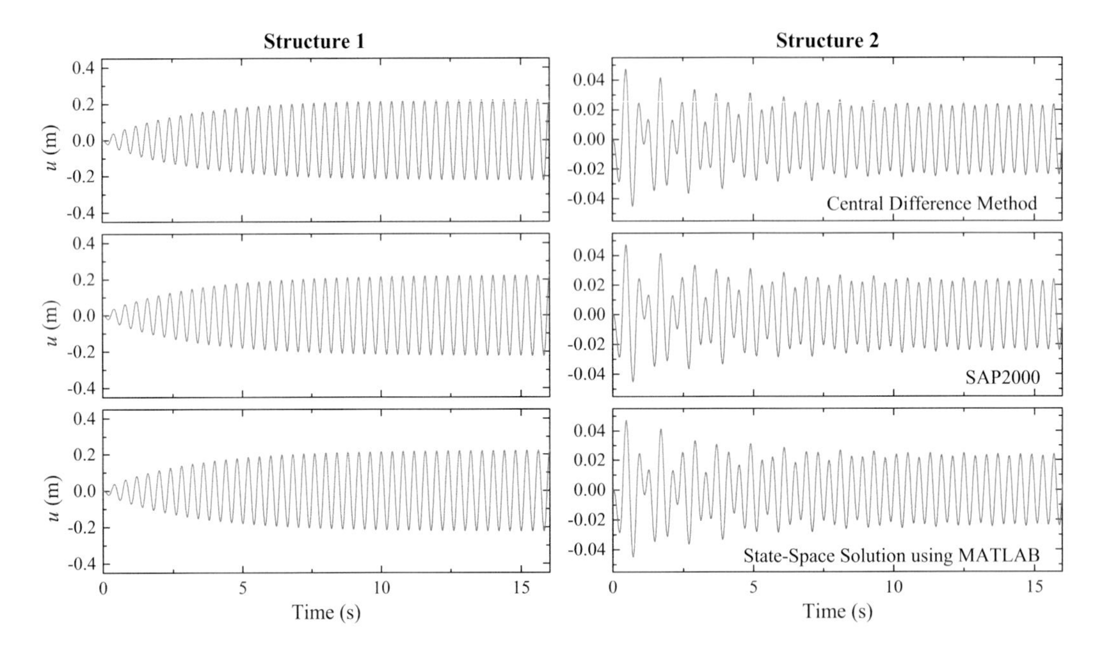

FIGURE 6.35 Displacement time history plots of the structures connected with friction damper (Case 3).

in the peak response quantities of the connected structures relative to the unconnected response quantities is presented in Table Ex.6.2.4.

Response time histories of the displacement and force of the friction damper together with the response of the connected structures (Case 3) are presented in Figure 6.37.

- **Force-deformation curves of the dampers**
  The force-deformation curve of the friction damper is given in Figure 6.38.

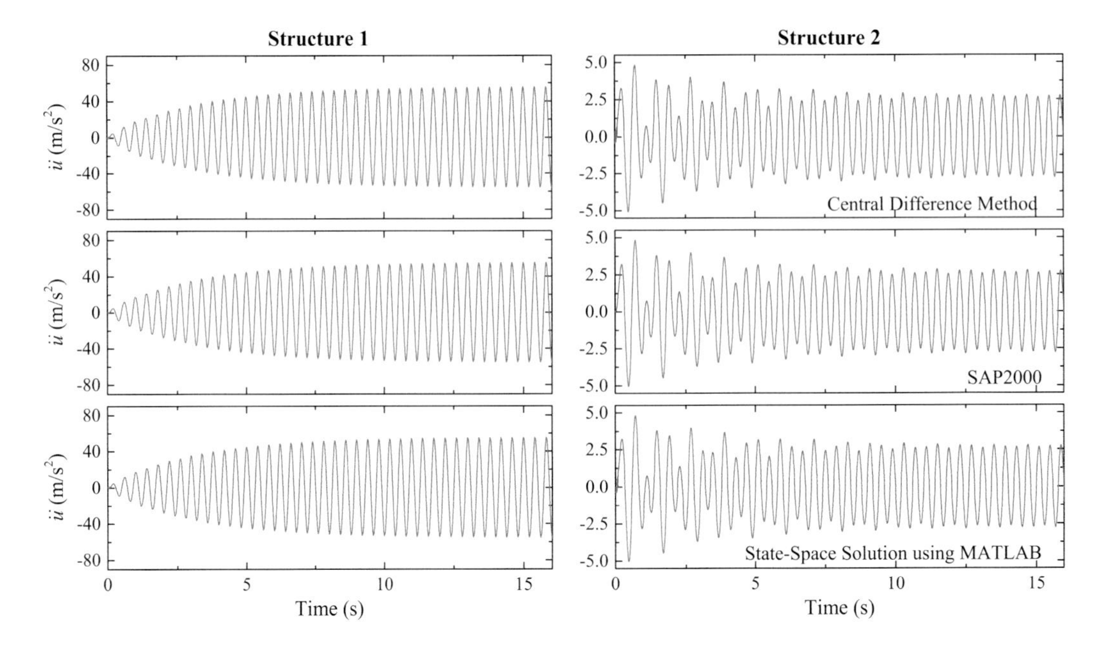

FIGURE 6.36 Acceleration time history plots of the structures connected with friction damper (Case 3).

TABLE EX.6.2.2
The Displacement and Absolute Acceleration Response History Data for the Structures Connected Using Friction Damper (Case 3)

		Solution Using Central Difference Method				SAP2000 Solution				MATLAB Solution – State-Space Method			
		Structure 1		Structure 2		Structure 1		Structure 2		Structure 1		Structure 2	
Time (s)	$\ddot{u}_g$ (m/s²)	u (m)	$\ddot{u}_{Abs}$(m/s²)	u (m)	$\ddot{u}_{Abs}$(m/s²)	u (m)	$\ddot{u}_{Abs}$(m/s²)	u (m)	$\ddot{u}_{Abs}$(m/s²)	u (m)	$\ddot{u}_{Abs}$(m/s²)	u (m)	$\ddot{u}_{Abs}$(m/s²)
0	0.0000	0.00000	0.00000	0.00000	0.00000	0.00000	0.00000	0.00000	0.00000	0.00000	0.00000	0.00000	0.00000
0.01	0.4693	–0.00001	–0.65509	–0.00001	0.36432	–0.00001	0.00248	–0.00001	0.00230	–0.00001	0.61676	–0.00001	–0.40700
0.02	0.9271	–0.00007	0.59244	–0.00007	–0.44231	–0.00006	0.01452	–0.00006	0.01414	–0.00006	0.63489	–0.00006	–0.39879
0.03	1.3620	–0.00023	–0.58507	–0.00023	0.39800	–0.00021	0.04293	–0.00021	0.04238	–0.00021	0.67830	–0.00021	–0.37991
0.04	1.7634	–0.00052	–0.49974	–0.00052	0.43671	–0.00048	0.09392	–0.00048	0.09308	–0.00049	0.75671	–0.00049	–0.34661
0.05	2.1213	–0.00098	0.85695	–0.00098	–0.32343	–0.00093	0.17263	–0.00093	0.17149	–0.00094	–0.34798	–0.00094	0.52176
0.06	2.4271	–0.00164	–0.18818	–0.00164	0.57238	–0.00156	0.28304	–0.00156	0.28157	–0.00158	1.04920	–0.00158	–0.22485
0.07	2.6730	–0.00251	0.04653	–0.00252	0.67342	–0.00241	0.42777	–0.00241	0.42596	–0.00243	0.04763	–0.00243	0.68499
0.08	2.8532	–0.00360	1.56468	–0.00361	–0.02075	–0.00348	0.60824	–0.00348	0.60555	–0.00351	1.55393	–0.00351	–0.01804
0.09	2.9631	–0.00492	0.68494	–0.00494	0.94214	–0.00478	0.82575	–0.00478	0.81859	–0.00481	0.66254	–0.00481	0.93555
0.1	3.0000	–0.00645	1.08193	–0.00647	1.10807	–0.00628	1.09008	–0.00629	1.05492	–0.00632	1.04879	–0.00632	1.09183
0.11	2.9631	–0.00816	1.52387	–0.00820	1.29152	–0.00798	1.46966	–0.00799	1.26222	–0.00802	1.47993	–0.00802	1.26600
0.12	2.8532	–0.01002	1.99760	–0.01009	1.49004	–0.00982	1.93214	–0.00986	1.45191	–0.00987	1.94401	–0.00990	1.45565
0.13	2.6730	–0.01197	2.48796	–0.01211	1.70010	–0.01176	2.41343	–0.01187	1.65367	–0.01181	2.42624	–0.01191	1.65747
0.14	2.4271	–0.01393	2.97780	–0.01424	1.91741	–0.01372	2.89649	–0.01398	1.86364	–0.01377	2.90979	–0.01402	1.86744
0.15	2.1213	–0.01584	3.44849	–0.01641	2.13701	–0.01563	3.36303	–0.01615	2.07710	–0.01568	3.37634	–0.01618	2.08079
0.16	1.7634	–0.01761	3.88061	–0.01858	2.35341	–0.01742	3.79386	–0.01832	2.28875	–0.01746	3.80666	–0.01835	2.29223
0.17	1.3620	–0.01917	4.25459	–0.02070	2.56070	–0.01900	4.16958	–0.02044	2.49287	–0.01904	4.18132	–0.02046	2.49602
0.18	0.9271	–0.02045	4.55141	–0.02269	2.75275	–0.02030	4.47121	–0.02244	2.68343	–0.02034	4.48139	–0.02246	2.68613
0.19	0.4693	–0.02136	4.75337	–0.02450	2.92330	–0.02125	4.68101	–0.02427	2.85427	–0.02128	4.68913	–0.02428	2.85641
0.2	0.0000	–0.02184	4.84471	–0.02606	3.06621	–0.02178	4.78309	–0.02586	2.99929	–0.02180	4.78870	–0.02587	3.00075
0.21	–0.4693	–0.02184	4.81234	–0.02732	3.17558	–0.02183	4.76415	–0.02715	3.11263	–0.02184	4.76680	–0.02715	3.11324
0.22	–0.9271	–0.02132	4.64636	–0.02821	3.24592	–0.02136	4.61382	–0.02808	3.18863	–0.02135	4.61330	–0.02807	3.18838
0.23	–1.3620	–0.02023	4.34061	–0.02869	3.27236	–0.02033	4.32568	–0.02860	3.22245	–0.02031	4.32170	–0.02858	3.22118
0.24	–1.7634	–0.01858	3.89308	–0.02871	3.25074	–0.01874	3.89713	–0.02866	3.20975	–0.01871	3.88959	–0.02863	3.20740

(Continued)

TABLE EX.6.2.2 (*Continued*)

The Displacement and Absolute Acceleration Response History Data for the Structures Connected Using Friction Damper (Case 3)

		Solution Using Central Difference Method				SAP2000 Solution				MATLAB Solution – State-Space Method			
		Structure 1		Structure 2		Structure 1		Structure 2		Structure 1		Structure 2	
Time (s)	$\ddot{u}_g$ (m/s²)	u (m)	$\ddot{u}_{Abs}$ (m/s²)	u (m)	$\ddot{u}_{Abs}$ (m/s²)	u (m)	$\ddot{u}_{Abs}$ (m/s²)	u (m)	$\ddot{u}_{Abs}$ (m/s²)	u (m)	$\ddot{u}_{Abs}$ (m/s²)	u (m)	$\ddot{u}_{Abs}$ (m/s²)
0.25	−2.1213	−0.01637	3.30613	−0.02822	3.17782	−0.01659	3.32999	−0.02823	3.14709	−0.01654	3.31889	−0.02819	3.14358
–	–	–	–	–	–	–	–	–	–	–	–	–	–
0.3	−3.0000	0.00174	−1.32117	−0.01767	2.00372	0.00128	−1.20657	−0.01798	2.03592	0.00138	−1.23201	−0.01788	2.02664
0.4	0.0000	0.03725	−8.68924	0.02922	−3.49547	0.03708	−8.60150	0.02871	−3.39873	0.03718	−8.62864	0.02876	−3.40389
0.5	3.0000	0.00233	0.46552	0.04451	−4.68640	0.00303	0.28882	0.04464	−4.70097	0.00290	0.32332	0.04449	−4.68756
0.6	0.0000	−0.04987	11.83031	−0.00474	1.16329	−0.04958	11.71512	−0.00413	1.05635	−0.04976	11.76263	−0.00428	1.07086
0.7	−3.0000	−0.00485	0.03576	−0.04488	4.83549	−0.00572	0.25447	−0.04477	4.82664	−0.00561	0.22556	−0.04465	4.81545
0.8	0.0000	0.06080	−14.53641	−0.01854	1.16892	0.06046	−14.40858	−0.01904	1.26466	0.06067	−14.46371	−0.01877	1.23981
0.9	3.0000	0.00566	−0.13139	0.02135	−2.57812	0.00669	−0.38919	0.02114	−2.55949	0.00661	−0.36717	0.02115	−2.56056
1	0.0000	−0.07081	17.00135	0.01542	−0.95877	−0.07045	16.87204	0.01572	−1.03539	−0.07065	16.92495	0.01548	−1.01191
1.1	−3.0000	−0.00524	−0.06882	−0.00307	0.72674	−0.00650	0.24704	−0.00291	0.71338	−0.00638	0.21570	−0.00308	0.72885
1.2	0.0000	0.08019	−19.30996	0.00726	−1.23774	0.07983	−19.17922	0.00710	−1.17452	0.08001	−19.22976	0.00716	−1.18061
1.3	3.0000	0.00460	0.31487	0.01011	−1.33004	0.00605	−0.04619	0.01020	−1.34048	0.00592	−0.01367	0.01027	−1.34739
1.4	0.0000	−0.08907	21.50349	−0.02144	2.69172	−0.08869	21.36981	−0.02115	2.61620	−0.08887	21.41850	−0.02113	2.61474
1.5	−3.0000	−0.00467	−0.38244	−0.03155	3.41982	−0.00631	0.02620	−0.03173	3.43867	−0.00613	−0.01982	−0.03166	3.43288
1.6	0.0000	0.09752	−23.59377	0.01191	−1.83125	0.09713	−23.45878	0.01140	−1.73379	0.09730	−23.50684	0.01145	−1.73907
1.7	3.0000	0.00517	0.34046	0.04155	−4.46812	0.00703	−0.12305	0.04153	−4.46821	0.00679	−0.06317	0.04142	−4.45782
1.8	0.0000	−0.10540	25.54019	0.00796	−0.11928	−0.10499	25.40322	0.00848	−0.21776	−0.10516	25.45026	0.00830	−0.20060
1.9	−3.0000	−0.00541	−0.35645	−0.02994	3.39330	−0.00747	0.15559	−0.02974	3.37508	−0.00718	0.08225	−0.02972	3.37334
2	0.0000	0.11270	−27.33962	−0.01451	0.82891	0.11229	−27.20139	−0.01484	0.90844	0.11244	−27.24649	−0.01464	0.88926

TABLE EX.6.2.3
Comparison of the Peak Response Quantities of the Unconnected and Connected Structures Using Different Response Analysis Methods

		Structure 1		Structure 2	
Structure	**Analysis Method**	**Displacement (m)**	**Acceleration (m/s²)**	**Displacement (m)**	**Acceleration (m/s²)**
Uncontrolled/ unconnected	Analytical	0.30359	74.90839	0.04866	4.77627
	SAP2000	0.30319	74.80871	0.04855	4.75863
	State-space	0.30250	74.78020	0.04860	4.76620
Friction damper connected (Case 1)	Central difference	0.30307	74.96583	0.04864	4.78867
	SAP2000	0.30283	74.71509	0.04853	4.75962
	State-space	0.30190	74.61990	0.04850	4.76620
Friction damper connected (Case 2)	Central difference	0.30276	74.89025	0.04862	4.78881
	SAP2000	0.30251	74.63511	0.04851	4.76009
	State-space	0.30160	74.53980	0.04850	4.76620
Friction damper connected (Case 3)	Central difference	0.22501	56.19804	0.04740	5.08181
	SAP2000	0.22471	55.69126	0.04725	5.04927
	State-space	0.22410	55.89190	0.04720	5.04170

TABLE EX.6.2.4
Percentage Reduction in the Peak Response Quantities of the Connected Structures as Compared to the Unconnected Response Quantities

		Structure 1		Structure 2	
Structure	**Analysis Method**	**Displacement Reduction (%)**	**Acceleration Reduction (%)**	**Displacement Reduction (%)**	**Acceleration Reduction (%)**
Friction damper connected (Case 1)	Central difference	0.17	−0.08	0.04	−0.26
	SAP2000	0.12	0.13	0.04	−0.02
	State-space	0.20	0.21	0.21	0.00
Friction damper connected (Case 2)	Central difference	0.27	0.02	0.08	−0.26
	SAP2000	0.22	0.23	0.08	−0.03
	State-space	0.30	0.32	0.21	0.00
Friction damper connected (Case 3)	Central difference	25.88	24.98	2.59	−6.40
	SAP2000	25.88	25.56	2.68	−6.11
	State-space	25.92	25.26	2.88	−5.78

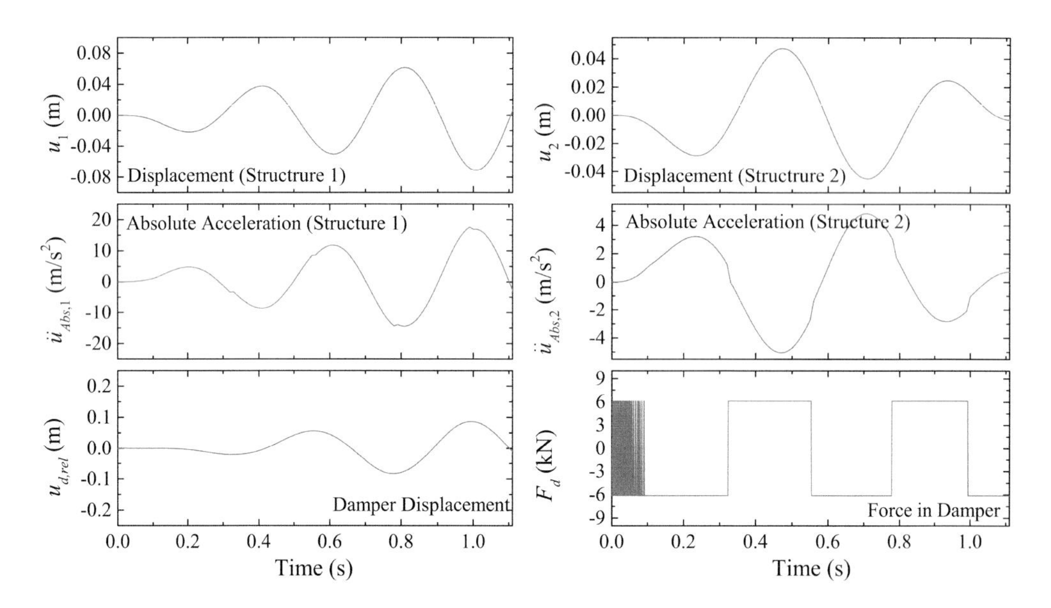

FIGURE 6.37 Response time histories of the structures connected with friction damper (Case 3).

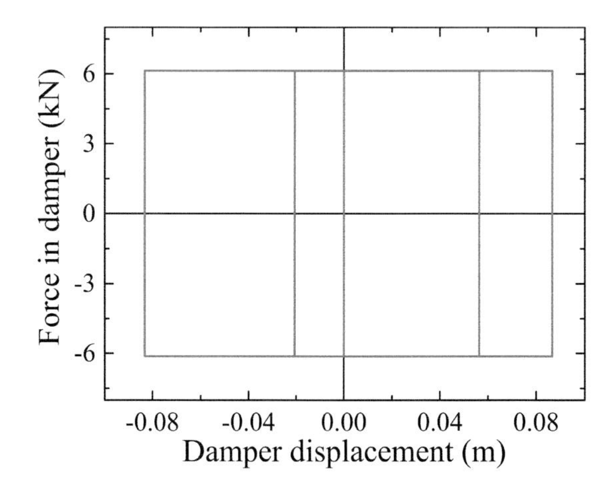

FIGURE 6.38 Force-deformation curve friction damper (Case 3).

REFERENCES

Heysami, A. (2015). Types of dampers and their seismic performance during an earthquake. *Current World Environment*, 10(Special-Issue1), 1002–1015.

Keightley, W. O. (1977). Building damping by Coulomb friction. In Sixth World Conference on Earthquake Engineering, New Delhi, India.

Mayes, R. L., and Mowbray, N. A. (1974). The effect of coulomb damping on multi degree of freedom elastic structures. *Earthquake Engineering and Structural Dynamics*, 3(3), 275–286.

Pall, A., and Pall, T. (2004). Performance-based design using pall friction dampers-an economical design solution. In 13th World Conference on Earthquake Engineering, Vancouver, BC, Canada, 70(7), 576–571.

Pall, A. S., and Marsh, C. (1982). Response of friction damped braced frames. *Journal of Structural Engineering*, 108(9), 1313–1323.

Pall, A. S., Marsh, C., and Fazio, P. (1980). Friction joints for seismic control of large panel structures. *Journal of Prestressed Concrete Institute*, 25(6), 38–61.

Patil, V. B., and Jangid, R. S. (2009). Double friction dampers for wind excited benchmark building. *International Journal of Applied Science and Engineering*, 7(2), 95–114.

7 Metallic Dampers

7.1 INTRODUCTION

Civil structures are commonly subjected to various types of dynamic and environment loadings, such as wind, traffic, and earthquakes. In particular, earthquake events cause severe damage to building and bridge structures. To prevent or mitigate such damage, various structural vibration control techniques have been developed and utilized skillfully in different structures. Passive hysteresis devices are divided into metallic and friction dampers, whose energy dissipation is independent of the loading rate. Metallic dampers dissipate energy through the inelastic deformation of their constitutive material. These dampers have stable hysteretic behavior, rate independence, and resistance against ambient temperature. Furthermore, they are reliable and provide the most effective and economical mechanism of energy dissipation through the yielding of metals. Many variables are typically considered in the study of yielding metallic dampers, such as its strength, stiffness, total energy absorbed, damping ratio, fatigue strength, deformation capacity ratio, dissipated energy to weight ratio, force to weight ratio, and the construction cost. Photo 7.1 shows metallic dampers installed in buildings, whereas Photo 7.2 shows various types of metallic dampers used for different applications.

7.2 X-SHAPED METALLIC DAMPER (ADAS DEVICE)

The conventional bracing system may not be adequate for the mitigation of dynamic loading. Added damping and added stiffness (ADAS) dampers are strongly recommended for use in moment resisting frames, particularly in chevron bracing system configuration in concrete and steel frames, to provide additional damping and stiffness to structures. Along with the conventional bracing, the X-type configuration of metallic dampers is commonly implemented. The ADAS dampers have stable hysteretic behavior, high lateral stiffness, high damping, long-term reliability, and they can be easily incorporated into structures. ADAS damper consists of X-shaped steel plates connected in parallel to the base plate using bolted connections. Figure 7.1 shows a typical X-shaped (ADAS) damper, also known as an X-plate metallic damper.

Working Mechanism of X-Shaped Metallic Damper

X-plate metallic dampers are made of multiple thin metallic plates and are intrinsically used along with the appropriate bracing system in buildings and bridges. Their combination with bracing system causes energy dissipation through flexural yielding deformation when such system is subjected to earthquake, wind, or any other lateral loading. The combined assembly of dampers and the bracing system is known as a device-brace assembly. Several cycles of yielding deformation are well sustained by the device-brace assembly, providing additional damping, thus causing a high level of energy dissipation. Each plate is characterized with double curvature, and such multiple plates are arranged parallel to each other. The number of plates used depends on the required amount of energy that needs to be dissipated in the given system. The plates used for manufacturing the X-shaped dampers are made of mild steel, lead, or any other metallic alloy, which exhibits large deformation characteristics. The advantage of using X-plate dampers in structures is that they show constant strain variation over the height of the structure, which ensures simultaneous yielding, and it may remain uniform over the full height of the damper. The damper restricts the behavior of the structure to the linear elastic range.

DOI: 10.1201/9781315269269-7

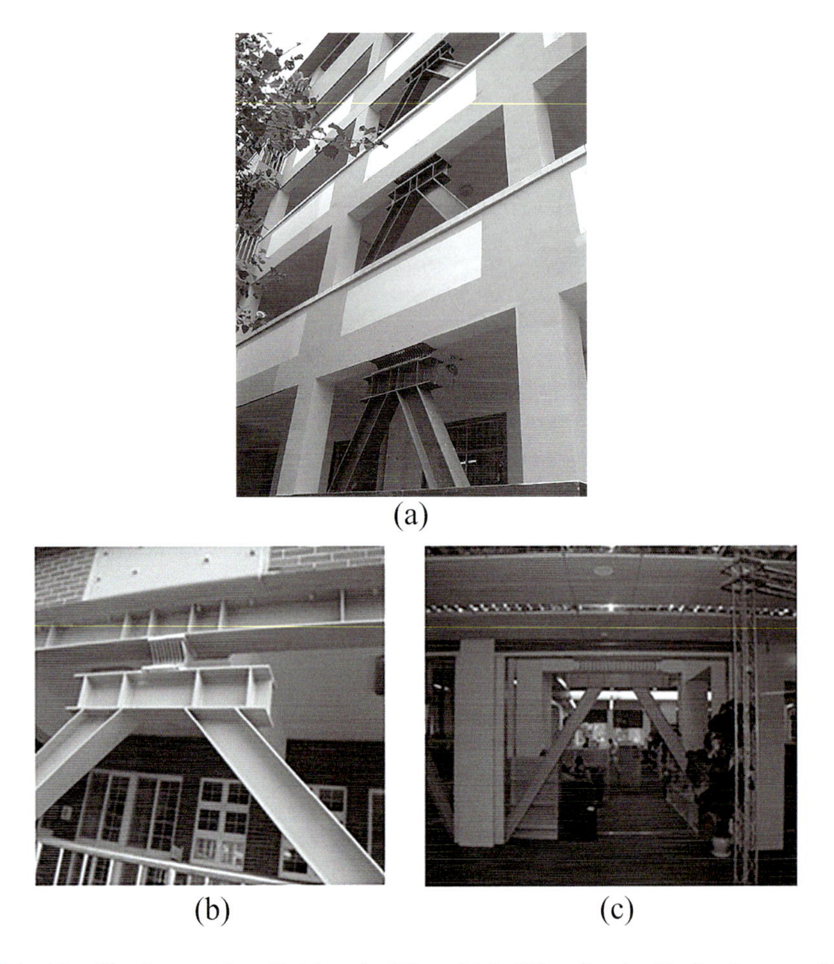

PHOTO 7.1 Metallic dampers installed in a building: (a) building fitted with the damper, (b) metallic damper detail, and (c) internal building frame with metallic damper. (Teruna et al. 2015.)

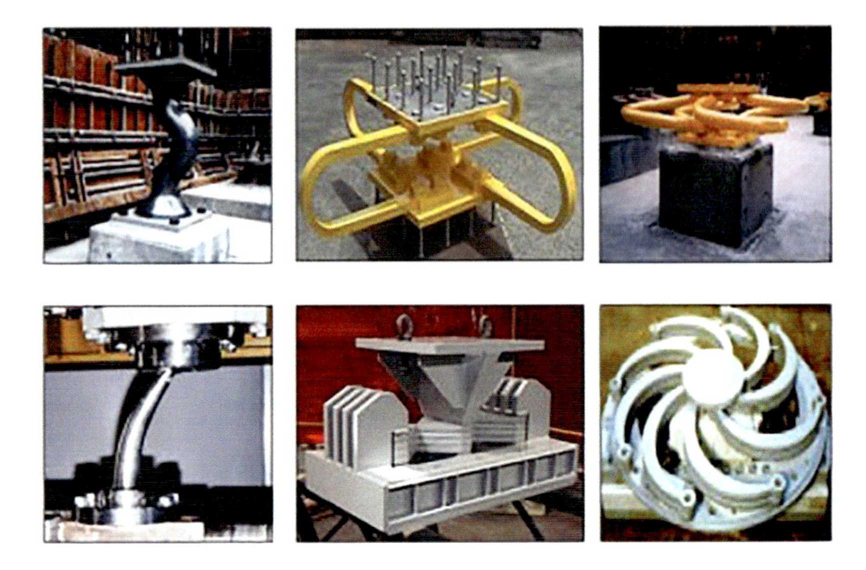

PHOTO 7.2 Various types of metallic dampers. (Heysami 2015.)

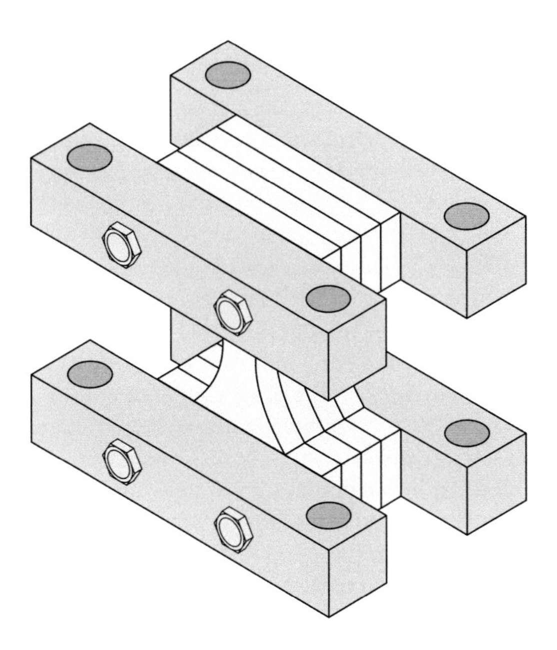

FIGURE 7.1 X-shaped added damping and added stiffness damper.

Mathematical Model of X-Plate Damper

Like friction dampers, metallic dampers are also modeled as displacement-dependent devices. During cyclic loading, the X-plate damper exhibits a stable hysteresis loop and enables a considerable amount of energy dissipation. The transition of the loop from its initial stiffness k_d to its limiting value is very smooth. For such smooth behavior, Wen's hysteresis model is often used to achieve the desired force-deformation characteristics. The restoring force in the X-plate damper is given by,

$$f_d = \alpha k_d u_p + (1-\alpha) k_d q Z \tag{7.1}$$

where Z is expressed by Equation (7.2).

$$q\frac{dZ}{dt} = A\dot{u}_p - \beta\left|\dot{u}_p\right| Z|Z|^{n-1} - \gamma\dot{u}_p|Z|^n \tag{7.2}$$

In Equation (7.2), A, β, γ, and n are dimensionless Wen's model parameters, and α is the post-to-pre yield stiffness ratio for the X-plate damper, Z is a function governed by Equation (7.2) that has a value that varies from −1 to +1, u_p and $\dot{u}_p$, respectively, are the displacement and velocity of the system where X-plate damper is installed. The Wen's model parameters influence only the geometry of the hysteresis loop. However, the geometrical properties of the X-plate damper control the seismic performance of the whole damping system. The rate of hysteresis energy dissipated by an X-plate damper at time T is given by Equation (7.3).

$$\frac{dE_h}{dT} = (1-\alpha) k_d Z\dot{u}_p \tag{7.3}$$

The total hysteresis energy dissipated is given by Equation (7.4).

$$E_h = (1-\alpha) k_d \int_0^T Z\dot{u}_p dT \tag{7.4}$$

7.3 TYLER'S YIELDING STEEL BRACING SYSTEM

Inelastic deformation of metals is an effective mechanism for energy dissipation. Traditionally, seismic-resistant design of structures depends on the ductility of structural members to dissipate energy imparted to the structure due to the dynamic excitations. To increase the seismic capabilities of concentrically braced frames, uniquely designed buckling restrained braces (BRB) are employed. However, the core of BRB, which extends beyond its sleeve, is likely to buckle and eventually fracture at its ends, as seen in Photo 7.3. This could lead to the failure of the connection. Hence, the ductility and energy dissipating capacities of BRB are not fully utilized.

To ensure the efficient utilization of the ductility and energy dissipating capacities, a steel yielding brace system (YBS) is developed, which is a highly ductile brace frame that can be used as an alternative to BRB. Further, a YBS has a high potential to provide a stiffer structure with increased ductility. The YBS is a highly elastic stiffened lateral force resisting system with a full and symmetrical hysteresis loop. It is characterized by a post-yield strength increase at large drifts. The main concern in seismic design is controling the peak lateral displacement during design basis earthquake (DBE). Under DBE, a system with high elastic stiffness shows excellent performance, as it is based on the equal displacement approximation. When drifts are large, an increase in post-yield strength ensures the prevention of potential collapse. Inelastic deformations are likely to concentrate at one of the stories of a building, where YBS with low post-yield stiffness is installed. Thus, premature collapse is localized in only one story of the building due to the increased concentration of elastic deformation. Figure 7.2 shows the steel YBS mechanism.

YBSs are generally provided along with ADAS devices to increase the seismic stability of structures. In structural design, the bracing members are considered rigid that provide large stiffness. When ADAS devices are employed along with the YBS, they are designed so that the column shear forces are drastically reduced. Considering that F_T is the maximum shear force developed at a story during an earthquake, and F_D is the design shear force in a column, the force F_R, to be resisted by ADAS dampers is given by Equation (7.5).

$$F_R = \left(F_T - F_D\right) \tag{7.5}$$

PHOTO 7.3 Out-of-plane buckling of gusset plate of BRB connection. (Gray 2012.)

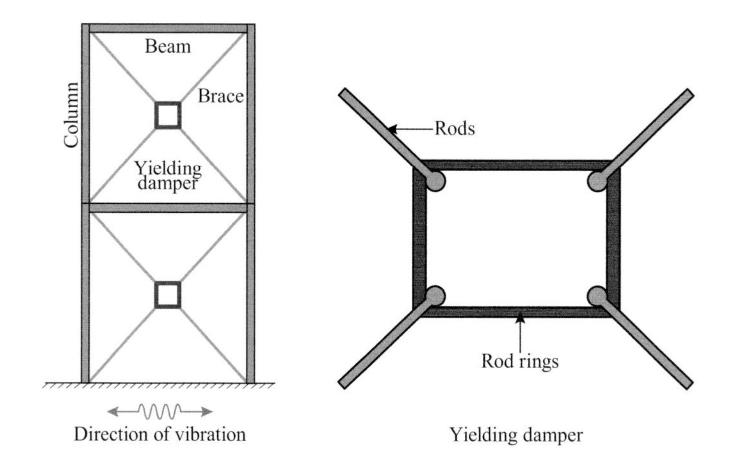

FIGURE 7.2 Yielding brace system.

Cast Steel Yielding Brace

The cast steel YBS is a hysteretic damper developed at the University of Toronto to enhance the seismic performance of braced frames. In this system, shown in Photo 7.4, the cast steel connector dissipates seismic energy through the inelastic flexural yielding of triangular fingers. The device prevents the tensile yielding and inelastic buckling of traditional braces. In addition, the damper provides a symmetrical hysteresis loop with increased energy dissipation, and the hysteresis loop of the cast steel YBS is presented in Figure 7.3.

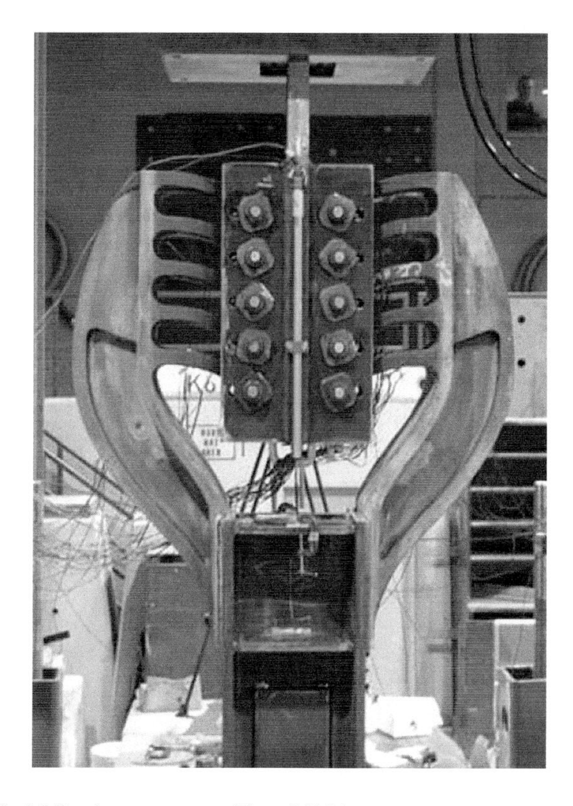

PHOTO 7.4 Cast steel yielding brace system. (Gray 2012.)

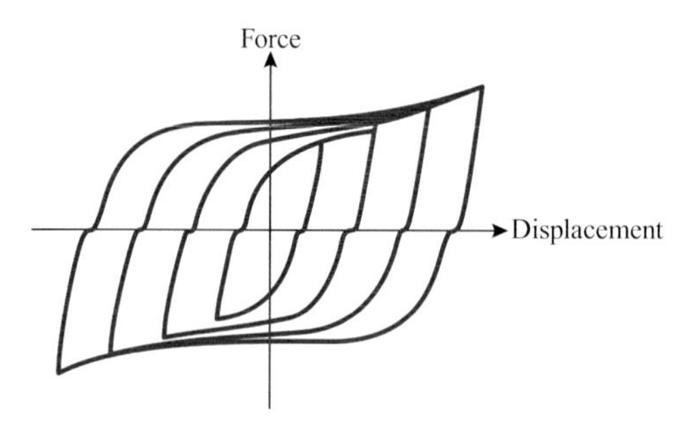

FIGURE 7.3 Hysteresis loop of cast steel yielding brace system.

7.4 STEEL DAMPER

Steel dampers are fabricated from mild steel plates with different geometrical shapes, such as straight, concave, or convex. The performance of steel dampers has been verified experimentally by several quasi-static cyclic tests. The resisting force offered by the dampers is proportional to the displacement. The specimen with a convex shape shows stable hysteretic behavior with desirable energy dissipation capabilities and ductility. Metallic dampers are capable of dissipating a large amount of input energy due to the hysteresis behavior of steel. It is shown in Figure 7.4 that the hysteretic energy is dissipated through: (i) axial yielding of the steel braces/BRB, (ii) flexural yielding of the ADAS/triangular-plate ADAS (TADAS) devices, or (iii) shear yielding of the shear links/link beams.

The steel damper exhibits stable hysteretic behavior, has long-term reliability, and can be easily incorporated into the structure. Various steel dampers such as bar type, loop type, portal type, ring type, and plate type are widely used. Figure 7.5 shows the hysteresis loop of a shear yielding damper.

7.4.1 Bar-Type Steel Damper

Bar dampers are primarily designed to be installed as beam-column connections at the corner of the building frame. Bar-type steel damper comprises a number of solid steel bars, provided in horizontal and vertical directions, as shown in Figure 7.6. The horizontal bars attached to the column in the structural frame dissipate seismic energy by flexural yielding. The vertical bars attached to the beam bottom dissipate energy by axial and shear yielding. As shown in Figure 7.6, the ends of the vertical bars are welded to the top and bottom plates to provide rigidity. For horizontal bars connected to the column, rigidity is provided through bolted connections. The damper may fail under cyclic loading, as the back-and-forth movement of the damper develops large axial compressive force in the bars. Due to cyclic loading, the mechanical properties of bar damper, such as energy dissipation rate, stiffness, and strength, degrade considerably. The parallel steel bars have common end support conditions.

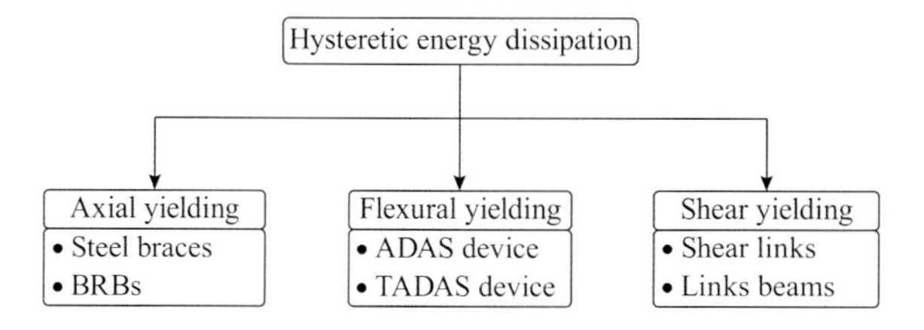

FIGURE 7.4 Dissipation of hysteretic energy.

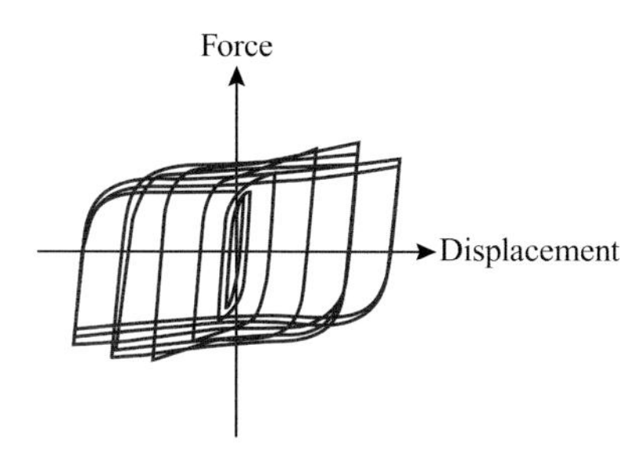

FIGURE 7.5 Hysteresis loop of shear yielding damper (aluminum/steel shear link).

When a horizontal deformation is imposed to the frame by applying force *P*, each solid bar of the damper will be under in-plane shear force, as shown in Figure 7.7, dissipating seismic energy by flexural yielding.

The main advantage of a bar damper is that it is a low-cost metallic damping system fabricated from ordinarily existing steel material without any complex assembling technique. Therefore, the damper is widely used in practice.

7.4.2 Loop-Type Steel Damper

The concept of a loop-type steel damper was first proposed by Kelly et al. (1972). In the loop-type steel metallic damper, also known as hysteretic steel damper, the seismic energy dissipation occurs through hysteresis behavior happening due to the plastic deformation and yielding of metallic components. The dampers are placed at selected locations of the main structure and are rigidly connected to it. They can be easily replaced if damaged after a large seismic event. The loop-type steel damper is typically designed with sufficient elastic strength and stiffness to prevent inelastic deformation under service conditions. Good energy dissipation capability and resistance to low cycle fatigue are the key characteristics of loop-type steel dampers.

7.4.3 Portal-Type Steel Damper

Ghabussi et al. (2020) used steel-curved dampers to improve the seismic performance of portal frame structures. Steel-curved dampers have been quite effective in increasing the strength of portal

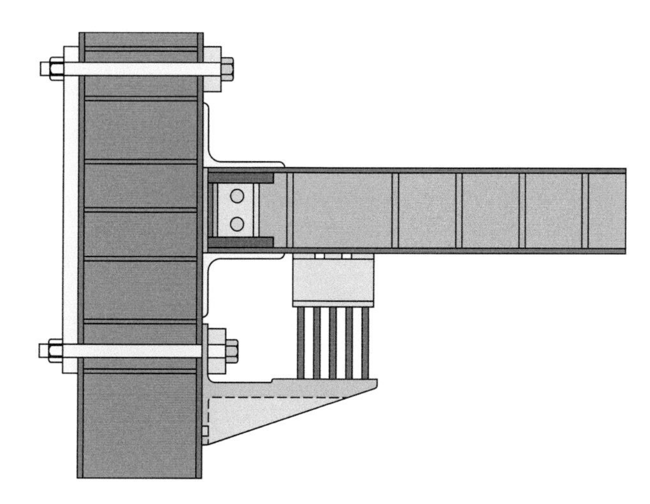

FIGURE 7.6 Bar-type damper installed in a frame.

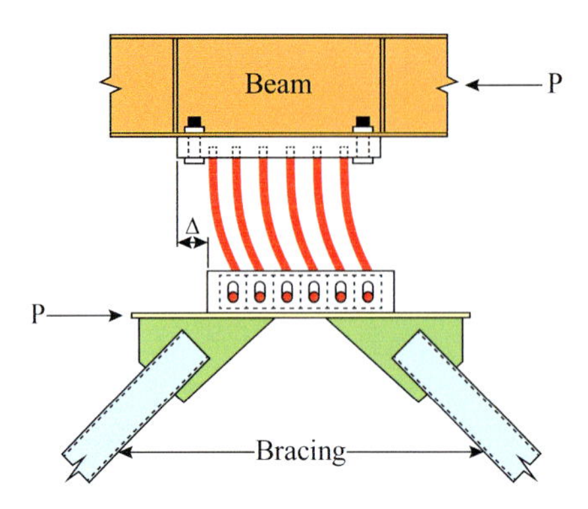

FIGURE 7.7 Bar damper connected to bracing system.

frames. The dampers are connected to portal frames using hinges made of steel plates, with specified geometries as per the precise design requirements. Also, the dampers are placed equidistant from the centroid of the frame and the axis of loading. Steel-curved dampers can easily bend under the action of external loads. The yielding of dampers at an early stage makes it possible for them to control the seismic damage of structural members. Steel-curved dampers can be used with different angles, i.e., 30°, 60°, 90°, and 120° between their two ends on pitched roof symmetric and mono-pitch portal frames. In countries where steel is very costly and large size steel sections are not readily available, providing built-up sections or increasing the member size is not feasible. Further, in earthquake-prone countries, reducing structural weight is essential to minimize the magnitude of the earthquake-induced inertial force considerably. In such cases, using steel-curved dampers can lead to considerable savings in the structural weight. However, portal-type steel dampers cannot be used for long-span portal frames, such as industrial buildings and hangars. Figure 7.8 shows the steel-curved damper attached to the portal frame, and Figure 7.9 shows the mono-pitch portal frame equipped with steel-curved dampers.

7.4.4 Ring-Type Steel Damper

The base shear demands of a structure during earthquakes are usually reduced by using concentrically braced frames. However, it is observed that concentrically braced frames fail to

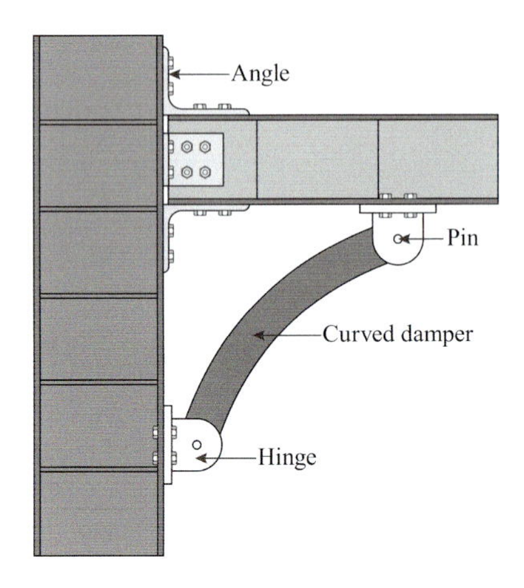

FIGURE 7.8 Steel-curved damper attached to a portal frame.

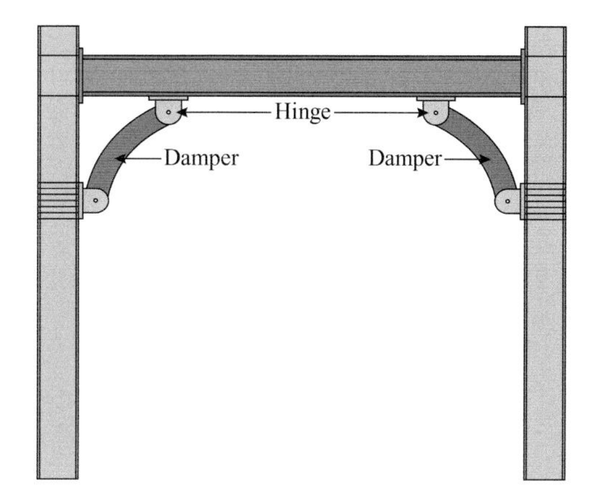

FIGURE 7.9 Mono-pitch portal frame equipped with steel-curved dampers.

provide the desired ductility at various locations in the structure. For improving the ductility and energy dissipating capability of concentrically braced frames, ring-type steel dampers are installed in frames. Azandariani et al. (2020) used steel ring dampers along with concentric frames, as shown in Figure 7.10. Further, Figure 7.11 depicts the general geometry of the steel ring damper.

Extensive parametric studies have been carried out to study the effect of thickness, length, and diameter of the steel ring on the behavior of steel ring dampers. The ultimate capacity of these dampers can be evaluated using the equations of equilibrium or principle of virtual work, after the formation of plastic hinges. Figure 7.12(a) and (b) show the plastic hinge formation due to tensile and compressive forces, respectively. Figure 7.12(c) shows the free body diagram of a one-quarter portion of steel ring damper of symmetrical geometry when subjected to symmetrical loading.

The relationship obtained considering the equilibrium of moments about a vertical axis is presented in Equation (7.6). Furthermore, the bending moment for a rectangular section is given in Equation (7.7).

$$\sum M = -\frac{P}{2} \times \frac{D}{2} + 2 \times M_P = 0 \qquad (7.6)$$
$$P = \frac{8M_P}{D}$$

$$M_P = Z_P F_y = \left(\frac{Lt^2}{4}\right) F_y \qquad (7.7)$$

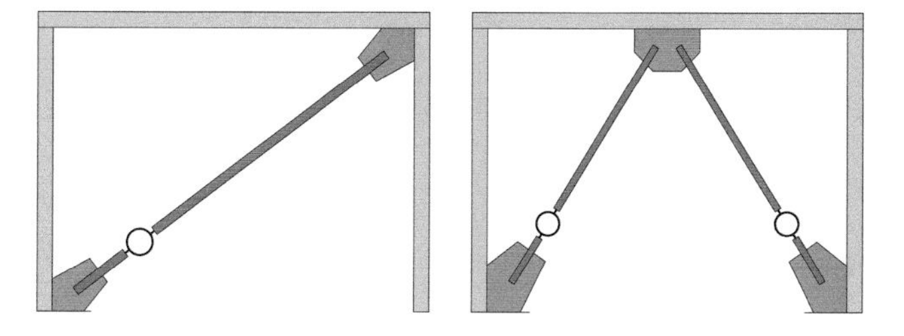

FIGURE 7.10 Location of steel ring dampers in concentrically braced steel frame system.

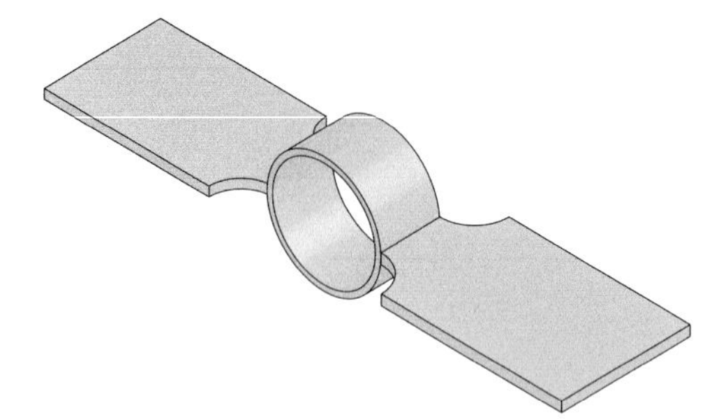

FIGURE 7.11 Geometry of steel ring damper.

where M_P and Z_P are the plastic moment and plastic section modulus, respectively; L and t are the length and thickness of the ring plate, respectively; and F_y is the yield stress of the ring plate. Moreover, the maximum bending moment in the plastic hinge is expressed as

$$M_{Pr} = C_{Pr} R_y Z_P F_y \tag{7.8}$$

where C_{Pr} is the coefficient that includes the effect of strain hardening factors, local constraints, and reinforcing factors; R_y is the ratio of expected yield stress to minimum yield stress, which is usually taken as 1. From Equations (7.6)–(7.8), the yield strength of the steel ring damper is expressed as,

$$P_y = 2C_{Pr} DLF_y \left(\frac{t}{D}\right)^2 \tag{7.9}$$

Figure 7.13 represents the performance of steel ring dampers in tension and compression.

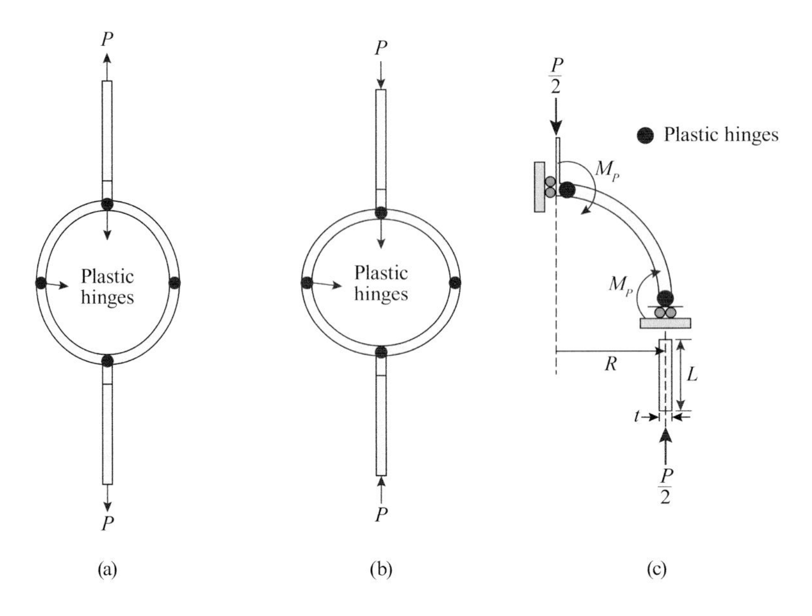

FIGURE 7.12 Plastic hinge formation mechanisms in steel ring damper. (a) Tension, (b) compression, and (c) free body diagram of a quarter steel ring.

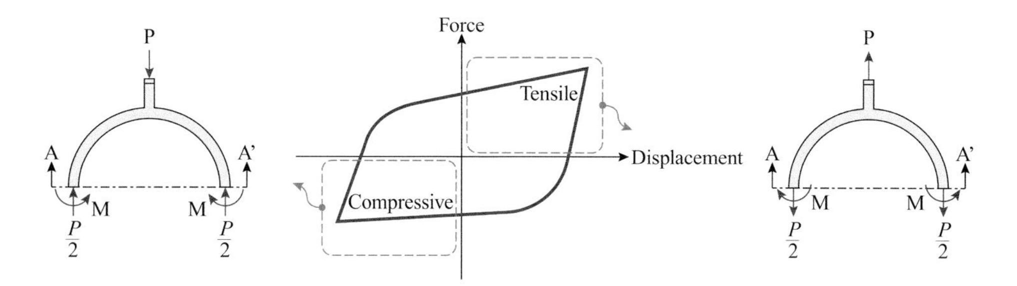

FIGURE 7.13 Performance of steel ring dampers in tension and compression.

7.4.5 Plate-Type Steel Damper

Steel plate metallic dampers are among the most widely used types of metallic dampers that are popular among researchers because of their hysteresis plastic deformation, versatility in shape, and good energy dissipation. Steel plate dampers have the advantage of simple structure, reliable performance, easy manufacturing, and easy method of implementation. Kelly et al. (1972) first used rectangular steel plate metal yielding damper. Further, Skinner et al. (1974) used a U-shaped steel plate damper in structures to enhance their seismic performance. Later, Tyler (1978) introduced tapered steel plates. The first application of steel dampers is a government building in New Zealand (1980). Whittaker et al. (1991) used triangular steel plates, and Tsai et al. (1993) employed X-shaped steel plates; both are widely used among steel plate dampers.

7.5 DUAL-PIPE DAMPER AND INFILLED-PIPE DAMPER

Two passive earthquake energy dissipation devices, i.e., dual-pipe damper and infilled-pipe damper, were introduced and investigated by Maleki and Mahjoubi (2013, 2014). The dual-pipe damper is made of two pipes welded at selected locations, as shown in Photo 7.5. The mechanism of energy dissipation is through the flexural yielding of the pipe. At large lateral displacement, the damper is subjected to tension, leading to increased stiffness and strength. In comparison with the single-pipe dampers, the double-pipe dampers show superior performance. It is reported that the dampers exhibit satisfactory ductility, energy absorption capacity, and stable hysteresis loops under cyclic quasi-static tests. Figure 7.14 shows the hysteresis loop of a double-pipe damper.

PHOTO 7.5 Dual-Pipe damper. (Courtesy of Shervin Maleki.)

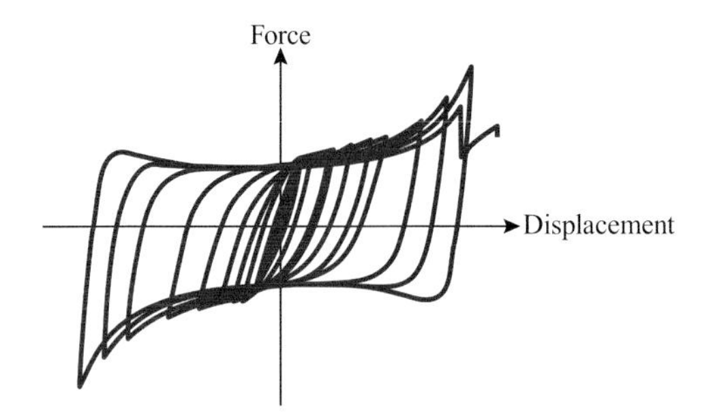

FIGURE 7.14 Hysteresis loop of double-pipe damper.

The infilled-pipe damper consists of two welded pipes that have two smaller pipes inside them, as shown in Photo 7.6. The spaces between the pipes are filled with metal such as lead or zinc. The energy dissipation mechanism in the device is: (i) the plastification of the outer pipes, inner pipes, and infilled metal, and (ii) the friction between metals.

Figure 7.15 shows the force-displacement loop of an infilled-pipe damper filled with lead.

7.6 TRIANGULAR METALLIC DAMPER

The performance of tall, slender, and flexible buildings is always a point of concern for researchers and structural designers because of the large inter-story deformations under dynamic excitations. Response control devices with stable force-displacement relationships and high energy dissipating capability could be suitable to mitigate the undesirable response of such structures. A steel triangular added damping added stiffness (TADAS) device is an economical energy dissipation solution to enhance the earthquake performance of such structures. TADAS exhibits the required elastoplastic constitutive behavior that is a typical characteristic of metallic dampers. TADAS device consists of several triangular plates welded to a common base plate, as shown in Figure 7.16(a). The device primarily uses mild steel plates, but devices with copper and stainless steel have also been developed. Due to the low-cost and simple manufacturing process of mild steel, TADAS made with mild steel is preferred. A large amount of heat is generated in the manufacturing process of TADAS while welding the plates. Special heat treatment is required to release the residual stresses developed in

PHOTO 7.6 Infilled-pipe damper. (Courtesy of Shervin Maleki.)

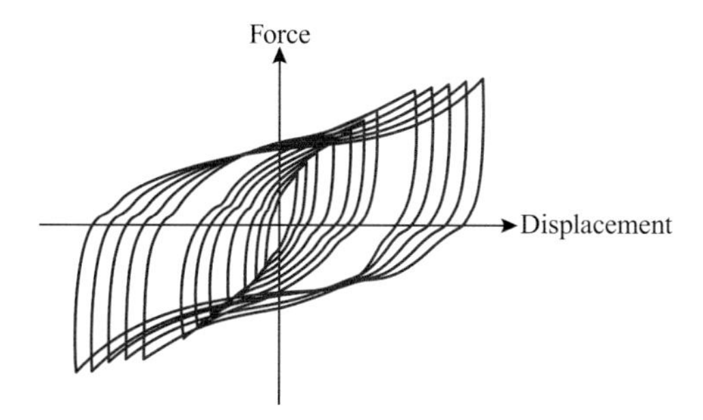

FIGURE 7.15 Hysteresis loop of infilled-pipe damper.

triangular plates of TADAS. Experimental results indicate that TADAS can sustain a large number of yielding reversals without any stiffness or strength degradation. Figure 7.16(b) illustrates the force-displacement loop of the TADAS device.

One end of the TADAS device is fixed to the structure, and the other end is pinned, as shown in Figure 7.16(a). The pinned end generates smaller axial forces in TADAS dampers. Both ends of ADAS devices or shear panels are fixed, which generate enormous axial forces. Therefore, TADAS are preferred to ADAS and shear panels.

The typical hysteresis loop of the TADAS damper is shown in Figure 7.16(b). Photo 7.7 shows the TADAS device installed in the test machine. Figure 7.17(a) shows the welded connection details of the TADAS device, and Figure 7.17(b) shows an individual steel triangular plate.

TADAS dampers can also be made using PTFE sliding plates placed inside the steel block, as shown in Figure 7.18. The variation of bending moment over the height (h) of the TADAS device is triangular. The moment is zero at the pinned end and $(P \times h)$ at the fixed end, where P is the lateral force acting at the pin. For a triangular plate of curvature, of the shape of a section of a circle, the displacement u at the pinned end is given by Equation (7.10), which depends on the strain ε, length L, and thickness t of the plate.

$$u = \frac{t}{2\varepsilon}\left[1 - cos\left(\frac{2L\varepsilon}{t}\right)\right] \tag{7.10}$$

The initial stiffness (k) of the TADAS damper, considering the number of plates (N) and their dimensions $(b, t \text{ and } h)$, is given by Tsai et al. (1993) as,

$$k = \frac{NEbt^2}{6h^3} \tag{7.11}$$

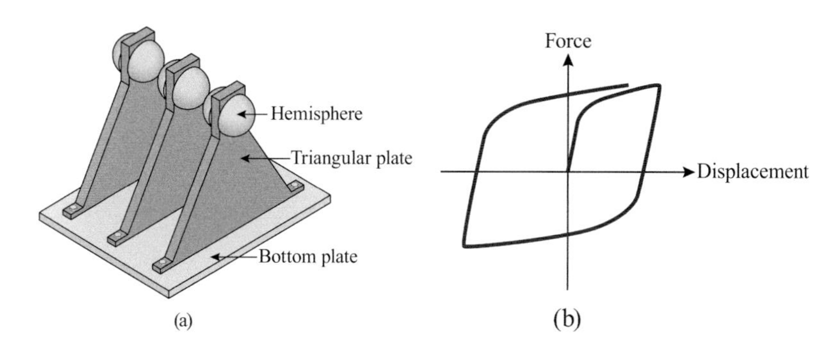

FIGURE 7.16 Schematic diagram and hysteretic loop of a TADAS device. (a) Schematic and (b) hysteresis loop.

PHOTO 7.7 TADAS devices installed in test machine. (Zemp et al. 2017.)

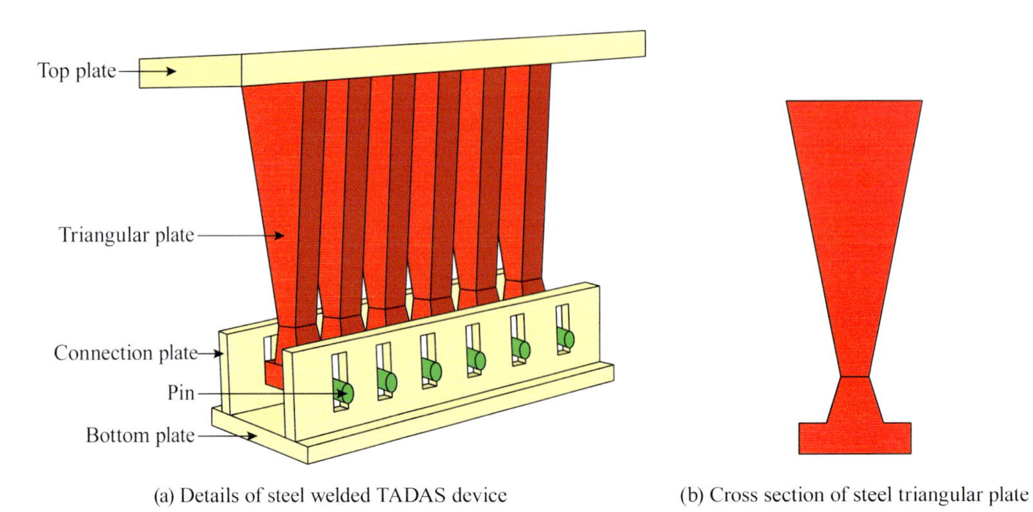

FIGURE 7.17 Triangular-plate added damping and added stiffness device: (a) details of steel welded TADAS device and (b) steel triangular plate.

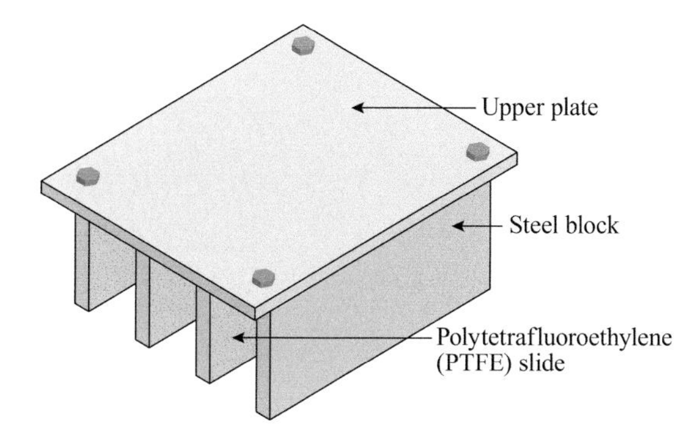

FIGURE 7.18 TADAS damper with PTFE slides.

The yield stress (F_y) of the TADAS damper is given by Equation (7.12), where f_y is the yield stress of the metal used for the plates.

$$F_y = \frac{Nf_y bt^2}{6h} \tag{7.12}$$

7.7 DUAL FUNCTION METALLIC DAMPER

The concept of dual function metallic damper was first proposed and justified by Li and Li (2013). It is called a dual function damper because it exhibits two characteristics: high initial stiffness and high energy dissipation capability. The initial stiffness of the damper can be increased by making its bearing exterior surface plane. The energy dissipation is increased by changing the shape of the metallic plate. Strip metallic damper, single round hole metallic damper, double round hole metallic damper, and double X-shaped metallic damper are some examples of dual function dampers.

7.7.1 Strip Metallic Damper

Figure 7.19 shows a strip metallic damper and its hysteresis loop. The hysteresis loop shows that the strip metallic damper has large initial stiffness, good deforming capability, and good energy-dissipating capability. It is important to note that a small vertical load causes large deformation of the whole damper system, making it unstable. Due to out-of-plane buckling in the strip region, the strength of the strip metallic damper abruptly reduces.

7.7.2 Round-Hole Metallic Damper

There are two types of round-hole metallic dampers, namely, single-hole metallic damper, and double-hole metallic damper. In Figure 7.20, a single round hole metallic damper, and its

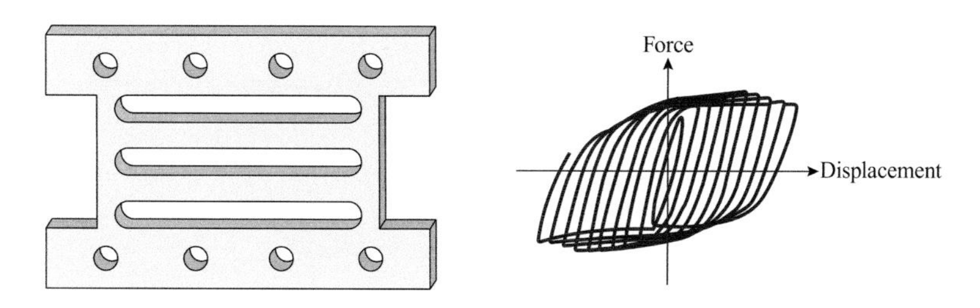

FIGURE 7.19 Strip metallic damper and its hysteresis loop.

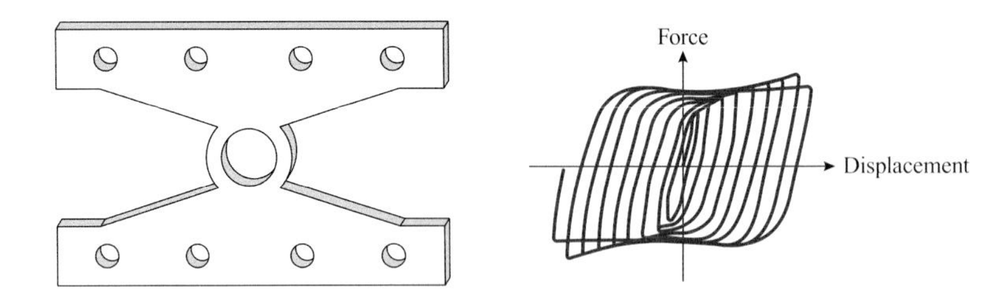

FIGURE 7.20 Single round hole metallic damper and its hysteresis loop.

force-deformation characteristics are shown. The round-hole metallic damper exhibits good energy dissipating capability and also has high initial stiffness.

Figure 7.21 shows a double round hole metallic damper and its hysteresis loop. The damper exhibits large initial stiffness at a very early stage of loading. However, at large deformation, the section near the round hole is likely to develop cracks. The hysteresis loop shows jaggedness; it is not as smooth as that of the single round hole damper.

7.7.3 Double X-Shaped Metallic Damper

Figure 7.22 shows the double X-shaped metallic damper and its hysteresis loop. This damper solves the buckling problem of strip metallic damper when it is subjected to small vertical loads. The hysteresis loop of double X-shaped metallic damper is very smooth and shows that the damper has large stiffness at the initial stage and large energy-dissipation capacity.

7.8 METALLIC YIELDING-FRICTION DAMPER

Metallic yielding dampers take advantage of the inelastic deformation of metallic materials under time-dependent cyclic loading for dissipating the input seismic energy. Generally, metallic yielding dampers are categorized into two types.

i. Shear-type metallic yielding damper
   The shear-type metallic yielding damper consists of metallic plates with a large number of openings, subjected to in-plane deformations wherein the seismic energy is resisted by metallic yielding plates. Typical examples of shear-type dampers are the honeycomb damper and the single round hole damper, shown in Figure 7.23 and Figure 7.20, respectively.
ii. Bending-type metallic yielding damper
   The bending-type metallic yielding damper utilizes flexural deformations of metallic dampers under out-of-plane bending. Examples of bending-type dampers are ADAS

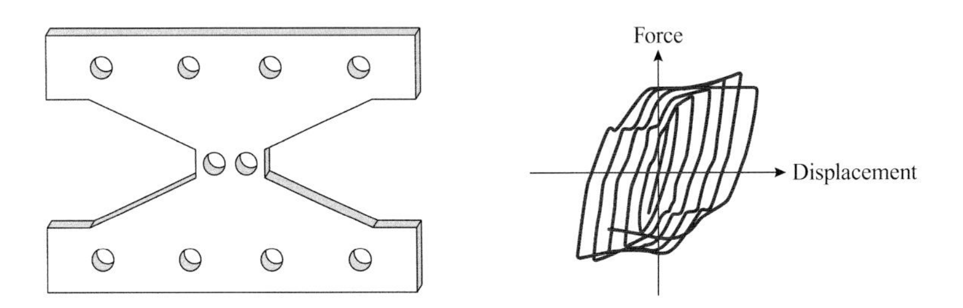

FIGURE 7.21 Double round hole metallic damper and its hysteresis loop.

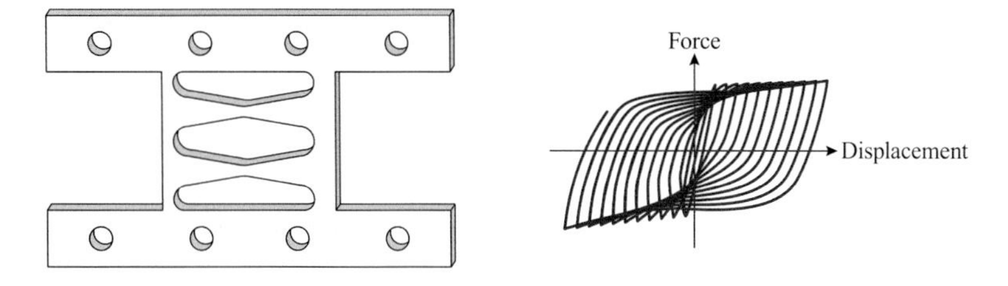

FIGURE 7.22 Double X-shaped metallic damper and its hysteresis loop.

and triangular metallic dampers. The metallic yielding-friction damper is a combination of metallic yielding damper and metallic friction damper, wherein the metallic plate is an important component in both the metallic and yielding dampers. Components of the metallic yielding-friction damper are shown in Figure 7.24(a), and Figure 7.24(b) shows the plane of yielding and friction plates.

The upper and lower connection plates are for fixing the yielding plate and the friction plate. Friction is generated by the interaction of sliding surfaces between the lower connection plate and copper plate. The energy dissipation of the friction part occurs at an early stage, compared to that of the yielding part, by adjusting the friction force less than that of the yielding plate part. In this case, the upper plate is subjected to horizontal force, and thus, the fixed plate is fastened. A steel bar is welded to the lower connection plate to avoid damage to the yielding plate under the pulling force generated by the relative deformation between the upper and the lower connection plates. This allows free up and down movement of steel plates, as shown in Figure 7.24(b). Copper and steel friction plates are used in the metallic yielding-friction damper, as friction between steel and steel leads to an unstable sliding force. Photo 7.8 shows the full-scale metallic yielding-friction damper installed in a building.

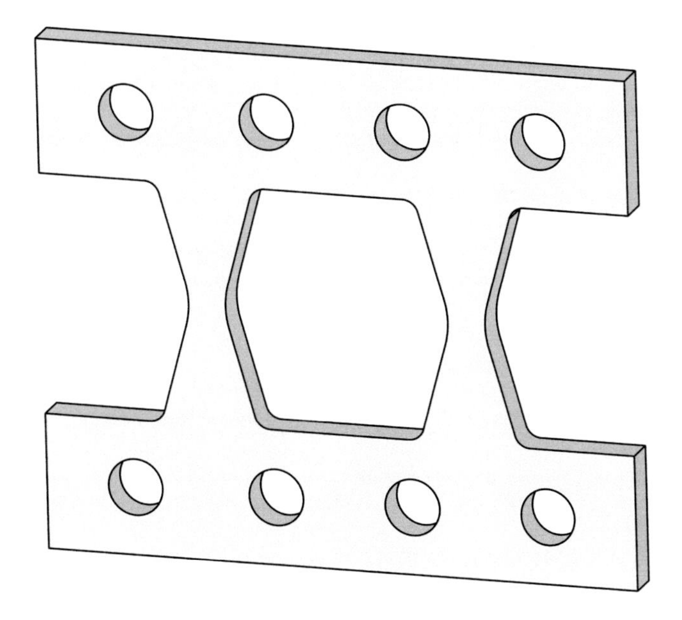

FIGURE 7.23 Honeycomb damper.

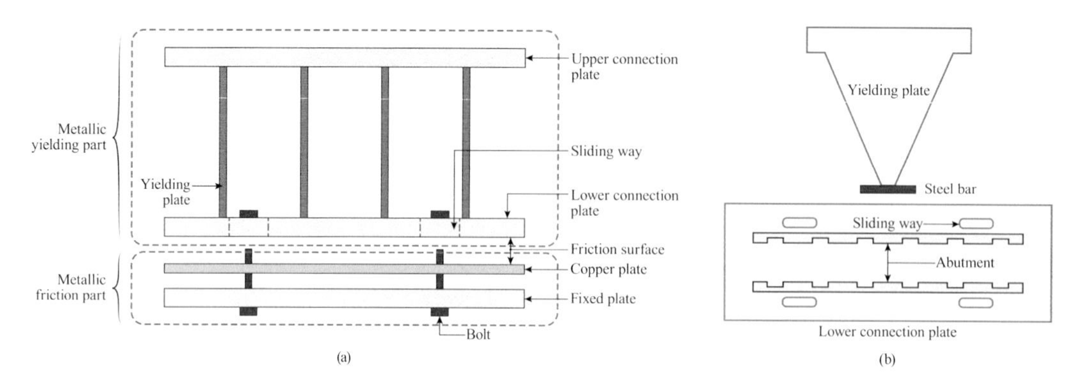

FIGURE 7.24 Metallic yielding-friction damper. (a) Components and (b) plane of yielding plate and friction plate.

PHOTO 7.8 Full-scale metallic yielding-friction damper installed in a building. (Courtesy of Hong-Nan Li.)

REFERENCES

Azandariani, M. G., Abdolmaleki, H., and Azandariani, A. G. (2020). Numerical and analytical investigation of cyclic behavior of steel ring dampers (SRDs). *Thin-Walled Structures*, 151, 106751.

Ghabussi, A., Marnani, J. A., and Rohanimanesh, M. S. (2020). Improving seismic performance of portal frame structures with steel curved dampers. *Structures*, 24, 27–40.

Gray, M. G. (2012). *Cast Steel Yielding Brace System for Concentrically Braced Frames*. Canada: University of Toronto

Heysami, A. (2015). Types of dampers and their seismic performance during an earthquake. Current World Environment, 10(Special-Issue 1), 1002–1015.

Kelly, J. M., Skinner, R. I., and Heine, A. J. (1972). Mechanisms of energy absorption in special devices for use in earthquake resistant structures. Bulletin of the New Zealand Society for Earthquake Engineering, 5(3), 63–88.

Li, G., and Li, H. N. (2013). Experimental study and application in steel structure of 'Dual Functions' metallic damper. *Advanced Steel Construction*, 9(3), 247–258.

Li, H., Li, G., and Wang, S. (2014). Study and Application of Metallic Yielding Energy Dissipation Devices in Buildings. In Tenth US National Conference on Earthquake Engineering, July 21–25. Anchorage, Alaska, United States.

Maleki, S., and Mahjoubi, S. (2013). Dual-pipe damper. *Journal of Constructional Steel Research*, 85, 81–91.

Maleki, S., and Mahjoubi, S. (2014). Infilled-pipe damper. *Journal of Constructional Steel Research*, 98, 45–58.

Skinner, R. I., Kelly, J. M., and Heine, A. J. (1974). Hysteretic dampers for earthquake-resistant structures. *Earthquake Engineering and Structural Dynamics*, 3(3), 287–296.

Teruna, D. R., Majid, T. A., and Budiono, B. (2015). Experimental study of hysteretic steel damper for energy dissipation capacity. *Advances in Civil Engineering*, 2015, 1–12.

Tsai, K.-C., Chen, H.-W., Hong, C.-P., and Su., Y.-F. (1993). Design of steel triangular plate energy absorbers for seismic-resistant construction. *Earthquake Spectra*, 9(3), 505–528.

Tyler, R. G. (1978). Tapered steel energy dissipators for earthquake resistant structures. Bulletin of the New Zealand Society for Earthquake Engineering, 11(4), 282–294.

Whittaker, A. S., Bertero, V. V., Thompson, C. L., and Alonso, L. J. (1991). Seismic testing of steel plate energy dissipation devices. *Earthquake Spectra*, 7(4), 563–604.

Zemp, R., Urrutia, R. C., Rendel, M., Cavalla, G., and de la Llera, J. C. (2017). Design, testing and implementation of TADAS devices in three RC buildings with shear walls and coupling beams. In Proceedings of the 16th World Conference on Earthquake Engineering.

8 Re-centering Devices

8.1 INTRODUCTION

Re-centering is the capability of any device to return to its original position and configuration after the occurrence of any seismic event. It is the most important aspect for the devices that possess low restoring capability. Lack of re-centering may lead to major damages and collapse of structures, during strong earthquake ground motions. On removal of the applied load, residual deformation of re-centering devices is negligible, and they retain their indigenous configuration. Devices that fall under re-centering capability group are pressurized fluid damper, preloaded spring friction damper, self-centering seismic-resistant system, and shape memory alloy (SMA) damper. These are described in the following sections.

8.2 PRESSURIZED FLUID DAMPER

Pressurized fluid damper was first employed by Tsopelas and Constantinou (1994), in order to provide both; damping and re-centering capability, prominently for a base isolation system. A schematic of pressurized fluid restoring damper is shown in Figure 8.1.

The resistance in the damper is due to the various physical phenomena listed below.

1. Resistance developed because of preload induced in the device due to initial pressurization of fluid. The resistive force $\left(F_p\right)$ due to this phenomenon is given mathematically as a function of the relative displacement in the damper (u) by Equation (8.1).

$$F_p = F_0\left[1 - exp\left(-\delta|u|\right)\right]sgn(u). \tag{8.1}$$

2. Resistance offered by the stiffness of damper linked with the compressibility of the fluid (silicone oil) in the device. Mathematically, the resistive force $\left(F_c\right)$ is expressed as,

$$F_c = k_0 u. \tag{8.2}$$

3. Resistance offered by the seal friction $\left(F_f\right)$, expressed mathematically as,

$$F_f = \left[F_m + \alpha k_0 |u|\right]Z. \tag{8.3}$$

4. Passing of fluid (silicone oil) through the orifice induces damping in the system. The damping force $\left(F_{dr}\right)$ can be expressed mathematically as,

$$F_{dr} = F_d\, sgn(\dot{u}). \tag{8.4}$$

Combining Equations (8.1)–(8.4) gives the total resisting force,

$$F = F_p + F_c + F_f + F_{dr}$$

$$\therefore F = F_0\left[1 - exp\left(-\delta|u|\right)\right]sgn(u) + k_0 u + \left[F_m + \alpha k_0 |u|\right]Z + F_d\, sgn(\dot{u}). \tag{8.5}$$

DOI: 10.1201/9781315269269-8

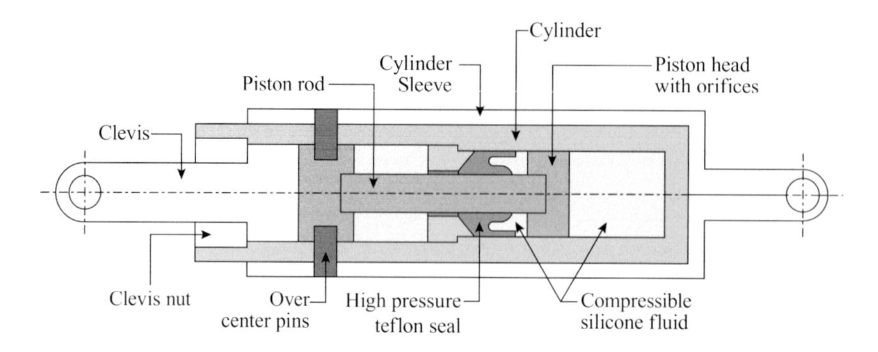

FIGURE 8.1 Pressurized fluid restoring damper. (Tsopelas and Constantinou 1994.)

In Equation (8.5), the mathematical terms F_0, k_0, F_m, and F_d indicate the preload, stiffness, seal friction force, and damping force of fluid respectively. Internal variable Z can be evaluated by Equation (8.6).

$$u\dot{Z} + \gamma|u|Z|Z| + \beta\dot{u}Z^2 - \dot{u} = 0. \tag{8.6}$$

The damping force depends on the following two conditions.

1. Product of displacement and velocity is greater than zero, i.e., $(u\dot{u} > 0)$.

$$F_d = F_{dp}\left[1 - exp\left(-\varepsilon_p|\dot{u}|\right)\right]. \tag{8.7}$$

2. Product of displacement and velocity is less than zero, i.e., $(u\dot{u} < 0)$.

$$F_d = F_{dm}\left[1 - exp\left(-\varepsilon_m|\dot{u}|\right)\right]. \tag{8.8}$$

Also,

$$\delta = \delta_0 \; exp\left(-\delta_1|\dot{u}|\right) \tag{8.9}$$

where F_{dp}, F_{dm}, α, β, γ, δ, δ_0, δ_1, ε_p, and ε_m are modal parameters.

From Equations (8.7) and (8.8), it is seen that there is a bias in the damping force provided by the damper. On loading, high damping is obtained while unloading results in lower damping. The typical force-displacement response obtained by Tsopelas and Constantinou (1994) using the above model is as shown in Figure 8.2.

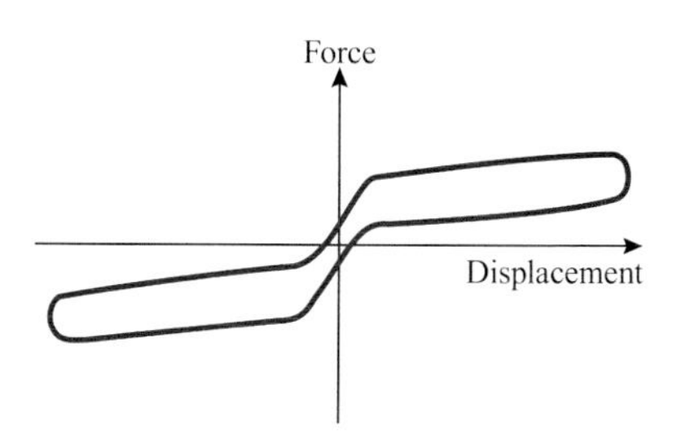

FIGURE 8.2 Force-displacement curve of pressurized fluid damper. (Tsopelas and Constantinou 1994.)

8.3 PRELOADED SPRING FRICTION DAMPER

Preloaded spring friction damper (Nims et al. 1993) consists of steel compression wedges embedded in thin cylinder along with an internally preloaded spring. A schematic of preloaded spring friction damper is shown in Figure 8.3.

The force-displacement loop for damper with preloaded spring has the characteristic double flag shape, and that with zero preload has triangular shape, as shown in Figure 8.4(a) and (b), respectively.

8.3.1 Mathematical Model of Preloaded Spring Friction Damper

Main components of spring friction damper include internal spring, compression and friction wedges, internal stops, and a cylinder accommodating all the components. Tension and compression gaps are the variable parameters of the damper, along with the numbers and shapes of compression and friction wedges, spring constant, and spring pre-compression. The wedges and spring are accommodated inside the core of the damper, as seen in Figure 8.3.

During operation, the to and fro movement of spring causes compressive force in the spring to act on friction and compression wedges. This causes a normal force to act on the cylindrical walls, which is proportional to the compressive force in the spring. Slip force in the damper is determined by the normal force and friction coefficient between friction wedge and cylindrical wall of the damper as given in Equation (8.10).

$$F_{sf} = \mu A F_s = \mu A k_s \Delta_s \tag{8.10}$$

where F_{sf} is the slip force in the damper, μ is the coefficient of friction, A is the constant relating slip force to the normal force acting on the cylindrical wall, F_s is the force in the spring, k_s is the spring constant, and Δ_s is the spring deformation.

The initial slip force can be altered by adjusting spring as per the requirement, to achieve appropriate deformation of spring. The slip force is always kept less than spring force, i.e., $\left(F_{sf} < F_s\right)$, so that the spring will always return to its initial position unless and until it is controlled by an external force. Thus, if both the tension and compression gaps are initially zero, the device will align itself in the center position.

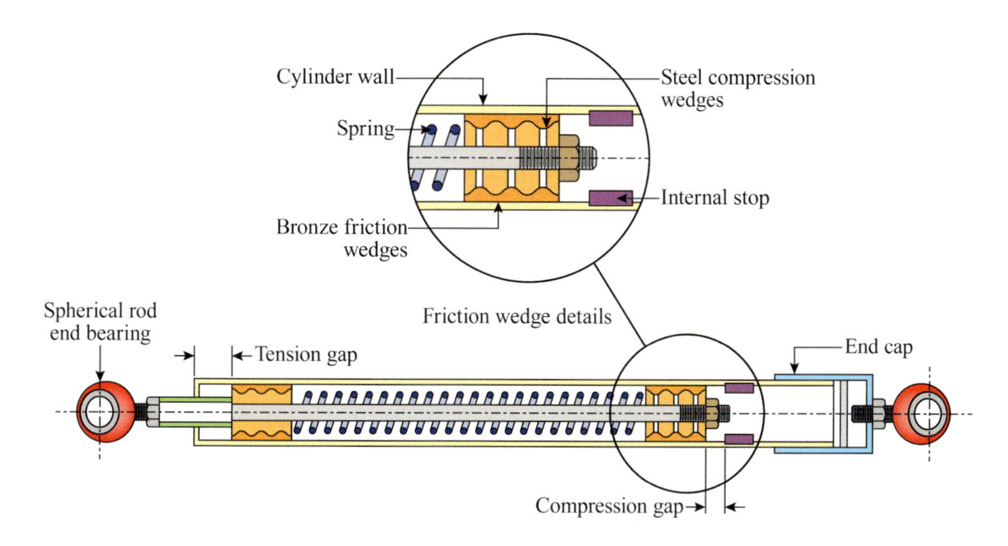

FIGURE 8.3 Schematic diagram of preloaded spring friction damper. (Nims et al. 1993.)

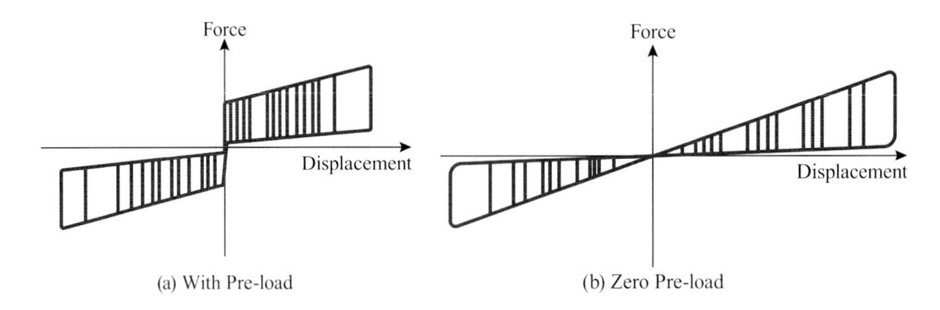

FIGURE 8.4 Force-displacement loops for spring-friction damper: (a) with pre-load, and (b) with zero pre-load. (Nims et al. 1993.)

8.4 SELF-CENTERING SEISMIC-RESISTANT SYSTEM

The current design methodology allows the structure to be designed to respond beyond elastic limit and, at the end, allows it to exhibit inelastic ductile response at specified locations. For this purpose, the specified regions are checked for ductility and energy dissipation capabilities.

Figure 8.5 shows the idealized force-displacement response of a linear elastic system, and of a system representing a yielding structure of equal mass, and equal initial stiffness. The maximum seismic-induced force in the yielding system is significantly lower than that of the linear elastic system. The maximum displacement of the yielding system can be smaller, comparable, or larger than that of the elastic system, depending on the natural period and on the strength of the yielding system. The area of the loop represents the energy dissipated per cycle through hysteretic yielding.

Designs that aim at inelastic response are highly recommended from the low initial cost; however, they have two major drawbacks, namely, (i) the regions in the principal lateral force resisting system are damaged in moderately strong earthquakes and need repair; or damaged beyond repair in strong earthquakes, and (ii) current design approaches are based on the principle that large energy dissipation capacity is necessary to mitigate the effects induced by earthquakes. This has very often led to the conception that a good structural system should be characterized by a fat

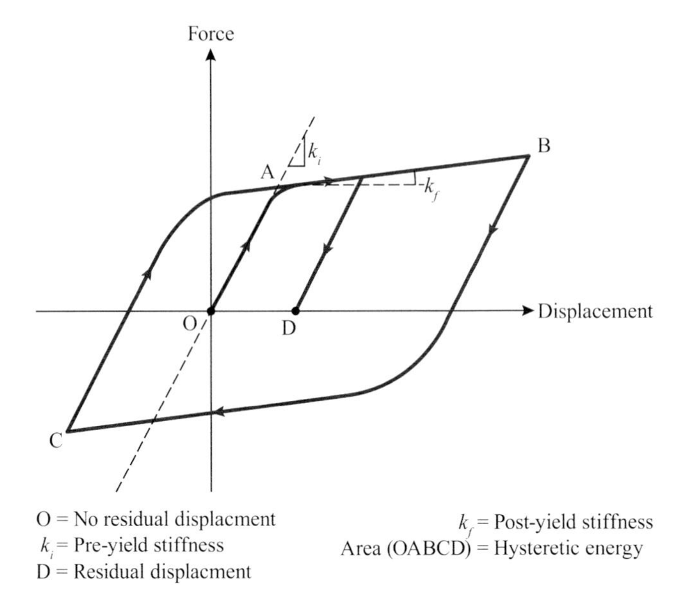

FIGURE 8.5 Idealized force-displacement relation of yielding system. (Christopoulos et al. 2002.)

hysteresis loop. As a large fraction of the input energy is expected to be dissipated by hysteresis, significant residual displacements could be expected in a structure after an earthquake, as illustrated in Figure 8.5. Excessive residual deformations can even result in the total damage of a structure due to geometric nonlinearity.

In order to overcome these drawbacks, innovative and economical self-centering earthquake-resistant systems are required, which would (i) incorporate the nonlinear characteristics of yielding structures and, thereby, limit the inertial force induced by seismic excitation and provide additional damping characteristics, (ii) incorporate self-centering properties allowing the structural system to return to its original position after an earthquake, and (iii) reduce or eliminate cumulative damage to the main structural elements.

Figure 8.6 shows the representative flag-shaped force-displacement relationship of a typical self-centering system. In this system, the amount of energy required to be dissipated is relatively less as compared to that of the yielding system shown in Figure 8.5. The most important point is that the structure returns to the origin of the graph, i.e., zero-force and zero-displacement point, at the end of every cycle of seismic loading.

8.4.1 Applications of Self-Centering System

Numerous experimental works have been performed by researchers to enhance the self-centering capabilities of structures. The self-centering capabilities are checked and enhanced, for various applications, namely, RC structures, steel structures, masonry walls, and bridges. These applications are briefly described in the following sections.

i. Reinforced concrete structures

Priestley and Tao (1993) proposed the use of self-centering precast concrete moment resisting frame (MRF) systems, prestressed with partially unbonded tendons as the primary lateral force resisting system in seismically prone areas. This primary lateral force resisting system proved very effective in energy dissipation during strong earthquakes.

Stanton and Nakaki (2002) developed post-tensioned (PT) split rocking wall system, as shown in Figure 8.7. In this system, the wall panels are split to allow rocking of the individual panels about their individual bases. The self-weight of the panels provides re-centering forces. If the weight is not sufficient to completely re-center the wall panels, unbonded PT tendons connecting the wall to its foundation can be installed. Energy dissipation can be introduced by grouting reinforcing bars into vertical ducts at the edges of the wall, so that

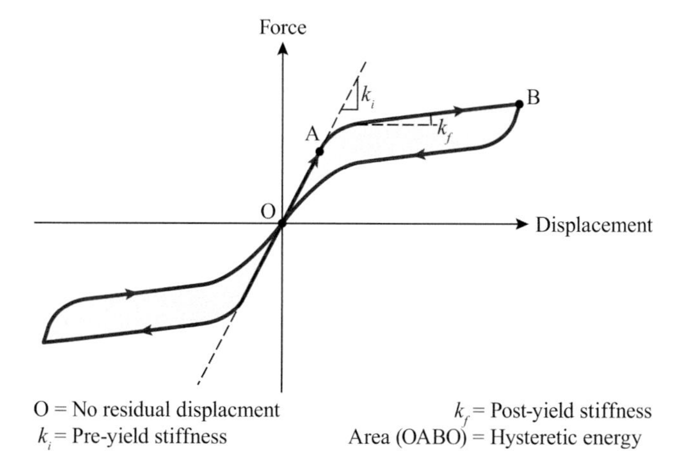

FIGURE 8.6 Idealized force-displacement relation of self-centering structural system. (Christopoulos et al. 2002.)

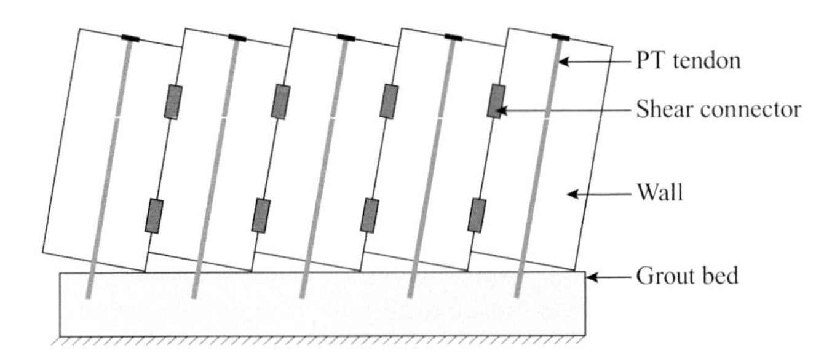

FIGURE 8.7 Post-tensioned split rocking wall system. (Stanton and Nakaki 2002.)

they yield cyclically in tension and compression during an earthquake. An alternative is to place ductile shear connectors between the wall panels, which deform cyclically in shear as the walls rock back and forth.

Restrepo (2002) extended the self-centering concept proposed by Stanton and Nakaki (2002), to RC cantilever walls by using a wall system prestressed with unbonded tendons and conventional reinforcement for energy dissipation, as shown in Figure 8.8. The main advantages of this hybrid jointed wall systems are the large lateral displacement capacity, the lack of structural damage associated with large displacements, and the ability to return to the original position upon unloading.

ii. Steel structures

Ricles et al. (2001) developed hybrid PT connection for steel MRF as shown in Figure 8.9. The connection consists of high-strength steel strands that run along the web of the beam and anchor to the exterior column flange at the end of the frame. In addition, seat and top angles are bolted to both column and beam. Shear resistance is provided by a combination of friction at the beam-column interface and also by the steel angles. The system is designed so that the steel angles are the only yielding elements. Therefore, only the steel angles would need to be replaced after a major earthquake. Additional benefits of this connection include (i) no field welding required, (ii) use of conventional materials and skills, and (iii) equal initial stiffness as that of the conventional welded connections.

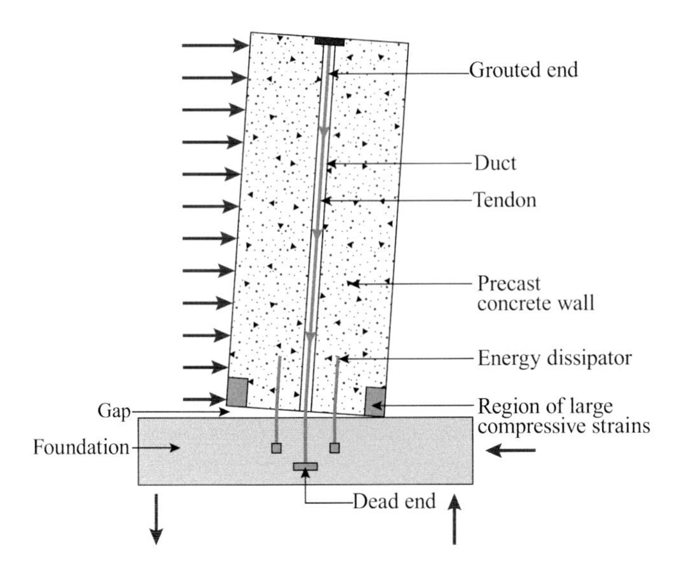

FIGURE 8.8 Hybrid reinforced concrete cantilever wall system. (Restrepo 2002.)

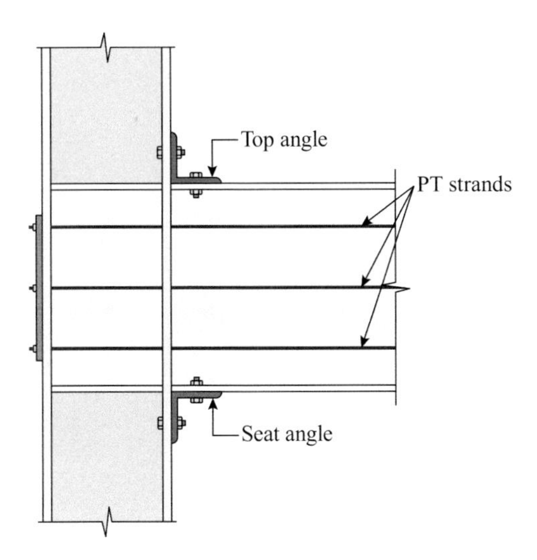

FIGURE 8.9 Hybrid posttensioned connections for steel frames. (Ricles et al. 2001.)

iii. Confined masonry walls

Toranzo et al. (2004) proposed the use of self-centering systems in combination with traditional methods of construction. Low-cost energy dissipaters were used with rocking confined masonry walls and were tested on shake table at the University of Canterbury, New Zealand. The wall built with brick masonry infill and slabs, as shown in Figure 8.10(a), was tested. The columns were designed for strain control to ensure that small shear distortions would occur in the wall panel, while the wall rocked at the foundation. Energy dissipation

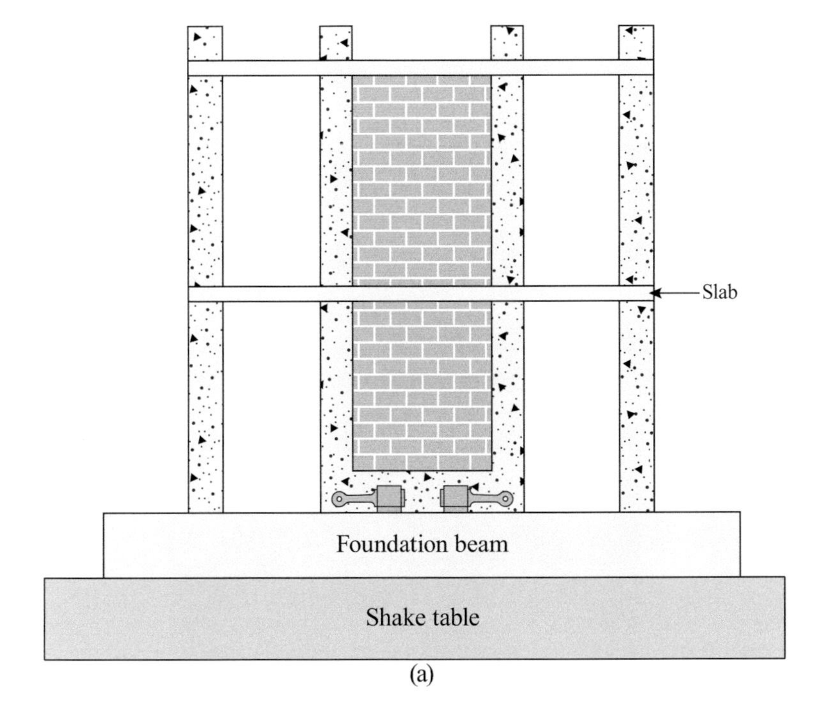

FIGURE 8.10 Rocking confined masonry wall system. (a) Rocking confined masonry system on shake table. (b) Flexural energy dissipation devices for rocking confined masonry system. (Toranzo et al. 2004.) (*Continued*)

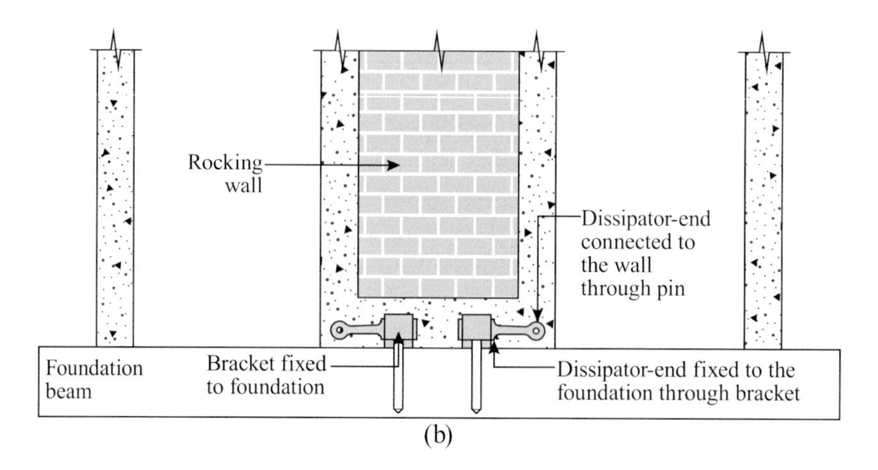

FIGURE 8.10 *(Continued)*

devices, in the form of tapered levers, designed to yield in bending with constant curvature, were installed at the wall toes, as shown in Figure 8.10(b). The seismic response of the unit was excellent and the masonry infill maintained its integrity throughout, without development of cracks.

iv. Bridge structures

The rocking concept combined with energy dissipation devices was effectively implemented in the design of the stepping rail bridge over the south Rangitikei River in New Zealand. This combination is conceptually similar to that of the self-centering system. The railroad bridge, shown in Photo 8.1, has been in operation since 1981. It is 70 m tall,

PHOTO 8.1 Application of self-centering system on the Rangitikei railway bridge, New Zealand (Christopoulos et al. 2002). (Courtesy of Nigel Priestley)

with six spans of prestressed concrete hollow-box girder, and overall length of 315 m. The base isolation is mainly designed to allow the sideways rocking of pairs of slender RC piers. Torsional-beam dampers are provided to limit the amount of rocking. The weight of the bridge, which allows its re-centering, is transmitted to the foundation through thin laminated rubber bearings. It is documented that the bridge is able to re-center itself, as the seismic vibrations pass away.

8.5 SHAPE MEMORY ALLOY DEVICES

Passive control systems proved to be very efficient in protecting the structures from earthquake-induced vibrations and the generated forces. Though the efficiency of passive control systems, such as base isolation devices, mass dampers, and viscous dampers, is up to the mark, there is still a requirement of passive devices that possess unique characteristics of innovative advanced materials and use them to achieve enhanced energy dissipation capability. Additionally, the traditional passive devices suffer from various drawbacks associated with ageing, durability, installation complexities and routine maintenance. This promoted the researchers to employ a novel and recent class of material such as shape memory alloy (SMA), in passive isolation system.

The concept of SMA was first introduced to the world in 1930. In 1980, SMA got much more attention when the construction industry started using smart materials for improved performance during seismic events. SMA-based dampers have potential to withstand severe earthquakes efficiently. They can reduce the permanent deformations and plastic deformations in the structure and can dissipate large amount of seismic energy. SMAs are known to have two unique characteristics, a shape memory effect and super-elasticity. The shape memory effect is the ability of the alloys to revert to their initial shape upon being heated until they enter their phase transformation temperature. Super-elasticity is where the alloys exhibit comparatively large recoverable strain. Different families of SMAs have different appropriate applications due to their different ranges of transformation temperature.

The SMAs exhibit two different crystalline phases: martensite and austenite. The phase transformation of the SMAs is a result of induction of the reversible temperature or the intensity of stresses. At low temperatures and high stresses, the martensite phase is stable, whereas at high temperature and low stresses, the austenite phase is stable. The structure of austenite phase is a cubic crystal. At a higher temperature, SMAs exhibit the super-elasticity. However, upon loading, stress-induced martensite is formed and upon unloading, the martensite reverts to austenite at a lower stress level, resulting in the hysteretic behavior, as shown in Figure 8.11. These properties make SMAs ideal candidates for use in seismic energy dissipation devices in structural control engineering.

The damping capacity of SMAs comes from two mechanisms, namely, martensite variations reorientation and stress-induced martensitic transformation.

Buehler and Wiley (1965) proposed a new composition having 53–57% nickel by weight, which showed an unusual effect. The remaining portion composed of titanium, giving a series of nickel-titanium alloys. It was seen that there were severally deformed specimen of alloys having a residual strain of 8–15%, which regained the original shape after a thermal cycle. Further, this effect came to be known as the shape memory effect, and the alloys exhibiting it were named SMAs. Nitinol is used as trade name to commercialize the innovative nickel-titanium alloy composition. The energy dissipation capacity of nickel–titanium SMA wires in the martensite phase is significantly higher than that in the austenite phase. As research progressed, it was found that at sufficiently high temperatures, the SMAs also possess the property of super-elasticity, which helps the materials in the recovery of large deformations during mechanical loading and unloading cycles, performed at constant temperature. This super-elasticity is efficiently used in SMA-based dampers.

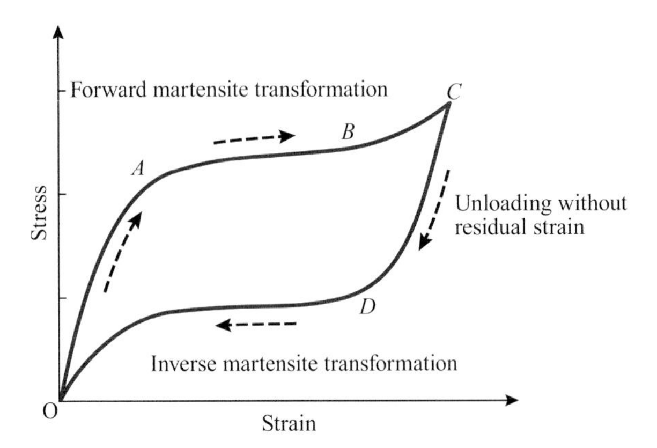

FIGURE 8.11 Super-elasticity effect of SMAs and total strain recovery. (Choi et al. 2013.)

Several advantages of implementing SMAs to mitigate the seismic response of structures are listed below.

1. SMAs are capable of sustaining a large amount of strain energy when subjected to cyclic loading, without causing any permanent deformation of the structure. SMAs allow manufacturing of supplemental energy dissipating devices that possess the self-centering capabilities.
2. Nitinol (nickel-titanium alloy) possesses good fatigue resistance under strain cycles having large amplitudes and requires little maintenance. Due to this property, the performance of SMAs is outstanding during the most devastative earthquakes.
3. Nitinol alloys have very good corrosion resistance, and considerable resistance to degradation caused due to ageing.
4. SMA materials show greater durability and are considered as reliable devices for long-standing usage.

8.5.1 Mathematical Model of SMA-Based Device

Various models have been proposed for modeling super-elasticity property of SMAs, which can simulate their hysteresis behavior. Over the time, advanced models have been developed to represent precise behavior of SMAs. Ozdemir (1976) was the first to model hysteretic properties of SMAs. His model was modified by Graesser and Cozzarelli (1991), named as modified Ozdemir model. Wilde et al. (2000) modified the Graesser and Cozzarelli (1991) model to include the strain hardening effect, which means strengthening of metal by plastic deformation due to dislocation in the crystal structure of the material. The model of SMA wire, proposed by Choi et al. (2013), is an improved version of Grasser and Cozzarelli model, which takes into account the strain hardening effect. A schematic of SMA-based damper is shown in Figure 8.12.

The differential equations of the model of SMA wire proposed by Choi et al. (2013) are given as,

$$\dot{F} = K_0\left[\dot{u} - |\dot{u}|\left|\frac{F-B}{B_c}\right|^{n-1}\left(\frac{F-B}{B_c}\right)\right] \tag{8.11}$$

$$B = K_0\alpha\left\{\begin{array}{l} u_{in} + f_T|u|^c \operatorname{erf}(au)\left[\Delta(-u\dot{u})\right] \\ +f_m\left[u - u_{mf}sgn(u)\right]^m \times\left[\Delta(u\dot{u})\right]\left[\Delta\left(|u| - u_{mf}\right)\right] \end{array}\right\} \tag{8.12}$$

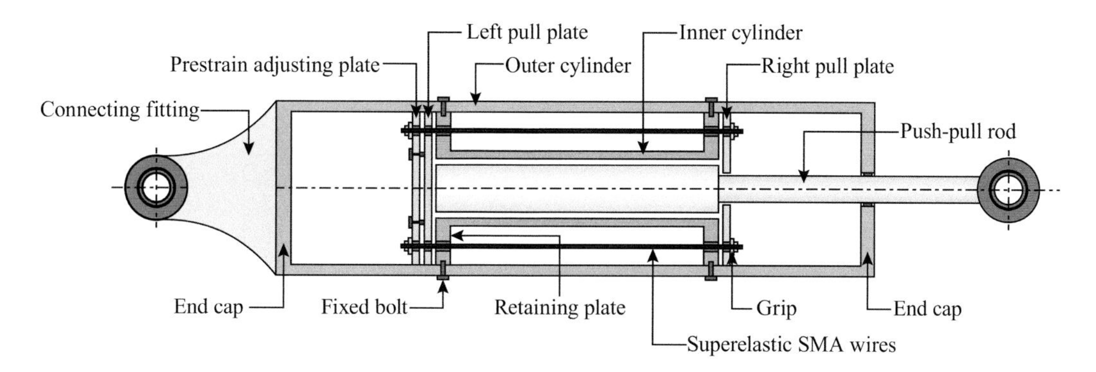

FIGURE 8.12 Schematic of SMA-based damper. (Choi et al. 2013.)

where F is restoring force, u is the relative inter-story displacement, $\dot{u}$ is the relative inter-story velocity, B_c is the force triggering the forward transformation from austenite to martensite phase, $\dot{F}$ is the time derivative of the restoring force, n is the constant that controls the sharpness of the transition between elastic to plastic phases, α is the constant controlling the slope of stress strain curve in the inelastic range, B is the back force, f_T is the constant to control the type and size of hysteresis, c is the slope of unloading stress plateau, a is the constant controlling the amount of elastic recovery during unloading, f_m and m are the material constants, controlling the martensitic hardening curve, and u_{mf} is the martensitic finish strain.

In Equation (8.12), u_{in} is the inelastic displacement, given as,

$$u_{in} = u - \frac{F}{K_0} \tag{8.13}$$

and $\Delta(u)$ is the unit step function, defined by Equation (8.14),

$$\Delta(u) = \begin{cases} 1 & u \geq 0 \\ 0 & u < 0 \end{cases} \tag{8.14}$$

erf (u) is the error function, defined as follows,

$$\text{erf}(u) = \frac{2}{\sqrt{\pi}} \int_0^u e^{-t^2} dt. \tag{8.15}$$

sgn (u) is the signum function, defined as,

$$sgn(u) = \begin{cases} 1 & u > 0 \\ 0 & u = 0 \\ -1 & u < 0 \end{cases} \tag{8.16}$$

Applications of SMA in Construction

Since its development from early 1960, SMAs have been successfully used in various fields such as medical, robotics, aerospace engineering, and in automobile industries. SMAs have also found to be useful in building and infrastructure sectors and in repairing and strengthening architectural heritage structures. Some of the examples of applications of SMAs in construction fields are described in the following section.

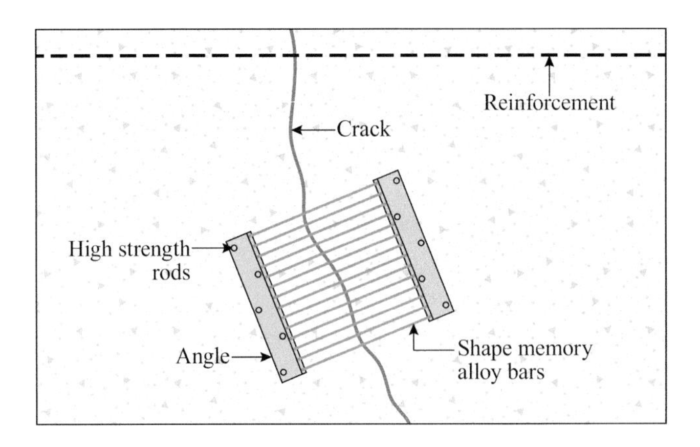

FIGURE 8.13 The SMA alloy bars used to strengthen the bridge girder.

- The first field implementation using the shape memory effect was for post-tensioning of a concrete structure on a highway bridge in Michigan. It had suffered cracks due to insufficient shear resistance. To strengthen the bridge girder, iron–manganese–silicon–chromium SMA rods with a diameter of 10·4 mm were mounted crossing the cracks at both faces of the web, as shown in Figure 8.13. Each rod was heated by electrical current, resulting in a reduction of the crack width by 40%.

  Excessive movement in bridge supports leads to unseating of bridge decks, which is a major cause of infrastructure failure during an earthquake. Reduction of the excessive movement of bridge decks depends on the provision of bridge restrainers. The experimental and analytical outcomes have revealed that bridge restrainers incorporating SMAs can reduce the movement of bridge decks during an earthquake and limit the residual deformation after an earthquake. However, the price of SMAs is the major barrier for them to be adopted, as most of the research work has been focused on nickel–titanium SMAs.
- SMAs are also used in repairing and strengthening works of architectural heritage structures. A nickel-titanium alloy damper developed under the EU-funded Istech project, shown in Photo 8.2, is used to ensure two-way super-plasticity under movement. Mounting of these SMAs inside the structure largely depends on their usage, whether they are being used to prevent large deformations or just to prevent out-of-plane collapse of building structures.

PHOTO 8.2 SMA alloy developed under the EU Istech Project. (Chang and Araki 2016)

PHOTO 8.3 Bell Tower of San Giorgio Church, Trignano, Italy. (Chang and Araki 2016)

- SMA devices are used to prevent the damage that can occur in structure due to earthquake ground vibrations. Although the SMA devices cannot prevent the tower or a slender structure developing cracks during an earthquake, they can certainly prevent the excessive structural deformations and collapse. A number of SMA devices were implemented in the restoration of the bell tower of San Giorgio Church shown in Photo 8.3 at Trignano in Italy. The tower was heavily damaged by an earthquake in 1996. SMA devices were connected with steel bars in series inside the bell tower, to restrain horizontal movement during an earthquake.
- Bell Tower of Badia Fiorentina, Italy, found similar application of large numbers of SMAs to enhance the seismic performance. In 2006, 18 SMA-based devices were used to improve the seismic performance of the tower. The installed SMA devices prevent the excessive deformation of the tower. Photo 8.4 shows Bell Tower of Badia Fiorentina, Firenze, Italy.

PHOTO 8.4 Bell Tower of Badia Fiorentina, Firenze, Italy.

PHOTO 8.5 Basilica of St Francis, Assisi, Italy.

- The mechanism of strengthening the heritage buildings during earthquake using the SMA devices is found in various Italian buildings such as the Basilica of St Francis of Assisi (Photo 8.5) and the San Serafino Church, Foligno (Photo 8.6). In Basilica of St Francis, a total of 47 SMA devices were installed to repair the facade, which was damaged during 1997 earthquake.

  Experiences on different projects indicate that, compared with high damping capacity and SMA effect, super-elasticity of SMAs is extremely important when improving the structural behavior in an earthquake or repairing mechanical damage of a structure after an earthquake, since damping can be provided by alternative mechanisms such as rubber or low-yield steel.
- Base isolation systems with SMAs
  A number of base isolation systems have been developed using nickel–titanium SMAs, with the objective of combining a conventional rubber bearing system with SMA. The

PHOTO 8.6 San Serafino Church, Foligno, Italy.

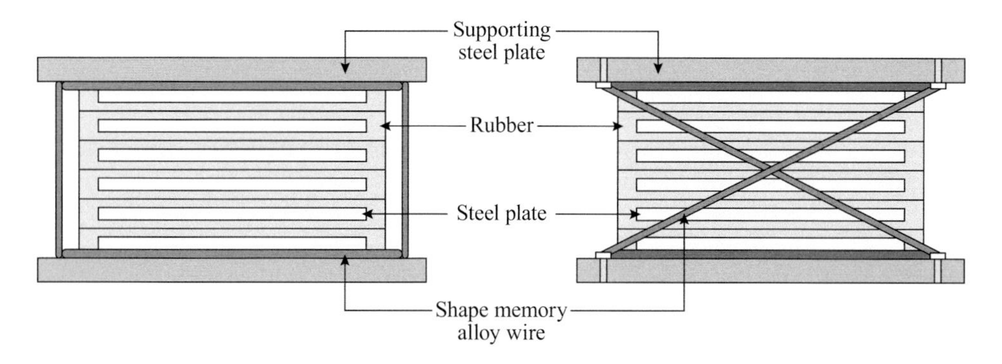

FIGURE 8.14 Shape memory alloy – rubber bearing. (Das and Mishra 2014)

advantage of this combined system is that it can significantly reduce the movement of superstructures during an earthquake. After an earthquake, the super-elasticity of the SMA will bring the rubber bearing system back to its original position; however, during the earthquake, large deformation is possible in the base-isolation system. To limit the large deformation of a conventional base-isolation system, Das and Mishra (2014) proposed an SMA-rubber bearing, shown in Figure 8.14, and demonstrated that it has better efficiency, compared to a conventional system. The shortcoming of this system is the material cost and the cost of construction, for implementation in an existing structure.

Although the SMAs yield during an earthquake, they facilitate the structure in recovering from deformation after an earthquake. Integration of SMAs in a bracing system can effectively resolve the pinching in the hysteresis loop of a structure after it undergoes a large deformation and demonstrates strong re-centering capacity. Previous experiences with concentrically braced steel frames suggest that this system has limited ductility and energy dissipation capacity due to buckling of the bracing. To tackle this limitation, several projects used SMAs connected with bracing of which, the rigidity is much higher than that of SMAs. Research has shown that braces with SMAs are effective in limiting inter-story drifts and residual drifts during an earthquake.

- Limitation

  One of the main barriers to SMAs being adopted in construction is the cost of the material, as materials needed in construction are in large amounts. The cost of producing SMAs involves the raw materials, processing, heat treatment, and machining. To enhance the market uptake of SMAs for use in construction, it is essential to develop low-cost and high-performance SMA products.

REFERENCES

Buehler, W. J., and Wiley, R. C., (1965). Nickel-base alloys, U.S. Patent 3,174,851.

Chang, W. S., and Araki, Y. (2016). Use of shape-memory alloys in construction: A critical review. *Proceedings of the Institution of Civil Engineers-Civil Engineering*, 169(2), 87–95.

Choi, E., Park, S. H., Cho, B. S., and Hui, D. (2013). Lateral reinforcement of welded SMA rings for reinforced concrete columns. *Journal of Alloys and Compounds*, 577, S756–S759.

Christopoulos, C., Filiatrault, A., and Folz, B. (2002). Seismic response of self-centring hysteretic SDOF systems. *Earthquake Engineering & Structural Dynamics*, 31(5), 1131–1150.

Das, S., and Mishra, S. K. (2014). Optimal performance of buildings isolated by shape-memory-alloy-rubber-bearing (SMARB) under random earthquakes. *International Journal for Computational Methods in Engineering Science and Mechanics*, 15(3), 265–276.

Graesser, E. J., and Cozzarelli, F. A. (1991). Shape-memory alloys as new materials for aseismic isolation. *Journal of Engineering Mechanics*, 117(11), 2590–2608.

Nims, D. K., Richter, P. J., and Bachman, R. E. (1993). The use of the energy dissipating restraint for seismic hazard mitigation. *Earthquake Spectra*, 9(3), 467–489.

Ozdemir, H. (1976). *Nonlinear transient dynamic analysis of yielding structure*. Ph.D. Dissertation, University of California.

Priestley, M. N., and Tao, J. R. (1993). Seismic response of precast prestressed concrete frames with partially debonded tendons. *PCI journal*, 38(1), 58–69.

Restrepo, J. I. (2002). New generation of earthquake resisting systems. The First Federation International du Béton (FIB) Congress, Paper E-268, Osaka, Japan.

Ricles, J. M., Sause, R., Garlock, M. M., and Zhao, C. (2001). Posttensioned seismic-resistant connections for steel frames. *Journal of Structural Engineering*, 127(2), 113–121.

Stanton, J. F., and Nakaki, S. D. (2002). Design guidelines for precast concrete seismic structural systems, PRESSS Report No. 01/03-09. UW Report No. SM 02-02. The University of Washington and The Nakaki Bashaw Group, Inc.

Toranzo, L. A., Restrepo, J. I., Carr, A. J., and Mander, J. B. (2004). Rocking confined masonry walls with hysteretic energy dissipators and shake-table validation. 13th World Conference on Earthquake Engineering.

Tsopelas, P., and Constantinou, M. C. (1994). NCEER-Taisei corporation research program on sliding seismic isolation systems for bridges: Experimental and analytical study of a system consisting of sliding bearings and fluid restoring force/damping devices (Report No. NCEER-94-0014). National Center for Earthquake Engineering Research, Buffalo, NY..

Wilde, K., Gardoni, P., and Fujino, Y. (2000). Base isolation system with shape memory alloy device for elevated highway bridges. *Engineering Structures*, 22(3), 222–229.

9 Miscellaneous Response Control Devices

9.1 WIRE ROPE ISOLATORS

Underground structures, such as hydrological project control rooms, oil and natural gas storage facilities, and large ammunition dumps, are significant contributors to the nation's economy. Their failure and any interruption of their operation lead to significant loss. It is important to protect these structures from external vibrations caused due to seismic activity or from the blast-induced ground motion (BIGM), blast, and transmitted shocks. Depending upon the intensity of loading, and geological considerations, the damage may range from minimal to catastrophic. It is also possible that though the structure may survive complete damage, the effect of explosion may render it operationally unserviceable.

Similar to earthquake response reduction, BIGM control can also be achieved by the three common vibration-suppression approaches; i.e., isolation, energy dissipation, and tuned vibration absorption. Traditional isolators and dampers used for seismic hazard mitigation may not meet the requirement of response reduction up to the desired level in specific situations, i.e., when applied against BIGM that are characterized by very high peak acceleration values.

Wire rope isolator (WRI), also known as wire rope spring or metal cable spring, is one of the most effective devices for such applications (Balaji et al. 2015; Demetriades et al. 1993). Due to their omnidirectional nature and ruggedness, WRIs have found wide-ranging applications in the protection of sensitive military equipment and heavy industrial machineries. In the event of a blast or explosion, significant amount of energy is transmitted to the ground in the form of shock and vibration. WRIs are widely used in shock and vibration isolation, due to their capability to absorb impact energy very efficiently. They have also been found to be useful for seismic protection of sensitive equipment in buildings.

Salient Features of WRI

WRIs are essentially a type of spring dampers, consisting of stranded stainless-steel wire ropes, held between rugged metal retainer bars. The spring function is provided by the elasticity, inherent in the flexing of the cable performed into a loop. The damping results from the relative friction between the cables, individual wires, and strands. Additionally, these isolators show great resistance to ageing and creep and are not affected by harsh environmental conditions, chemical agents, and temperature variations. The peculiarity and effectiveness of the WRI over conventional isolation devices lies in their ability to provide isolation in all three planes and all directions, which makes them exclusive. The behavior of WRI is strongly affected by its configuration and design and is dependent on its various characteristics, such as diameter of wire rope, diameter of coil, length of cable, direction of force applied, and the number of wire strands. The competence of a WRI depends on its stiffness, which in turn depends upon the material and geometric properties. WRIs are available in various shapes and sizes, some of which are shown in Figure 9.1.

Mathematical Model of WRI

Structural systems very often show nonlinear behavior under severe excitations like earthquake or explosions. In such cases, the resulting restoring forces become highly nonlinear and show significant hysteresis, which is indicative of the fact that the force cannot be represented as a function of

DOI: 10.1201/9781315269269-9

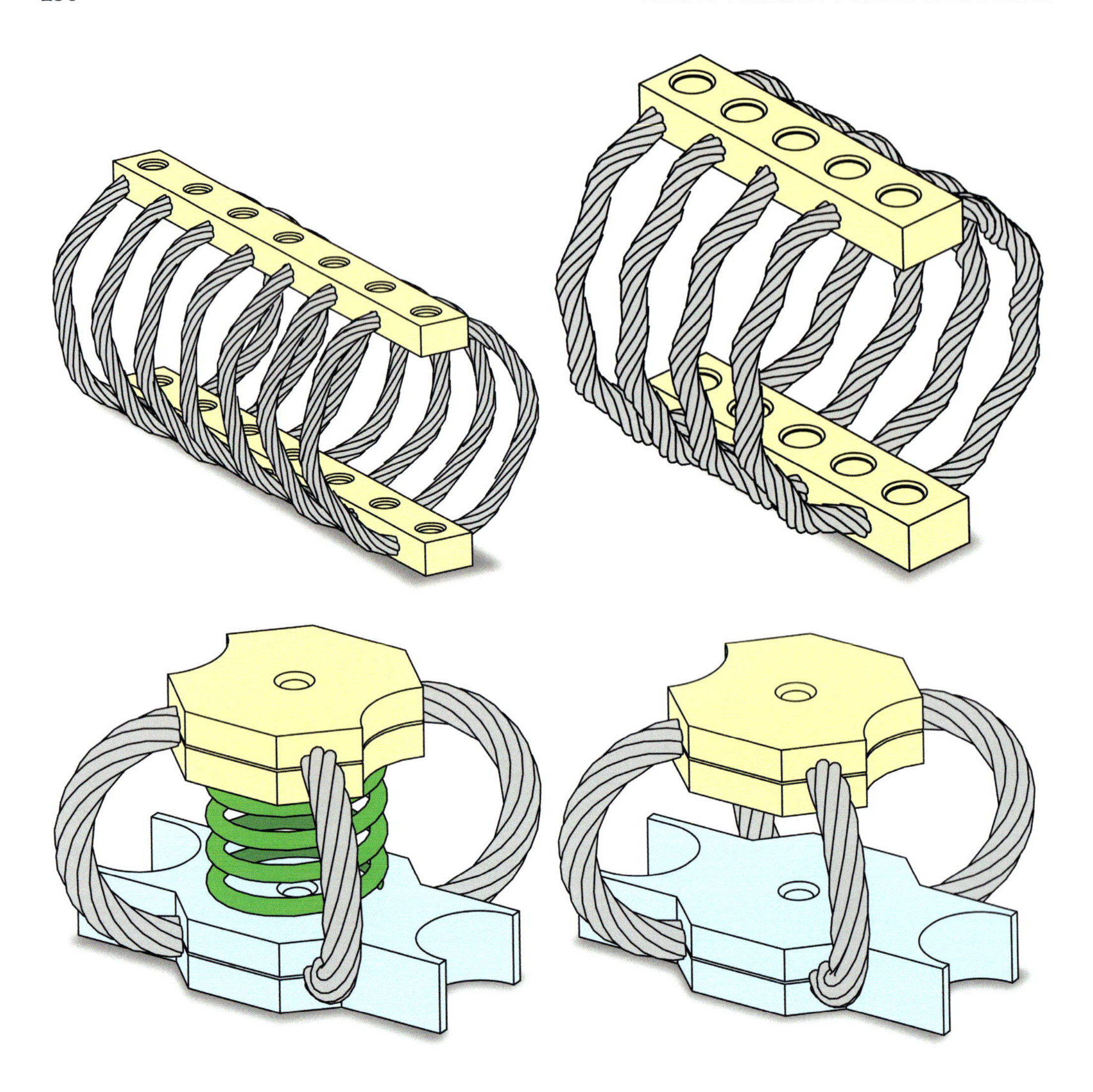

FIGURE 9.1 Typologies of wire rope isolators: (a) helical, (b) half-helical, (c) polycal, and (d) spring polycal.

instantaneous displacement and velocity. The Bouc-Wen model is used for describing the dynamic behavior of a WRI in larger displacement amplitudes. The model can describe the nonlinear hysteretic systems, i.e., the response characteristics of hardening and softening systems, due to its versatility and mathematical tractability. The model considers forced vibrations of a nonlinear system with hysteresis, under periodic excitation. The equation of motion for such system is represented as,

$$\ddot{u}(t) + z = F(t) \tag{9.1}$$

$$\dot{z} + \gamma|\dot{u}|z + \beta\dot{u}|z| = \alpha\dot{u} \tag{9.2}$$

where z is the intrinsic force, u is the displacement, t is the time, and $F(t)$ is the excitation force.

$$\ddot{u}(t) + F_r(u,\dot{u}) = F(t) \tag{9.3}$$

$$F_r(u,\dot{u}) = \bar{g}(u,\dot{u}) + z(u) \tag{9.4}$$

$$\dot{z} = -\beta|\dot{u}|z^n - \gamma\dot{u}|z|^n + \alpha\dot{u};\ n = 1,\ 3,\ 5\ ... \tag{9.5}$$

$$\dot{z} = -\beta|\dot{u}|z^{n-1}|z| - \gamma\dot{u}z^n + \alpha\dot{u};\ n = 2,\ 4,\ 6\ ... \tag{9.6}$$

Equations (9.3)–(9.6) represent the equations of motion. The restoring force (F_r), represented in Equation (9.4), in nonlinear hysteretic model, is decomposed in two parts, $\bar{g}(u,\dot{u})$ and $z(u)$, where $\bar{g}$ is a nonlinear non-hysteretic polynomial, which is a function of the instantaneous displacement and velocity, and $z(u)$ is the hysteretic component. This model allows analytical treatment; for instance, analytical calculation of the Jacobean. It can be seen that Equation (9.2) is a special case of Equation (9.5), where $n = 1$.

$$\dot{z}(t) = \alpha\dot{u}(t) - \beta|\dot{u}(t)|z(t)|z(t)|^{n-1} - \gamma\dot{u}(t)|z(t)|^n \tag{9.7}$$

Equation (9.7), representing the Bouc-Wen model, is constituted from four parameters, i.e., α, β, γ and n. The parameters α, β, γ and n govern the shape and size of the hysteresis loop. α, β and n are positive real numbers, whereas γ may be a positive or negative real number. The parameter β controls the stiffness change as the sign of $\dot{u}$ changes. The shape of the hysteresis loop is determined by varying the combinations of β and γ, with different stiffness characteristics. The parameter α controls the slope of the loop, at $z = 0$. Therefore, with the other parameters being constant, the variation in α alters the height and thickness of the loop. The yielding exponent n is a dimensionless quantity, which governs the smoothness of transition from linear to nonlinear range. Increasing the value of n leads to further elastoplastic behavior. As stated by Wen, $n \in \mathbb{N}$. However, n is a real positive number. For the case, $0 < n < 1$, the term tends to infinity, when $z(t)$ approaches zero; leading to numerical errors. Hence, Equation (9.7) is rewritten in the following form,

$$\dot{z}(t) = \alpha\,\dot{u}(t) - \beta\left|\dot{u}(t)\right| z(t)\left|z(t)\right|^{n-1} - \gamma\,\dot{u}(t)|z(t)|^n \tag{9.8}$$

The non-hysteretic term is essential to describe the restoring force of actual hysteretic vibration isolators. From Equation (9.8), the non-hysteretic term can be easily derived as follows.

$$\frac{\partial z}{\partial u} = \alpha - \left\{\gamma + \beta\, sgn\left[\dot{u}(t)\right] sgn\left[z(t)\right]\right\}|z(t)|^n \tag{9.9}$$

The restoring force (F_r) has been decomposed into z and g, in Equation (9.4), but if $(F_r) = z$, then $g = 0$. It is clear from Equation (9.9) that $\left(\frac{\partial F_r}{\partial u}\right)$ i.e., $\left(\frac{\partial z}{\partial u}\right)$ varies only with z and the sign of $\dot{u}(t)$. Hence, it is independent of $u(t)$. Subsequently, $\left.\frac{\partial F_r}{\partial u}\right|_{F_r=0} = \left.\frac{\partial z}{\partial u}\right|_{z=0} = \alpha$. This implies that for any specified value for $(F_r) = (F_z)$, the hysteretic curves corresponding to various excitations and response levels have the same slope at the points where $(F_r) = (F_z)$ and identical sign of $\dot{u}$, which conflicts with the experimental hysteresis loops. This inconsistency may be avoided by the inclusion of a nonhysteretic term $\bar{g}$ in the equation of restoring force. When $\bar{g}$ represents a linear spring, $\bar{g} = ku(t)$. This leads to

$$\frac{\partial F_r}{\partial u} = k + \left.\frac{\partial z}{\partial u}\right|_{F_r - ku} = k + \alpha - \left\{\gamma + \beta sgn(\dot{u}) sgn(F_r - ku)\right\}\left|(F_r - ku)\right|^n \tag{9.10}$$

The parameter α controls the slope of the loop at $z = 0$, as seen from Table 9.1.

TABLE 9.1
Parameters Governing the Slope of the Hysteresis Loop

$\frac{\partial z}{\partial u}$	$z>0$	$z=0$	$z<0$
$\dot{u}$	$\alpha-(\gamma+\beta)\lvert z\rvert^n$	α	$\alpha-(\gamma-\beta)\lvert z\rvert^n$
$\dot{u}$	$\alpha-(\gamma-\beta)\lvert z\rvert^n$	α	$\alpha-(\gamma+\beta)\lvert z\rvert^n$

- Modeling of the cyclic behavior of WRI
  The cyclic behavior of WRI is well represented for shear cases, (since the hysteresis loop in shear-roll is almost identical in positive and negative directions) by the Equation (9.4). In this model, the restoring force of an elastoplastic model with hardening can be expressed as,

$$F_r=\alpha\left(\frac{F_y}{u_y}\right)u+(1-\alpha)\left(\frac{F_y}{u_y}\right)z \tag{9.11}$$

$$z=A-\beta sgn(u)|z|^{n-1}z-\gamma|z|^n \tag{9.12}$$

where F_r is the restoring force; u is the displacement; α is the post-yield stiffness ratio; u_y and F_y, respectively, are the displacement at yield and yield force; A is the hysteretic amplitude; and z is an internal variable governed by the differential Equation (9.2). The energy loss ratio (ELR), and effective stiffness (K_{eff}) of WRI can be evaluated experimentally.

Energy Loss Ratio (ELR) of WRI

The ELR of WRI is the amount of energy dissipated due to friction between the wire strands, expressed by Equation (9.13). The ELR provides information on the damping capability of the WRI; i.e., a higher value of ELR indicates higher damping.

$$\text{ELR}=\frac{A_{loop}}{\pi\left(\frac{F_{max}-F_{min}}{2}\right)\left(\frac{u_{max}-u_{min}}{2}\right)} \tag{9.13}$$

where, A_{loop} is the total area of the hysteresis loop in positive (tension) and negative (compression) directions, F_{max} and F_{min} are the maximum and minimum values of force, and u_{max} and u_{min} are the maximum and minimum values of displacement, respectively. The geometric characteristics of WRI do not have any significant impact on the ELR; and the ELR largely remains unaffected by the change in wire rope diameter and number of turns. WRIs are reported to have the highest damping, when loaded under smaller amplitude levels. With increasing amplitude of vibrations, the ELR decreases for all orientations of WRI, which indicates reduction in damping and reduced area of the hysteresis loop.

Effective Stiffness of WRI

The effectiveness of WRIs depends primarily on the stiffness. The effective stiffness (K_{eff}) is a dynamic property, which is strongly governed by the geometric properties of WRI, such as rope diameter, number of coils, displacement amplitude, and material properties. The spring orientation affects the effective stiffness in shear-roll. With increase in wire rope diameter, (K_{eff}) increases, due to increase in the number of contact points between individual wire strands. (K_{eff}) is doubled, if the number of turns is doubled. Various experimental studies have shown that WRIs are stiffer in tension-compression as compared to in shear-roll. Stiffer isolators have higher damping effectiveness, i.e., increased energy dissipation capability. For tension-compression, the effective stiffness is expressed by Equation (9.14). It decreases for higher displacement values, on account of the increase in individual slip between wire strands.

$$K_{eff} = \left(\frac{F_{max} - F_{min}}{u_{max} - u_{min}} \right) \tag{9.14}$$

Energy Dissipation Capability of WRIs

The various geometrical characteristics, such as wire rope diameter, number of turns, orientation of cables, loosening and tightening of strands, govern the hysteresis behavior and affect the shape of hysteresis curve of WRI. However, it is independent of the frequency of loading. The components of the restoring force are a nonhysteretic elastic force and a hysteretic damping force. The cyclic behavior of WRI for tension-compression, and shear-roll is different. The amplitude of displacement significantly affects the shape of hysteresis curves. Rubbing and sliding friction between the individual strands provide inherent damping due to flexural hysteresis. The wires and cables possess inherent capability to dampen the vibration due to static hysteresis resulting from the sliding friction between cable strands. The viscous damping in WRI is negligible. The damping in WRI is maximum, when loaded under smaller displacement amplitudes in vertical direction.

As seen from Figure 9.2(a), the hysteresis curves under vertical, i.e., tension-compression loading are asymmetrical, indicating variable stiffness in tension and compression modes. Asymmetry can be attributed to the nonlinear contact between the steel-stranded wires for higher loads. Asymmetry of curves results due to different numbers of strands tightening and relaxing during tension and compression in different arrangements and hardening in tension and softening in compression. As seen in Figure 9.2(a), initially, for smaller displacements, the curves are symmetrical in both directions; however, at larger amplitudes, the asymmetrical nature becomes evident due to the nonlinear contact between stranded wires. Since the hysteresis force-deformation curves in tension-compression modes are asymmetric, they cannot be represented accurately by the classical Bouc-Wen model. Furthermore, the hardening behavior in tension is because of the number of contact points increasing significantly, as the wire strands get closer, resulting into increased frictional resistance. This eventually leads to a stiffer isolator.

While in compression mode, as the strands tend to move apart, the number of contact points decreases, leading to decrease in friction and, hence, the softening behavior. Hardening behavior is observed at tension, and softening under compression. With increasing amplitude, this effect is seen more prominently.

As seen from Figure 9.2(b), tests in the horizontal direction, (shear-roll) exhibit symmetric behavior of the hysteresis due to circular section of the coil, and equal number of coil's turns from the center of the isolator. The dissipated energy increases progressively as the deformation increases.

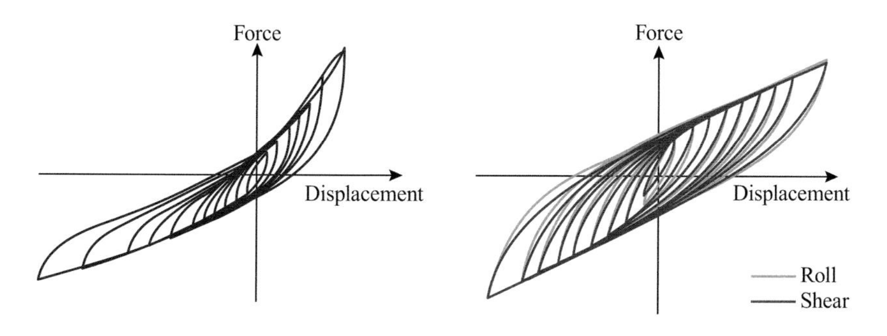

FIGURE 9.2 Hysteresis curves of wire rope isolators: (a) tension-compression and (b) shear-roll.

Unlike in tension-compression, the hysteresis plots in shear-roll are symmetric in positive and negative directions because the restoring force mechanism is the same and the stiffness of the WRI remains equal for positive and negative shear.

9.2 BRETELLE DAMPER

Bretelle dampers are used in the form of clamps that are meant for installation with conductor loops. Bretelle dampers not only exhibit damping function but also help in preventing conductor rupture and are designed to carry high slipping load. The conductor channel of the Bretelle damper is grooved, whereas, the channel for the conductor is serrated in order to protect the conductor without affecting the clamp function. Bretelle dampers are typically installed on most suspension clamps in 225- and 400-kV lines. They are composed of a small length of the same conductor of the line, attached to the conductor, by special fixing blocks. Positioning of dampers is same at each suspension tower of the line. The tension in the cables does not affect the performance of the dampers.

Figure 9.3 shows a typical Bretelle damper. There are two types of Bretelle clamps, (i) open type, where it is a continuous channel for the conductor (multiple loops), and (ii) closed type, in which the closed end is intended to optimal corona protection. All internal spaces are provided to take up the clamping of the dampers, which are installed as per the specific requirements.

9.3 SPACER DAMPER

Transmission lines are laid along long distance in order to transmit signals from one place to the other. When cables of the transmission lines are required to span along very long distance, they are supported by pylons on adjacent spans. In this case, wind may tend to cause significant vibration of the cables. Several forms of vibration, i.e., aeolian vibration, sub-conductor oscillations, galloping, and wind-sway occur in such cases. The most dominant is the Aeolian vibration. Aeolian vibration, shown in Figure 9.4 is characterized as low-amplitude vibration with high frequency, generally ranging from 20 to 150 Hz.

It is the most important type of vibration, because it causes fatigue failure of conductor strands or the system used for protection of conductors. Aeolian vibration can occur in relatively light winds, with velocity of 2 to 24 km/h, resulting in the formation of eddies at the end of the cable. When the frequency of eddies generated in the cable becomes equal to the natural frequency of the cable, the forces generated result in the fatigue failure. Nowadays, a transmission line consists of several sub-conductors for each pole, and at each position; where the sub-conductors are grouped in bundles. However, such type of transmission lines are subjected to aeolian vibration. Due to the flow of wind over the windward conductors, eddies and vortices are formed, between the sub-conductors, which are placed adjacent to each other. In order to prevent the failure of conductor and protect the cables of transmission lines from wind effect, spacer dampers are widely used. Photo 9.1 shows the spacer

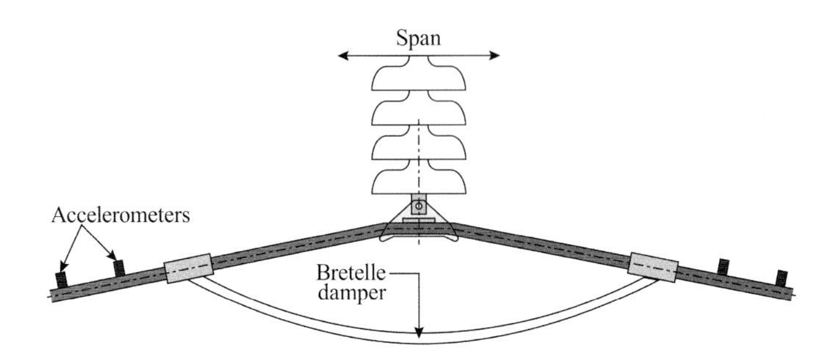

FIGURE 9.3 Bretelle damper.

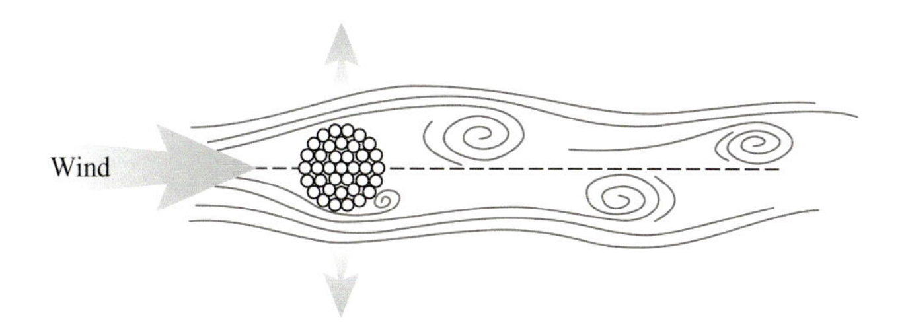

FIGURE 9.4 Aeolian vibration.

dampers, used to prevent aeolian vibration. For effective functioning, installation of exact number of spacer dampers is done along each span precisely. The spacer dampers protect the conductors from high-speed winds, and also from galloping, by preventing the clashing of sub-conductors within the span length. Such spacers are flexible in nature, so that they can take care of high short circuit current loads in compression and tension to protect the conductors and cables from damage. For double conductor system, flexible spacers with high energy capabilities are required.

General Information About Spacer Damper

A spacer damper essentially consists of conductor clamps, a connecting frame with damping elements made of highly elastic macro-molecular elastomers, characterized by high energy-absorbing capabilities that can damp harmful wind vibrations and also have required flexibility and stiffness in all directions. The damping elements also possess the capability of resisting UV rays and ozone attacks. Generally, the clamps of the dampers are made of high-grade cast aluminum, and high-strength screws. There are two types of spacer dampers, rigid spacer and flexible spacer. Rigid spacers are used where conductor connections are very short, and where it is not feasible to use flexible spacers. Rigid spacers are readily available in market for double, triple, and quadruple bundles; however, other designs may be manufactured according to the specific requirements. Rigid spacers equipped with counterweights to create tower clearance are used for jumpers. Figure 9.5 shows such type of spacer dampers.

Flexible dampers are used as horizontal and vertical twin bundles made of aluminum. In order to impart flexibility, these dampers are provided with moving clamp bodies connected by means of spacing piece as shown in Figure 9.6. However, the clamp bodies are guided over steel bodies and

PHOTO 9.1 Dampers used to prevent aeolian vibration.

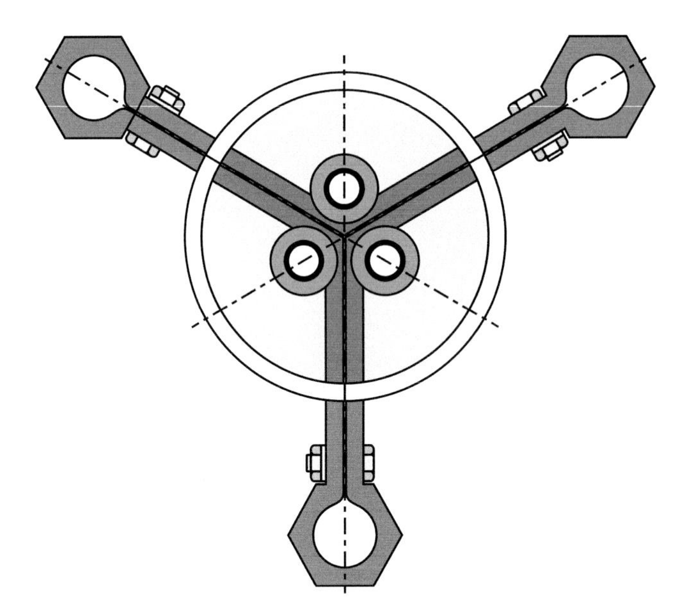

FIGURE 9.5 Rigid spacer dampers for jumpers with counterweights.

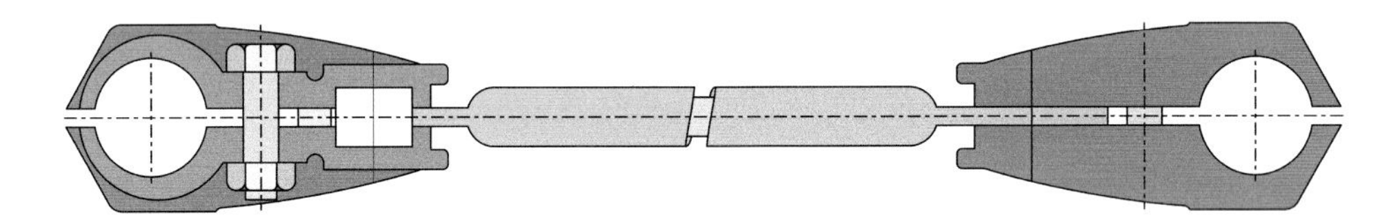

FIGURE 9.6 Flexible spacer dampers.

FIGURE 9.7 Flexible spacer damper with neoprene washers.

are capable of moving only in one direction. Stainless sleeves are provided with the bore in the link so as to provide the optimum bolt connection. Flexible spacer dampers, provided with neoprene washers are shown in Figure 9.7. They are used in the form of horizontal and vertical twin bundles.

REFERENCES

Balaji, P. S., Rahman, M. E., Moussa, L., and Lau, H. H. (2015). Wire rope isolators for vibration isolation of equipment and structures – a review. IOP Conference Series: Materials Science and Engineering, 78 (01).

Demetriades, G. F., Constantinou, M. C., and Reinhorn, A. M. (1993). Study of wire rope systems for seismic protection of equipment in buildings. *Engineering Structures*, 15(5), 321–334.

10 Advances in Passive Control Strategies

Hybrid Control

10.1 INTRODUCTION

During earthquakes, structural damages could be caused because of excessive displacements and large forces induced on structural members. In view of the extensive damage of structures during earthquakes, scientists and engineers are exerting significant efforts to mitigate their consequences. Hence, the current research is focused on identifying more effective solutions and enhancing existing strategies for enhanced structural protection. Despite substantial improvements in retrofit programs and new design approaches, such as better detailing, dissipating the seismic energy through inelastic deformations without complete collapse, and considering higher loads and forces, earthquakes are still causing significant damage to structures, property, and the occupants around the world. This calls for strengthened research effort in the area and the implementation of state-of-the-art techniques to protect structures against the undesirable effects of various hazards, such as earthquakes and wind. In this chapter, the implementation of one of the promising approaches, i.e., passive hybrid response control strategies, for vibration response control of structures is discussed.

10.2 PASSIVE CONTROL SYSTEMS

A passive structural control system provides control forces according to the local motion and the dynamic characteristics of the control devices at the point where they are connected to the structure, without demanding for any source of external power supply. Once designed and implemented to satisfy specific requirements, passive control systems are unable to adapt to the changes in structural properties and the stochastic nature of external excitations. In this type of system, the mass/damping/stiffness or a combination of any two or all properties are modified by adding some components into the structure. The passive control devices are actuated by the movements of the structure, which in turn leads to the exertion of the control force depending on their dynamic characteristics. Three common types of passive control devices are given as:

1. Base isolators
    - i. Laminated rubber bearings (LRBs)
    - ii. Lead rubber bearings
    - iii. Pure friction (PF) system
    - iv. Friction pendulum system (FPS)
    - v. Resilient-friction base isolator (R-FBI) system
    - vi. Électricité de France
2. Energy dissipation devices (dampers)
    - i. Fluid viscous dampers (VDs)
    - ii. Viscoelastic dampers (VEDs)
        - a. Viscoelastic solids
        - b. Viscoelastic fluids
    - iii. Friction dampers (FDs)

DOI: 10.1201/9781315269269-10

iv. Hysteretic dampers
 a. Metallic dampers
3. Tuned dampers
 i. Tuned mass dampers (TMDs)
 ii. Tuned liquid dampers (TLDs)

In this chapter, passive hybrid control strategies that incorporate multiple passive control devices are discussed.

10.3 HYBRID CONTROL SYSTEMS

Hybrid vibration control systems are vibration response control strategies that implement two or more response control techniques simultaneously on a single structure. Equipping a structure with multiple control techniques helps in delivering enhanced control performance. The implementation of passive isolation bearings in combination with other passive, semi-active, or active control devices is a typical example of hybrid control strategy.

10.4 HYBRID BASE ISOLATION SYSTEM FOR BUILDINGS AND BRIDGES

Hybrid base isolation (BI) systems can be implemented for the mitigation of vibrations in buildings as well as bridges. Examples of various types of hybrid BI systems for application to buildings and bridges are shown in Figure 10.1(a) and (b).

Hybrid control systems can be designed by combination of base isolators and passive dampers, as shown in the figure.

Seismic BI is the most widely used earthquake-protective scheme for civil engineering structures. However, base-isolated structures are susceptible to large isolator displacement, and serious damages can occur in the base-isolated structures, especially under cases of near-fault excitations. The large isolator displacement could compromise the stability of the structure. Furthermore, it poses a challenge to the proper functioning of various service fittings crossing the isolation level/layer. In order to reduce the seismic impact due to severe earthquakes, the performance of BI system could be improved by adding supplemental passive, semi-active, and active control devices; creating hybrid structural control.

Elastomeric and sliding bearings are the most commonly used base isolators. One of the common elastomeric bearings is the lead rubber bearing (N-Z system). Lead rubber bearings combine the function of isolation and energy dissipation in a single compact system and also support the weight of the superstructure. Further, the N-Z system provides restoring force. Sliding bearings allow mobility and provide energy dissipation through friction. Spherical sliding bearings provide a restoring mechanism to the structure, after it is displaced by seismic horizontal forces, because of their curvature. However, despite the inherent restoring force in the passive isolators, under very severe ground motions, the superstructure may experience large displacements, which may lead to failure of isolation bearings. For this reason, supplementary dampers are used in combination with seismic isolators to reduce the displacement response of the base-isolated structure without significant compromise to the effectiveness of the BI system in reducing the superstructure acceleration and shear forces.

The examples presented in the following sections demonstrate the application of an appropriate combination of base isolators and passive supplemental dampers to mitigate seismic response of buildings and bridges, subjected to real earthquake ground motions.

EXAMPLES

Example 10.1 Hybrid Response Control of a Building Using Base Isolation and Tuned Mass Damper

A building is isolated by lead rubber bearing for protection against dynamic loads, such as earthquakes. The superstructure of the building is modeled by lumping the total mass of

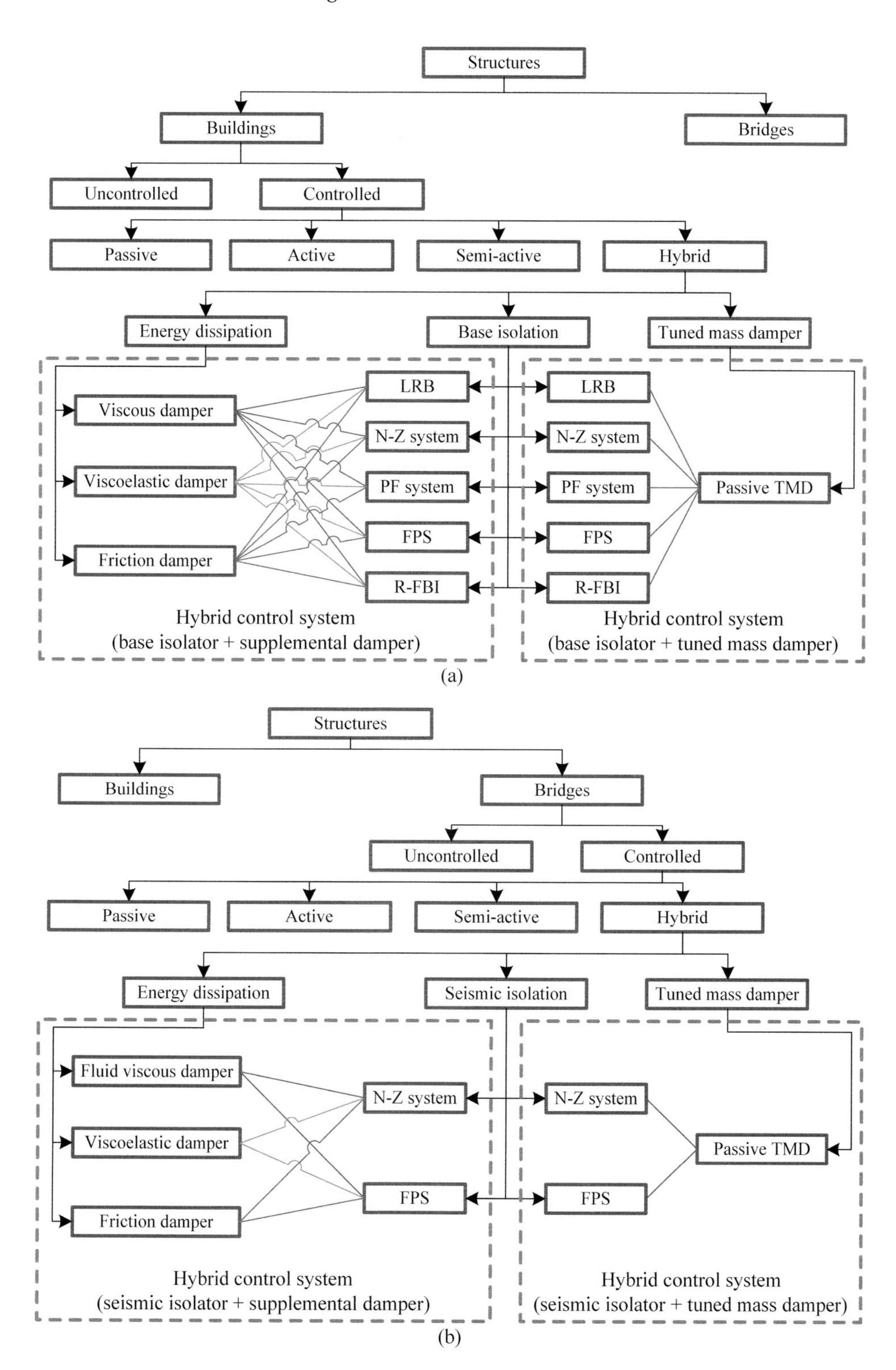

FIGURE 10.1 Examples of hybrid base isolation systems for (a) buildings and (b) bridges.

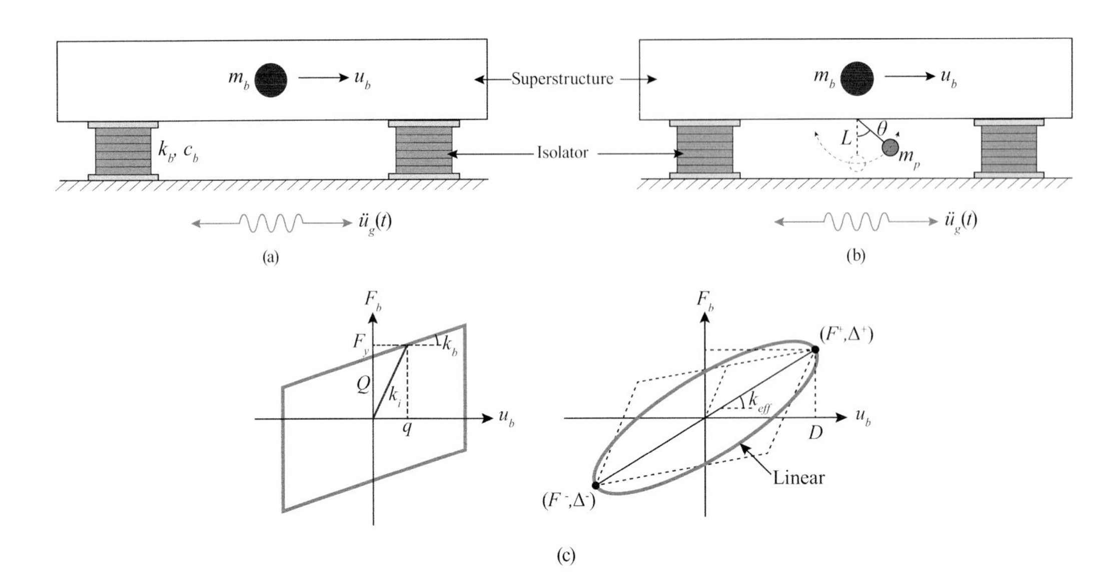

FIGURE 10.2 The idealized: (a) mathematical model of the base-isolated building, (b) mathematical model of the base-isolated building with the pendulum tuned mass damper, and (c) bilinear and equivalent linear force-deformation curves of the isolation system.

the structure as a single mass (m) of 6,000 ton. The idealized mathematical model of the base-isolated building with the lumped superstructure is shown in Figure 10.2(a). A hybrid structural response control strategy of BI and pendulum TMD is also implemented as an alternative control approach as shown in Figure 10.2(b). The lead rubber bearing can be modeled using a bilinear force-deformation behavior depicted in Figure 10.2(c). The yield displacement (q), the yield strength (F_y), the initial stiffness (k_i), and the post-yield stiffness (k_b) of the isolation system are given as 2.5 cm, 4,120 kN, 164,808 kN/m, and 37,900 kN/m, respectively. Alternatively, the bilinear isolation system can also be approximately modeled as an equivalent linear system using the effective stiffness (k_{eff}) and effective damping ratio (β_{eff}) as shown in Figure 10.2(c). The values of (k_{eff}) and (β_{eff}), respectively, can be obtained using Equations (10.1) and (10.2).

$$k_{eff} = k_b + \frac{Q}{D} \tag{10.1}$$

$$\beta_{eff} = \frac{4Q(D-q)}{2\pi k_{eff} D^2} \tag{10.2}$$

where D is the design isolation displacement given as 0.4 m and $\left(Q = F_y - qk_b\right)$ is the characteristic strength of the isolation system. Furthermore, the time period of the simple pendulum is given as $T_p = 2\pi\sqrt{\frac{L}{g}}$; where $L = 1$ m is the length of the pendulum, $g = 9.81\ \text{m/s}^2$ is gravitational acceleration, and $m_p = 300\,\text{t}$ is the mass of the pendulum. In addition, both the building models shown in Figure 10.2(a) and (b) are subjected to ground acceleration of $\left[\ddot{u}_g = sin(\pi t) + sin(2\pi t)\right]$ for a 30-s duration.

a. Perform a time history analysis for the base-isolated building and determine the peak values of the displacement of the superstructure (u_b) and absolute acceleration of the superstructure ($\ddot{u}_{b,Abs}$).
b. Perform a time history analysis for the base-isolated building with the pendulum TMD and determine the peak values of the displacement of the superstructure (u_b) and absolute acceleration of the superstructure ($\ddot{u}_{b,Abs}$). Also, compare the peak values of the response quantities for the two response control approaches.

SOLUTION

First, the data given regarding the base-isolated structure is provided.

- **Given Data**

  Structure
  Total mass $m = 6{,}000{,}000$ kg

  Base Isolation System
  Yield displacement $q = 2.5$ cm
  Yield strength $F_y = 4{,}120$ kN
  Initial stiffness $k_i = 164{,}808$ kN/m
  Post-yield stiffness $k_b = 37{,}900$ kN/m
  Design displacement $D = 0.4$ m

  Pendulum Tuned Mass Damper
  Mass $m_p = 300{,}000$ kg
  Length $L = 1$ m

- **Modeling the Base Isolator Using the Equivalent Linear Approach**
  The characteristic strength of the isolation system can be calculated as:

$$Q = F_y - qk_b \tag{10.3}$$

Therefore, the characteristic strength of the isolation system is evaluated and presented subsequently.

Characteristic strength $Q = 3{,}172.72$ kN

The effective stiffness of the isolation system can be calculated as:

$$k_{eff} = k_b + \frac{Q}{D} \tag{10.4}$$

Therefore, the effective stiffness of the isolation system is evaluated and presented subsequently.

Effective stiffness $k_{eff} = 45{,}831.1$kN/m

The effective time period of the isolation system can be calculated as:

$$T_{eff} = 2\pi\sqrt{\frac{m}{k_{eff}}} \tag{10.5}$$

Therefore, the effective time period of the isolation system is evaluated and presented subsequently.

Effective time period $T_{eff} = 2.2734$ s

The angular frequency

$$\omega_{eff} = \frac{2\pi}{T_{eff}} = 2.764 \text{ rad/s} \tag{10.6}$$

The effective damping ratio of the isolation system can be calculated as:

$$\beta_{eff} = \frac{4Q(D-q)}{2\pi k_{eff} D^2} \tag{10.7}$$

Therefore, the effective damping ratio of the isolation system is evaluated and presented subsequently.

Effective damping ratio $\beta_{eff} = 0.10323$

The angular frequency

$$C_{eff} = 2\beta_{eff} m\omega_{eff} = 3,425.7\,\text{kNs/m} \tag{10.8}$$

- **Pendulum TMD**
  The time period of the pendulum TMD can be calculated as:

$$T_p = 2\pi\sqrt{\frac{L}{g}} \tag{10.9}$$

Therefore, the time period of the pendulum TMD is evaluated and presented subsequently.

Time period of the pendulum $T_p = 2.01$ s

To model the pendulum as a spring-mass system, the stiffness can be obtained based on the time period as:

$$k_p = m_p\left(\frac{2\pi}{T_p}\right)^2 = 2,943\,\text{kN/m} \tag{10.10}$$

a. **Time History Analysis of the Base-Isolated Structure**
   The time history response quantities of the base-isolated structure under the base are evaluated.

- **Base Excitation**
  The base excitation is given as:

$$\ddot{u}_g = sin(\pi t) + sin(2\pi t) \tag{10.11}$$

- **Equation of Motion**
  The equation of motion for the base-isolated structure, which is idealized as an SDOF structure, can be written as:

$$m\ddot{u}_b + c_{eff}\dot{u}_b + k_{eff}u_b = -m\ddot{u}_g, \tag{10.12}$$

$$m\ddot{u}_b + c_{eff}\dot{u}_b + k_{eff}u_b = -m\left[sin(\pi t) + sin(2\pi t)\right] \tag{10.13}$$

b. **Time History Analysis of the Base-Isolated Structure with the Pendulum TMD**
   The vibration response of the base-isolated building with the pendulum is evaluated subsequently.

   The mass matrix and the stiffness matrix of the system, respectively, are given as:

$$\begin{bmatrix} m & 0 \\ 0 & m_p \end{bmatrix} = \begin{bmatrix} 6,000,000 & 0 \\ 0 & 300,000 \end{bmatrix}\text{kg and} \tag{10.14}$$

$$\begin{bmatrix} k_{eff}+k_p & -k_p \\ -k_p & k_p \end{bmatrix} = \begin{bmatrix} 48,774.1 & -2,943 \\ -2,943 & 2,943 \end{bmatrix}\text{kN/m} \tag{10.15}$$

The damping constant of the pendulum damper, c_p, is taken as 0. The damping matrix of the combined system can be given as:

$$\begin{bmatrix} c_{eff}+c_p & -c_p \\ -c_p & c_p \end{bmatrix} = \begin{bmatrix} 3,425.7+0 & -0 \\ -0 & 0 \end{bmatrix}\text{kNs/m} = \begin{bmatrix} 3,425.7 & 0 \\ 0 & 0 \end{bmatrix}\text{kNs/m.} \tag{10.16}$$

The matrix form of the equations of motion for the base-isolated structure with the pendulum can be expressed as:

$$\begin{bmatrix} m & 0 \\ 0 & m_p \end{bmatrix}\begin{Bmatrix} \ddot{u}_b \\ \ddot{u}_p \end{Bmatrix}+\begin{bmatrix} c_{eff}+c_p & -c_p \\ -c_p & c_p \end{bmatrix}\begin{Bmatrix} \dot{u}_1 \\ \dot{u}_2 \end{Bmatrix}+\begin{bmatrix} k_{eff} & 0 \\ 0 & k_p \end{bmatrix}\begin{Bmatrix} u_b \\ u_p \end{Bmatrix}=-\begin{bmatrix} m & 0 \\ 0 & m_p \end{bmatrix}\begin{Bmatrix} 1 \\ 1 \end{Bmatrix}\ddot{u}_g \quad (10.17)$$

- **Solution of the Governing Equations of Motion**

  The equations of motion given in Equations (10.13) and (10.17) can be solved numerically to obtain the time history response quantities of the structures. There are various methods that can be implemented to calculate the response of the structures numerically. Some of the common methods include the central difference method, Newmark's-β method, Wilson-θ method, Hilber-Hughes-Taylor (HHT) method, state-space method, and modal time history analysis. For this problem, the central difference method is used.

  The bilinear force-deformation model of the isolation system can also be used. The time history analysis of the models of (a) the base-isolated structure and (b) the base-isolated structure with the pendulum is run on SAP2000 considering the bilinear force-deformation relationship.

  The plot of the time histories of the response quantities for the base-isolated structure and the base-isolated structure with the pendulum are given in Figure 10.3(a) and (b). Furthermore, the comparison of the peak values of the response quantities are also presented in Table 10.1.

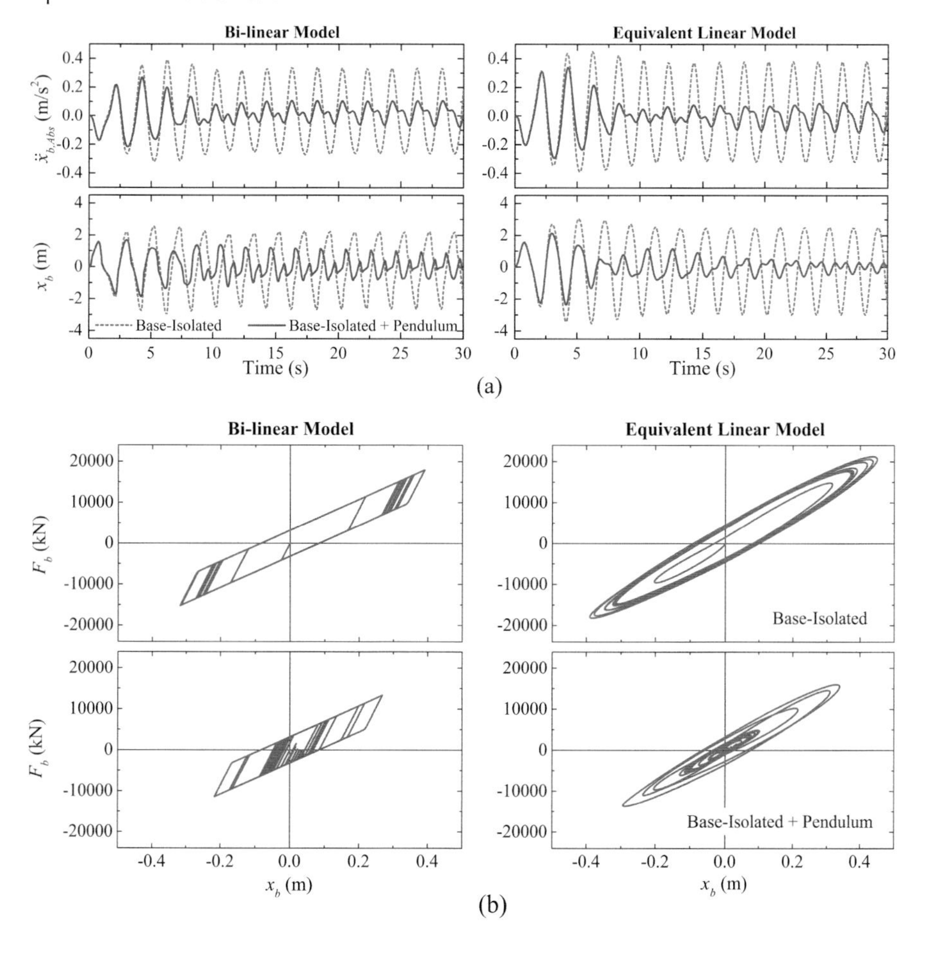

FIGURE 10.3 Comparison of response time histories and force-deformation curves of the isolator for the bilinear and equivalent linear modeling approaches: (a) response time histories of the building and (b) force-deformation curves of the isolator.

TABLE 10.1
Percentage Reduction in the Peak Response Quantities of the Connected Structures as Compared to the Unconnected Response Quantities

	Bilinear		Equivalent Linear	
Structure	**Peak Displacement (m)**	**Peak Acceleration (m/s^2)**	**Peak Displacement (m)**	**Peak Acceleration (m/s^2)**
Base-Isolated	0.3904	2.9934	0.4495	3.5398
Base-Isolated + Pendulum	0.2680	1.8634	0.3384	2.3571
Percentage Difference	31.35	37.75	24.72	33.41

Example 10.2 Hybrid Response Control: Base Isolation, Supplemental Damper, and Tuned Mass Damper

The fixed-base building frame shown in Figure 10.4(a) has a mass (m) of 10,000 kg, and the total lateral stiffness of the columns (k) is 2,470 kN/m. The damping ratio of the superstructure (ξ_s) is 0.02, and the base excitation is given as $\ddot{u}_g = 3sin(5\pi t)$. Answer the following by performing time history analysis under the base excitation ($\ddot{u}_g$) acting for a total duration of 10 s. Use a time step (Δt) of 0.001 s for the time history analysis. For all the cases, evaluate the properties of the control devices and construct the mass, the stiffness, and the damping matrices. Calculate the peak values of the displacement and absolute acceleration response quantities of the top floor of the building shown in Figure 10.4(a), considering the hybrid control strategy of either: (i) BI with VD (BI + VD), or (ii) BI with TMD (BI + TMD). The properties of the BI system, the TMD, and the VD can be obtained based on the data shown in respective figures.

Compare the peak values of the response quantities of the building frame controlled by the two hybrid control strategies with that of the uncontrolled building. Figure 10.4(a) shows the schematic diagram of the fixed-base building frame, whereas Figure 10.4(b) details the properties of the building. Further, Figure 10.4(c) shows the building frame isolated by LRB.

SOLUTION

The solution of the problem is presented as follows.

1. **Hybrid control strategy: BI with VD**
   In the hybrid control strategy involving BI and VD shown in Figure 10.4(d), there are two lateral (translational) degrees of freedom at the base mass and at the top floor of the building. The damping constant of the VD can be combined with the damper of the superstructure in constructing the damping matrix.

   The mass and the stiffness matrices remain the same as that of the base-isolated building.

   Mass matrix,

$$[M] = \begin{bmatrix} m_b & 0 \\ 0 & m_t \end{bmatrix} = \begin{bmatrix} 6.67 & 0 \\ 0 & 10 \end{bmatrix} \times 1,000\,\text{kg}$$

   Stiffness matrix,

$$[K] = \begin{bmatrix} k_b + k_t & -k_t \\ -k_t & k_t \end{bmatrix} = \begin{bmatrix} 2.63 & -2.47 \\ -2.47 & 2.47 \end{bmatrix} \times 10^6\,\text{N/m}$$

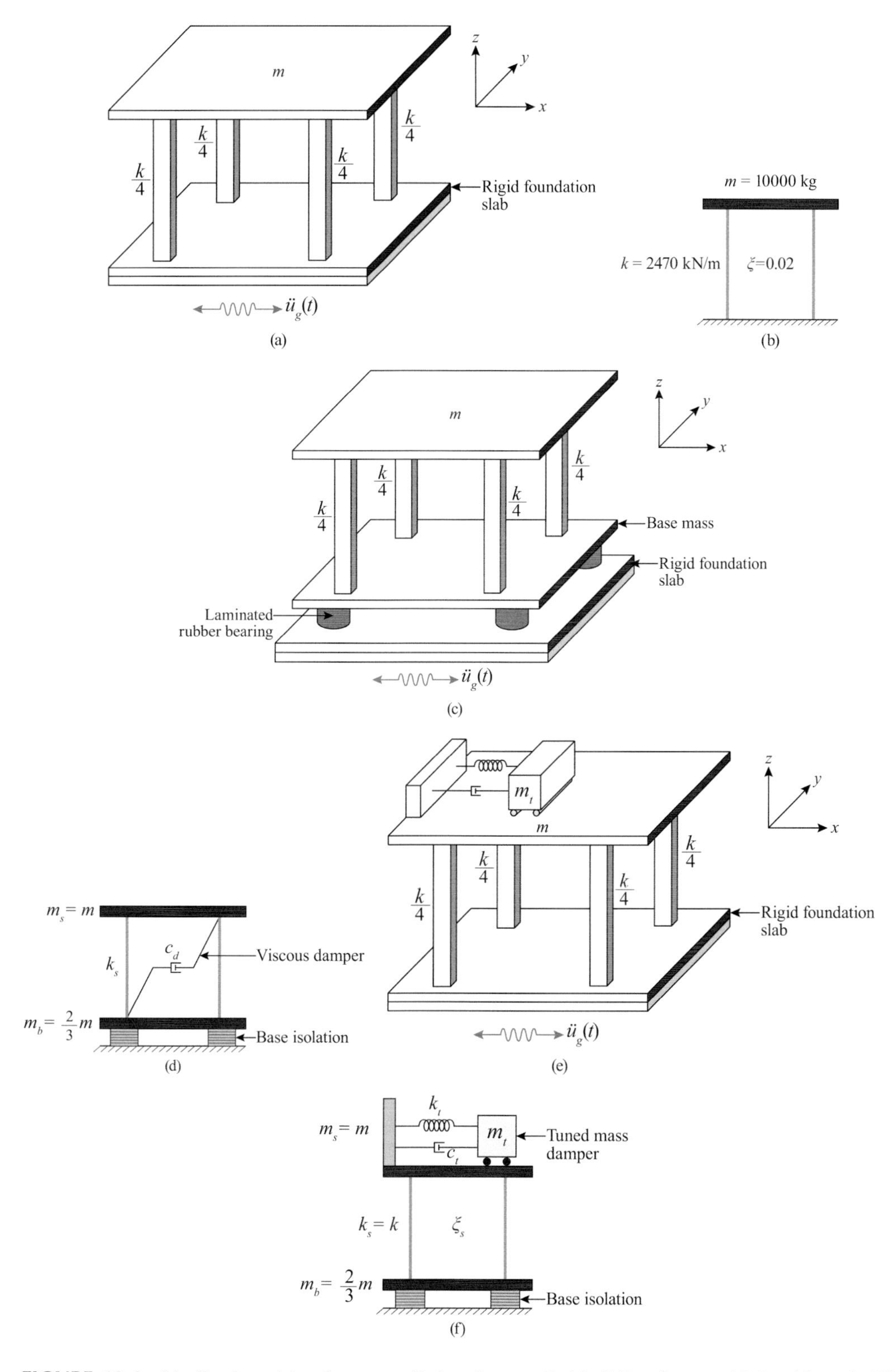

FIGURE 10.4 Idealized models of uncontrolled and controlled building frames: (a) fixed-base building frame, (b) properties of fixed-base building, (c) building isolated by LRB, (d) building with base isolation + viscous damper, (e) building with TMD, and (f) building with base isolation + TMD.

Damping matrix,

$$[C]=\begin{bmatrix} c_b+(c_s+c_d) & -(c_s+c_d) \\ -(c_s+c_d) & (c_s+c_d) \end{bmatrix}=\begin{bmatrix} 21.76 & -11.29 \\ -11.29 & 11.29 \end{bmatrix}\times 1{,}000\,\text{Ns/m}.$$

The matrix form of the equations of motion for the base-isolated building with the VD can be written as:

$$\begin{bmatrix} m_b & 0 \\ 0 & m_s \end{bmatrix}\begin{Bmatrix} \ddot{u}_b \\ \ddot{u}_s \end{Bmatrix}+\begin{bmatrix} c_b+(c_s+c_d) & -(c_s+c_d) \\ -(c_s+c_d) & (c_s+c_d) \end{bmatrix}\begin{Bmatrix} \dot{u}_b \\ \dot{u}_s \end{Bmatrix}+\begin{bmatrix} k_b+k_s & -k_s \\ -k_s & k_s \end{bmatrix}\begin{Bmatrix} u_b \\ u_s \end{Bmatrix}=-\begin{bmatrix} m_b & 0 \\ 0 & m_s \end{bmatrix}\begin{Bmatrix} 1 \\ 1 \end{Bmatrix}\ddot{u}_g$$

2. **Hybrid control strategy: BI with TMD**
Figure 10.4(e) shows the building frame with TMD. Figure 10.4(f) shows the mathematical model of the base-isolated building frame with TMD.
The hybrid control strategy makes the numbers of degrees of freedom three: viz. at the base mass, at the top floor of the building, and at the TMD.
The mass, the stiffness, and the damping matrices are constructed as shown below.
Mass matrix,

$$[M]=\begin{bmatrix} m_b & 0 & 0 \\ 0 & m_s & 0 \\ 0 & 0 & m_t \end{bmatrix}=\begin{bmatrix} 6.667 & 0 & 0 \\ 0 & 10 & 0 \\ 0 & 0 & 0.2 \end{bmatrix}\times 1{,}000\,\text{kg}$$

Stiffness matrix,

$$[K]=\begin{bmatrix} k_b+k_s & -k & 0 \\ -k_s & k_b+k_t & -k_t \\ 0 & -k_t & k_t \end{bmatrix}=\begin{bmatrix} 2{,}634.49 & -2{,}470 & 0 \\ -2{,}470 & 2{,}591.40 & -49.4 \\ 0 & 49.4 & 49.4 \end{bmatrix}\times 10^3\,\text{N/m}$$

Damping matrix,

$$[C]=\begin{bmatrix} c_b+c_s & -c_s & 0 \\ -c_s & c_s+c_t & -c_t \\ 0 & -c_t & -c_t \end{bmatrix}=\begin{bmatrix} 16.76 & -6.29 & 0 \\ -6.29 & 6.92 & -0.63 \\ 0 & -0.63 & 0.63 \end{bmatrix}\times 1{,}000\,\text{Ns/m}$$

$$\begin{bmatrix} m_b & 0 & 0 \\ 0 & m_s & 0 \\ 0 & 0 & m_t \end{bmatrix}\begin{Bmatrix} \ddot{u}_b \\ \ddot{u}_s \\ \ddot{u}_t \end{Bmatrix}+\begin{bmatrix} c_b+c_s & -c_s & 0 \\ -c_s & c_s+c_t & -c_t \\ 0 & -c_t & c_t \end{bmatrix}\begin{Bmatrix} \dot{u}_b \\ \dot{u}_s \\ \dot{u}_t \end{Bmatrix}+\begin{bmatrix} k_b+k_s & -k_s & 0 \\ -k_s & k_s+k_t & -k_t \\ 0 & -k_t & k_t \end{bmatrix}\begin{Bmatrix} u_b \\ u_s \\ u_t \end{Bmatrix}$$

$$=-\begin{bmatrix} m_b & 0 & 0 \\ 0 & m_s & 0 \\ 0 & 0 & m_t \end{bmatrix}\begin{Bmatrix} 1 \\ 1 \\ 1 \end{Bmatrix}\ddot{u}_g$$

The solution for the governing equations of motion derived for the structures can be solved by implementing various methods, such as the central difference method and Newmark's β method. Here, the central difference method is implemented in Excel to obtain the response quantities. The peak values of the response quantities for all the cases are presented and compared in the following table.

- **Comparison of Response Quantities**
The comparison of the response quantities is given in Table 10.2.

TABLE 10.2
Comparison of Response Quantities for Base-Isolated Building Equipped with Either Viscous Damper or Tuned Mass Damper

	Response Value		Percentage Reduction with Respect to Uncontrolled	
Combination	**Displacement (m)**	**Acceleration (m/s²)**	**Displacement (m)**	**Acceleration (m/s²)**
BI + VD	0.0621	0.82	78.63	98.86
BI + TMD	0.0621	0.82	78.61	98.86

Example 10.3 Response Control of a Multistory Building Using Base Isolation and Supplemental Dampers

Determine the effectiveness of the hybrid BI systems for the response control of a five-story building frame under near-fault earthquake ground motions. Implement fifteen different hybrid BI types, obtained as a combination of five isolators (i.e., the LRB, lead-rubber bearing [N-Z system], Pure friction (PF) system, FPS, and Resilient friction base isolation (R-FBI)), and three dampers (i.e., the VD, VED, and FD). Also, compare the floor acceleration, the isolator displacement, the base shear, and other pertinent response quantities of the structure for: (a) the fixed-base building, (b) the base-isolated building, and (c) the base-isolated building equipped with the supplemental dampers.

a. Perform time history analysis under the near-fault earthquakes and compare the time histories of the pertinent response quantities for the uncontrolled and controlled building considering all cases.
b. Compare the force-deformation plots of the passive and hybrid BI systems considering all cases.
c. Compare the distribution of the peak values of the floor accelerations, floor displacements, and story shear over the height of the building for the different cases.
d. Compare the Fast Fourier Transform (FFT) plots of the top floor acceleration of the building for the different cases.
e. Plot time histories of the input energy, the energy dissipated by the base isolator, the energy dissipated by the damper in the uncontrolled and controlled building, considering all cases.
f. Evaluate and compare the percentage reductions in the response quantities when the hybrid isolation is used as compared to the use of only passive base isolators.

In this problem, the multistory shear model of a building is used in the investigation and the supplemental damper is connected at base level with one end attached to the ground and the other to the base mass. The seismic response of five-story building models is investigated under real earthquake ground motion records. The three types of idealized mathematical models of the multistory building considered are (a) the fixed-base, (b) base-isolated building, and (c) base-isolated building with supplemental damper connected at the base mass level. The three models are shown in Figure 10.5.

The following assumptions can be made in modeling the building:

1. The building is modeled as a shear frame considering only one lateral degree of freedom at each floor level.
2. The mass of the structure is lumped at each floor level and the floors are considered to be infinitely stiff in their own plane.

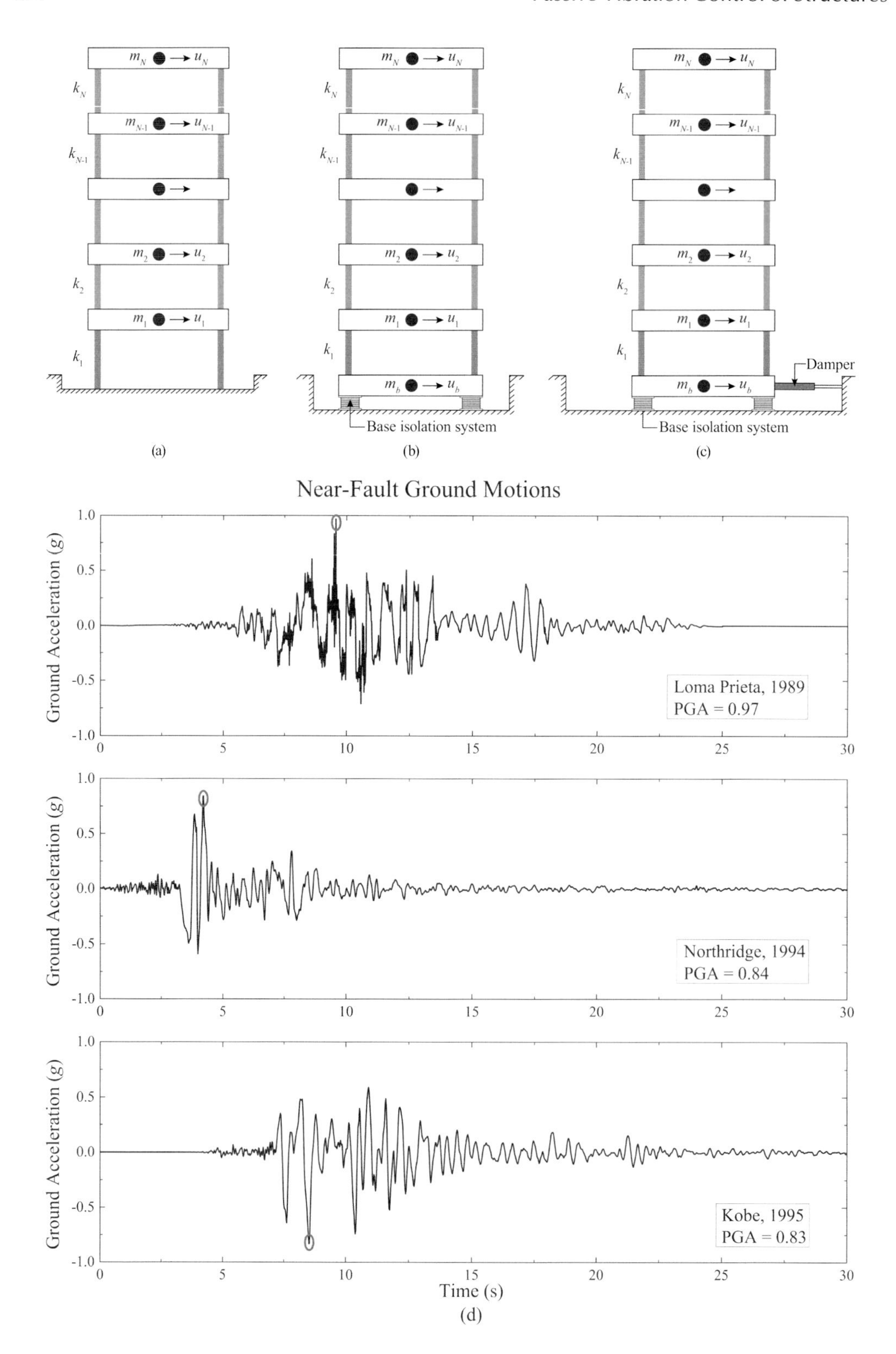

FIGURE 10.5 Mathematical models of the multistory building and time histories of ground motions used in the study: (a) fixed-base, (b) base-isolated, (c) base-isolation with damper, and (d) time histories of ground motions used in the study. (Zelleke et al. 2015.)

3. The columns providing the lateral stiffness are assumed to be axially inextensible.
4. The effects of soil-structure interaction are neglected.
5. Only one horizontal component of the earthquake ground motion is applied to the building at a time.
6. Taking in to account the fact that BI reduces the earthquake response of structures, helping the elastic limit not to be exceeded in the superstructure, the analysis is done under the reasonable assumption that the building frame remains within the elastic range under the excitations from the ground motions.

Details of Building Frame

The details of the building are given as follows. A uniform damping ratio (ξ_s) of 0.02 is taken for the superstructure. All floor masses are taken to be equal, and the floor stiffness is decided in such a way that a required value of fundamental time period of the fixed-base building (T) of 0.5 s is obtained. For this example, all stories are considered to have the same lateral stiffness and the base mass (m_b) of the base-isolated building model is taken to be equal to the floor masses.

Parameters of Base Isolators

The LRB, lead-rubber bearings (N-Z system), PF system, FPS, and R-FBI are the isolators studied. Modeling of these isolators is done using specific parameters that affect the response of the isolated building. LRB is modeled using isolation time period (T_b) and damping ratio (ξ_b), whereas the N-Z system is modeled using isolation time period (T_b), damping ratio (ξ_b), normalized yield strength (F_0), and isolator yield displacement (q). The friction coefficient (μ) is used to characterize the PF system and the FPS, with FPS having stiffness (k_b) in addition to the friction coefficient. Moreover, the R-FBI system is modeled using similar parameters as FPS with an additional damping parameter (ξ_b) incorporated.

The five isolation systems are modeled with specific parameters and paired with the three types of dampers and the responses are compared. Here, (ξ_b) value of 0.15; (T_b) of 2.5 s; (F_0) and (μ) values of 0.075; and (q) of 2.5 cm are used to model the five isolation systems accordingly.

Parameters of Supplemental Dampers

The passive dampers used in this study (i.e., VD, VED, and FD) are attached at the base level between the ground and the base mass to control the large isolator displacement that occurs during earthquake excitation. These dampers can be used to dissipate the earthquake energy and mitigating the large isolator displacement without excessively affecting other response quantities of the building, such as floor accelerations and story shear.

VD is a velocity-dependent damping device and the parameter used in modeling is the damping coefficient (c_d), which is obtained in terms of the external damping ratio (ξ_d) added to the system. For this example, (ξ_d) value of 20% is considered.

The force displacement model of VED is represented with the combined effect of the damper elastic force and the damping force. The elastic force is represented with the damper stiffness (k_d), which can be obtained from the relative external damper stiffness (k') while damping coefficient of damper (c_d) is used to obtain the damping force. (c_d) is expressed in terms of the external damping ratio (ξ_d) added to the system due to the supplemental viscoelastic device. The relative external damper stiffness value of 0.75 is used, whereas the damping coefficient value considered here is 0.2.

FDs are displacement-dependent devices and are modeled in this example by specifying the limiting friction force (F_L). The normalized limiting friction force $(F' = F_L/W)$, which is the ratio of the limiting friction force to the total weight of the base-isolated building, is used to define the FD. The value of (F') used in this example is 0.07. The multistory building being studied is analyzed for all of the combinations of the five isolation systems and the three supplemental dampers.

Earthquake Ground Motion Data

Three near-fault earthquake ground motions are used in this example. The near-fault earthquake data used here are the 1989 Loma Prieta, the 1994 Northridge, and the 1995 Kobe ground motions. The peak ground accelerations (PGAs) and other relevant details of the selected ground motion data are shown in Table 10.3, whereas the time histories of the ground motions are depicted in Figure 10.5(d).

Equations of Motion and Response Analysis

The governing equations of motion are derived by taking equilibrium of forces at each degree of freedom during seismic excitation. The equations of motion for the base-isolated building with

TABLE 10.3
List of Near-Fault Ground Motions Used in the Study

Sr. No.	Earthquake	Event Year	Recording Station	Magnitude	PGA (g)
1	Loma Prieta	1989	Los Gatos Presentation Center (LGPC)	6.93	0.97
2	Northridge	1994	Sylmar – Hospital	6.69	0.84
3	Kobe, Japan	1995	Kobe Japan Meteorological Agency (KJMA)	6.90	0.83

supplemental damper are expressed in matrix form, which is of order of the number of degrees of freedom in the building ($N + 1$), and given as

$$\begin{bmatrix} m & \{r\}^T[M] \\ [M]\{r\} & [M] \end{bmatrix}\begin{Bmatrix} \ddot{u}_b \\ \{\ddot{U}\} \end{Bmatrix}+\begin{bmatrix} c & \{0\}^T \\ \{0\} & [C] \end{bmatrix}\begin{Bmatrix} \dot{u}_b \\ \{\dot{U}\} \end{Bmatrix}$$
$$+\begin{bmatrix} k & \{0\}^T \\ \{0\} & [K] \end{bmatrix}\begin{Bmatrix} u_b \\ \{U\} \end{Bmatrix}=-\begin{bmatrix} m & \{r\}^T[M] \\ [M]\{r\} & [M] \end{bmatrix}\begin{Bmatrix} 1 \\ \{0\} \end{Bmatrix}\ddot{x}_g \tag{10.18}$$

where $[M]$, $[C]$, and $[K]$ are the mass, damping, and stiffness matrices of the superstructure which are of order ($N \times N$), respectively; c and k are the damping coefficient and stiffness at the isolation level obtained by combining the contributions of the isolator and attached damper; u_b, $\dot{u}_b$, and $\ddot{u}_b$ are the displacement, velocity, and acceleration of the base mass relative to the ground, respectively; $m = (m_b + \Sigma_{j=1}^{N} m_j)$ is the total mass of the base-isolated building; m_j is the mass of jth floor; $\{U\} = \{u_1, u_2, \ldots, u_N\}^T$, $\{\dot{U}\}$, and $\{\ddot{U}\}$ are the vectors of floor displacements, velocities, and accelerations, respectively; $\{r\} = \{1,1,\ldots,1\}^T$ is vector of influence coefficients; and $\ddot{x}_g$ is the earthquake ground acceleration. $[M]$ and $[K]$ are given below in Equations (10.19) and (10.20), respectively, while $[C]$ is constructed by using a constant damping ratio for the superstructure.

$$[M]=\begin{bmatrix} m_1 & 0 & 0 & . & 0 & 0 \\ 0 & m_2 & 0 & . & 0 & 0 \\ 0 & 0 & m_3 & . & 0 & 0 \\ . & . & . & . & 0 & 0 \\ 0 & 0 & 0 & 0 & m_{N-1} & 0 \\ 0 & 0 & 0 & 0 & 0 & m_N \end{bmatrix} \tag{10.19}$$

$$[K]=\begin{bmatrix} k_1+k_2 & -k_2 & 0 & . & 0 & 0 \\ -k_2 & k_2+k_3 & -k_3 & . & 0 & 0 \\ 0 & -k_3 & k_3+k_4 & . & 0 & 0 \\ . & . & . & . & 0 & 0 \\ 0 & 0 & 0 & 0 & k_{N-1}+k_N & -k_{N-1} \\ 0 & 0 & 0 & 0 & -k_{N-1} & k_N \end{bmatrix} \tag{10.20}$$

Due to the nonlinear force-deformation behavior of the isolator and the difference between the damping of the isolation system, the superstructure, and the supplemental damper, the equation of motion cannot be solved using the classical modal superposition technique. Therefore, the equations of motion are numerically solved using the Newmark's method of step-by-step integration using a small time interval (Δt) with linear variation of acceleration.

Nonlinear time history analysis is carried out to obtain the response quantities of the base-isolated building equipped with the supplemental VD, VED, and FD. The time history analysis can be performed using any of the nonlinear time history analysis methods. The results of the analysis are presented subsequently.

Time Histories of Response Quantities

The time histories of top floor acceleration, isolator displacement, and story shear for the five-story base-isolated building modeled using each isolation system (i.e., LRB, N-Z, PF, FPS, and R-FBI) paired with viscous, viscoelastic, and FDs under the 1989 Loma Prieta, the 1994 Northridge, and the 1995 Kobe ground motions are plotted from Figures 10.6–10.20.

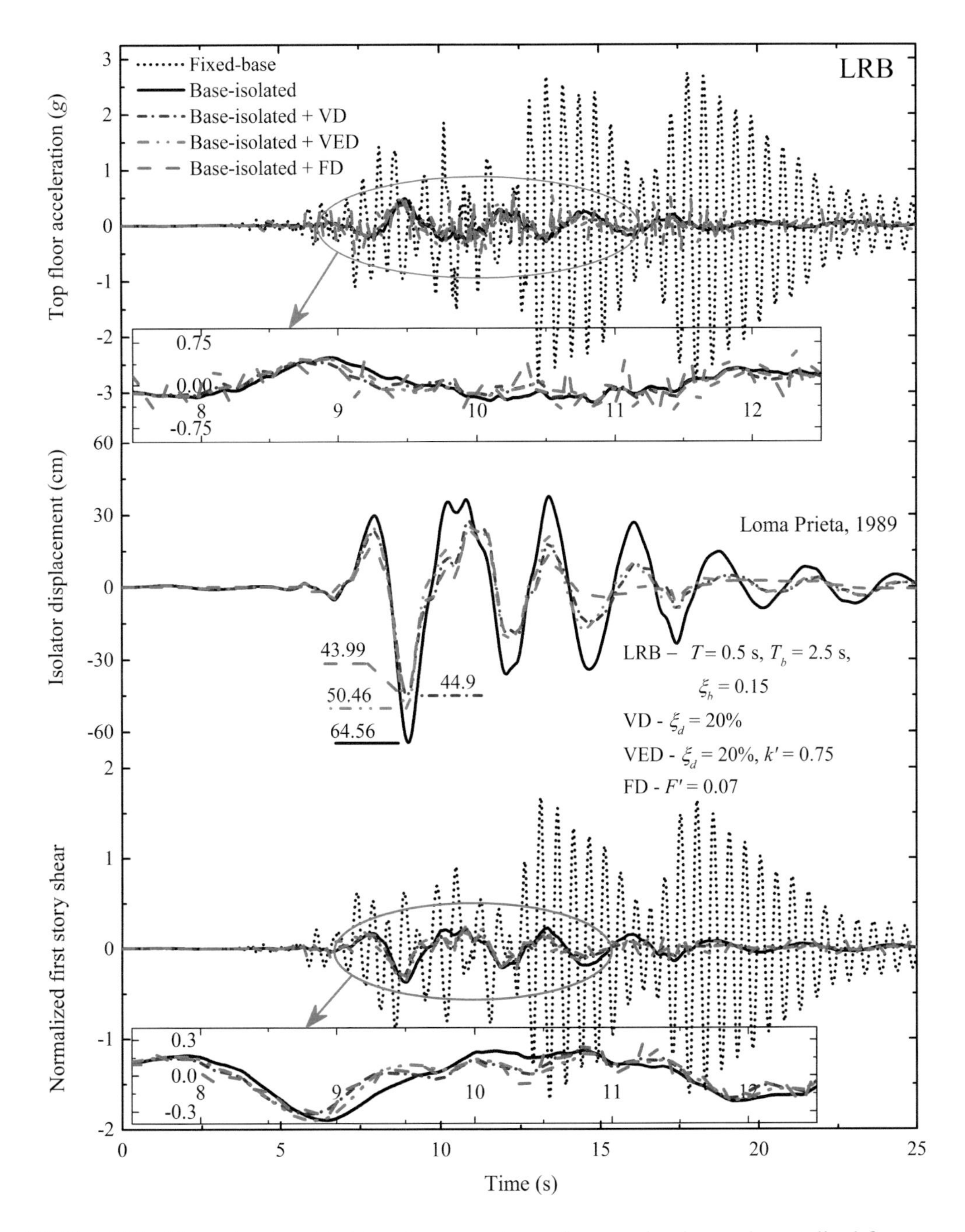

FIGURE 10.6 Time histories of isolator displacement, top floor acceleration, and normalized first story shear of five-story building isolated by LRB under Loma Prieta, 1989 ground motion.

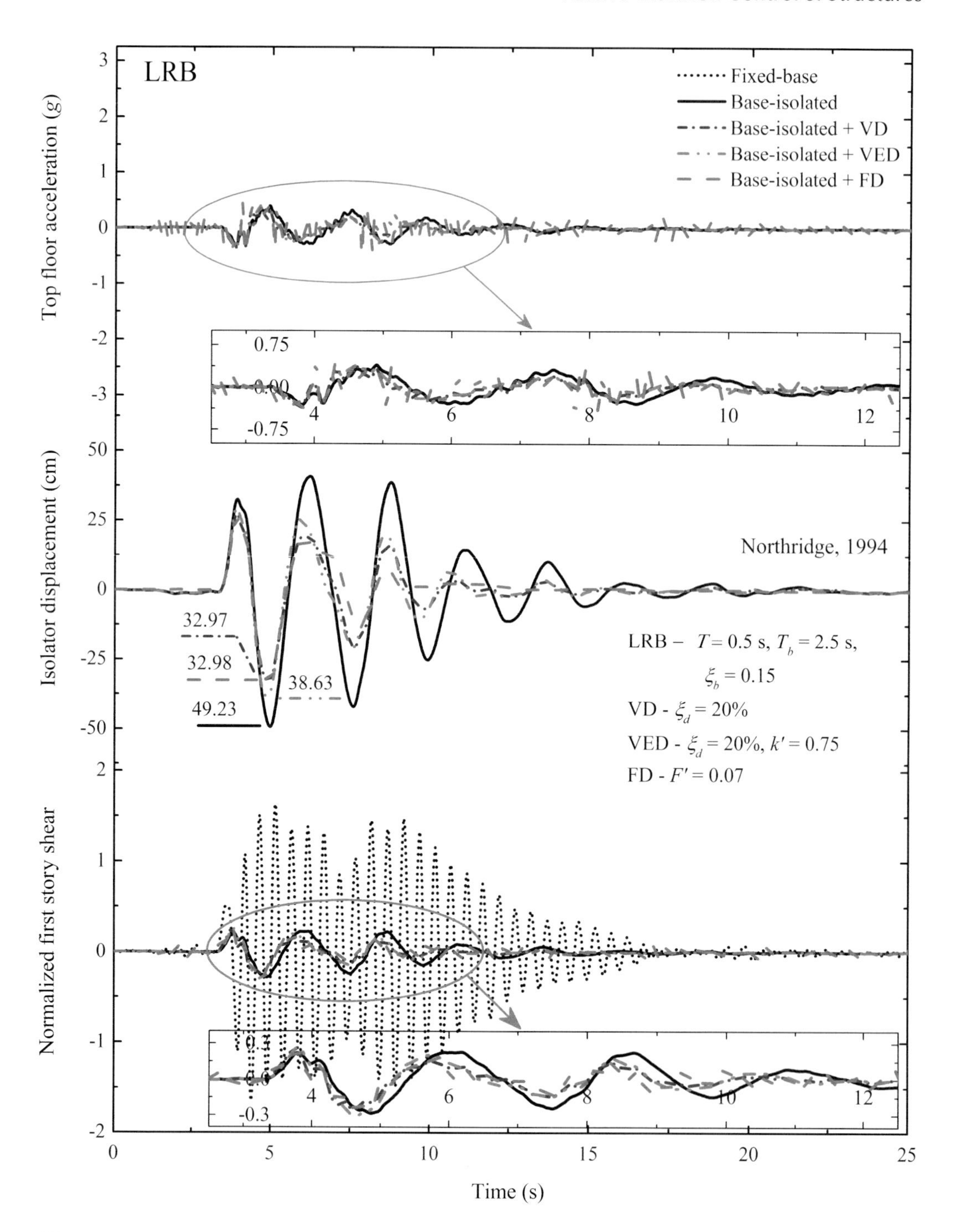

FIGURE 10.7 Time histories of isolator displacement, top floor acceleration, and normalized first story shear of five-story building isolated by LRB under Northridge, 1994 ground motion.

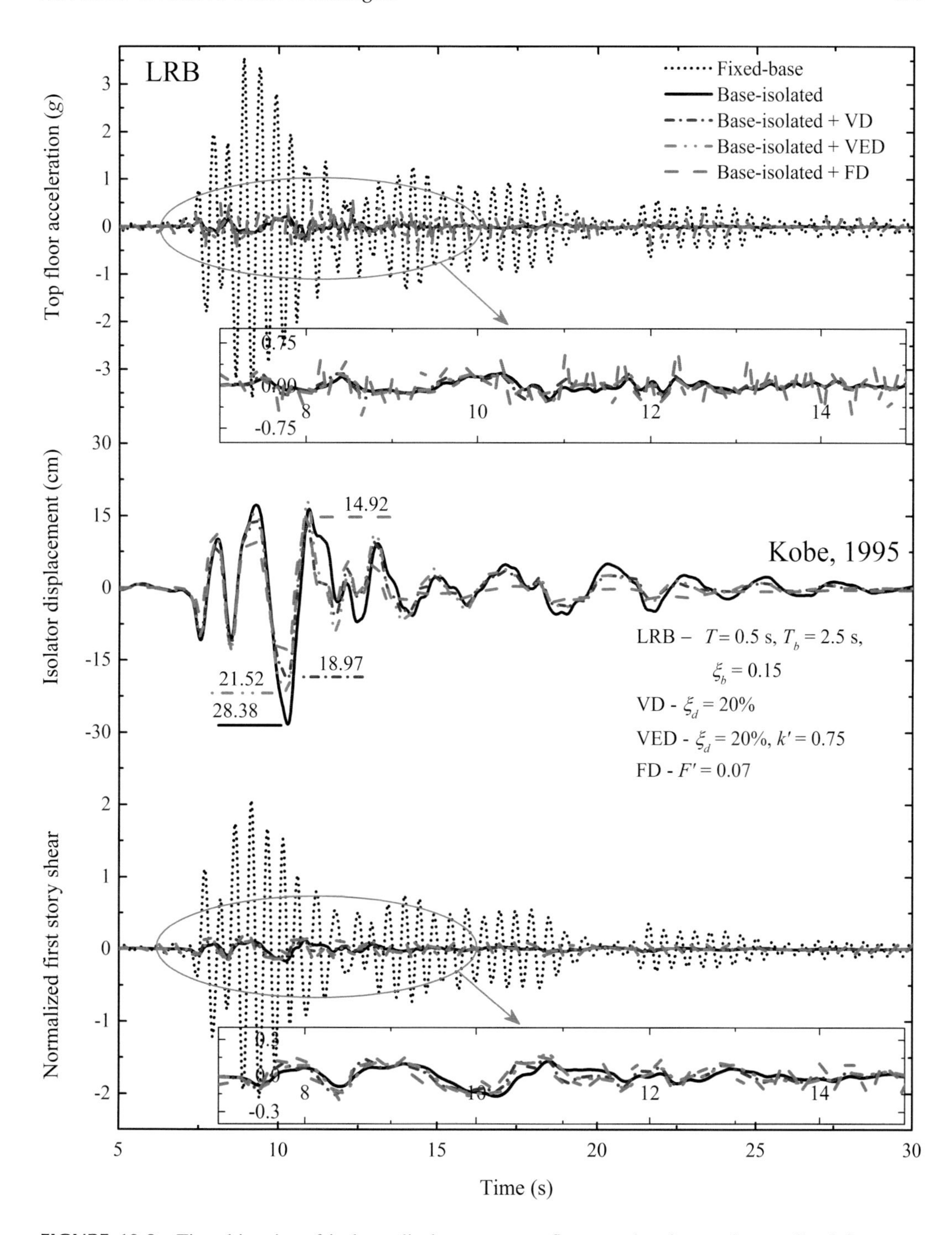

FIGURE 10.8 Time histories of isolator displacement, top floor acceleration, and normalized first story shear of five-story building isolated by LRB under Kobe, 1995 ground motion.

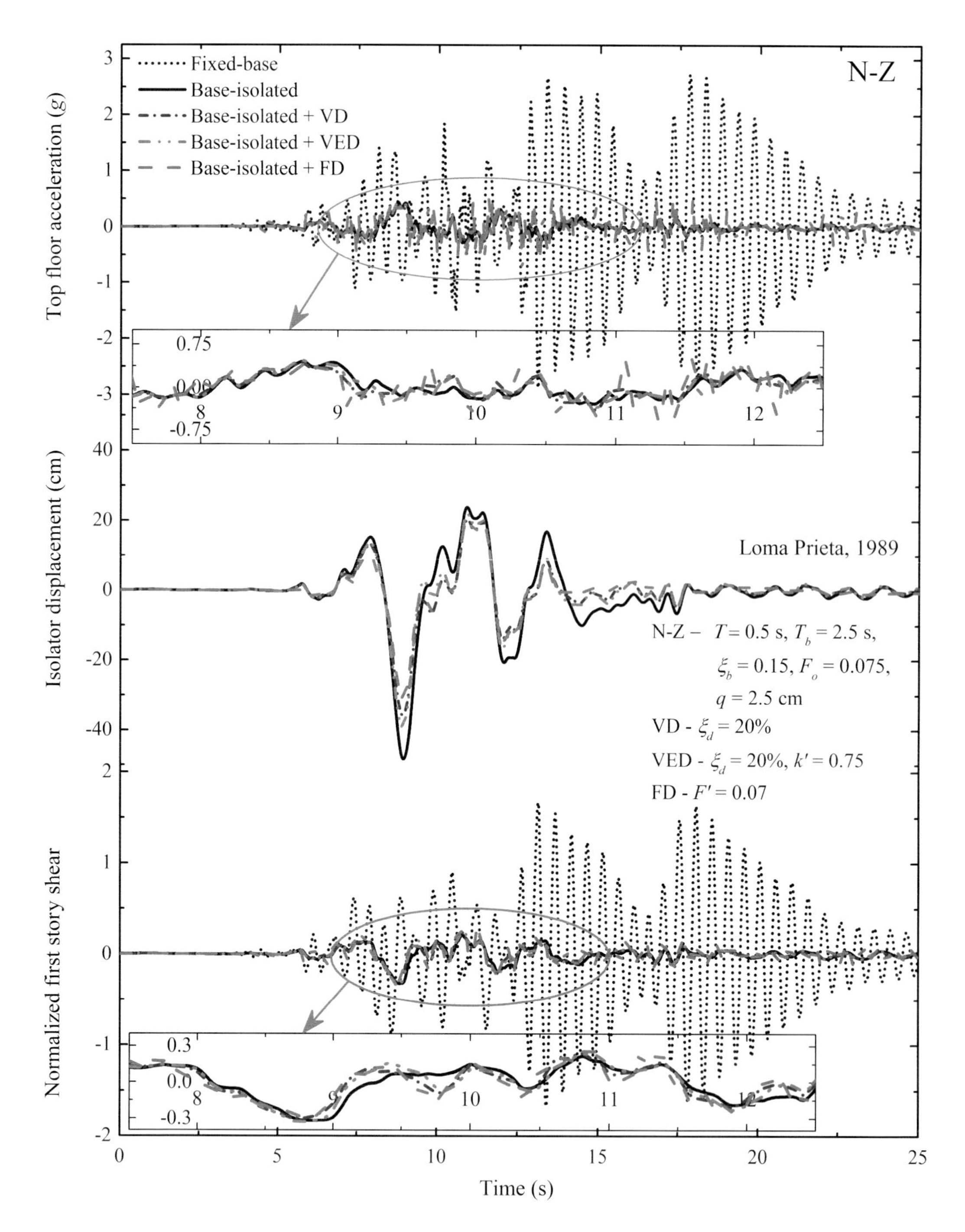

FIGURE 10.9 Time histories of isolator displacement, top floor acceleration, and normalized first story shear of five-story building isolated by N-Z system under Loma Prieta, 1989 ground motion.

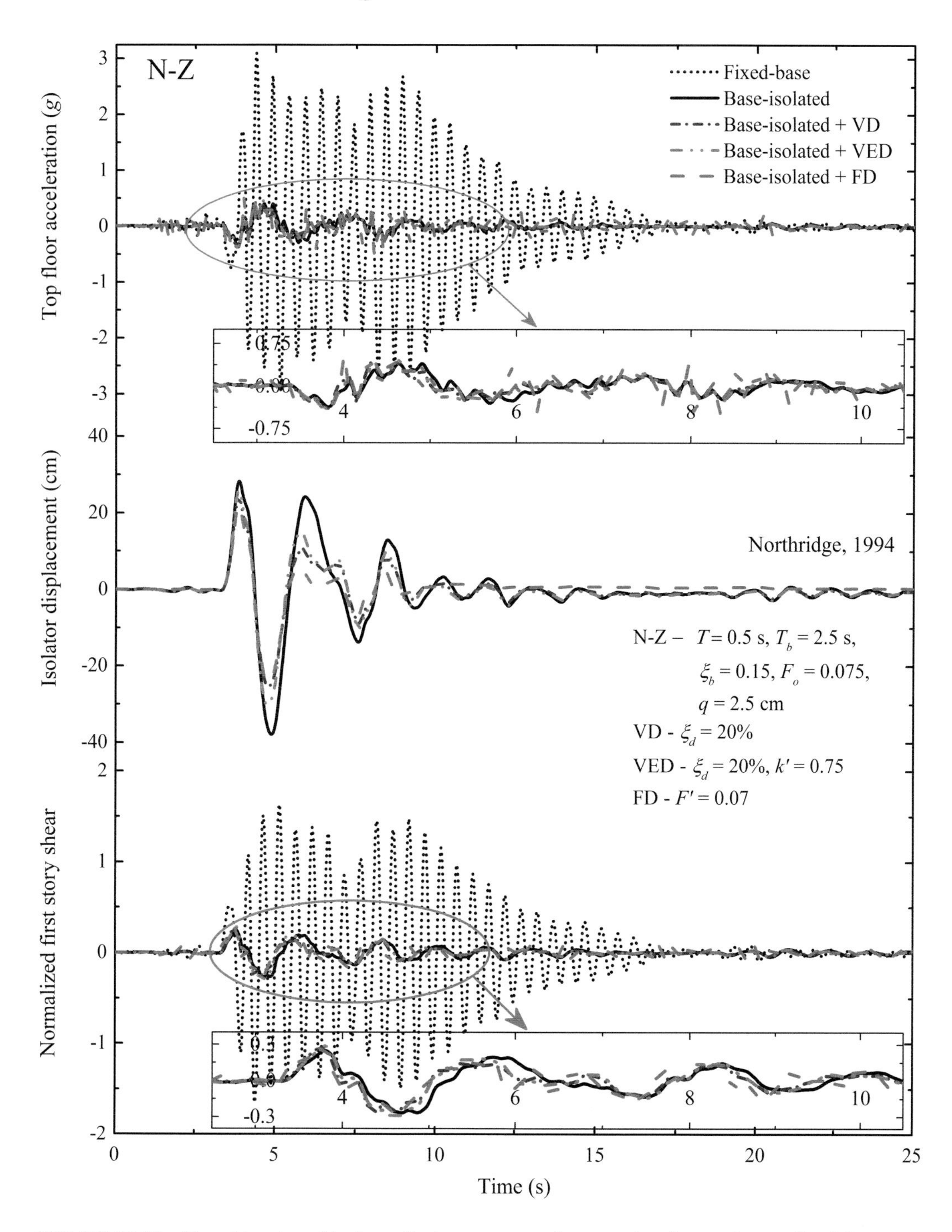

FIGURE 10.10 Time histories of isolator displacement, top floor acceleration, and normalized first story shear of five-story building isolated by N-Z system under Northridge, 1994 ground motion.

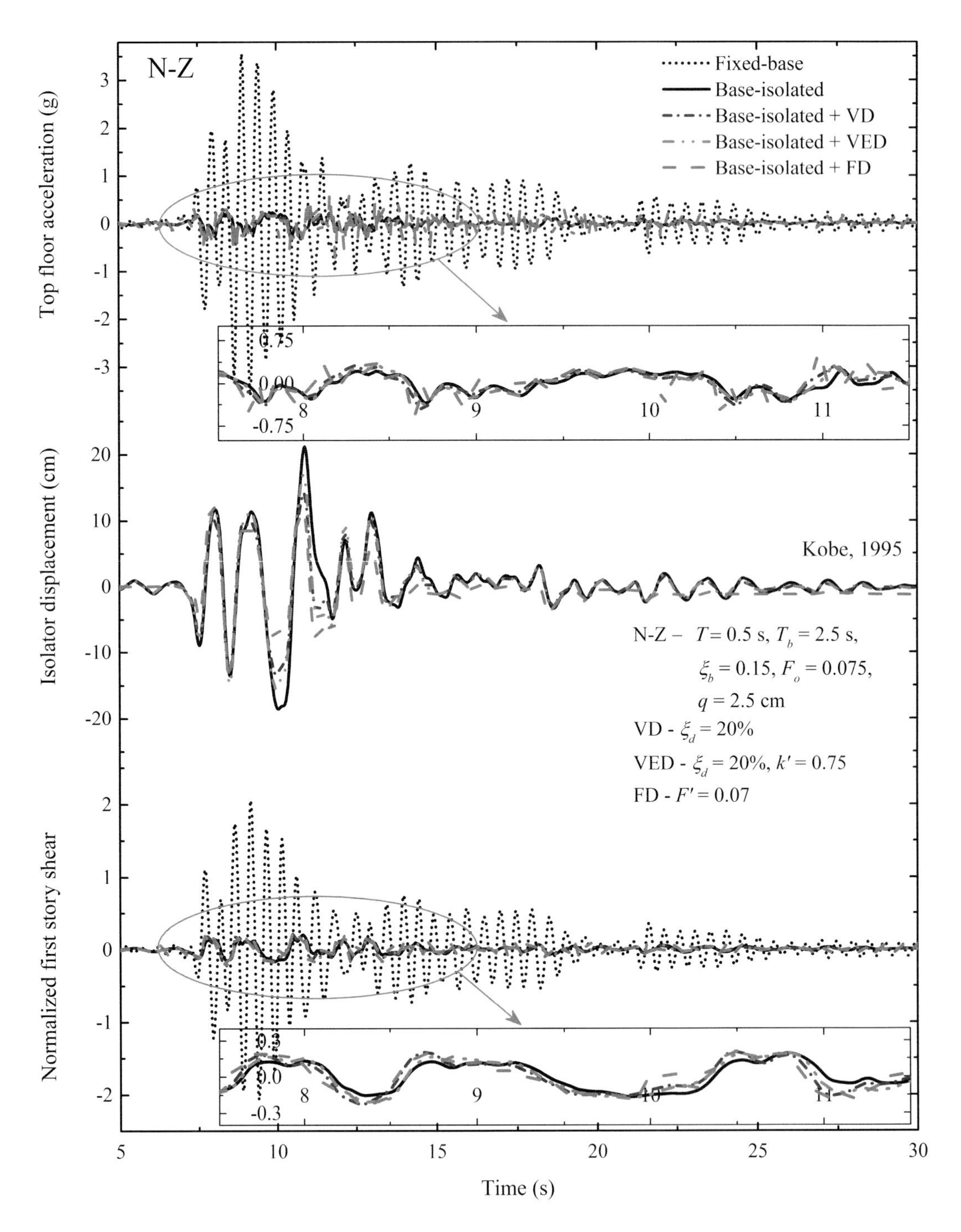

FIGURE 10.11 Time histories of isolator displacement, top floor acceleration, and normalized first story shear of five-story building isolated by N-Z system under Kobe, 1995 ground motion.

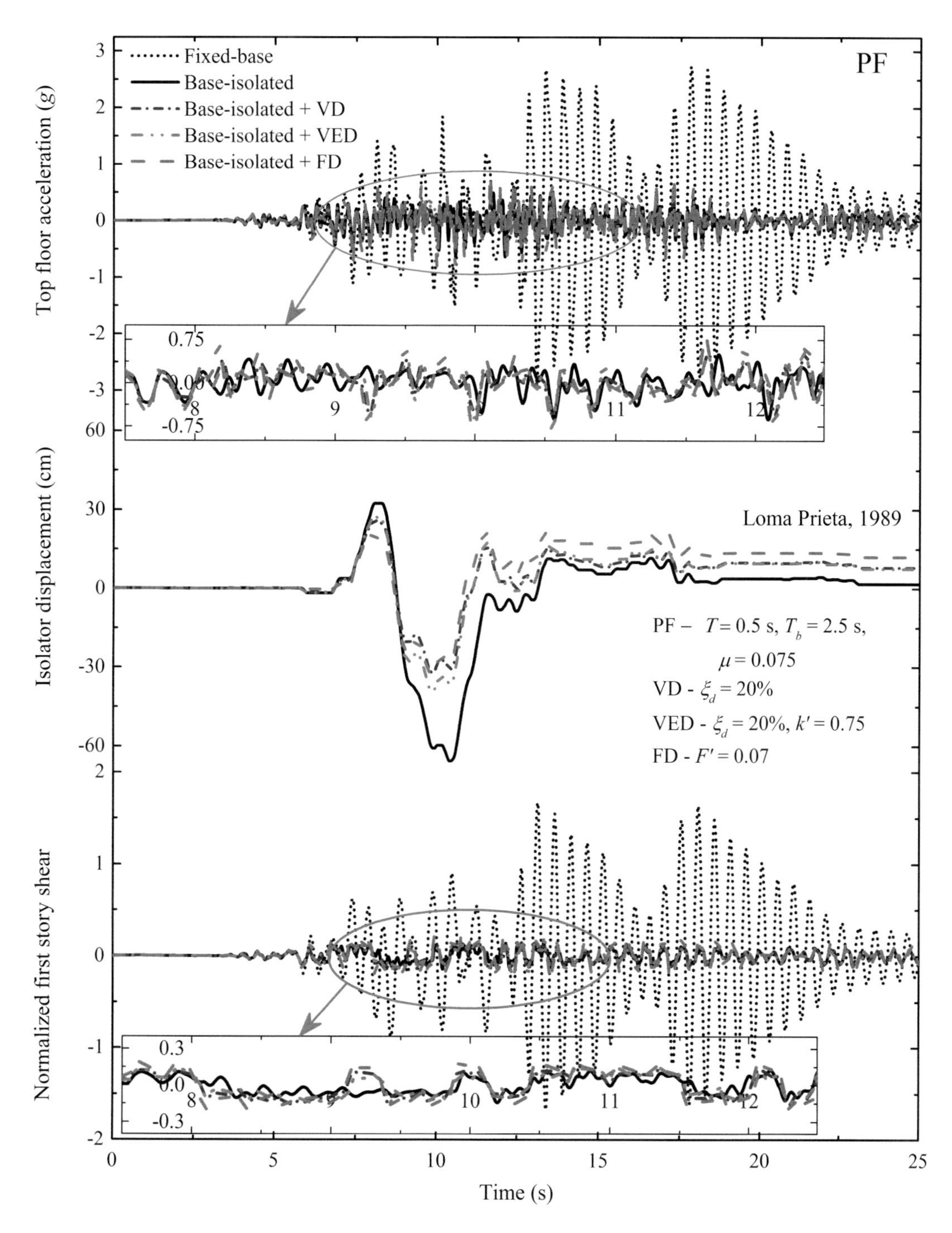

FIGURE 10.12 Time histories of isolator displacement, top floor acceleration, and normalized first story shear of five-story building isolated by PF system under Loma Prieta, 1989 ground motion.

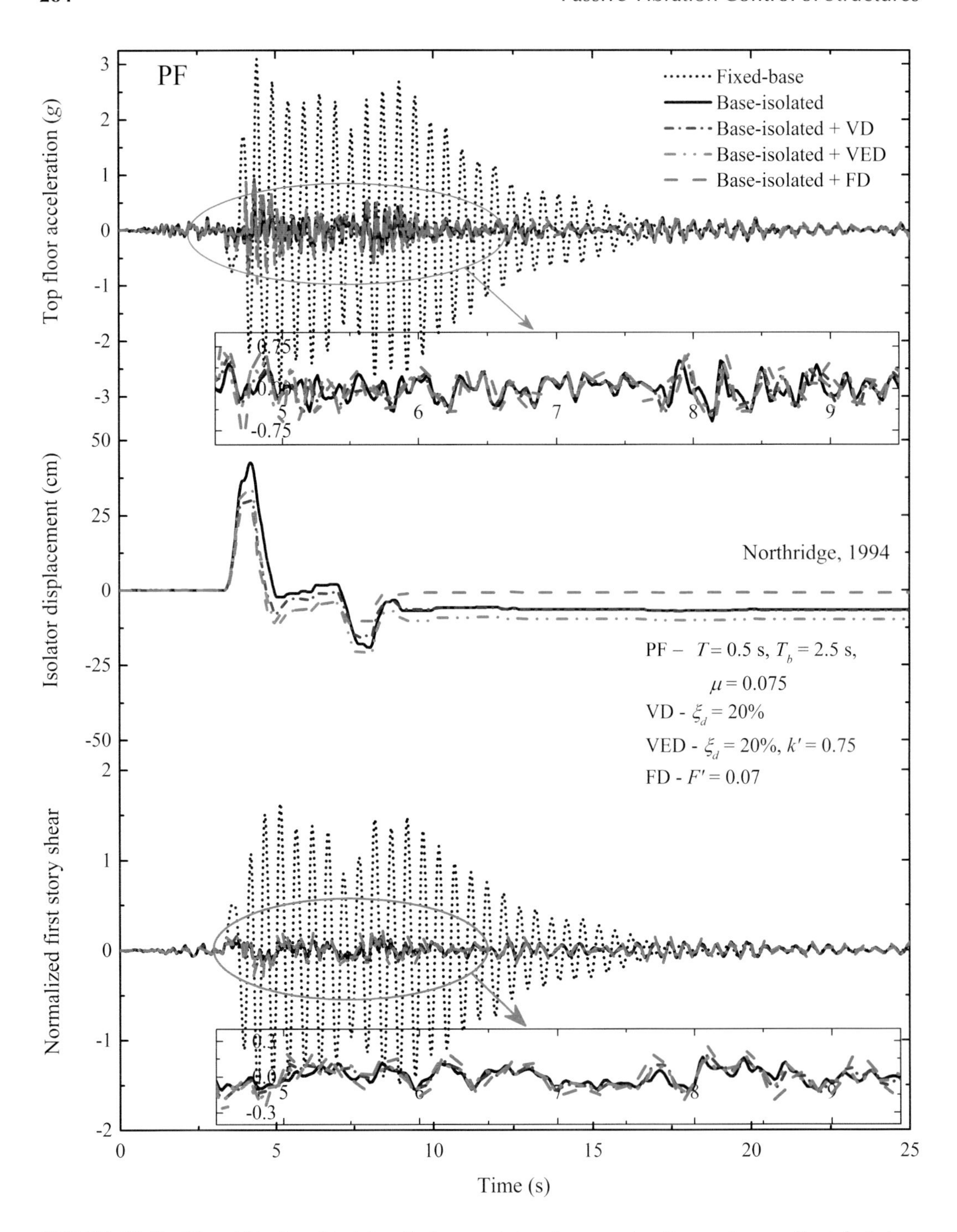

FIGURE 10.13 Time histories of isolator displacement, top floor acceleration, and normalized first story shear of five-story building isolated by PF system under Northridge, 1994 ground motion.

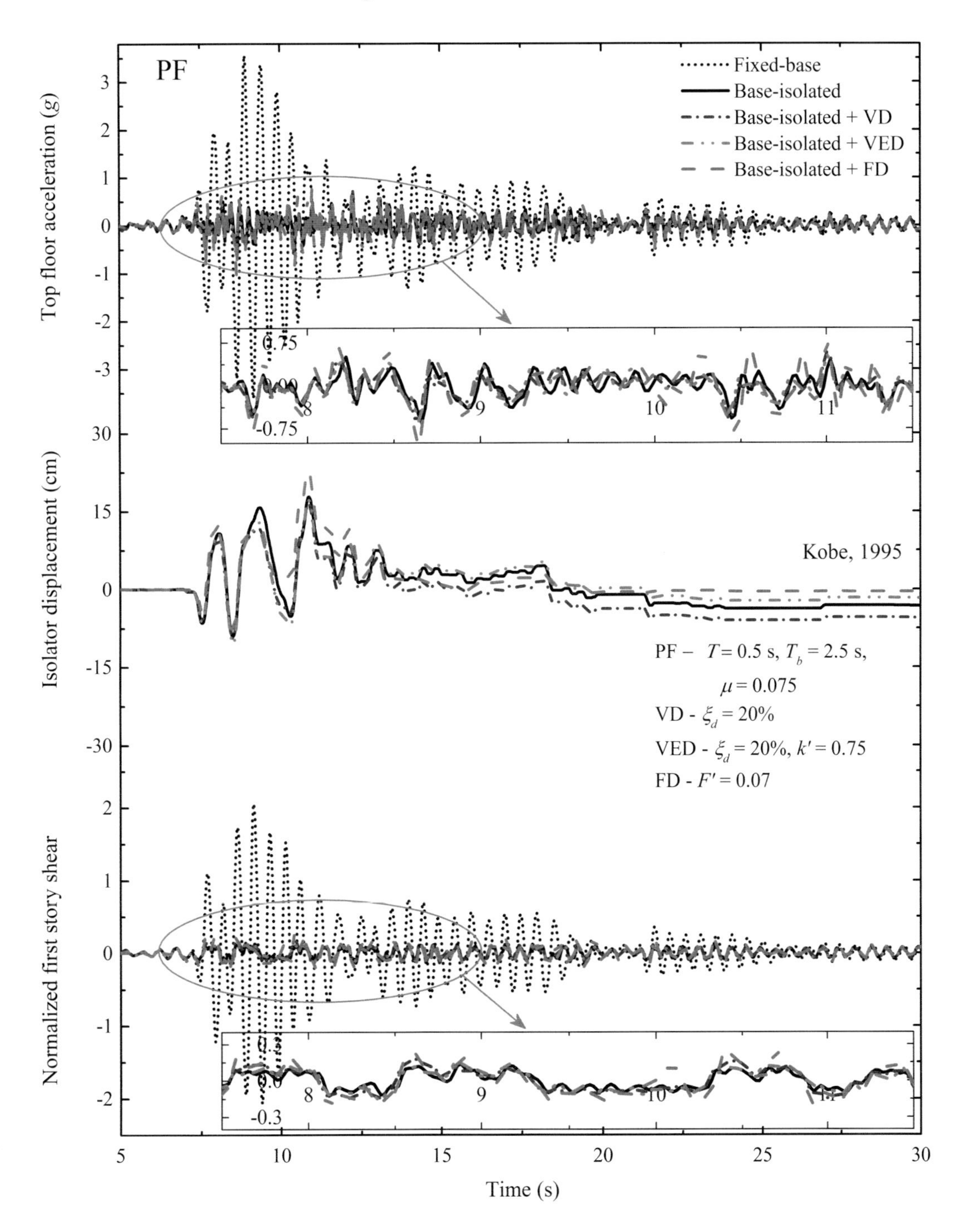

FIGURE 10.14 Time histories of isolator displacement, top floor acceleration, and normalized first story shear of five-story building isolated by PF system under Kobe, 1995 ground motion.

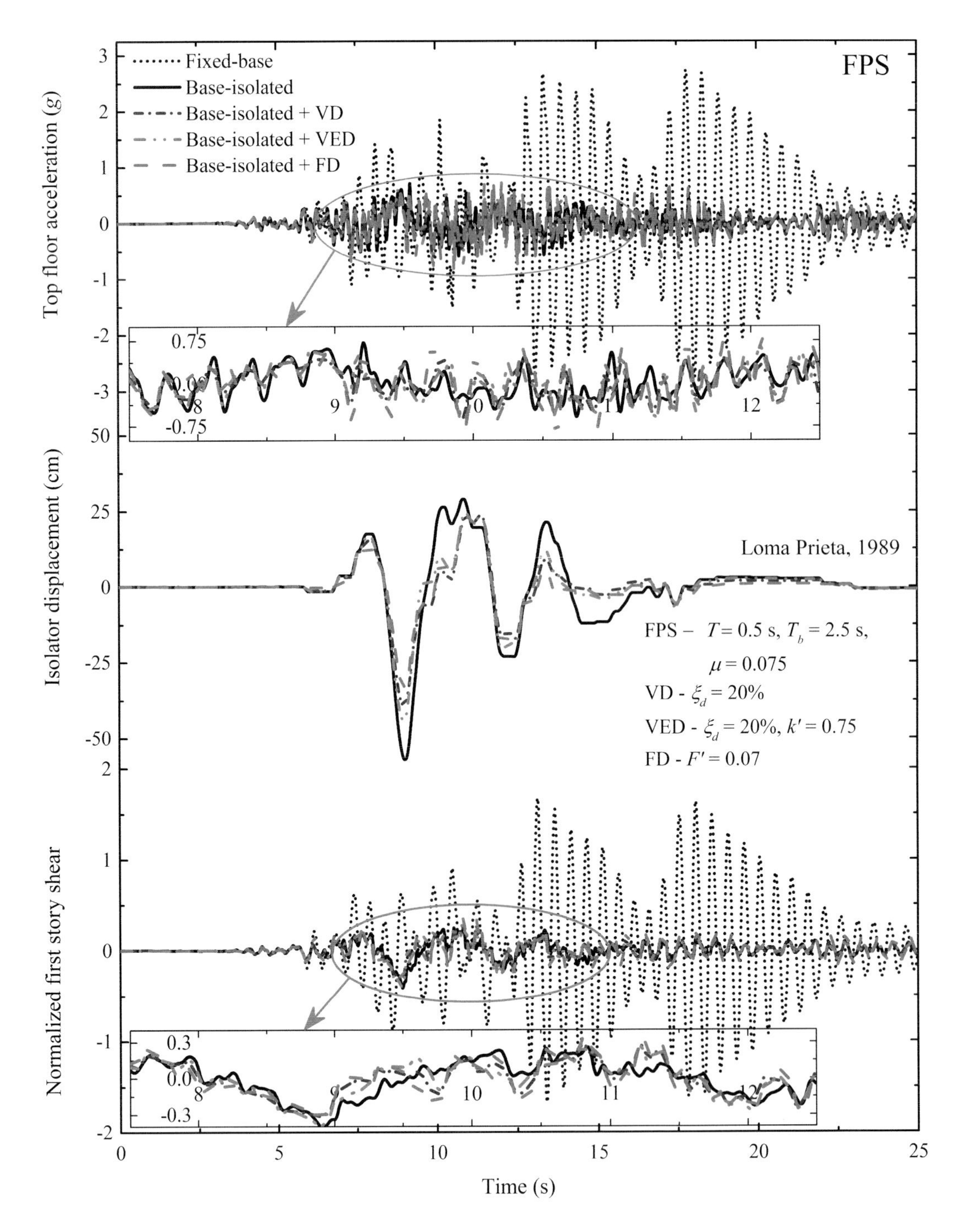

FIGURE 10.15 Time histories of isolator displacement, top floor acceleration, and normalized first story shear of five-story building isolated by FPS under Loma Prieta, 1989 ground motion.

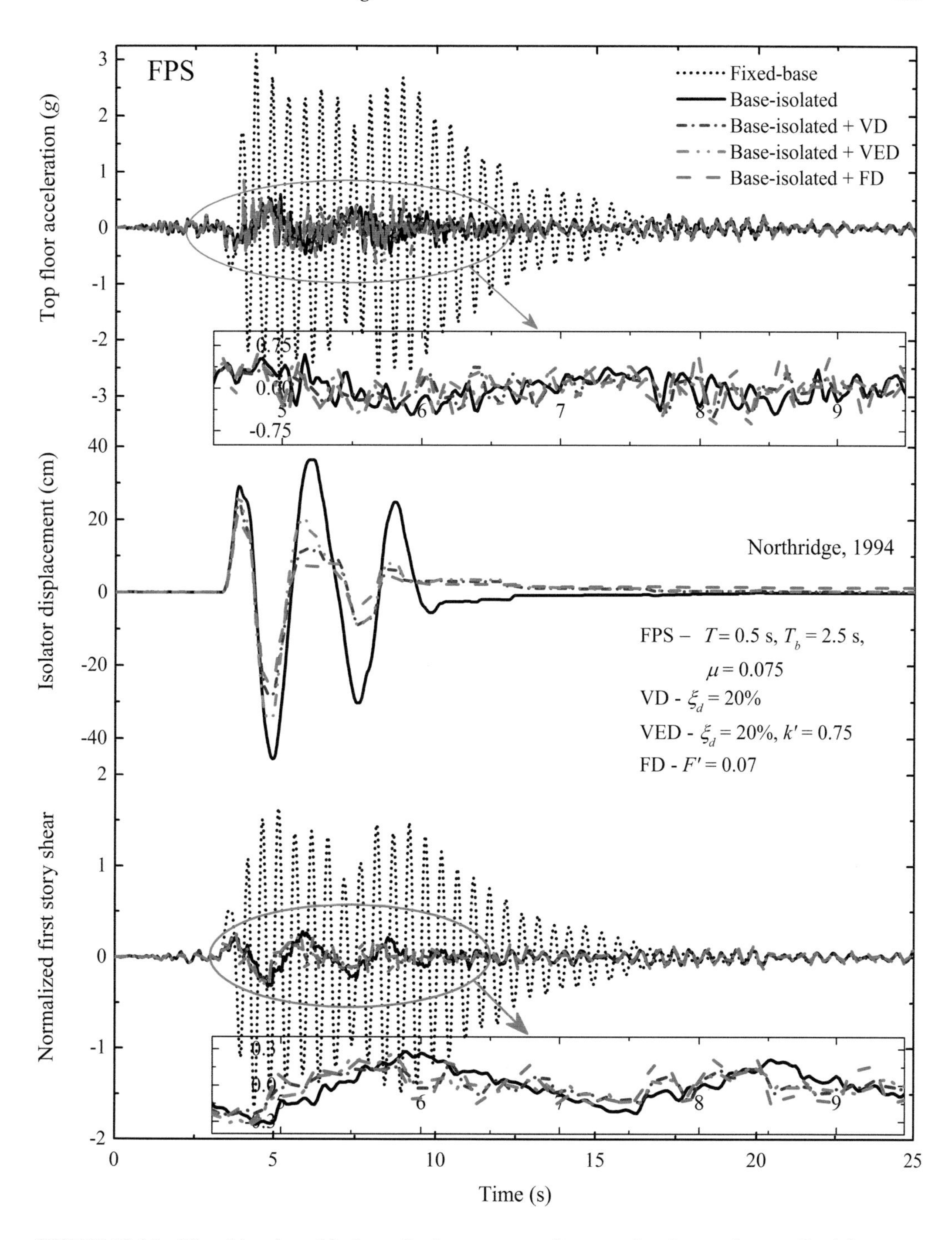

FIGURE 10.16 Time histories of isolator displacement, top floor acceleration, and normalized first story shear of five-story building isolated by FPS under Northridge, 1994 ground motion.

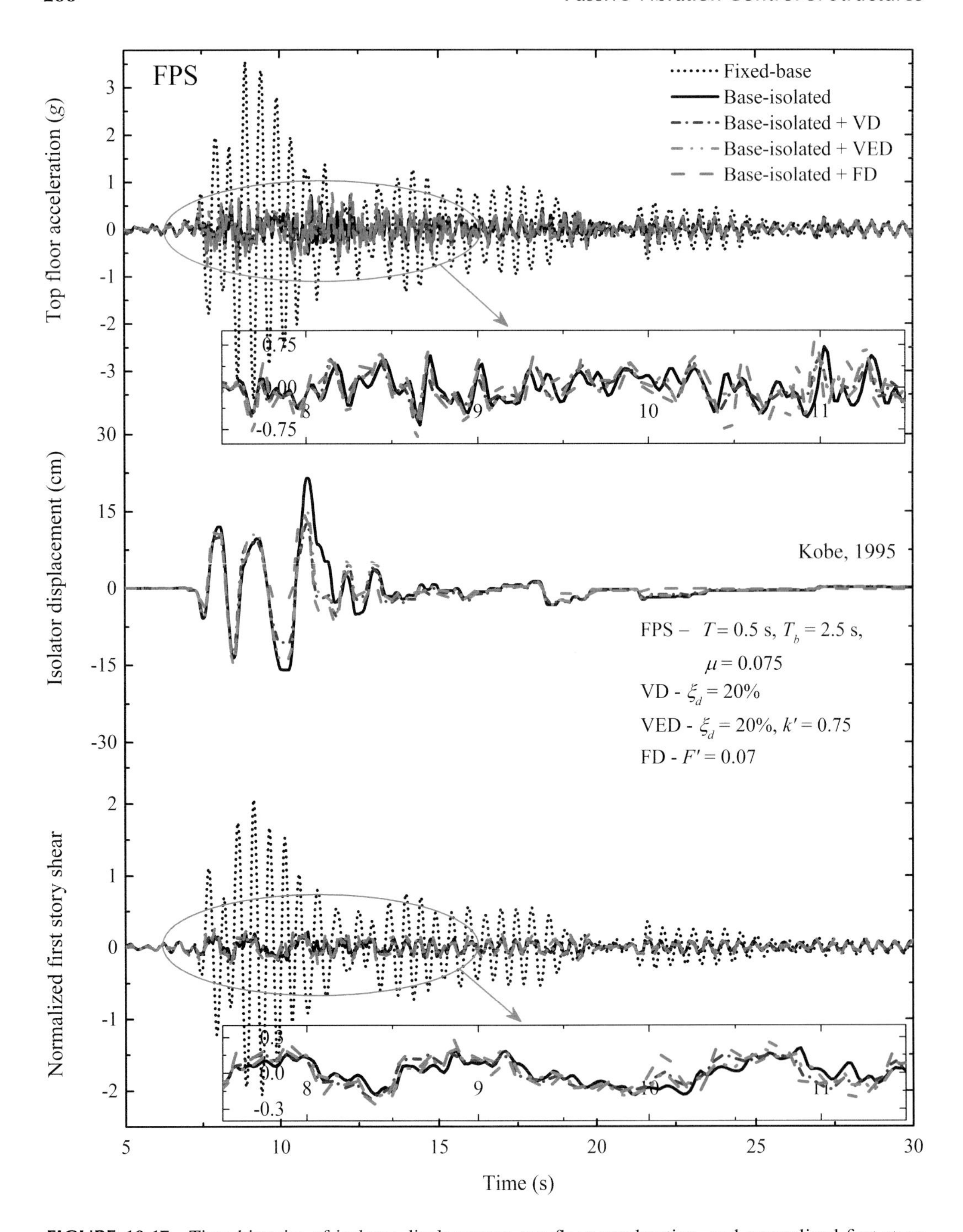

FIGURE 10.17 Time histories of isolator displacement, top floor acceleration, and normalized first story shear of five-story building isolated by FPS under Kobe, 1995 ground motion.

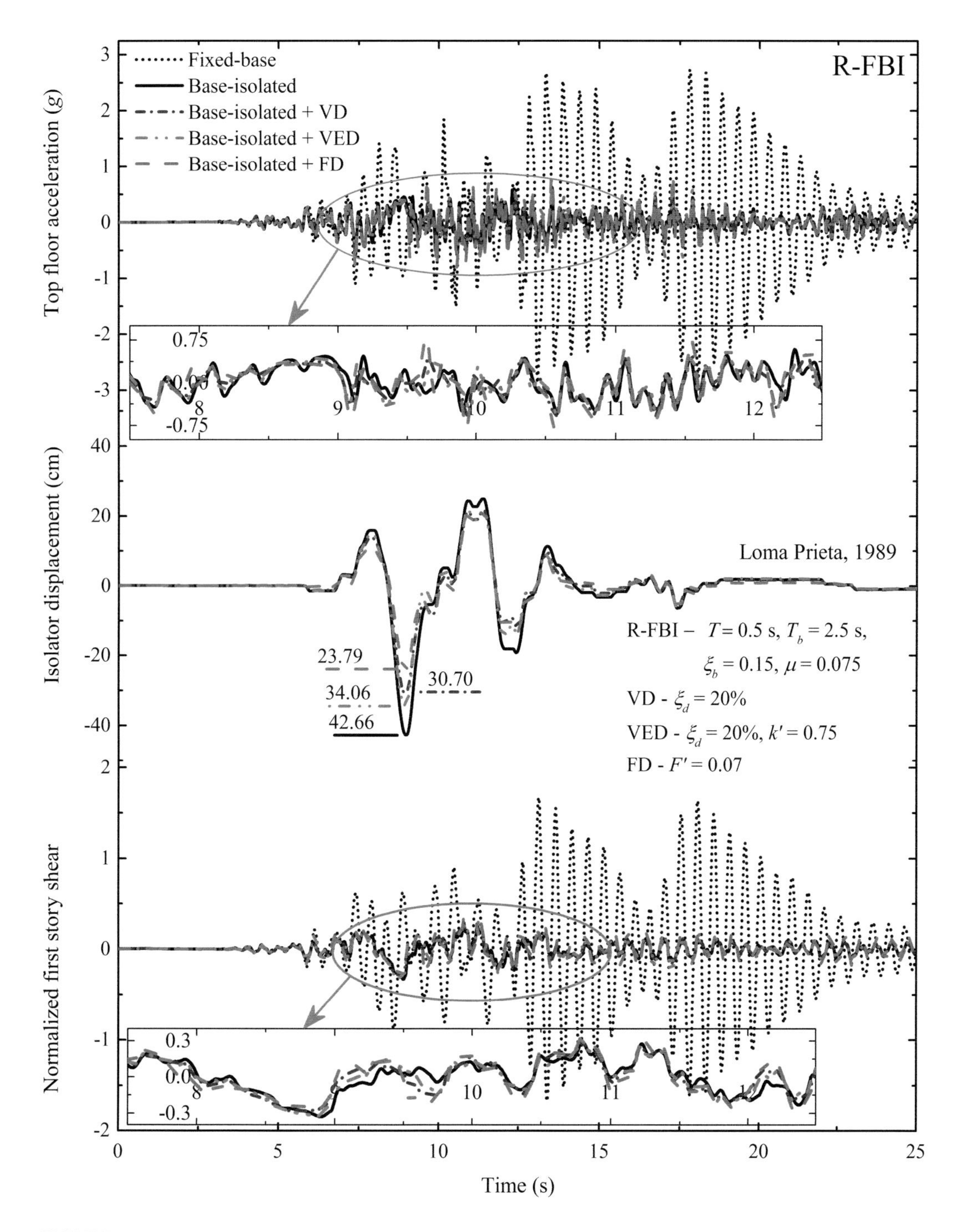

FIGURE 10.18 Time histories of isolator displacement, top floor acceleration, and normalized first story shear of five-story building isolated by R-FBI under Loma Prieta, 1989 ground motion.

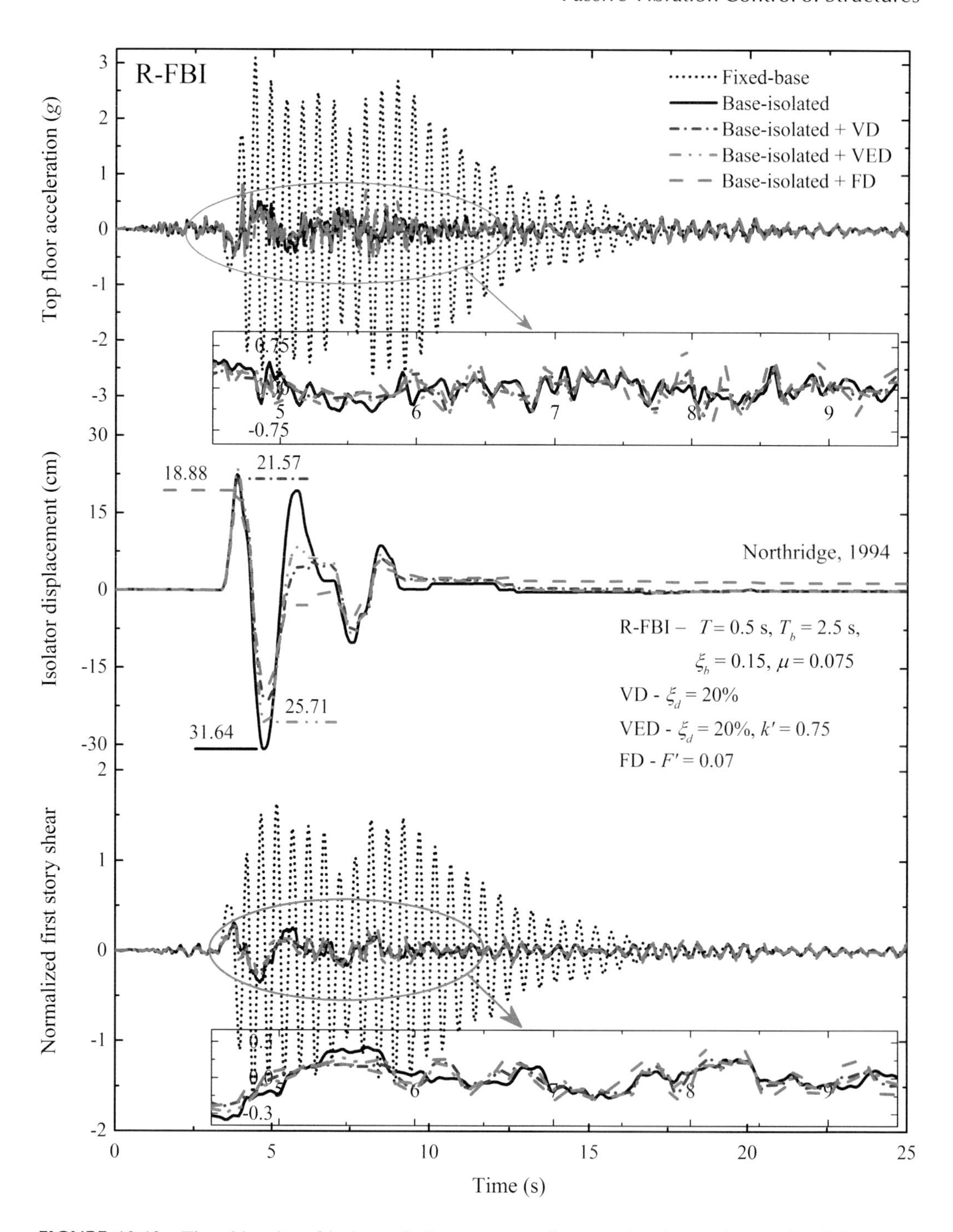

FIGURE 10.19 Time histories of isolator displacement, top floor acceleration, and normalized first story shear of five-story building isolated by R-FBI under Northridge, 1994 ground motion.

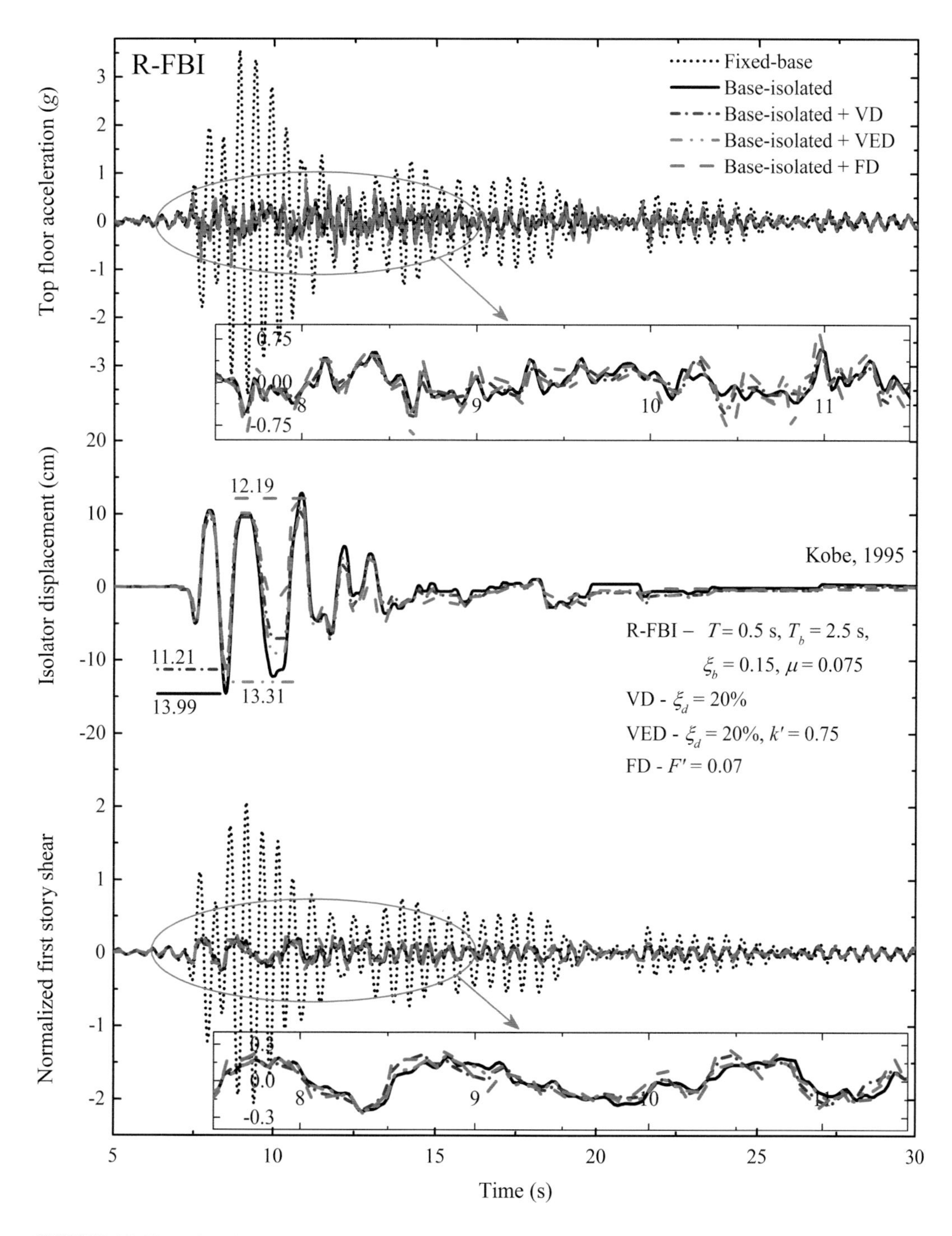

FIGURE 10.20 Time histories of isolator displacement, top floor acceleration, and normalized first story shear of five-story building isolated by R-FBI under Kobe, 1995 ground motion.

Force-Deformation Curves

Figures 10.21–10.25 show the force-deformation behavior of the five isolation systems, i.e., LRB, N-Z system, PF system, FPS, and R-FBI, and the three dampers, i.e., VD, VED, and FD, for the base-isolated building as well as the base-isolated buildings installed with the supplemental dampers under the three near-fault ground motions.

Distribution of Peak Response Quantities Over Building Height

Figures 10.26–10.30 show the distribution of the peak floor displacement, peak floor acceleration and peak story shear over the total height of the building for all isolation systems under the three near-fault ground motions.

FFTs of Top Floor Accelerations

In Figures 10.31–10.35, the comparisons of the FFT amplitude spectra of top floor acceleration are shown for the base-isolated building and base-isolated building with each damper type.

Time Histories of Energies

The time histories of input and dissipated energies per unit mass (E) of the base-isolated building with and without supplemental damper under near-fault ground motions are given in Figures 10.36–10.40. The energy plots show that the supplemental dampers dissipate a significant amount of the earthquake energy.

Summary of Peak Response Quantities

The summary of the peak values of the key response quantities of the multistory building under the near-fault ground motions is presented in Table 10.4.

Percentage Response Reduction due to Use of Supplemental Dampers

The percentage response reductions achieved for all cases are plotted in Figure 10.41 for near-fault ground motions. Average isolator displacement percentage reductions of 32%, 30%, 30%, 34%, and 25% are achieved for the multistory building isolated by LRB, N-Z system, PF system, FPS, and R-FBI, respectively, due to the use of VD having a damping ratio (ξ_d) of 20%. For the use of VED having a damping ratio (ξ_d) of 20% and relative external damper stiffness (k') of 0.75, average isolator displacement percentage reductions of 21%, 20%, 22%, 22%, and 15% are achieved

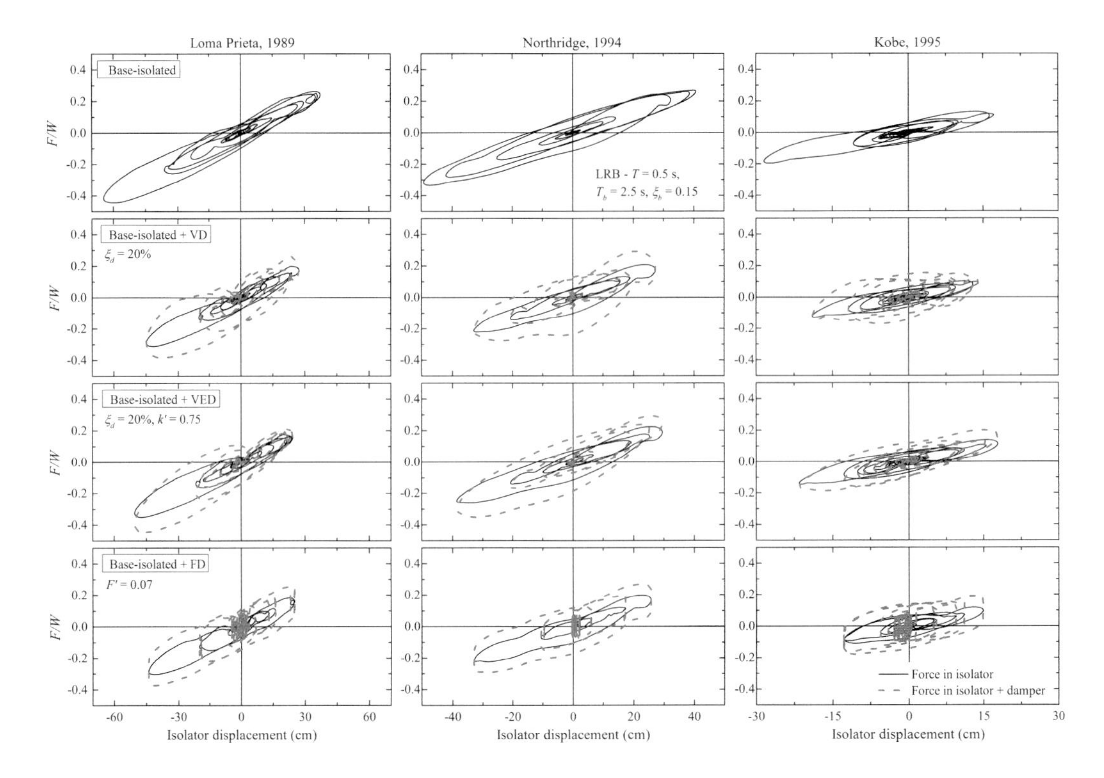

FIGURE 10.21 Comparison of force-deformation behavior of isolator for five-story building isolated by LRB with and without supplemental dampers under near-fault ground motions.

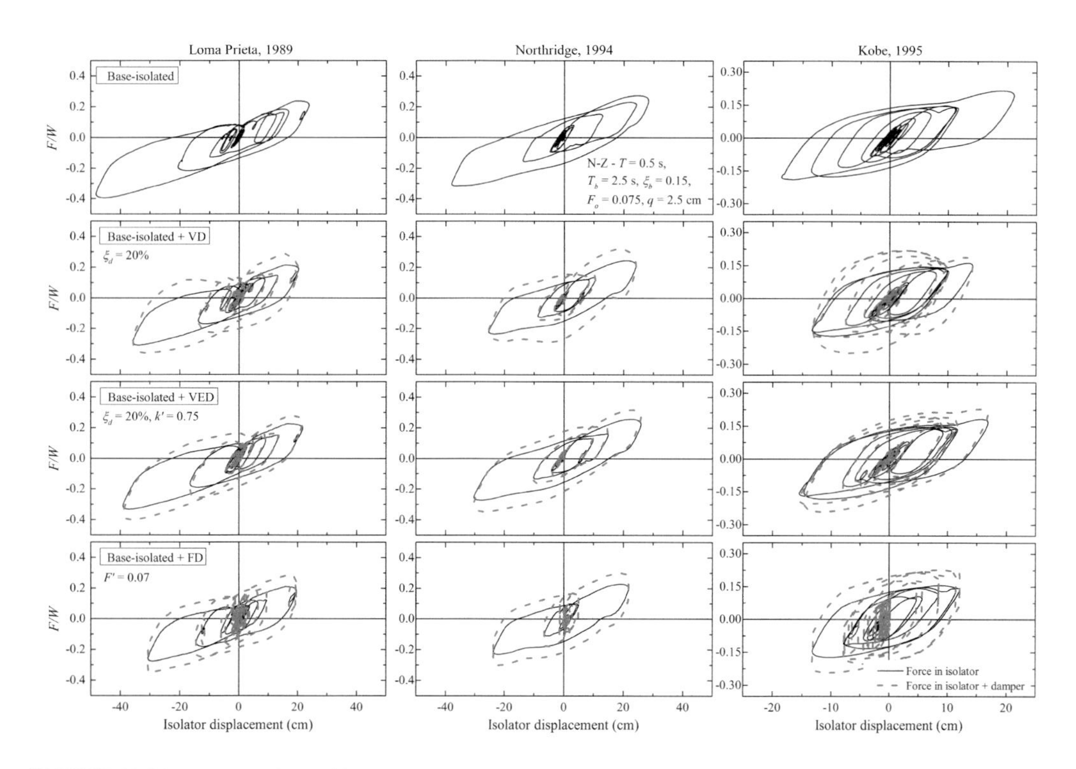

FIGURE 10.22 Comparison of force-deformation behavior of isolator for five-story building isolated by N-Z system with and without supplemental dampers under near-fault ground motions.

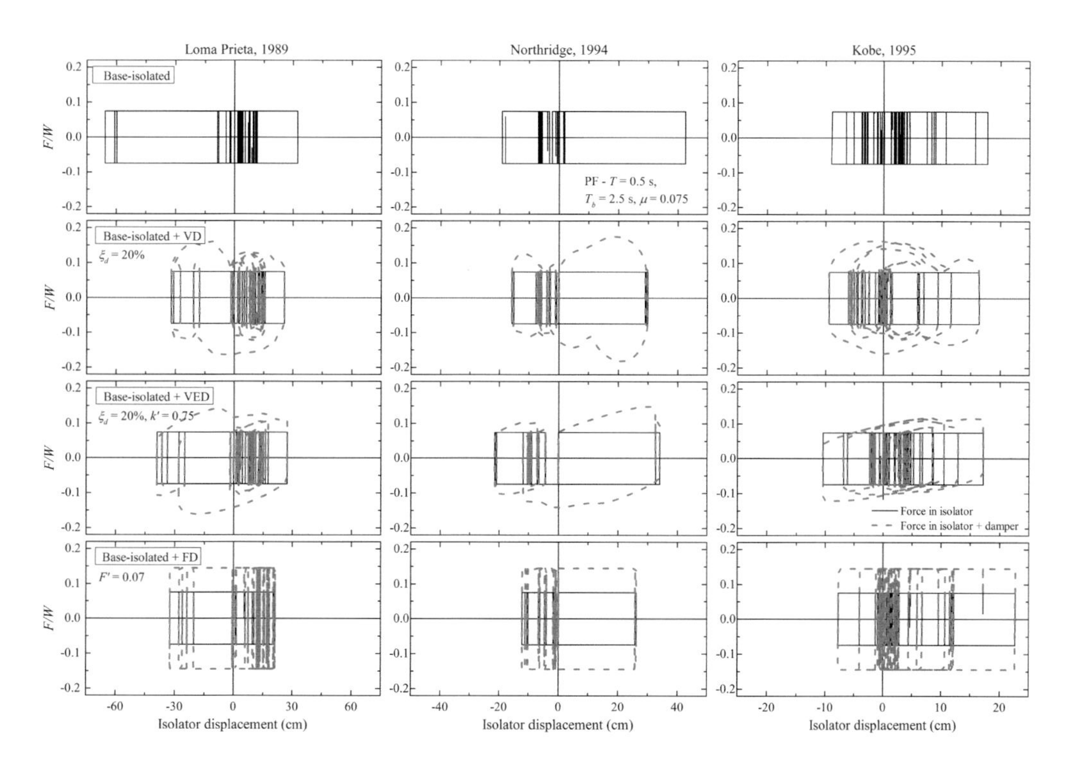

FIGURE 10.23 Comparison of force-deformation behavior of isolator for five-story building isolated by PF system with and without supplemental dampers under near-fault ground motions.

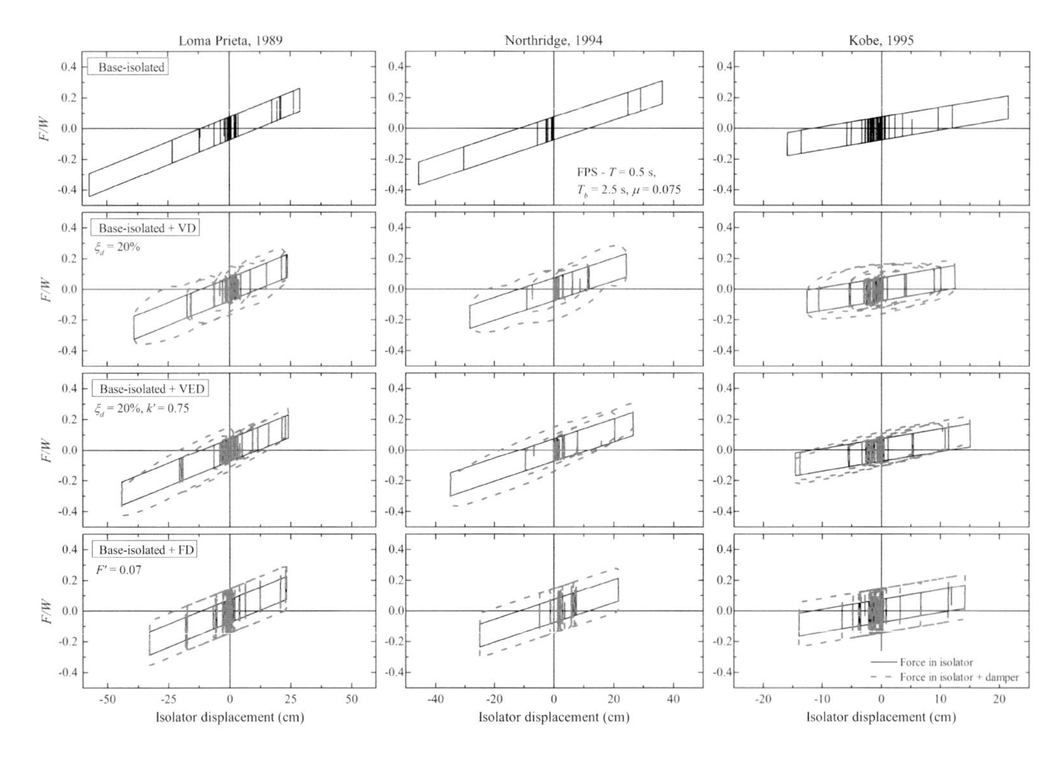

FIGURE 10.24 Comparison of force-deformation behavior of isolator for five-story building isolated by FPS with and without supplemental dampers under near-fault ground motions.

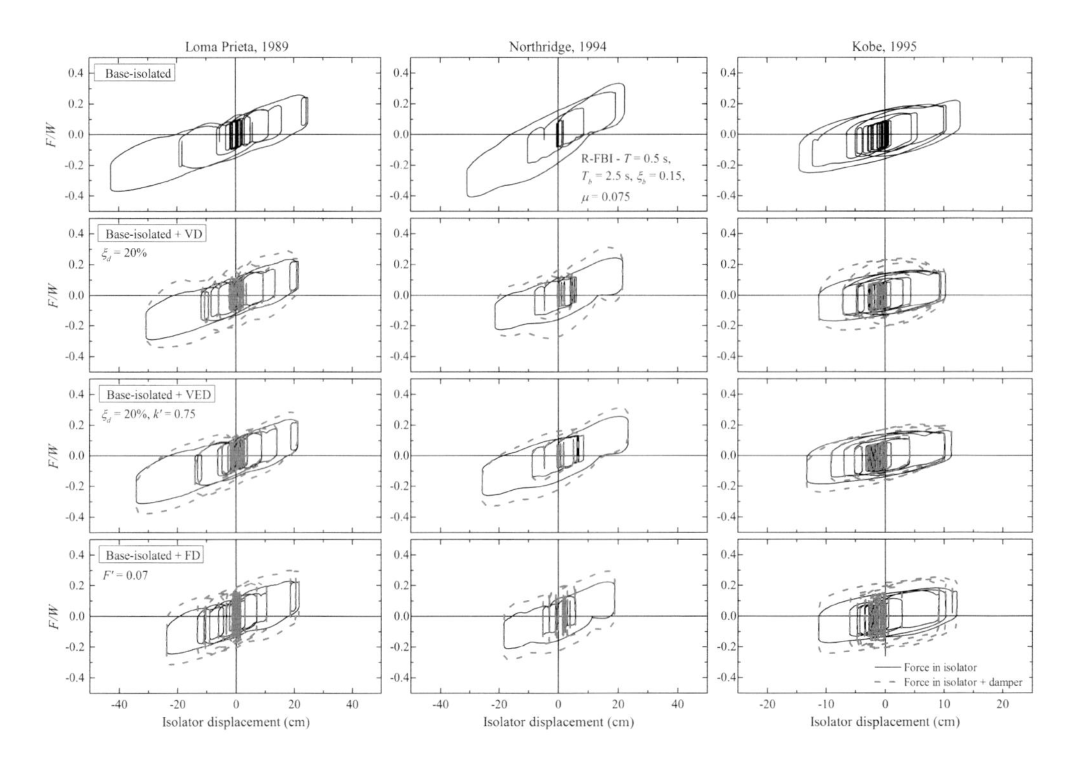

FIGURE 10.25 Comparison of force-deformation behavior of isolator for five-story building isolated by R-FBI with and without supplemental dampers under near-fault ground motions.

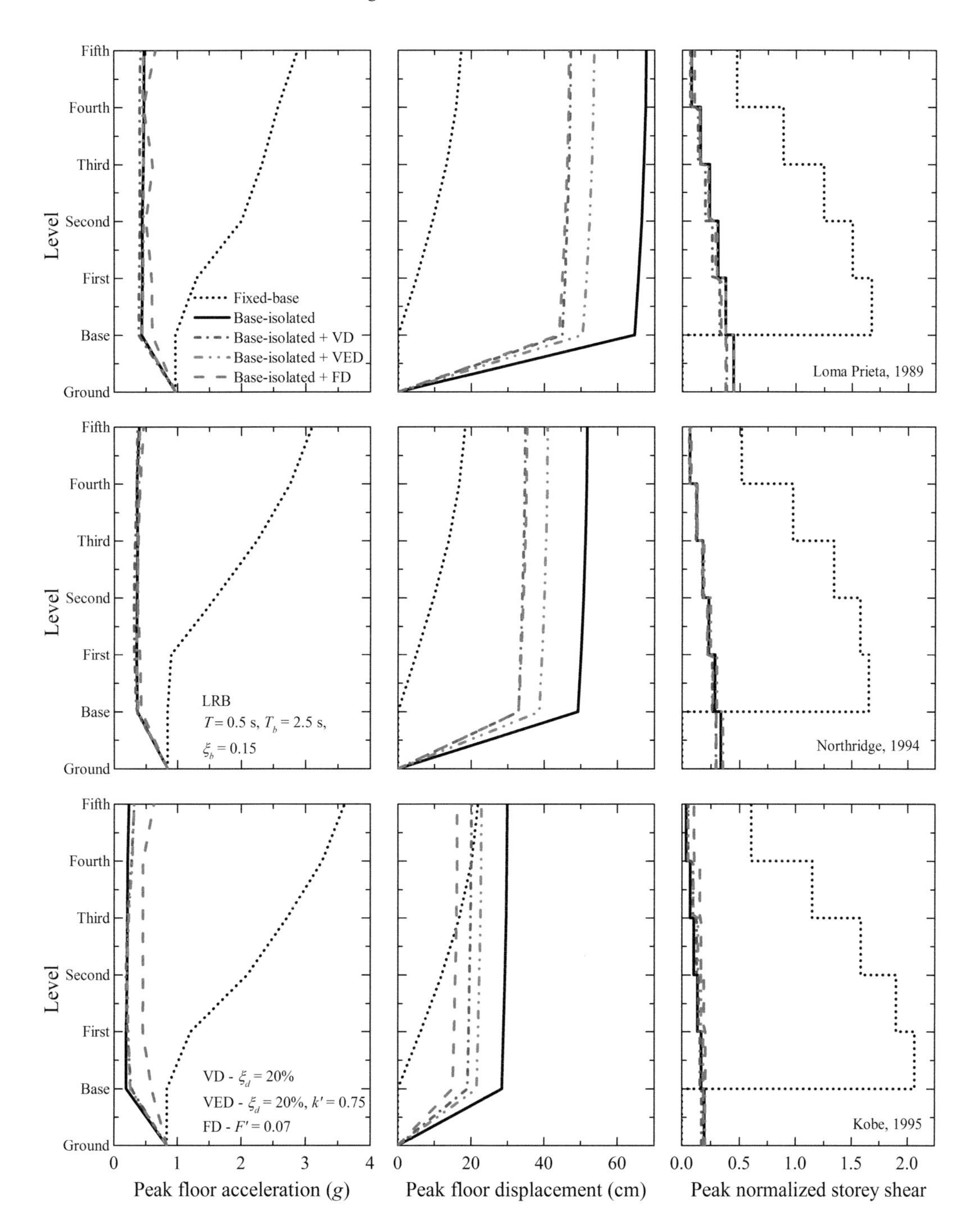

FIGURE 10.26 Distribution of peak floor acceleration, peak floor displacement, and peak normalized story shear over the height of five-story building isolated by LRB under near-fault ground motions.

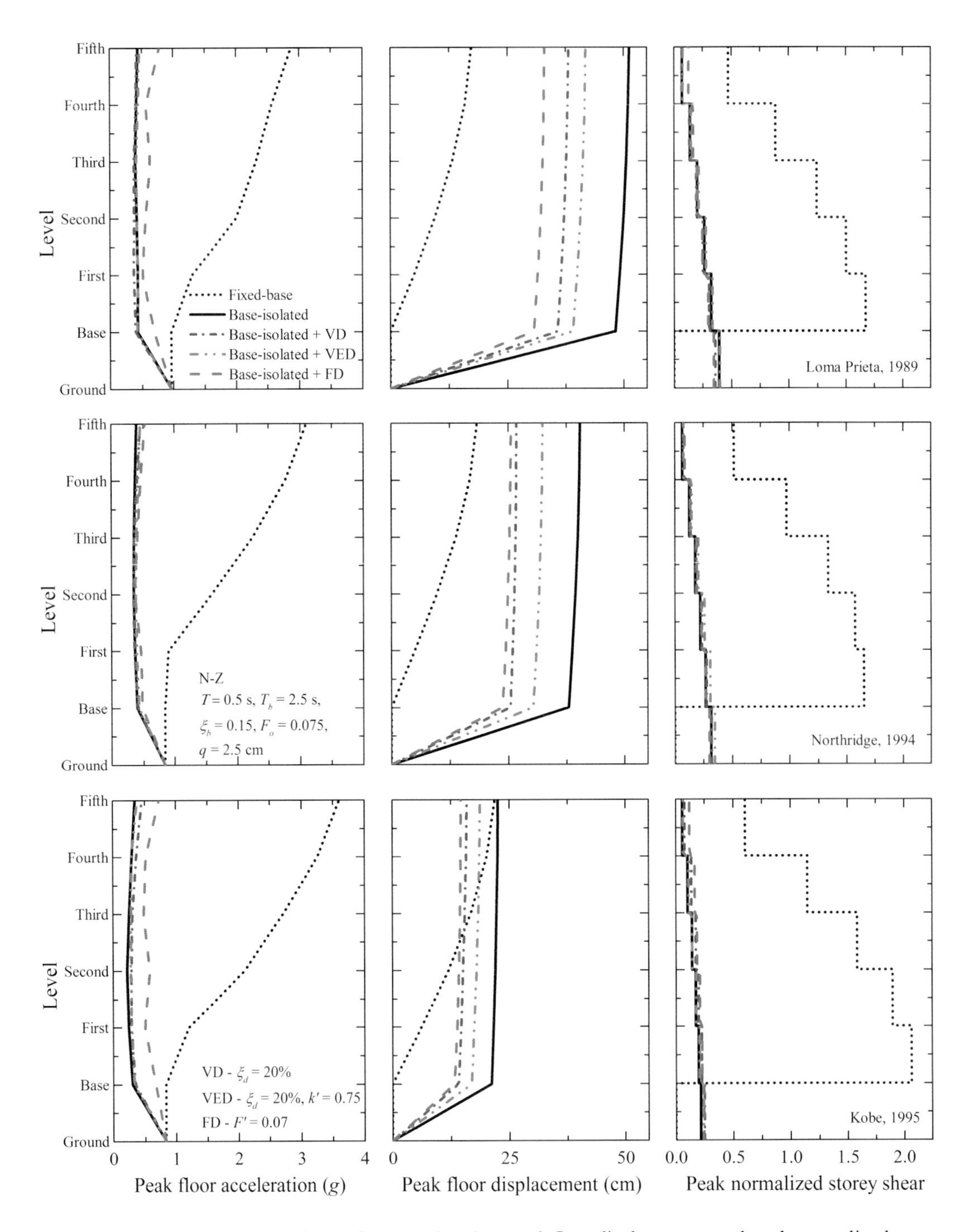

FIGURE 10.27 Distribution of peak floor acceleration, peak floor displacement, and peak normalized story shear over the height of five-story building isolated by N-Z system under near-fault ground motions.

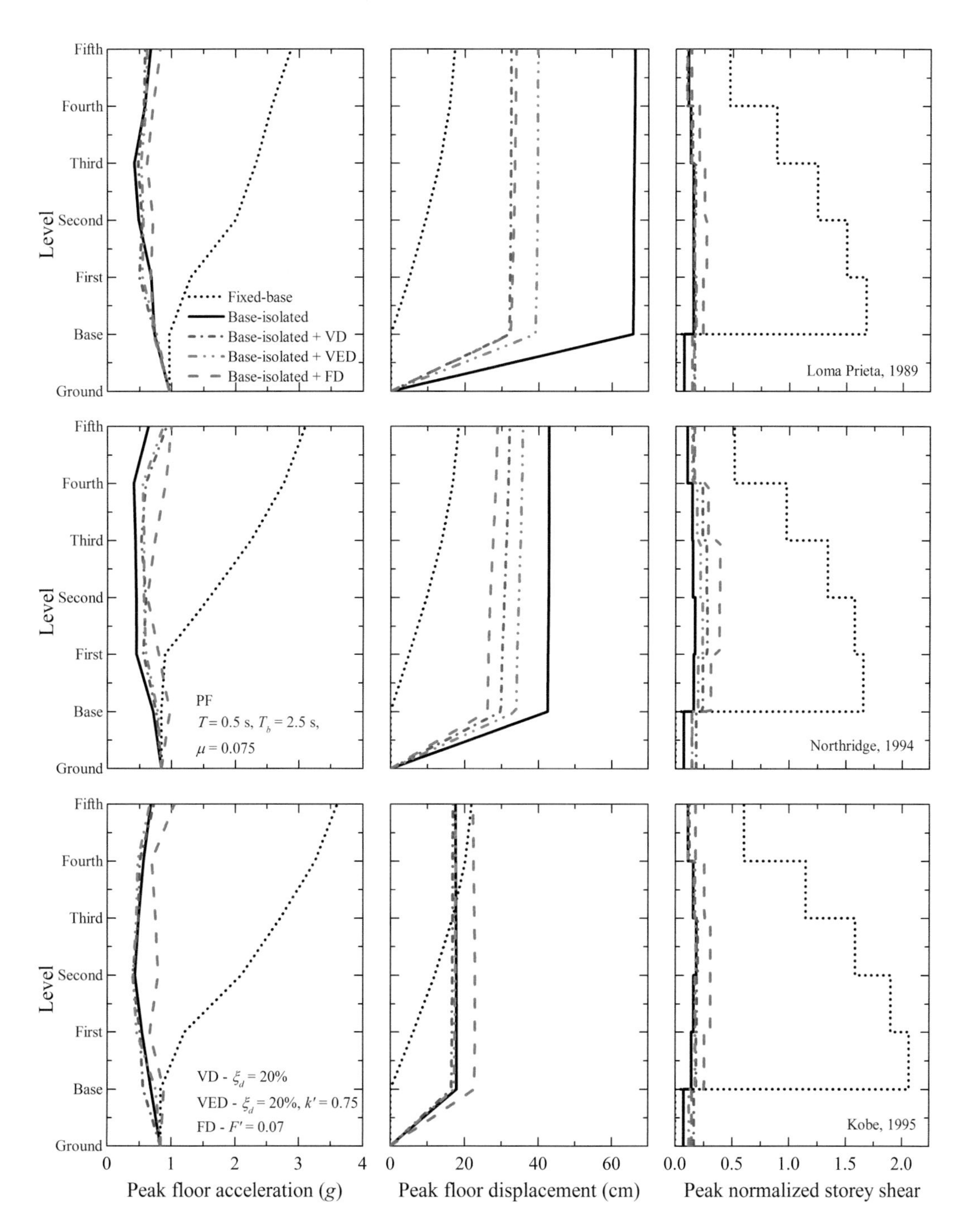

FIGURE 10.28 Distribution of peak floor acceleration, peak floor displacement, and peak normalized story shear over the height of five-story building isolated by PF system under near-fault ground motions.

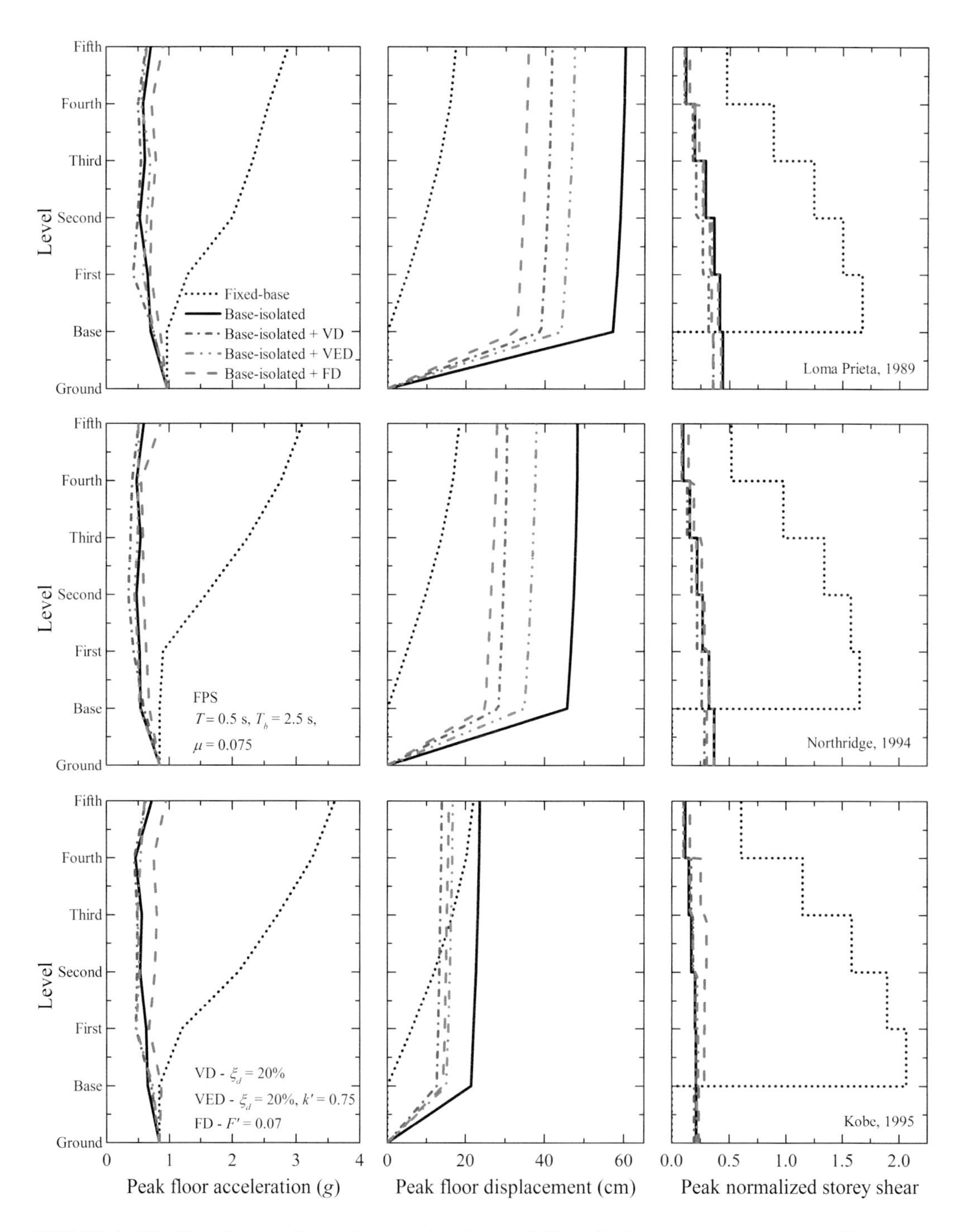

FIGURE 10.29 Distribution of peak floor acceleration, peak floor displacement, and peak normalized story shear over the height of five-story building isolated by FPS under near-fault ground motions.

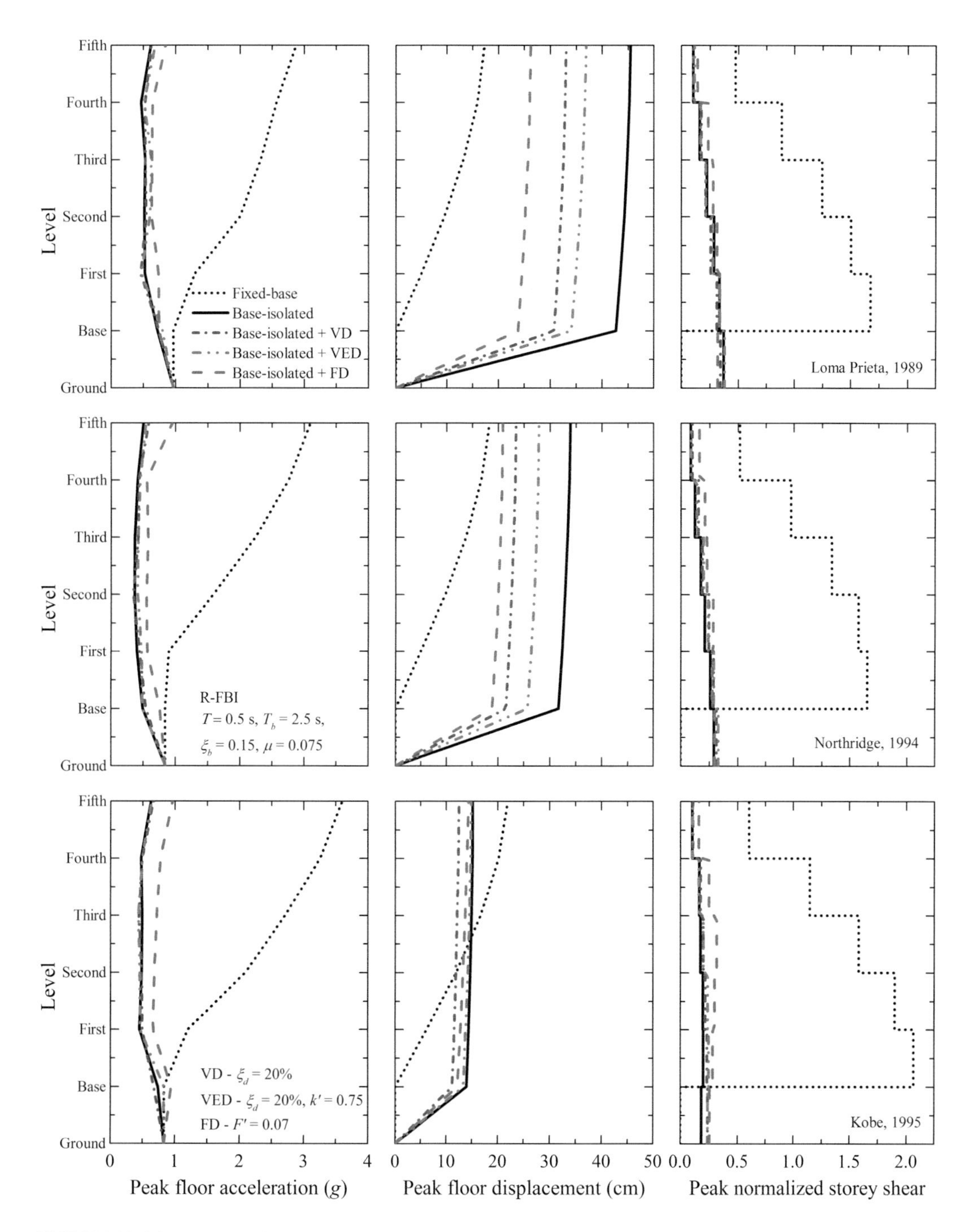

FIGURE 10.30 Distribution of peak floor acceleration, peak floor displacement, and peak normalized story shear over the height of five-story building isolated by R-FBI under near-fault ground motions.

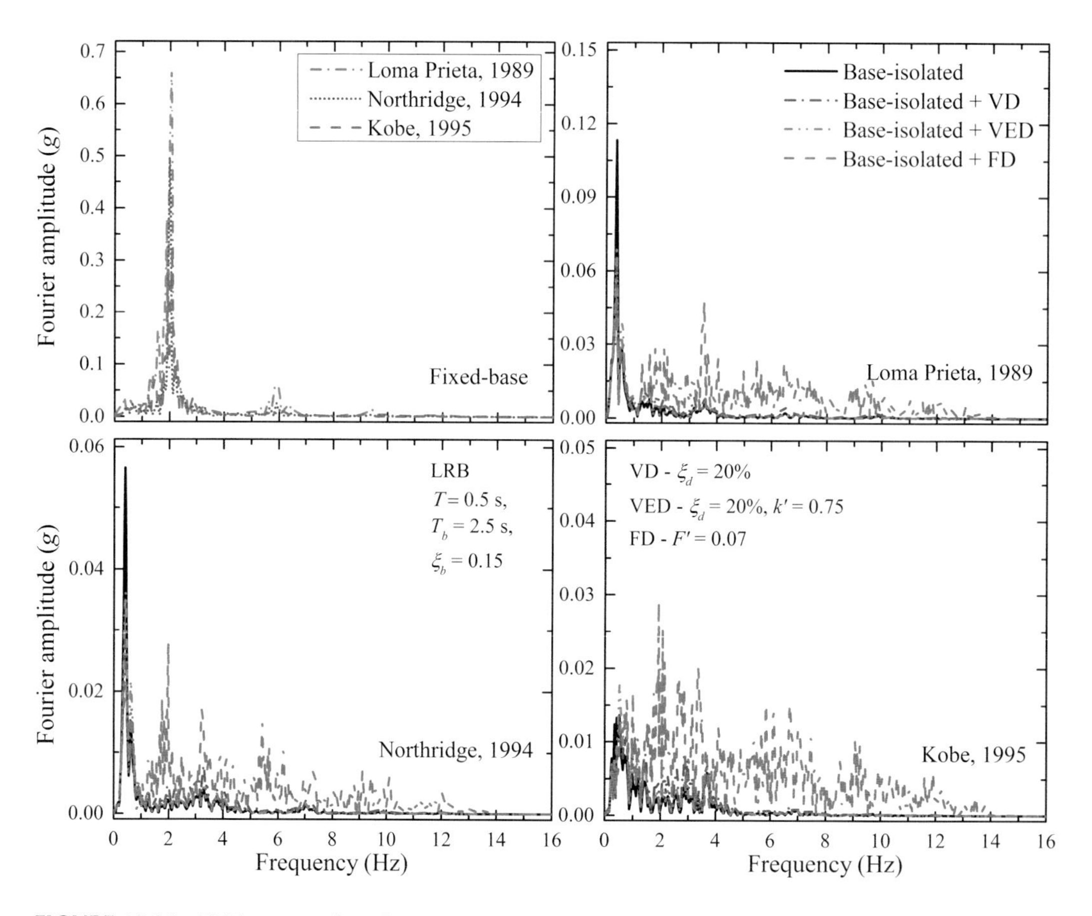

FIGURE 10.31 FFT spectra of top floor acceleration for five-story building isolated by LRB with and without supplemental dampers under near-fault ground motions.

for the multistory building isolated by LRB, N-Z system, PF system, FPS, and R-FBI, respectively. Further, average isolator displacement percentage reductions of 36%, 36%, 21%, 36%, and 31% are achieved for the multistory building isolated by LRB, N-Z system, PF system, FPS, and R-FBI, respectively, due to the use of FD having a normalized limiting friction force (F') of 0.07.

Example 10.4 Response Control of a Reinforced Concrete Bridge Using Base Isolation and Supplemental Dampers

In this example, the response control of a highway bridge using BI and supplemental dampers is presented. The bridge presented in Kunde and Jangid (2006) is considered here. In this example, the bridge model is considered to be equipped with supplemental dampers in addition to the base isolators as shown in Figure 10.42. The details of the bridge are given in Table 10.5. Lead rubber bearing (N-Z system) is used as isolation system, whereas the supplemental damper used in this control problem is VD. The isolation time period (T_b), normalized yield strength (F_0), yield displacement (q), and isolation damping ratio (ξ_b) of the N-Z system are 2.5 s, 0.05, 2.5 cm, and 5%, respectively. Furthermore, the damping ratio of the supplemental VD (ξ_d) is taken as 10%.

Perform dynamic response history analysis of the bridge equipped with the hybrid response control strategy (i.e., BI and supplemental damper) and determine the time histories of the deck acceleration, isolator displacement and pier base shear under the 1995 Kobe earthquake. Furthermore, compare the response quantities of the bridge controlled using the hybrid strategy with that of the bridge controlled using only BI.

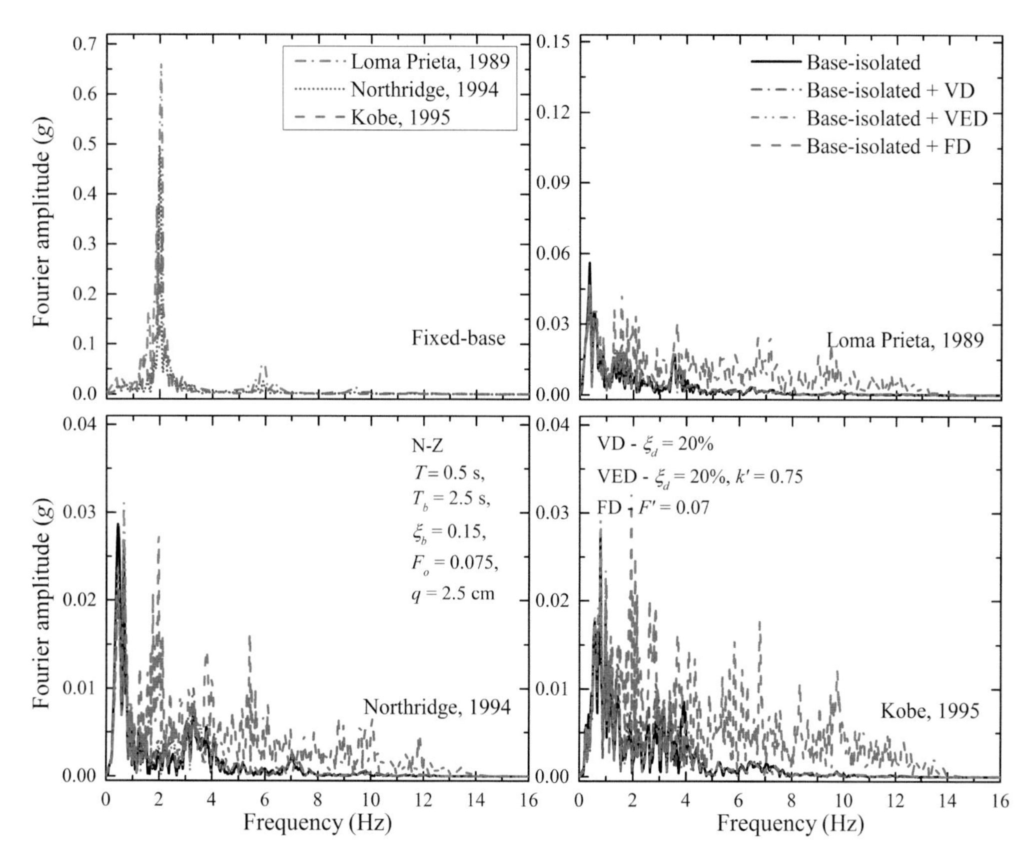

FIGURE 10.32 FFT spectra of top floor acceleration for five-story building isolated by N-Z system with and without supplemental dampers under near-fault ground motions.

SOLUTION

The equations of motion of the bridge are derived considering the bridge model shown in Figure 10.43. The model shows the forces acting on the bridge deck and on the pier.

For bidirectional excitation, the matrix form of the equations of motion of the base-isolated bridge equipped with supplemental VD is given in Equation (10.21).

$$
\begin{bmatrix} \begin{bmatrix} \hat{M}_p \end{bmatrix} & \{\mathbf{0}\} & \overleftrightarrow{\mathbf{0}} \\ \{\mathbf{0}\}^T & m_d & \\ \overleftrightarrow{\mathbf{0}} & \begin{bmatrix} \hat{M}_p \end{bmatrix} & \{\mathbf{0}\} \\ & \{\mathbf{0}\}^T & m_d \end{bmatrix}
\begin{Bmatrix} \{\ddot{X}_p\} \\ \ddot{x}_d \\ \{\ddot{Y}_p\} \\ \ddot{y}_d \end{Bmatrix}
+
\begin{bmatrix} \{\hat{C}_p\} & \{\mathbf{0}\} & \overleftrightarrow{\mathbf{0}} \\ \{\mathbf{0}\}^T & 0 & \\ \overleftrightarrow{\mathbf{0}} & \{\hat{C}_p\} & \{\mathbf{0}\} \\ & \{\mathbf{0}\}^T & 0 \end{bmatrix}
\begin{Bmatrix} \{\dot{X}_p\} \\ \dot{x}_d \\ \{\dot{Y}_p\} \\ \dot{y}_d \end{Bmatrix}
+
\begin{bmatrix} \{\hat{K}_p\} & \{\mathbf{0}\} & \overleftrightarrow{\mathbf{0}} \\ \{\mathbf{0}\}^T & 0 & \\ \overleftrightarrow{\mathbf{0}} & \{\hat{K}_p\} & \{\mathbf{0}\} \\ & \{\mathbf{0}\}^T & 0 \end{bmatrix}
\begin{Bmatrix} \{X_p\} \\ x_d \\ \{Y_p\} \\ y_d \end{Bmatrix}
$$

$$
-\begin{Bmatrix} \{r_{p,2}\} f_{bx,2} \\ f_{bx,1} + f_{bx,2} \\ \{r_{p,2}\} f_{by,2} \\ f_{by,1} + f_{by,2} \end{Bmatrix}
-\begin{Bmatrix} \{r_{p,2}\} f_{dx,2} \\ f_{dx,1} + f_{dx,2} \\ \{r_{p,2}\} f_{dy,2} \\ f_{dy,1} + f_{dy,2} \end{Bmatrix}
= -\begin{bmatrix} \begin{bmatrix} \hat{M}_p \end{bmatrix} & \{\mathbf{0}\} & \overleftrightarrow{\mathbf{0}} \\ \{\mathbf{0}\}^T & m_d & \\ \overleftrightarrow{\mathbf{0}} & \begin{bmatrix} \hat{M}_p \end{bmatrix} & \{\mathbf{0}\} \\ & \{\mathbf{0}\}^T & m_d \end{bmatrix}
\begin{bmatrix} \{r_p\} & 0 \\ 1 & \{\mathbf{0}\}^T \\ 0 & \{r_p\} \\ \{\mathbf{0}\}^T & 1 \end{bmatrix}
\begin{Bmatrix} \ddot{x}_g \\ \ddot{y}_g \end{Bmatrix}
\tag{10.21}
$$

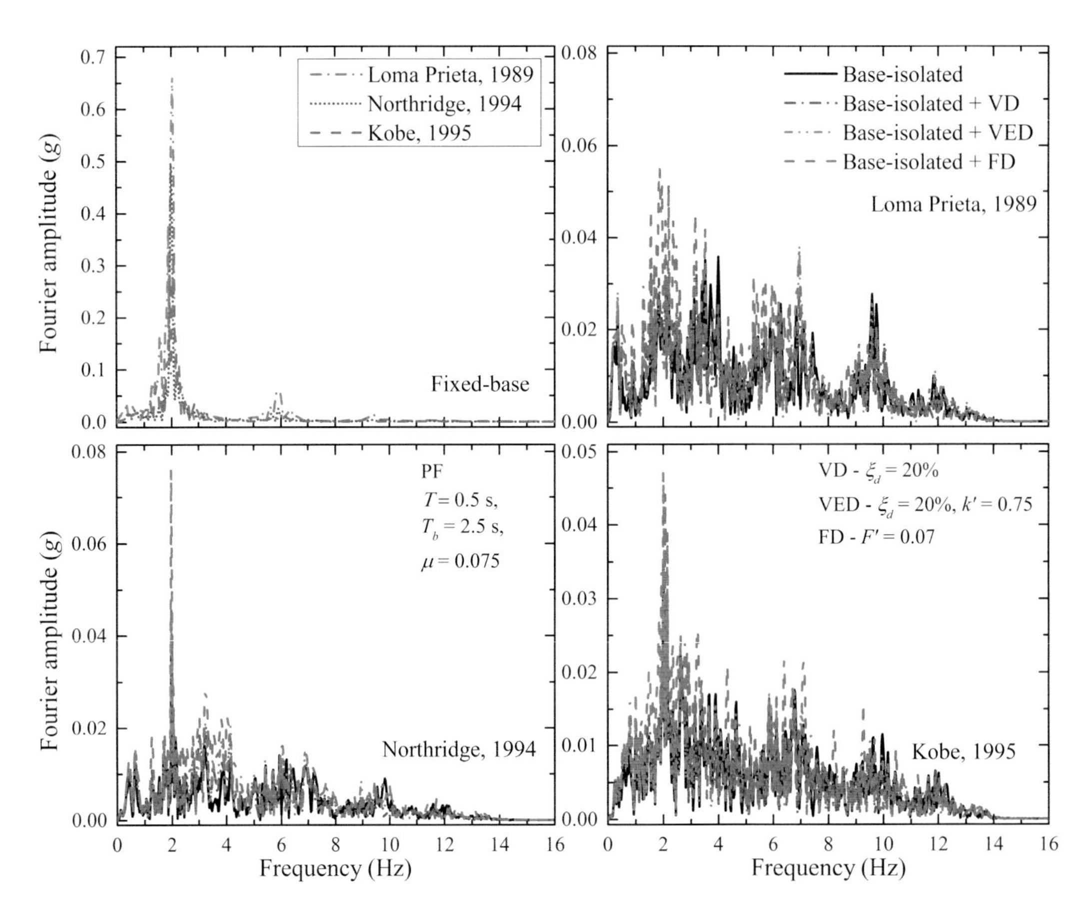

FIGURE 10.33 FFT spectra of top floor acceleration for five-story building isolated by PF system with and without supplemental dampers under near-fault ground motions.

where $f_{bx,1}$ and $f_{by,1}$ are the forces in the isolators located at the abutment in the longitudinal and transverse directions, respectively; $f_{bx,2}$ and $f_{by,2}$ are the forces in the isolators located at the pier in the longitudinal and transverse directions, respectively; $f_{dx,1}$ and $f_{dy,1}$ are the forces in the VDs located at the abutment in the longitudinal and transverse directions, respectively; and $f_{dx,2}$ and $f_{dy,2}$ are the forces in the VDs located at the pier in the longitudinal and transverse directions, respectively.

The time histories of the response quantities of the bridge under the bidirectional earthquake excitation are obtained by solving Equation (10.21) numerically. In Figure 10.44, the time histories of the North-South (NS) and the East-West (EW) components of the 1995 Kobe earthquake ground acceleration are shown. The time histories of the deck acceleration in the longitudinal and transverse direction for the: (a) uncontrolled bridge, (b) the base-isolated bridge, and (c) the base-isolated bridge equipped with VD are presented in Figure 10.45. Furthermore, the time histories of the normalized pier base shear in the longitudinal and transverse direction are presented in Figure 10.46. From the results presented in the figures, it can be seen that the deck acceleration response quantities obtained for the base-isolated bridges with and without the VD are significantly less than those of the uncontrolled bridge. The time histories of the isolator displacement for the: (a) the base-isolated bridge and (b) the base-isolated bridge equipped with VD are presented in Figure 10.47. It can be seen in the figure that the isolator displacement for the base-isolated bridge equipped with the VD is significantly smaller than that of the base-isolated bridge without the supplemental damper. The comparison of the peak values of all the response quantities is given in Table 10.6.

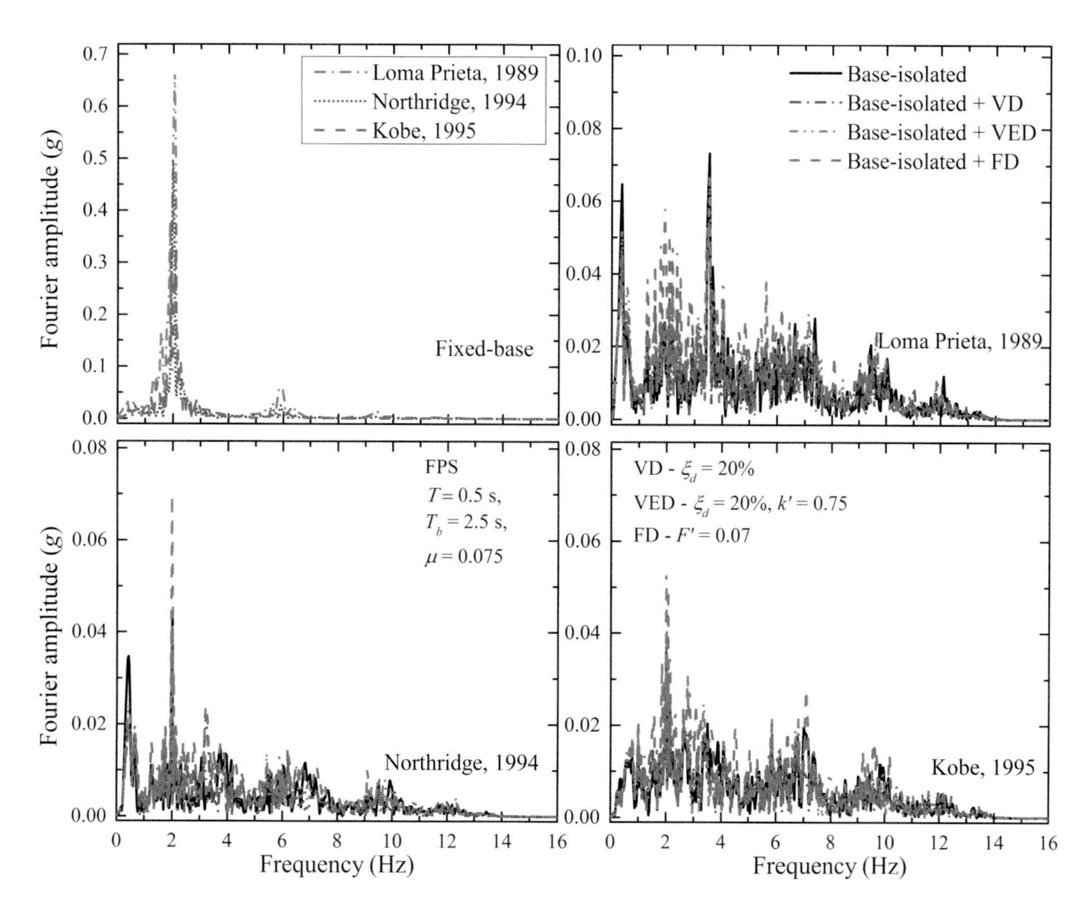

FIGURE 10.34 FFT spectra of top floor acceleration for five-story building isolated by FPS with and without supplemental dampers under near-fault ground motions.

Example 10.5 Hybrid Control of Benchmark Highway Bridge (Highway Benchmark 91/5 Overcrossing) Using Base Isolation and Supplemental Dampers

In this example, the response control of a benchmark highway bridge (Highway Benchmark 91/5 Overcrossing) using BI and supplemental fluid VDs is presented. The bridge response is computed by inducing ground motions applied simultaneously in the longitudinal as well as in transverse direction, i.e., fault normal component in transverse direction and fault parallel component in longitudinal direction. The elevation and plan view of the benchmark bridge are presented in Figure 10.48, whereas the cross-sectional elevation of the bridge is given in Figure 10.49. Furthermore, the details of the benchmark highway bridge are presented in Table 10.7.

Perform response history analysis of the benchmark bridge equipped with a hybrid BI strategy (BI + fluid VD [FVD]) under the 1994 Northridge, 1995 Kobe, and 1999 Chi-Chi earthquake ground motions. Furthermore, compare the response of the hybrid-controlled bridge with that of the uncontrolled and the seismic-isolated bridge models. The details of the three earthquake ground motions are presented in Table 10.8, whereas the response spectra of the ground motions for 2% of the critical damping are given in Figure 10.50.

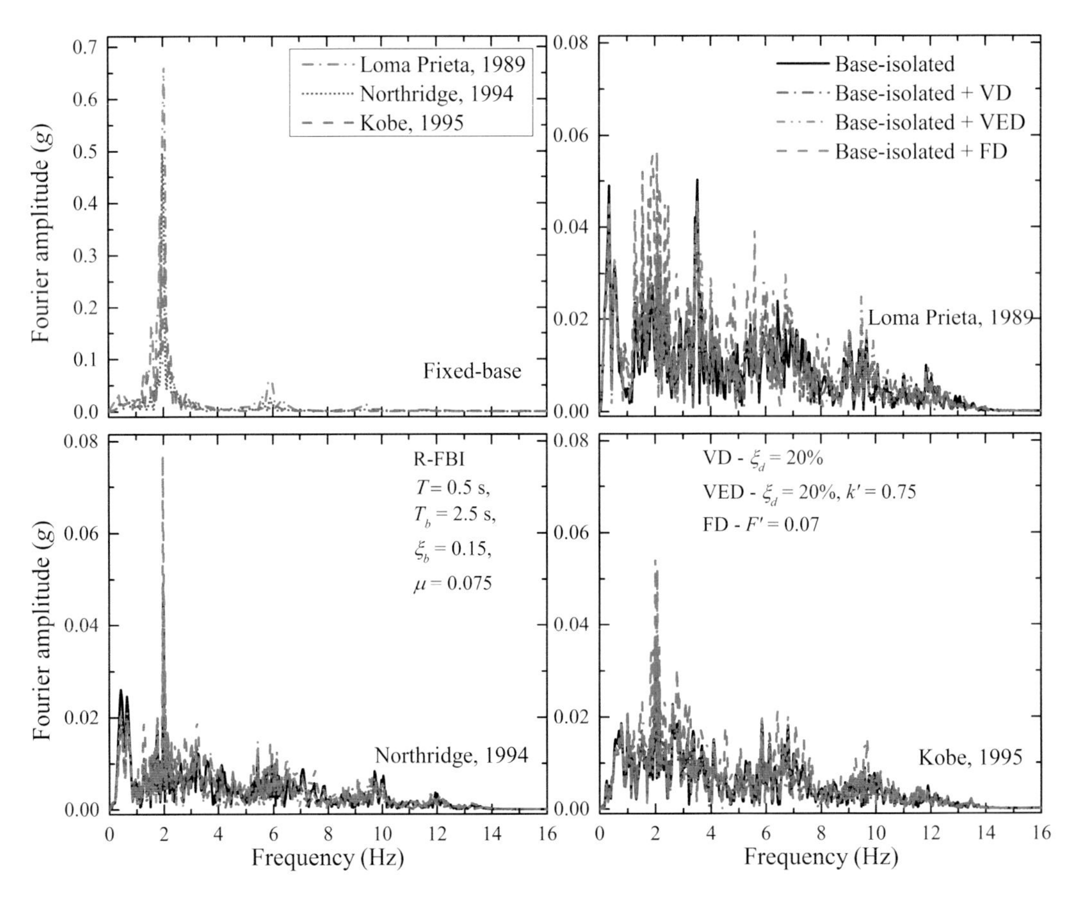

FIGURE 10.35 FFT spectra of top floor acceleration for five-story building isolated by R-FBI with and without supplemental dampers under near-fault ground motions.

SOLUTION

The matrix form of the equations of motion of the bridge can be given as follows.

$$[M]\{\ddot{u}(t)\}+[C]\{\dot{u}(t)\}+[K(t)]\{u(t)\}+\{F_d\}=-[M]\{r\}\{\ddot{u}_g(t)\} \tag{10.22}$$

where $[M]$, $[C]$, and $[K(t)]$, respectively, are the mass, damping, and stiffness matrices of the base isolated bridge; $\{F_d\}$ is the vector of the damper force; and $\{\ddot{u}_g\}=\{\ddot{x}_g,\ddot{y}_g\}^T$ the vector of earthquake ground accelerations in the longitudinal and transverse directions.

The response quantities of the bride can be obtained by solving the governing equations of motion. The time histories of the bearing displacement and the base shear under the 1994 Northridge earthquake are presented in Figures 10.51 and 10.52, respectively; whereas the time histories of the response quantities under the 1995 Kobe earthquake are presented in Figures 10.53 and 10.54, respectively. Furthermore, the time histories of the bearing displacement and the base shear, respectively, under the 1999 Chi-Chi earthquake are presented in Figures 10.55 and 10.56.

The figures compare the time histories of the response quantities for the: (a) uncontrolled bridge, (b) bridge isolated by N-Z system, and (c) bridge equipped with N-Z system and FVD, in both the longitudinal and transverse directions. The results show that the hybrid control strategy (seismic isolation + FVD) results in significant reduction of the isolator displacement response of the bridge as compared to that of the isolated bridge without the damper. However, the base shear is not significantly influenced by the installation of the FVD.

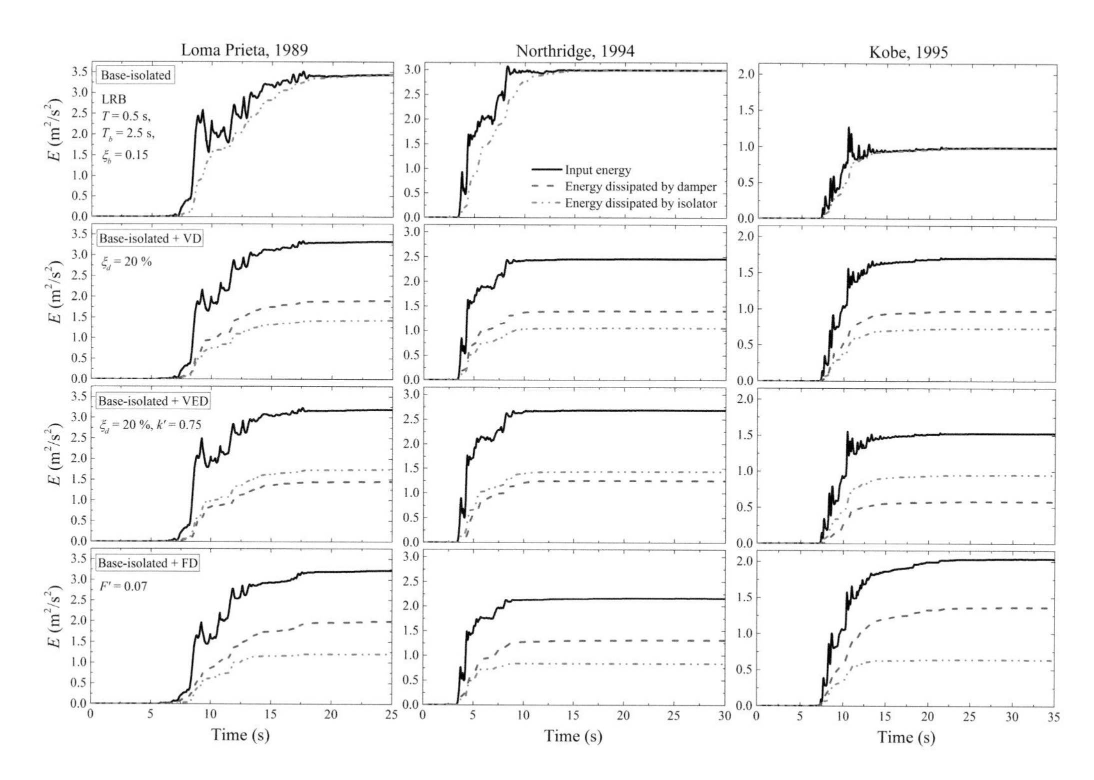

FIGURE 10.36 Time histories of input and dissipated energies per unit mass for five-story building isolated by LRB with and without supplemental dampers.

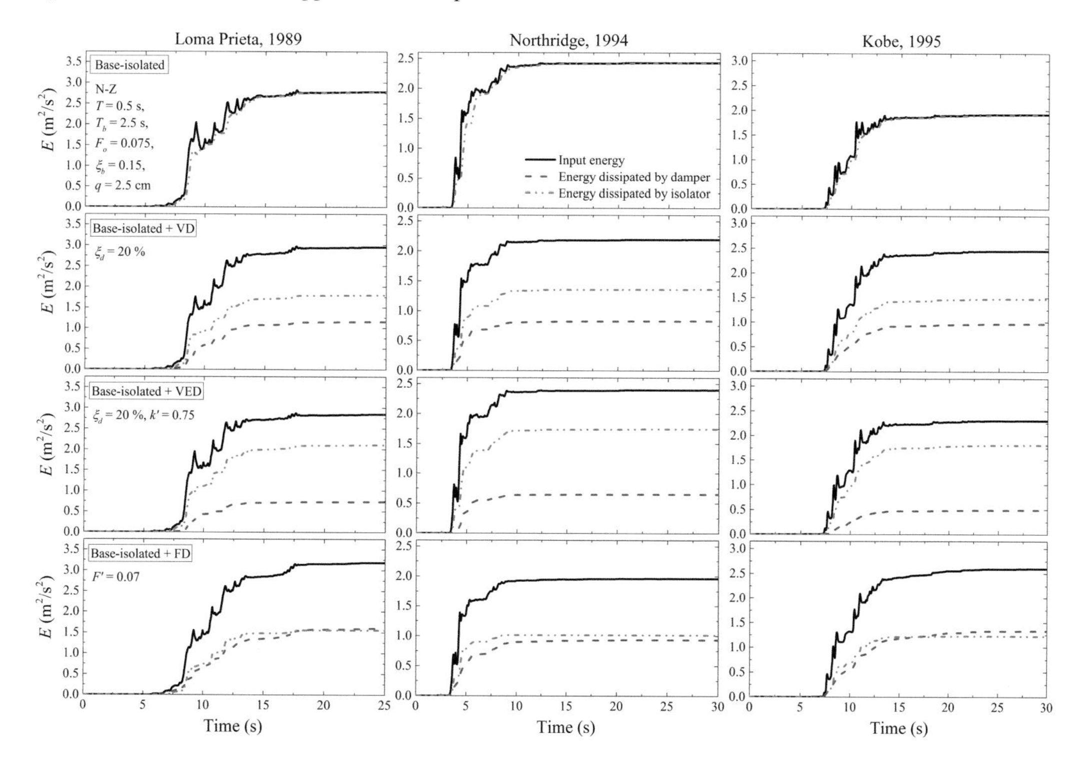

FIGURE 10.37 Time histories of input and dissipated energies per unit mass for five-story building isolated by N-Z system with and without supplemental dampers.

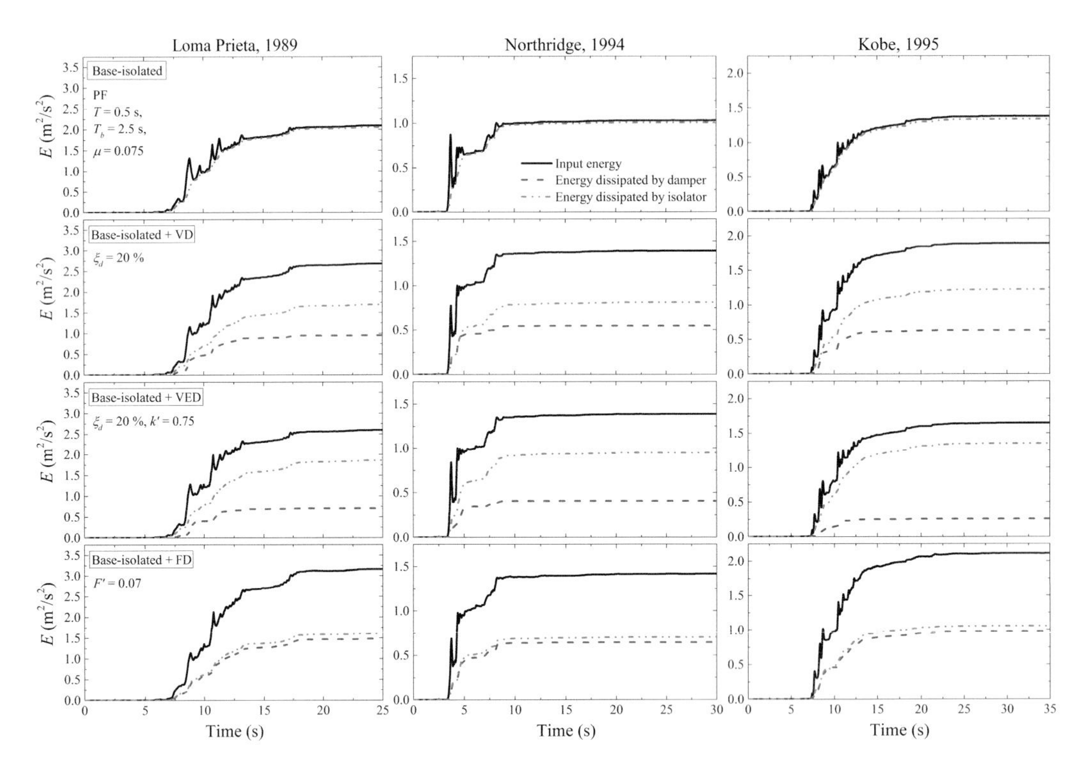

FIGURE 10.38 Time histories of input and dissipated energies per unit mass for five-story building isolated by PF system with and without supplemental dampers.

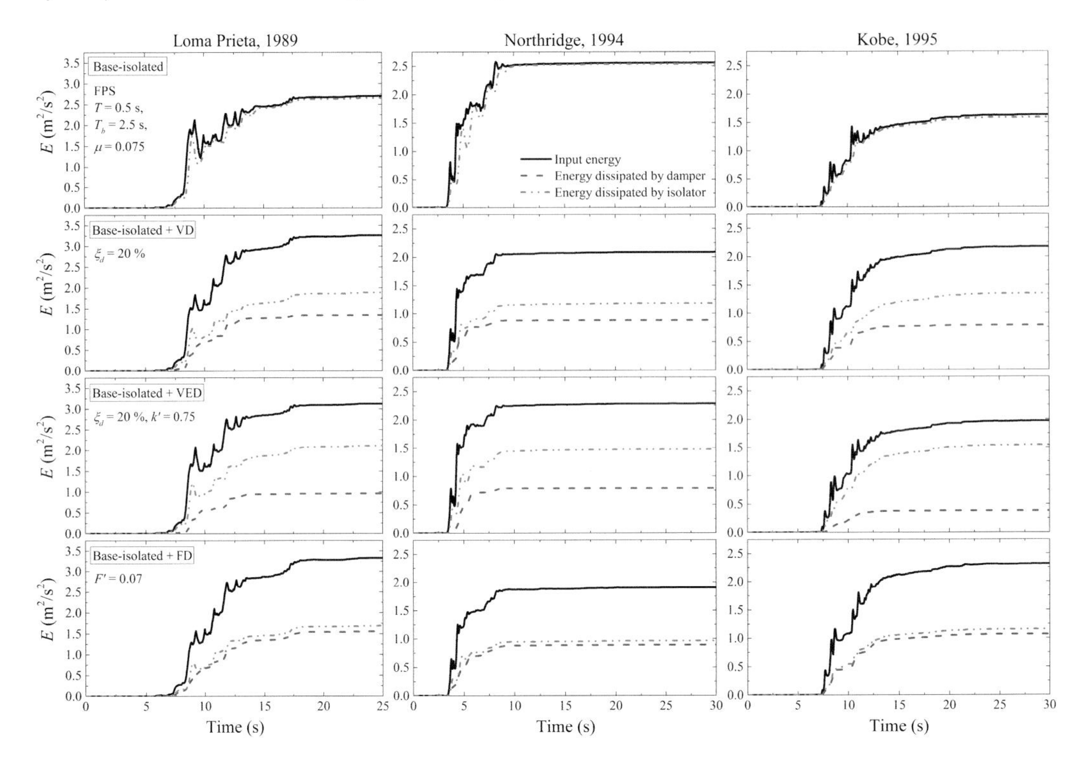

FIGURE 10.39 Time histories of input and dissipated energies per unit mass for five-story building isolated by FPS with and without supplemental dampers.

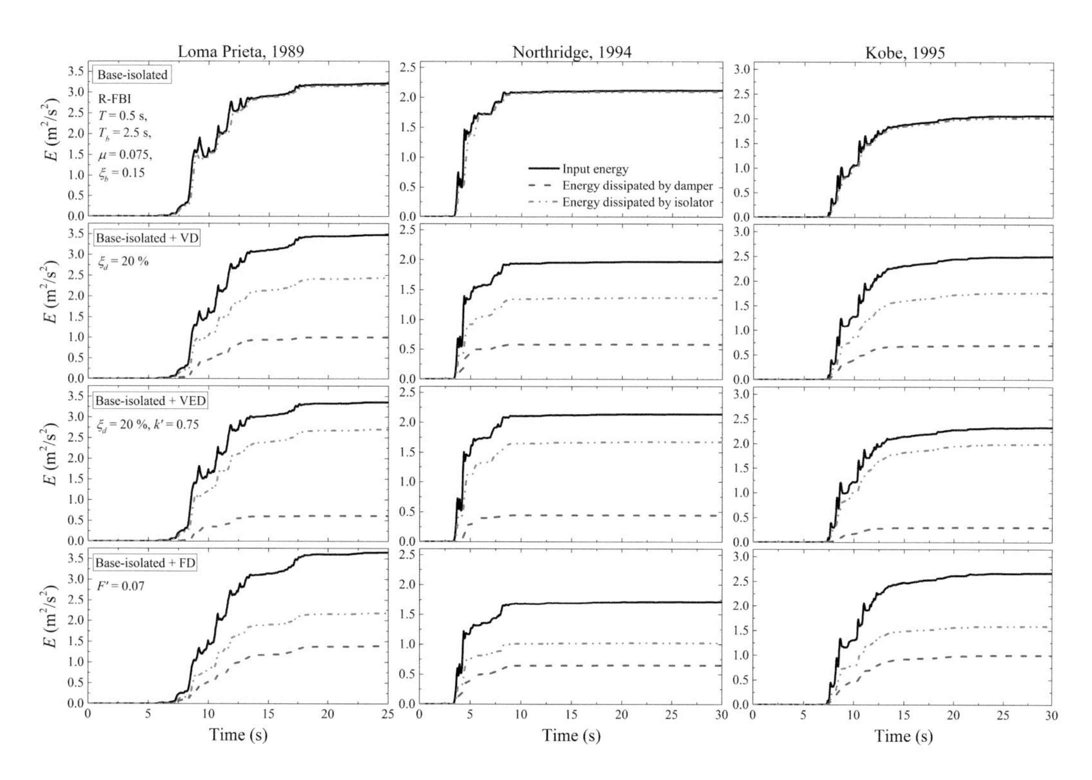

FIGURE 10.40 Time histories of input and dissipated energies per unit mass for five-story building isolated by R-FBI with and without supplemental dampers.

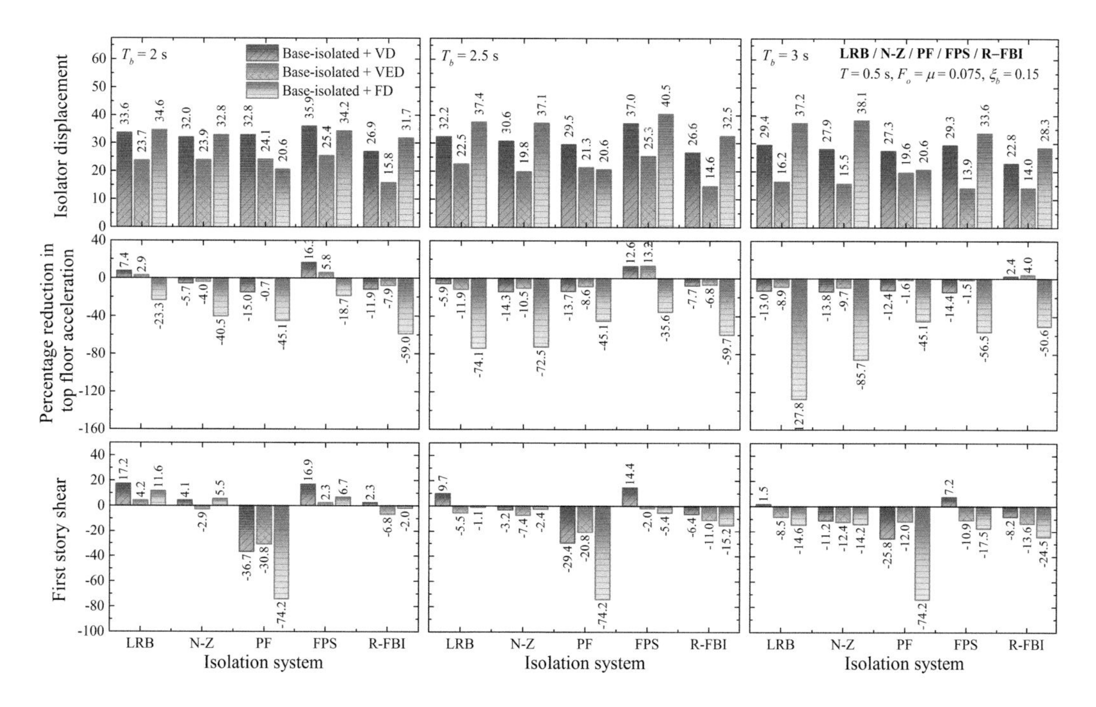

FIGURE 10.41 Comparison of percentage reductions in the isolator displacement, top floor acceleration, and first-story shear achieved due to supplemental dampers for various isolators under near-fault ground motions.

TABLE 10.4
Peak Responses of Five-Story Base-Isolated Building with and without Supplemental Damper under Near-Fault Ground Motions
$(T = 0.5\ s, T_b = 2.5\ s)$

Response under Near-Fault Ground Motions			Base-Isolated	Base-Isolated Building with			Fixed-Base
				VD ($\xi_d = 20\%$)	VED ($\xi_d = 20\%$, $k' = 0.75$)	FD ($F' = 0.07$)	
LRB	Loma Prieta, 1989	Isolator displacement (cm)	612.558	412.902	50.457	43.990	-
$\xi_b = 0.15$,		Top floor acceleration (g)	0.476	0.421	0.466	0.650	2.875
$T_b = 2.5$ s		Normalized first story shear	0.376	0.322	0.376	0.335	1.675
	Northridge, 1994	Isolator displacement (cm)	49.227	32.972	38.633	32.981	-
		Top floor acceleration (g)	0.396	0.368	0.404	0.463	3.098
		Normalized first story shear	0.284	0.259	0.301	0.266	1.652
	Kobe, 1995	Isolator displacement (cm)	28.378	18.975	21.518	112.921	-
		Top floor acceleration (g)	0.231	0.315	0.314	0.621	3.606
		Normalized first story shear	0.168	0.158	0.186	0.203	2.066
N-Z	Loma Prieta, 1989	Isolator displacement (cm)	48.231	35.776	39.049	30.667	-
$\xi_b = 0.15$,		Top floor acceleration (g)	0.438	0.421	0.461	0.776	2.875
$F_0 = 0.075$,		Normalized first story shear	0.324	0.305	0.336	0.298	1.675
$q = 2.5$ cm,	Northridge, 1994	Isolator displacement (cm)	37.986	25.421	30.334	23.703	-
$T_b = 2.5$ s		Top floor acceleration (g)	0.395	0.456	0.459	0.525	3.098
		Normalized first story shear	0.268	0.276	0.308	0.271	1.652
	Kobe, 1995	Isolator displacement (cm)	21.128	112.169	16.878	13.257	-
		Top floor acceleration (g)	0.348	0.456	0.382	0.723	3.606
		Normalized first story shear	0.199	0.224	0.206	0.228	2.066
PF	Loma Prieta, 1989	Isolator displacement (cm)	65.724	32.059	39.066	32.410	-
$\mu = 0.075$		Top floor acceleration (g)	0.675	0.591	0.617	0.829	2.875
		Normalized first story shear	0.155	0.170	0.179	0.237	1.675
	Northridge, 1994	Isolator displacement (cm)	42.459	29.864	33.987	26.191	-
		Top floor acceleration (g)	0.637	0.926	0.873	1.001	3.098
		Normalized first story shear	0.160	0.239	0.198	0.307	1.652
	Kobe, 1995	Isolator displacement (cm)	17.814	16.450	17.188	22.683	-
		Top floor acceleration (g)	0.676	0.730	0.658	1.048	3.606
		Normalized first story shear	0.142	0.183	0.175	0.252	2.066

(Continued)

TABLE 10.4 (*Continued*)
Peak Responses of Five-Story Base-Isolated Building with and without Supplemental Damper under Near-Fault Ground Motions
$(T = 0.5\ \text{s}, T_b = 2.5\ \text{s})$

Response under Near-Fault Ground Motions			Base-Isolated	Base-Isolated Building with			Fixed-Base
				VD ($\xi_d = 20\%$)	VED ($\xi_d = 20\%$, $k' = 0.75$)	FD ($F' = 0.07$)	
FPS	Loma Prieta, 1989	Isolator displacement (cm)	57.127	39.016	412.121	32.877	-
$\mu = 0.075$,		Top floor acceleration (g)	0.705	0.631	0.638	0.900	2.875
$T_b = 2.5$ s		Normalized first story shear	0.414	0.315	0.399	0.343	1.675
	Northridge, 1994	Isolator displacement (cm)	45.650	28.314	312.924	212.930	-
		Top floor acceleration (g)	0.591	0.508	0.507	0.860	3.098
		Normalized first story shear	0.324	0.260	0.327	0.320	1.652
	Kobe, 1995	Isolator displacement (cm)	21.434	12.583	15.075	112.221	-
		Top floor acceleration (g)	0.712	0.619	0.599	0.950	3.606
		Normalized first story shear	0.212	0.212	0.230	0.285	2.066
R-FBI	Loma Prieta, 1989	Isolator displacement (cm)	42.662	30.705	312.064	23.790	-
$\mu = 0.075$,		Top floor acceleration (g)	0.617	0.625	0.687	0.843	2.875
$\xi_b = 0.15$,		Normalized first story shear	0.333	0.307	0.328	0.334	1.675
$T_b = 2.5$ s	Northridge, 1994	Isolator displacement (cm)	31.644	21.569	25.706	18.884	-
		Top floor acceleration (g)	0.505	0.586	0.541	0.953	3.098
		Normalized first story shear	0.257	0.285	0.288	0.273	1.652
	Kobe, 1995	Isolator displacement (cm)	13.989	11.206	13.310	12.189	-
		Top floor acceleration (g)	0.627	0.664	0.639	0.964	3.606
		Normalized first story shear	0.201	0.234	0.247	0.280	2.066

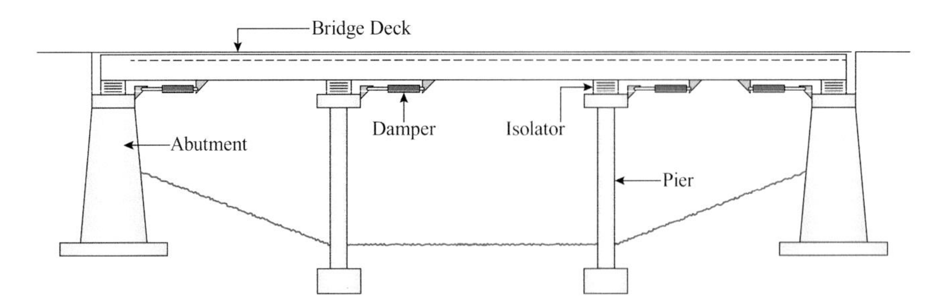

FIGURE 10.42 Concrete bridge equipped with base isolators and supplemental dampers.

TABLE 10.5
Dynamic Properties of the Bridge Considered Herein

Sr. No.	Bridge Property	Value
1	Span length (m)	3 at 30
2	Pier height (m)	8
3	Pier shape	Circular
4	Cross-sectional area of deck (m^2)	3.57
5	Cross-sectional area of pier (m^2)	4.09
6	Moment of inertia of deck in transverse direction (m^4)	2.08
7	Moment of inertia of pier in longitudinal direction (m^4)	0.64
8	Moment of inertia of pier in transverse direction (m^4)	0.64
9	Fundamental time period of pier in longitudinal direction (s)	0.1
10	Fundamental time period of pier in transverse direction (s)	0.1
11	Young's modulus of elasticity (GPa)	20.64
12	Mass density (kg/m^3)	2,400
13	Fundamental time period of non-isolated bridge in longitudinal direction (s)	0.45
14	Fundamental time period of non-isolated bridge in transverse direction (s)	0.45

Source: Kunde and Jangid (2006).

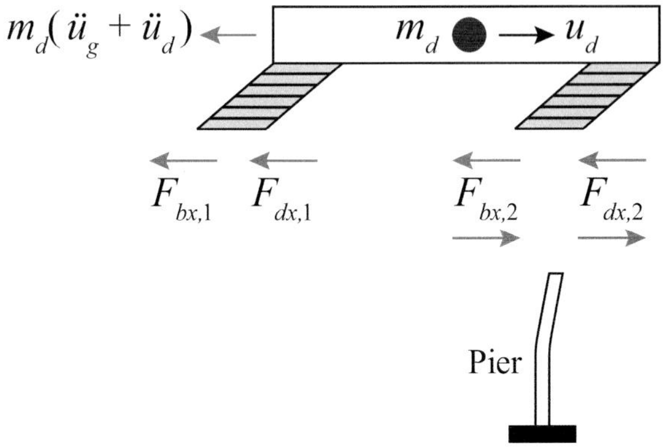

FIGURE 10.43 Forces acting on the bridge deck and the pier in the *x* direction.

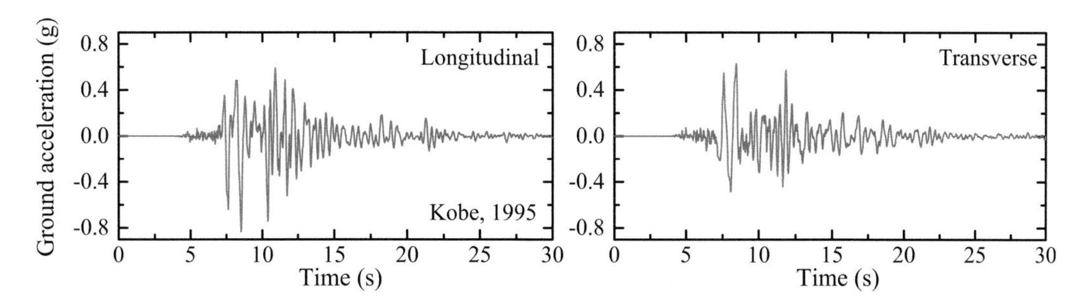

FIGURE 10.44 The NS and the EW components of the 1995 Kobe earthquake ground acceleration applied in the longitudinal and transverse directions of the bridge, respectively.

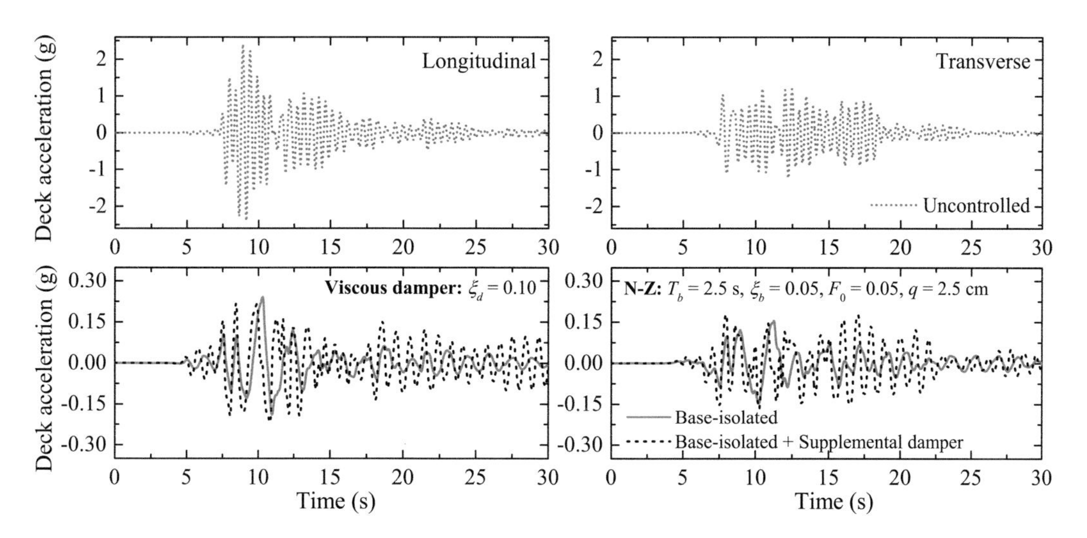

FIGURE 10.45 Comparison of time histories of deck acceleration under the 1995 Kobe earthquake.

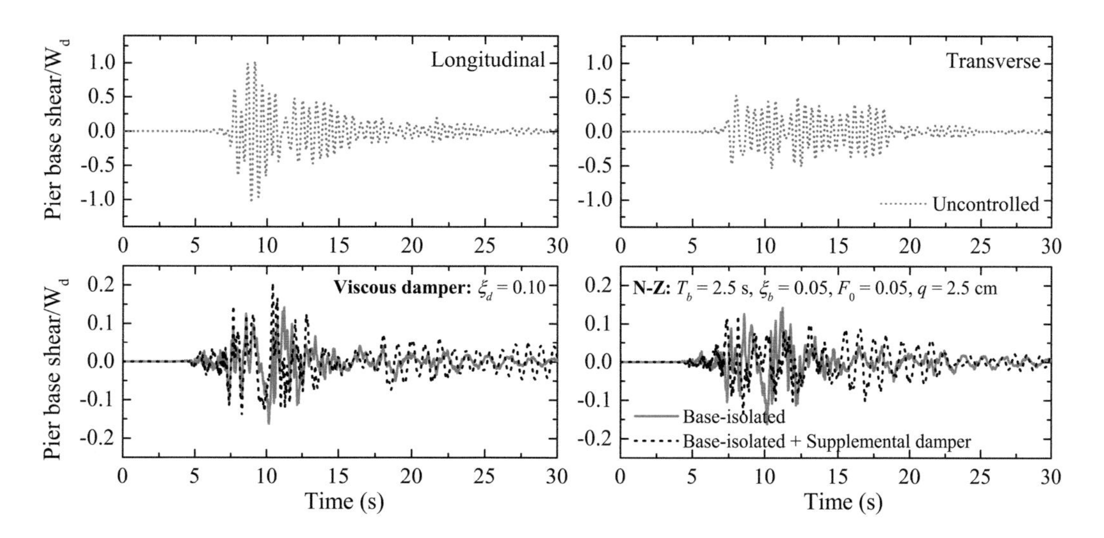

FIGURE 10.46 Comparison of time histories of normalized pier base shear under the 1995 Kobe earthquake.

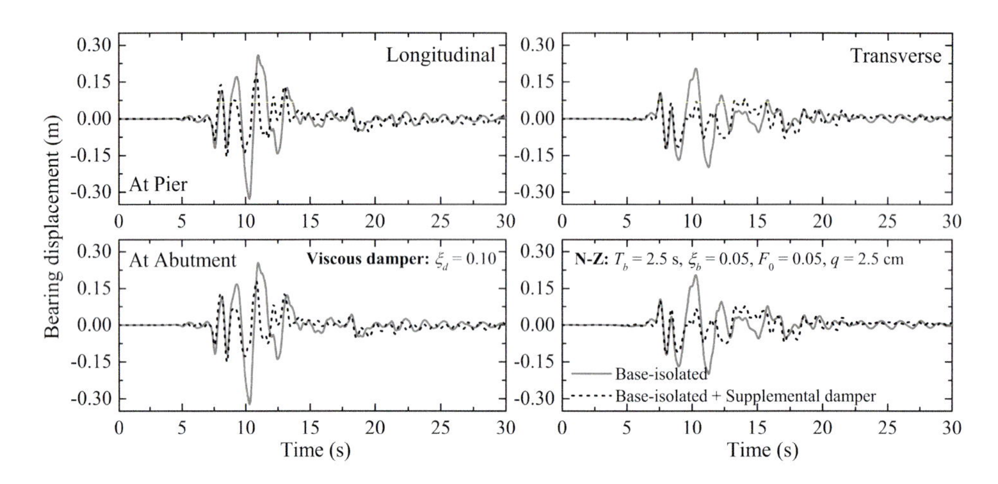

FIGURE 10.47 Comparison of time histories of isolator displacement under the 1995 Kobe earthquake.

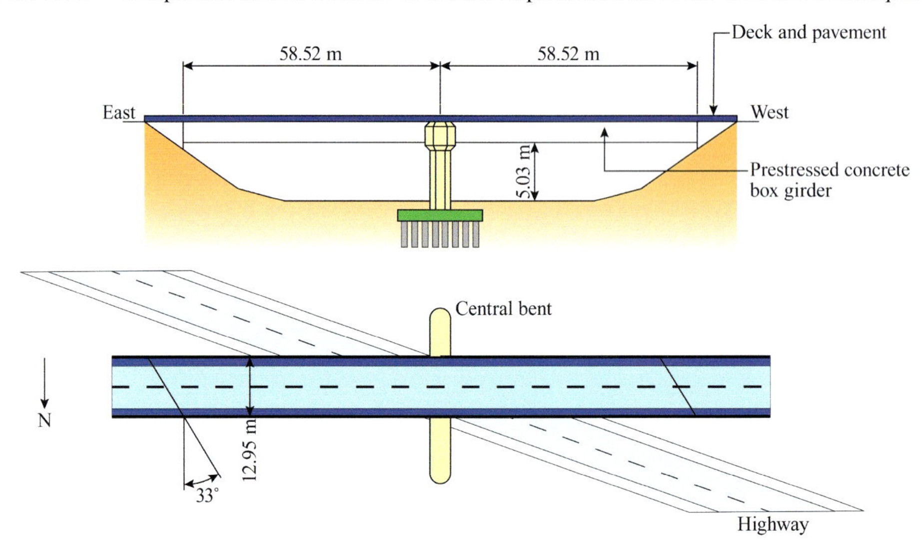

FIGURE 10.48 Elevation and plan view of the benchmark highway bridge. (Agrawal et al. 2006)

TABLE 10.6
Comparison of Peak Values of Response Quantities of the Bridge under the 1995 Kobe Earthquake

		Bridge		
Response Quantity	**Direction**	**Uncontrolled**	**Base-Isolated**	**Base-Isolated + Supplemental Damper**
Deck acceleration (g)	Longitudinal	2.406	0.241	0.218
	Transverse	1.230	0.155	0.179
Normalized pier base shear	Longitudinal	1.042	0.162	0.206
	Transverse	0.545	0.118	0.135
Isolator displacement at pier (m)	Longitudinal	-	0.328	0.185
	Transverse	-	0.205	0.124
Isolator displacement at abutment (m)	Longitudinal	-	0.323	0.179
	Transverse	-	0.202	0.119

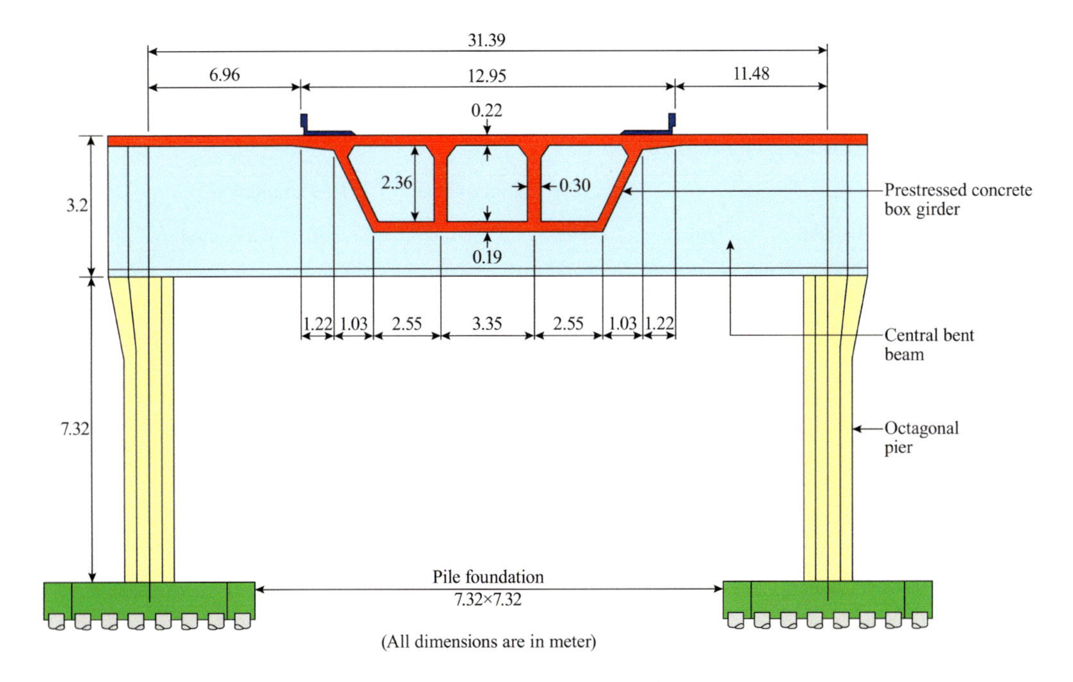

FIGURE 10.49 Cross-sectional view of the 91/5 highway overcrossing.

TABLE 10.7
Details of the Benchmark Highway Bridge

Bridge Property		Description/Value of Property
Type of bridge		Continuous, cast-in-situ prestressed concrete (PC) box girder bridge.
Number of spans		Two
Length of each span (m)		58.52
Number of abutments		Two skewed at 33°
Width of deck (m)	Along east span	13.95
	Along west span	15
Length and height of PC outrigger supporting deck (m)		31.4 and 6.9
Number of pile groups supporting PC outrigger		Two friction pile groups
Number of piles in each group		49
Depth of the bent beam (m)		3.2
Height of octagonal columns (m)		6.9
Total mass of the bridge (kg)		4,237,544
Total mass of the central bent (kg)		959,140
Total mass of piers (kg)		236,114
Natural time period (s)	Transverse direction	0.4
	Longitudinal direction	0.813
Type of isolator		Lead rubber bearing (N-Z system)
Number of N-Z bearings		Four N-Z bearings on each abutment and one N-Z bearing at each bent column located at the center
Yield displacement of lead plugs (m)		0.015
Pre-yield shear stiffness of bearings (N/m)		4.8×10^6
Post-yield shear stiffness of bearings (N/m)		0.6×10^6

TABLE 10.8
Details of the 1994 Northridge, 1995 Kobe, and 1999 Chi-Chi Earthquake Ground Motions

		EW Component		NS Component	
Earthquake	**Type**	**PGA (g)**	**PGV (m/s)**	**PGA (g)**	**PGV (m/s)**
Northridge, 1994	Near-fault	0.838	1.661	0.472	0.73
Kobe, 1995	Near-fault	0.509	0.373	0.503	0.366
Chi-Chi, 1999	Near-fault	1.157	1.147	0.417	0.456

Source: Talyan et al. (2021).
Note: PGA = Peak ground acceleration, PGV = Peak ground velocity

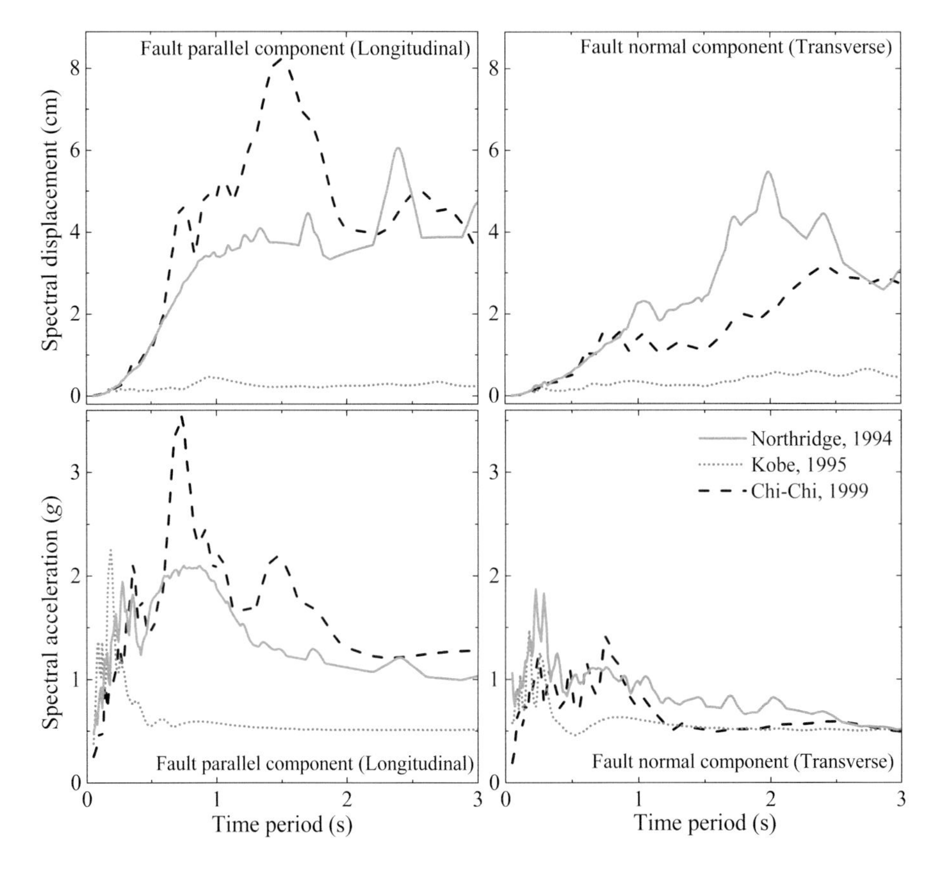

FIGURE 10.50 Response spectra for the 1994 Northridge, 1995 Kobe, and 1999 Chi-Chi earthquake ground motions.

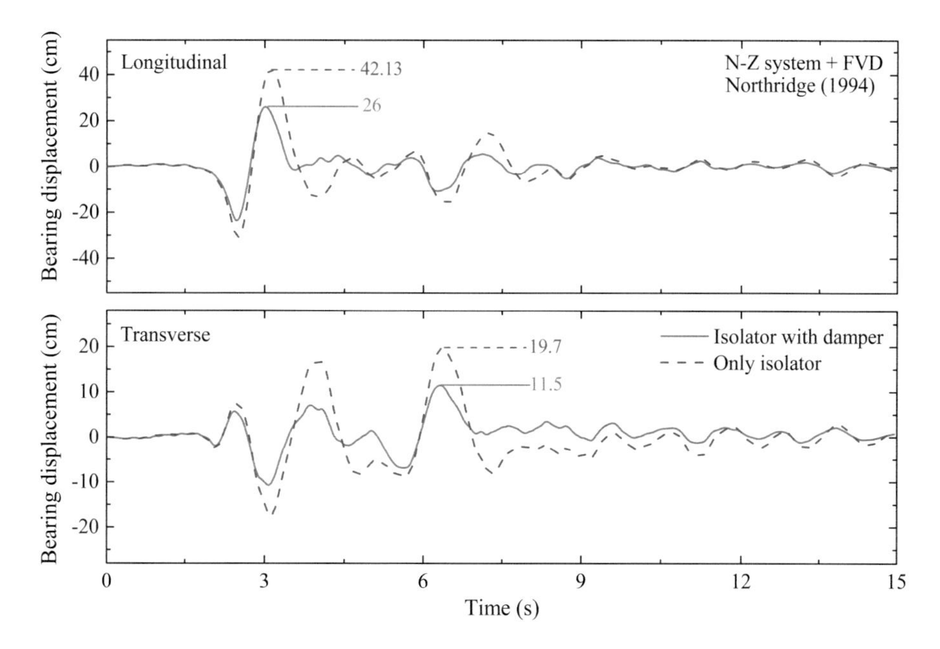

FIGURE 10.51 Comparison of the time histories of the bearing displacement for the: (a) uncontrolled bridge, (b) bridge isolated by N-Z system, and (c) bridge equipped with N-Z system and FVD under the 1994 Northridge earthquake ground motion.

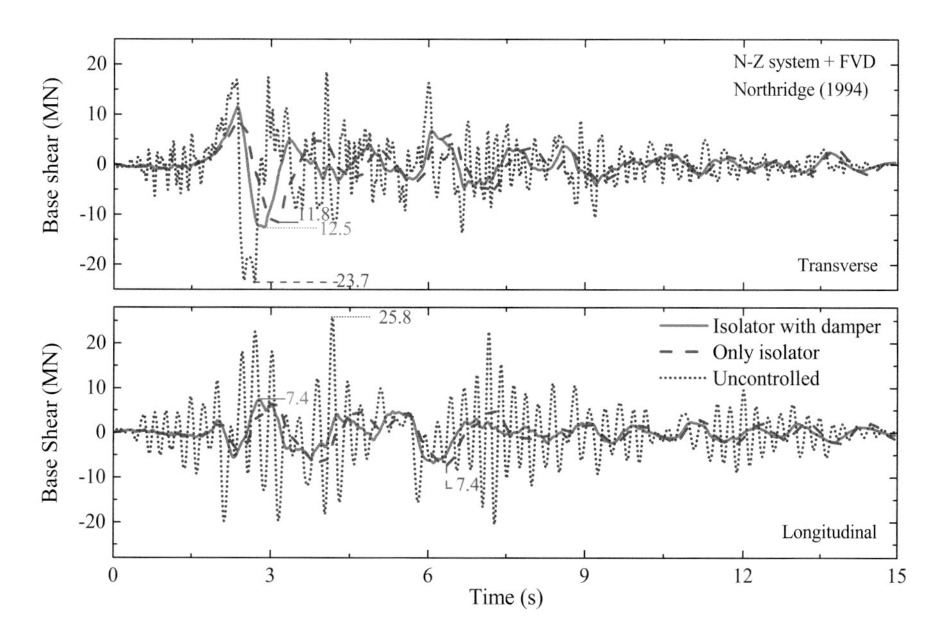

FIGURE 10.52 Comparison of the time histories of the base shear for the: (a) uncontrolled bridge, (b) bridge isolated by N-Z system, and (c) bridge equipped with N-Z system and FVD under the 1994 Northridge earthquake ground motion.

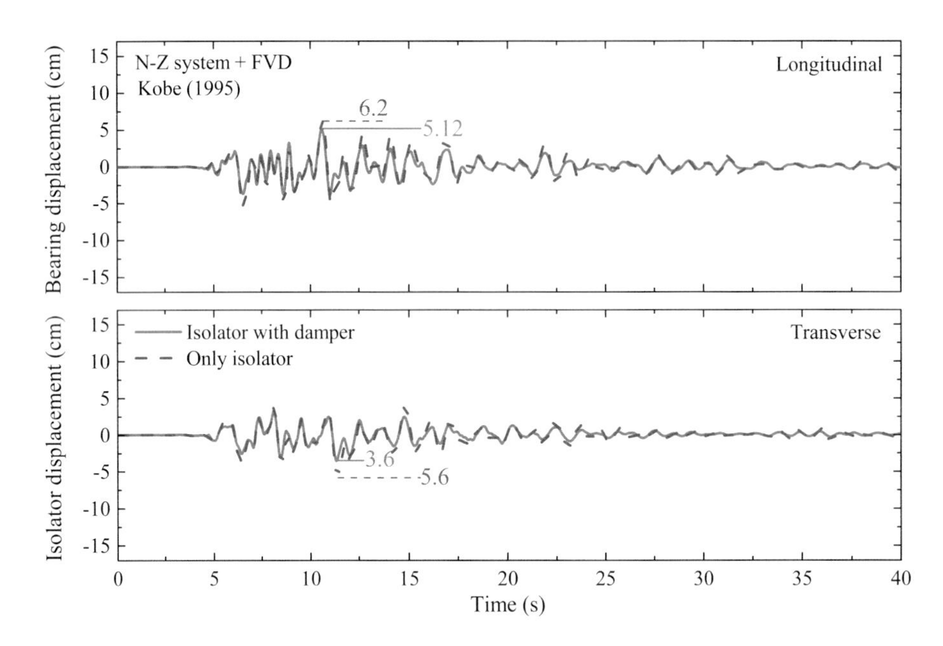

FIGURE 10.53 Comparison of the time histories of the bearing displacement for the: (a) uncontrolled bridge, (b) bridge isolated by N-Z system, and (c) bridge equipped with N-Z system and FVD under the 1995 Kobe earthquake ground motion.

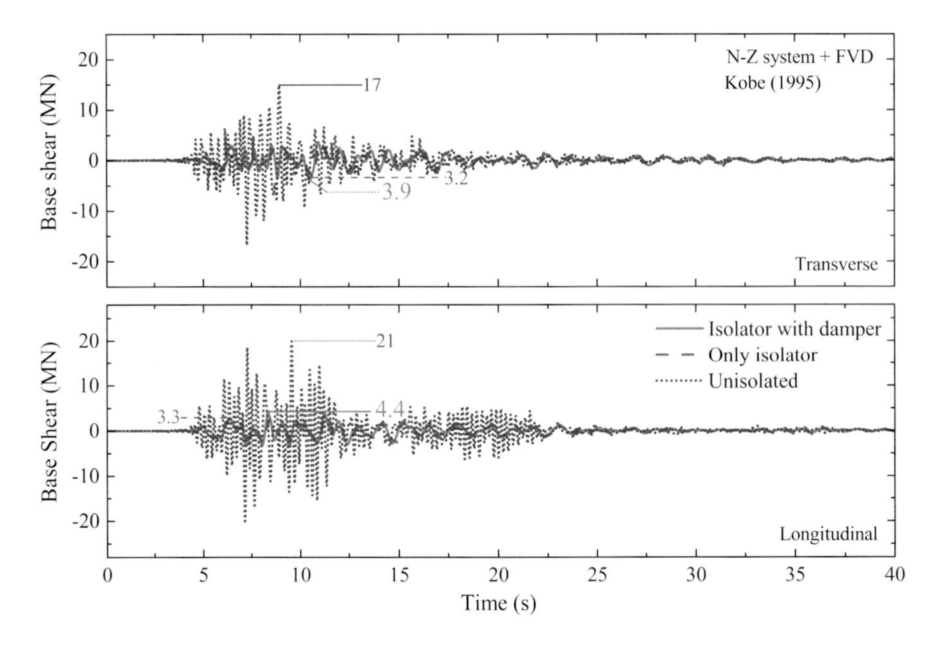

FIGURE 10.54 Comparison of the time histories of the base shear for the: (a) uncontrolled bridge, (b) bridge isolated by N-Z system, and (c) bridge equipped with N-Z system and FVD under the 1995 Kobe earthquake ground motion.

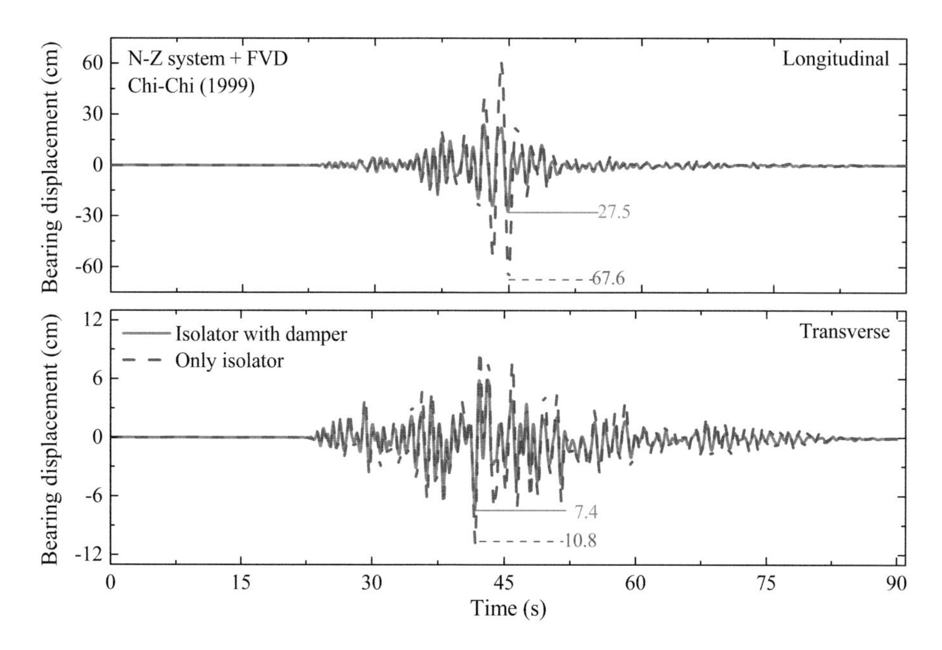

FIGURE 10.55 Comparison of the time histories of the bearing displacement for the: (a) uncontrolled bridge, (b) bridge isolated by N-Z system, and (c) bridge equipped with N-Z system and FVD under the 1999 Chi-Chi earthquake ground motion.

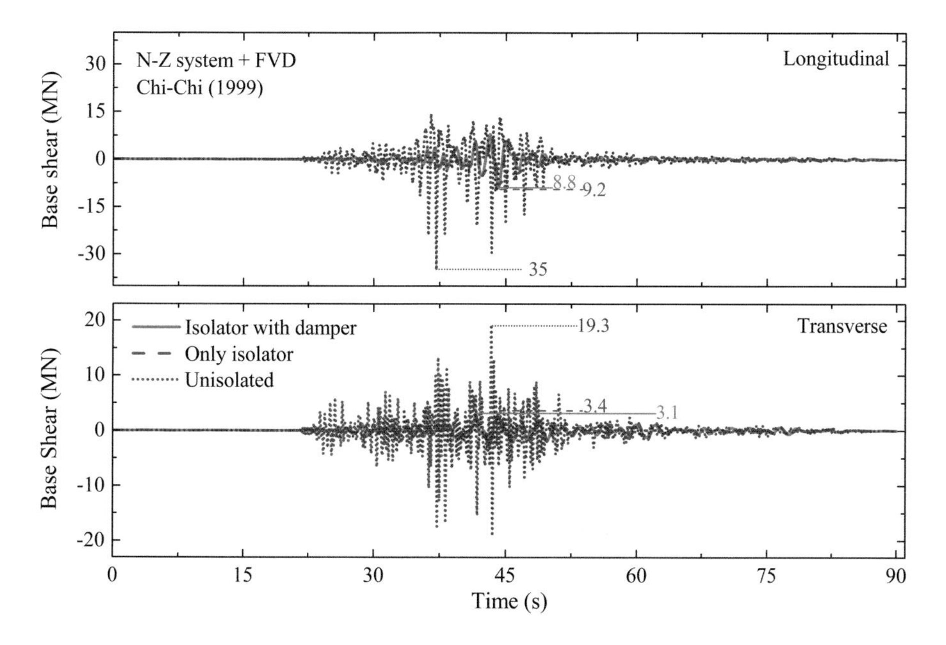

FIGURE 10.56 Comparison of the time histories of the base shear for the: (a) uncontrolled bridge, (b) bridge isolated by N-Z system, and (c) bridge equipped with N-Z system and FVD under the 1999 Chi-Chi earthquake ground motion.

REFERENCES

Agrawal, A. K., Tan, P., Nagarajaiah, S., and Zhang, J. (2006). Benchmark structural control problem for a seismically excited highway bridge. In Structures Congress 2006.

Kunde, M. C., and Jangid, R. S. (2006). Effects of pier and deck flexibility on the seismic response of isolated bridges. *Journal of Bridge Engineering*, 11(1), 109–121.

Talyan, N., Elias, S., and Matsagar, V. (2021). Earthquake response control of isolated bridges using supplementary passive dampers. *Practice Periodical on Structural Design and Construction*, 26(2), 04021002.

Zelleke, D. H., Elias, S., Matsagar, V. A., and Jain, A. K. (2015). Supplemental dampers in base-isolated buildings to mitigate large isolator displacement under earthquake excitations. Bulletin of the New Zealand Society for Earthquake Engineering, 48(2), 100–117.

11 Introduction to Benchmark Control Problems

11.1 INTRODUCTION

Structural response control is a subject of wide interest among structural engineers internationally. It was not possible, in general, to compare different control strategies directly because the controls were applied to different structures and under different loading conditions. This problem was addressed by NASA in their space structures program. The solution they adopted was to build a "test bed" structure installed with a number of sensors and actuators. Various control strategies could then be implemented and objective comparisons were made. International Association for Structural Control (IASC) implemented the "test bed" in software, where the software contained an accurate description of the structure together with the dynamics of sensors and actuators. By applying different control strategies to the "test bed" structure and measuring the response to a standardized set of earthquake or wind excitations, direct comparison of the effectiveness of the control strategies could be made. Also, the researchers have developed software and generously made their software available for researchers worldwide to implement their favorite control strategy on the "test bed" problem. Benchmark structural control problems can be used to evaluate the relative effectiveness and applicability of structural control devices and algorithms and to provide an analytical "test bed" for evaluation of various control design issues. Simulation programs in MATLAB® have been developed and are made available to facilitate comparison of the efficiency and merits of various control strategies. The benchmark problems are made as realistic as possible. SIMULINK models are developed to simulate the features and limitations of the structural control problems.

11.2 SEISMICALLY EXCITED 8-STORY BASE-ISOLATED BENCHMARK BUILDING

A benchmark problem on the seismic response control has been developed and presented as a test-bed structure for response control of seismically excited 8-story base-isolated building (Narasimhan et al., 2006). The benchmark base-isolated existing 8-story building is located in Los Angeles, California, United States. Passive control strategies have been developed for its seismic response control and the building is evaluated with a broad set of carefully chosen parameters and performance measures, under a suite of seven historic earthquake ground motions. Salient features of the building are presented in Table 11.1. Furthermore, Figure 11.1 shows the schematic diagram of the base-isolated building with its different components. Figure 11.2 shows the three-dimensional (3D) view of the building. Figure 11.3 displays the floor plan of the building, showing location of isolators.

11.2.1 Model of 8-Story Benchmark Building

The RC floor slabs are supported by metal decking and a grid of steel beams. Steel bracing is provided in the superstructure in the external frames located around the building perimeter. The RC base slab supports the steel superstructure, which is integral with RC beams and drop panels below the location of each column. The details of structural model are presented in Table 11.2.

The time period of the building for the first eight modes is presented in Table 11.3, for N-S and E-W directions.

DOI: 10.1201/9781315269269-11

TABLE 11.1
Salient Features of the 8-Story Base-Isolated Benchmark Building

Building Feature	Detail
Type of building	Base-isolated RC + steel building Irregular in plan and in elevation
Number of stories	8
Total length of building (m)	82.40
Total width of building (m)	54.30
Shape of building	L-shape – up to 6th story Rectangular shape – for 7th and 8th story
Total weight of the superstructure (kN)	20,200

TABLE 11.2
Details of Structural Model of the 8-Story Benchmark Building

Structural Model Feature	Detail/Value
Modeling of the building	3D linear elastic system
Type of floor slab and base mat	Rigid in its own plane
Degrees of freedom (DOF) of superstructure	24 (3 at each story)
Degrees of freedom (DOF) of isolation system	3
Number of modes of vibration considered	24
Superstructure damping ratio (%)	5
Number of isolators provided (LRB + FPS)	92

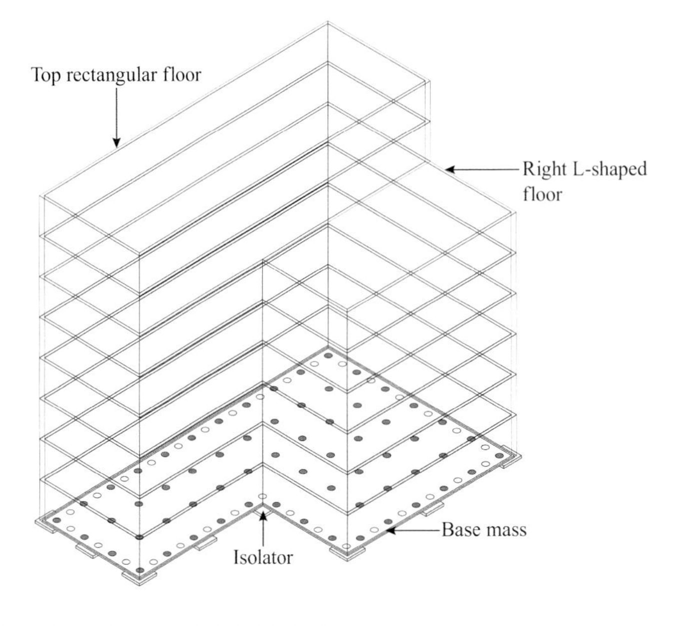

FIGURE 11.1 Schematic of the base-isolated building.

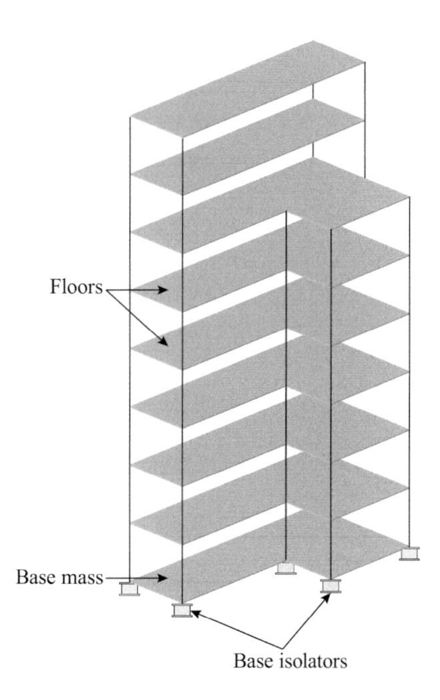

FIGURE 11.2 Three-dimensional view of the base-isolated building. (Sharma and Jangid, 2010)

Figure 11.4 shows the connection of isolation bearing with the foundation. Dampers are provided along with isolators, in order to control the excessive lateral displacement at the isolator level. One end of the damper is connected to the ground, and the other end to the base mat.

11.2.2 Governing Equations of Motion

A MATLAB-based nonlinear structural analysis tool is developed for vibration control simulation of the seismically excited building. This tool is written as S-function and is incorporated into

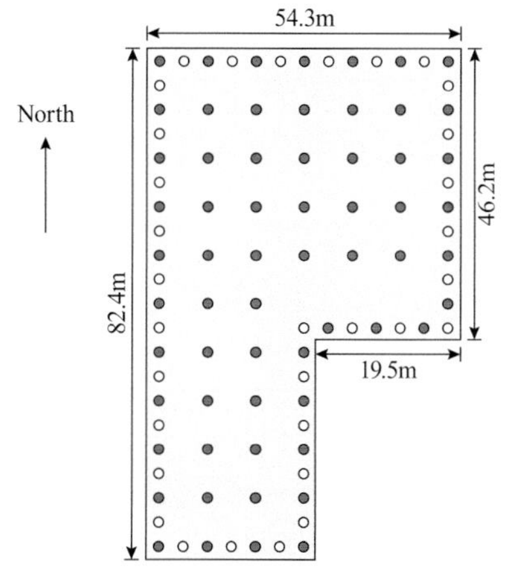

FIGURE 11.3 Floor plan of the base-isolated building. (Narasimhan et al., 2006)

TABLE 11.3
Natural Period of Building in First Eight Modes

Mode	North-South	East-West
1	0.78	0.89
2	0.27	0.28
3	0.15	0.15
4	0.11	0.11
5	0.08	0.08
6	0.07	0.07
7	0.06	0.06
8	0.05	0.06

the SIMULINK model. The S-function solves the incremental dynamic equations of motion using Newmark's method. The inputs to the S-function block in SIMULINK model are the seismic excitation and the control forces provided by the control devices. The nonlinear analysis tool remains the same for any control strategy employed.

Figure 11.5 shows a schematic of MATLAB/SIMULINK implementation for the building. The same block diagram can be used for any type of controller. The nonlinear dynamic behavior of isolators is modeled using the discrete bilinear Bouc-Wen model. The forces in the isolators are transferred to the center of the mass of the rigid base slab. Thus, the isolation bearings can be modeled individually or globally as lumped equivalent elements at the center of mass of the rigid base mat. The governing equations of motion of the superstructure are expressed as below.

$$[M]\{\ddot{u}\}+[C]\{\dot{u}\}+[K]\{u\}=-[M]\{r\}\left(\{\ddot{u}_g\}+\{\ddot{u}_b\}\right) \tag{11.1}$$

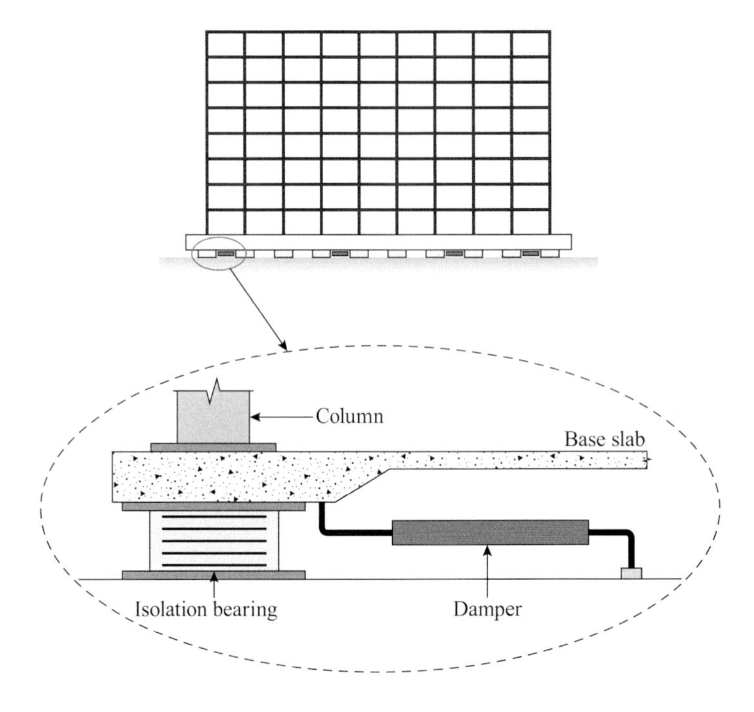

FIGURE 11.4 Isolators used in the base-isolated building. (Narasimhan et al., 2006.)

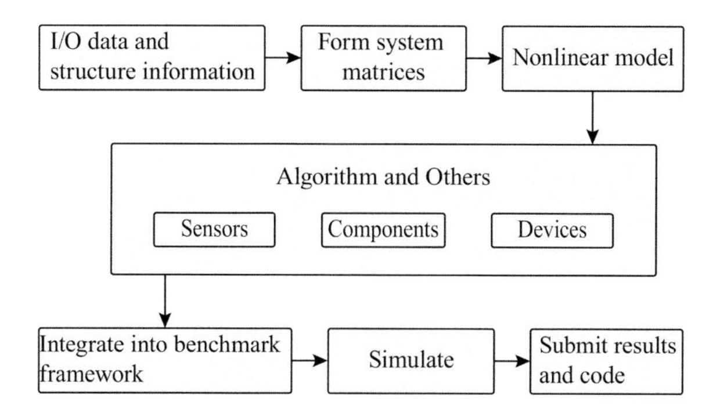

FIGURE 11.5 Schematic of MATLAB/SIMULINK implementation.

where $[M]$, $[C]$, and $[K]$, respectively, are the lumped mass, damping, and stiffness matrices of size 24 × 24, of the fixed-base building, considering 24 degrees of freedom (DOF). $\{r\}$ is the vector of influence coefficients of size 24 × 3; $\{\ddot{u}\},\{\dot{u}\},\text{and}\{u\}$ are the vectors of acceleration, velocity, and displacement response of the structure of size 24 × 1; $\{\ddot{u}_g\}$ and $\{\ddot{u}_b\}$ are the vectors of ground acceleration and base acceleration, respectively, of size 3 × 1.

Further, the governing equation of motion for the base mass is written as follows,

$$\{r\}^T[M]\{\ddot{u}\}+[C_b]\{\dot{u}_b\}+[K_b]\{u_b\}+\{f_b\}=-\left[\{r\}^T[M]\{r\}+[M_b]\right]\left\{\{\ddot{u}_g\}+\{\ddot{u}_b\}\right\} \quad (11.2)$$

where $[M_b],[C_b],\text{and}[K_b]$ are the diagonal mass, resultant damping, and the resultant stiffness matrices of the isolation bearings, $\{f_b\}$ is the vector of size 3 × 1, containing the nonlinear bearing forces; and $\{r\}^T$ represents the transpose of the influence coefficient matrix $\{r\}$. The combined governing equations of motion of the entire structure can be written as,

$$\begin{bmatrix} [M] & [M]\{r\} \\ \{r\}^T[M] & 0 \end{bmatrix}\begin{Bmatrix} \{\ddot{u}\} \\ 0 \end{Bmatrix}+\begin{bmatrix} [C] & 0 \\ 0 & [C_b] \end{bmatrix}\begin{Bmatrix} \{\dot{u}\} \\ \{\dot{u}_b\} \end{Bmatrix}+\begin{bmatrix} [K] & 0 \\ 0 & [K_b] \end{bmatrix}\begin{Bmatrix} \{u\} \\ \{u_b\} \end{Bmatrix}$$
$$+\begin{Bmatrix} 0 \\ \{f_b\} \end{Bmatrix}=-\begin{bmatrix} [M]\{r\} \\ \{r\}^T[M]\{r\}+[M_b] \end{bmatrix}\left\{\{\ddot{u}_g\}+\{\ddot{u}_b\}\right\} \quad (11.3)$$

11.2.3 Evaluation Criteria

To facilitate direct and systematic comparison of various control strategies, a set of eight normalized evaluation criteria is defined in the benchmark problem definition paper. The evaluation criteria are the dimensionless structural seismic response quantities, which consider the ability of the employed controller to reduce the peak and normalized response quantities over the entire time record, and the control requirements. Smaller values of evaluation criteria indicate superior performance. The merit of a controller is based on the criteria, expressed in terms of both the peak response quantities and the root mean square (RMS). The RMS displacement provides a measure of the required physical size of the device. The RMS velocity provides a measure of the control power required, and RMS absolute acceleration is a measure of the magnitude of the control forces generated. For each controller design, eight evaluation criteria are found for the defined seven earthquake records. For base-isolated benchmark building subjected to earthquake excitation, critical response values are related to the structural integrity of the building. The evaluation criteria are peak base shear force

(J_1), peak story shear (J_2), peak base displacement (J_3), peak story drift (J_4), peak absolute acceleration (J_5), peak compound isolator force at the center of base mat (J_6), RMS base displacement (J_7), and RMS absolute acceleration (J_8). The evaluation criteria are normalized by their corresponding fixed-base values.

11.2.4 Peak Response

The first six evaluation criteria, (J_1) to (J_6), are defined to measure the ability of the controller to reduce the peak response quantities of the base-isolated benchmark building.

1. Peak base shear force in the controlled structure, normalized by the corresponding shear in the uncontrolled structure

$$J_1(q) = \frac{\max_t \left\| V_0(t,q) \right\|}{\max_t \left\| V_0(t,q) \right\|}$$

2. Peak structure shear at the first story level in the controlled structure, normalized by the corresponding shear in the uncontrolled structure

$$J_2(q) = \frac{\max_t \left\| V_1(t,q) \right\|}{\max_t \left\| V_1(t,q) \right\|}$$

3. Peak base displacement or isolator deformation in the controlled structure, normalized by the corresponding displacement in the uncontrolled structure

$$J_3(q) = \frac{\max_{t,i} \left\| d_i(t,q) \right\|}{\max_{t,i} \left\| d_i(t,q) \right\|}$$

4. Peak inter-story drift in the controlled structure, normalized by the inter-story drift in the uncontrolled structure

$$J_4(q) = \frac{\max_{t,f} \left\| d_f(t,q) \right\|}{\max_{t,f} \left\| d_f(t,q) \right\|}$$

5. Peak absolute floor acceleration in the controlled structure, normalized by the corresponding acceleration in the uncontrolled structure

$$J_5(q) = \frac{\max_{t,f} \left\| a_f(t,q) \right\|}{\max_{t,f} \left\| \hat{a}_f(t,q) \right\|}$$

6. Peak force generated by all control devices normalized by the peak base shear in the uncontrolled structure

$$J_6(q) = \frac{\max_t \left\| \sum_k F_k(t,q) \right\|}{\max_t \left\| V_0(t,q) \right\|}$$

11.2.5 RMS Response

The second set of two evaluation criteria is based on RMS responses, over the entire simulation time of the earthquake.

7. RMS base displacement in the controlled structure normalized by the corresponding RMS base displacement in the uncontrolled structure

$$J_7(q) = \frac{\max_i \left\| \sigma_d(t,q) \right\|}{\max_t \left\| \sigma_{\hat{d}}(t,q) \right\|}$$

8. RMS absolute floor acceleration in the controlled structure normalized by the corresponding RMS base acceleration in the uncontrolled structure

$$J_8(q) = \frac{\max_f \left\| \sigma_a(t,q) \right\|}{\max_f \left\| \sigma_{\hat{a}}(t,q) \right\|}$$

In the evaluation criteria, $i = 1, \ldots, N_8$ represent the story number; $k = 1, \ldots, N_d$ is the device number, $f = 1, \ldots, N_f$; $q = 1, \ldots, 7$ is the earthquake number, $t = 0 \leq t \geq Tq$ is the time, $\langle \cdot \rangle$ is the inner product, and $\|\cdot\|$ is the vector magnitude incorporating N-S and E-W components. The evaluation criteria are summarized in Table 11.4.

The uncontrolled (fixed-base) values of the response quantities for the seven earthquakes are presented in Table 11.5. Fixed-base values of structural response quantities like base displacement, RMS base displacement, and isolation force do not exist. Therefore, for normalization, their values are taken as 1 for obtaining the performance indices.

TABLE 11.4
Summary of Evaluation Criteria

Peak Response	**Peak Response**
Peak base shear	Peak structure shear at the first story
$J_1(q) = \frac{\max_t \left\| V_0(t,q) \right\|}{\max_t \left\| V_0(t,q) \right\|}$	$J_2(q) = \frac{\max_t \left\| V_1(t,q) \right\|}{\max_t \left\| V_1(t,q) \right\|}$
Peak base displacement	Peak inter-story drift
$J_3(q) = \frac{\max_{t,i} \left\| d_i(t,q) \right\|}{\max_{t,i} \left\| d_i(t,q) \right\|}$	$J_4(q) = \frac{\max_{t,f} \left\| d_f(t,q) \right\|}{\max_{t,f} \left\| d_f(t,q) \right\|}$
Peak absolute floor acceleration	Peak force generated
$J_5(q) = \frac{\max_{t,f} \left\| a_f(t,q) \right\|}{\max_{t,f} \left\| \hat{a}_f(t,q) \right\|}$	$J_6(q) = \frac{\max_t \left\| \sum_k F_k(t,q) \right\|}{\max_t \left\| V_0(t,q) \right\|}$
RMS Responses	
RMS base displacement	RMS absolute floor acceleration
$J_7(q) = \frac{\max_i \left\| \sigma_d(t,q) \right\|}{\max_t \left\| \sigma_{\hat{d}}(t,q) \right\|}$	$J_8(q) = \frac{\max_f \left\| \sigma_a(t,q) \right\|}{\max_f \left\| \sigma_{\hat{a}}(t,q) \right\|}$

TABLE 11.5
Peak Structural Response Quantities of Fixed-Base Benchmark Building

Earthquake	Base Shear (kN)	Story Shear (kN)	Story Drift (m)	Abs. Acceleration (m/s^2)	RMS Abs Acceleration
Newhall	200,100	197,010	0.0658	23.9789	3.5437
Sylmar	199,130	169,270	0.0455	17.8316	2.9744
El-Centro	74,156	67,371	0.0247	7.8285	2.2349
Rinaldi	264,940	251,710	0.0913	27.5217	5.6909
Kobe	270,200	268,000	0.0865	28.1471	5.1098
Jiji	139,740	131,330	0.0372	11.7303	2.7056
Erzincan	135,480	134,720	0.0396	18.6368	2.914

11.2.6 Earthquake Ground Excitations

For passive control design strategy, the following seven earthquake records are used to estimate the evaluation criteria.

1. Northridge, 1994, Earthquake recorded at Newhall Fire Station
2. Northridge, 1994, Earthquake recorded at Sylmar Station
3. Northridge, 1994, Earthquake recorded at Rinaldi Station
4. Imperial Valley, 1940, Earthquake
5. Kobe, 1995, Earthquake
6. Jiji, 1999, Earthquake
7. Erzincan, 1992, Earthquake

The particulars of the seven earthquakes are given in Table 11.6. The two horizontal components of the earthquakes are designated as fault-normal (F-N) and fault-parallel (F-P). The earthquakes are applied bidirectionally to the benchmark building in horizontal plane; where the F-N and F-P components are applied in x and y directions, respectively.

11.2.7 Numerical Study

The control of seismic energy transmitted from ground to the building is one of the most effective techniques of seismic response control. In recent years, sliding-type isolators are widely used to add damping and to increase the flexibility of the structure. The most effective sliding isolator is friction pendulum system (FPS), due to its ease of installation and simple mechanism of restoring force by gravity action. FPS, variable friction pendulum isolator (VFPI), variable FPS (VFPS), and variable curvature FPS (VCFPS) are employed for response evaluation of the 8-story base-isolated benchmark building. The optimum parameters of the control devices are presented in Table 11.7.

Recently, hybrid isolation systems are being used to mitigate the seismic response of the structural systems. Example of such strategies is the use of hybrid control system which includes a combination of two passive devices or one passive and one active or semi-active device. It is proved that the hybrid systems perform better than the traditional isolation systems. In the study of the 8-story benchmark building, 31 laminated rubber bearings (LRB) are used along with 61 sliding isolators, that is, FPS, VFPI, VFPS, and VCFPS. Table 11.8 presents the maximum evaluation criteria for the seven selected earthquakes, when 92 sliding isolators are installed. Table 11.9 presents the maximum evaluation criteria for the hybrid isolation systems,

TABLE 11.6
Particulars of Earthquake Records

Earthquake	Station	Magnitude (M_w)	Time Step (s)	Designation	Direction	PGA (g)	PGV (cm/s)	PGD (cm)
Northridge (1994)	Newhall, Fire station	6.7	0.001	NORTHR/NWH 360	F-N	0.590	97.2	38.05
				NORTHR/NWH 90	F-P	0.583	75.5	17.57
Northridge (1994)	Sylmar – Olive View Med FF	6.7	0.001	NORTHR/SYL 360	F-N	0.843	139.6	32.68
				NORTHR/SYL 90	F-P	0.604	78.2	16.05
Northridge (1994)	Rinaldi (77), Receiving Station	6.7	0.001	NORTHR/RRS 228	F-N	0.838	166.1	28.78
				NORTHR/RRS 318	F-P	0.472	73.0	19.76
Imperial Valley (1940)	117 El Centro Array #9	7.0	0.001	IMPVALL/I-ELC180	F-N	0.313	29.8	11.32
				IMPVALL/I-ELC270	F-P	0.215	30.2	23.91
Kobe (1995)	0 KJMA	6.9	0.001	KOBE/KJM000	F-N	0.841	81.3	17.68
				KOBE/KJM090	F-P	0.631	74.3	19.95
Jiji (1999)	TCU068	7.6	0.001	CHICHI/TCU068-N	F-N	0.462	263.1	430.0
				CHICHI/TCU068-W	F-P	0.566	176.6	324.11
Erzincan (1992)	95 Erzincan	6.9	0.001	ERZIKAN/ERZ-N-S	F-N	0.5153	83.9	27.35
				ERZIKAN/ERZ-E-W	F-P	0.4955	64.3	22.78

Note: F-N = Fault-normal; F-P = Fault-parallel.

i.e., considering 31 LRBs and 61 sliding isolators. The location of LRB and sliding isolators is as shown in Figure 11.3.

11.3 SEISMICALLY EXCITED 20-STORY STEEL BENCHMARK BUILDING

The 20-story steel benchmark building was designed for the Los Angeles, California, region and proposed as a benchmark problem by Spencer et al. (1998). The goal of this study was to provide a clear basis to evaluate the efficacy of various structural control strategies. American Society of Civil

TABLE 11.7
Optimum Parameters of Control Devices

Control Device	Parameter	Value
Friction pendulum system (FPS)	μ	0.09
	T_b (s)	3.5
Variable friction pendulum system (VFPS)	μ_{max}	0.15
	μ_{min}	0.03
	fvf	1
	T_b (s)	3
Variable friction pendulum isolator (VFPI)	μ	0.09
	T_b (s)	3
Variable curvature friction pendulum system (VCFPS)	μ	0.07
	T_b (s)	1.5
Resilient-friction bearing isolator (R-FBI)	μ	0.09
	T_b (s)	2

Note: μ_{max} and μ_{min} are the maximum and minimum values of coefficient of friction, T_b is the base isolation period, and fvf is the frequency variation factor.

TABLE 11.8
Maximum Evaluation Criteria for the Seven Selected Earthquakes (with 92 Sliding Isolators)

Evaluation Criteria	Isolator	Newhall	Sylmar	El Centro	Rinaldi	Kobe	Jiji	Erzincan
J_1	FPS	0.22335	0.34907	0.38669	0.24894	0.15527	0.62692	0.48659
	VFPS	0.30217	0.33975	0.39717	0.26137	0.20817	0.55898	0.44449
	VFPI	0.19009	0.21831	0.3839	0.16057	0.13382	0.3657	0.35251
	VCFPS	0.21142	0.27768	0.38573	0.19574	0.14441	0.47830	0.43459
J_2	FPS	0.18933	0.35871	0.45887	0.21436	0.13560	0.56338	0.41803
	VFPS	0.27360	0.3965	0.41179	0.23530	0.18256	0.49152	0.38867
	VFPI	0.20634	0.23734	0.45766	0.14177	0.13722	0.32575	0.28086
	VCFPS	0.19852	0.28686	0.45846	0.17213	0.13276	0.44934	0.36045
J_3	FPS	0.15074	0.34939	0.03396	0.33404	0.14625	0.5186	0.34628
	VFPS	0.13296	0.2607	0.04317	0.30907	0.13138	0.35596	0.26367
	VFPI	0.16753	0.35755	0.03523	0.29331	0.14876	1.28048	0.39030
	VCFPS	0.15610	0.36130	0.03434	0.30744	0.14198	0.65174	0.35496
J_4	FPS	0.28147	0.39192	0.58575	0.16447	0.19985	0.499	0.38892
	VFPS	0.34415	0.51481	0.42897	0.19430	0.21808	0.54379	0.36619
	VFPI	0.29589	0.30245	0.58570	0.13624	0.20181	0.35883	0.32400
	VCFPS	0.28763	0.32854	0.58574	0.13954	0.20079	0.41038	0.34366
J_5	FPS	0.34363	0.54301	0.76546	0.25932	0.26065	0.41338	0.34181
	VFPS	0.48973	0.62218	0.63138	0.35603	0.37166	0.53049	0.44606
	VFPI	0.35068	0.56649	0.76555	0.25985	0.26302	0.51877	0.26313
	VCFPS	0.34439	0.55669	0.76548	0.2484	0.26195	0.40562	0.31031
J_6	FPS	0.2192	0.34089	0.14427	0.30175	0.20566	0.39557	0.3009
	VFPS	0.29605	0.33319	0.14157	0.33906	0.27181	0.35183	0.27764
	VFPI	0.18716	0.20991	0.14282	0.20484	0.17236	0.22499	0.21426
	VCFPS	0.2077	0.27103	0.14381	0.25094	0.19179	0.29391	0.26111
J_7	FPS	0.03275	0.06962	0.02419	0.06137	0.03024	0.13991	0.09305
	VFPS	0.03555	0.07633	0.02838	0.0618	0.02759	0.08874	0.06396
	VFPI	0.03616	0.06939	0.02398	0.07102	0.03289	0.46214	0.08118
	VCFPS	0.03337	0.06692	0.02413	0.06696	0.03022	0.14593	0.09593
J_8	FPS	0.36085	0.37658	0.60256	0.21909	0.27892	0.50697	0.3857
	VFPS	0.42784	0.49817	0.5483	0.26252	0.30836	0.50821	0.37383
	VFPI	0.3605	0.3916	0.60319	0.21544	0.27602	0.41469	0.35282
	VCFPS	0.36052	0.37799	0.60278	0.21074	0.27784	0.43963	0.35214

Engineers (ASCE) designed nonlinear benchmark problems considering 3-, 9-, and 20-story steel buildings as the benchmark structures. This section presents the 20-story steel benchmark building.

11.3.1 Model of the Seismically Excited 20-Story Steel Benchmark Building

The 20-story building used for the benchmark study was designed by Brandow and Johnston Associates. Although not actually constructed, this structure meets the seismic code specifications and represents a typical high-rise building designed for the Los Angeles, California region. The building provides a wider basis for the comparison of results by different structural control strategies. A detailed description of the 20-story benchmark building is presented in Table 11.10.

TABLE 11.9
Maximum Evaluation Criteria for the Seven Selected Earthquakes (31 LRB + 61 Sliding Isolators)

Evaluation Criteria	Isolator (LRB + Sliding Isolator)	Newhall	Sylmar	El Centro	Rinaldi	Kobe	Jiji	Erzincan
J_1	FPS	0.17697	0.26677	0.21891	0.17943	0.10746	0.70095	0.37896
	VFPS	0.25094	0.31742	0.33669	0.23506	0.17635	0.56747	0.43909
	VFPI	0.18519	0.23842	0.28359	0.1718	0.11013	0.47548	0.36836
	VCFPS	0.1608	0.20151	0.21538	0.15311	0.09699	0.47186	0.30161
J_2	FPS	0.15779	0.27008	0.22137	0.16476	0.09249	0.63333	0.3235
	VFPS	0.22674	0.33841	0.33548	0.21437	0.15723	0.49377	0.3704
	VFPI	0.19107	0.24135	0.34722	0.15442	0.1024	0.43027	0.3013
	VCFPS	0.14292	0.20002	0.22059	0.14117	0.08354	0.42252	0.25356
J_3	FPS	0.30319	0.52875	0.0945	0.43592	0.25697	0.97273	0.52677
	VFPS	0.18909	0.3828	0.0532	0.32966	0.15784	0.5386	0.29885
	VFPI	0.21305	0.44531	0.05281	0.34675	0.16541	0.92388	0.3872
	VCFPS	0.3131	0.50783	0.09514	0.4485	0.25497	1.23898	0.51329
J_4	FPS	0.18796	0.26639	0.27654	0.13262	0.08433	0.56377	0.29743
	VFPS	0.26764	0.41959	0.32359	0.16008	0.16206	0.43568	0.3477
	VFPI	0.25906	0.24142	0.46604	0.13242	0.14341	0.38652	0.32516
	VCFPS	0.17784	0.20679	0.2765	0.11019	0.08141	0.35592	0.24688
J_5	FPS	0.14678	0.177	0.37879	0.13972	0.10985	0.43208	0.15443
	VFPS	0.27371	0.47991	0.46463	0.28113	0.23478	0.37694	0.22521
	VFPI	0.26461	0.40455	0.63037	0.17583	0.19656	0.32476	0.21141
	VCFPS	0.13998	0.15508	0.37883	0.11086	0.1113	0.27018	0.13347
J_6	FPS	0.17194	0.25437	0.07992	0.23036	0.14603	0.44542	0.23814
	VFPS	0.24872	0.31017	0.13338	0.30385	0.22277	0.35699	0.26811
	VFPI	0.17456	0.23376	0.10615	0.21874	0.1451	0.3022	0.22623
	VCFPS	0.15654	0.19974	0.07863	0.19632	0.13137	0.30075	0.18638
J_7	FPS	0.07152	0.14872	0.02011	0.09106	0.04718	0.29251	0.16734
	VFPS	0.03982	0.07291	0.01809	0.0638	0.03183	0.16016	0.07858
	VFPI	0.04452	0.0967	0.02174	0.06866	0.03477	0.20326	0.106
	VCFPS	0.07724	0.13985	0.0201	0.09066	0.0475	0.2816	0.15341
J_8	FPS	0.19732	0.28028	0.31192	0.11476	0.13917	0.52035	0.29058
	VFPS	0.35556	0.33973	0.39647	0.17905	0.23733	0.51524	0.32136
	VFPI	0.29151	0.30946	0.46783	0.15737	0.21117	0.37773	0.29013
	VCFPS	0.19104	0.22638	0.31157	0.10761	0.13746	0.34297	0.24102

The column lines employ three-tier construction, i.e., monolithic column pieces are connected at every three levels beginning with the first level. Column splices, which are seismic (tension) splices to carry bending moment and uplift forces, are located on the 1st, 4th, 7th, 10th, 13th, 16th, and 18th levels at 1.83 m above the centerline of the beam-column joint. The column bases are modeled as pinned and are secured to the ground at the B-2 level. RC foundation walls and surrounding soil are assumed to restrain the structure at the ground level from horizontal displacement. The floor system comprises 248-MPa steel wide-flange beams acting compositely with the floor slab. Each frame resists 50% of the seismic mass associated with the entire structure. The seismic mass of the structure is due to the steel framing, floor slabs, ceiling/flooring, mechanical/electrical, partitions, roofing, and a penthouse located on the roof. The plan of the 20-story N-S moment-resisting frame is depicted in Figure 11.6(a) and the elevation in Figure 11.6(b).

TABLE 11.10
Particulars of the Seismically Excited 20-Story Steel Benchmark Building

Building Feature			Detail
Type of building			The lateral load-resisting system comprised of steel perimeter of moment-resisting frames
Number of basement levels			2
Height of levels in building w.r.t. ground floor		B-1 level	Level directly below the ground level
		B-2 level	Level below B-1
Floor-to-floor height for the two basement levels (m)			3.65
Height for the ground level (m)			5.49
Typical floor-to-floor height (for analysis purpose measured from center-of-beam to center-of-beam) (m)			3.96
Plan dimensions (m)			30.48 × 36.58
Total height of the building (m)			80.77
The interior bays of the structure contain simple framing with composite floors			6.10 m c/c in both directions Number of bays in the N-S direction = 5 Number of bays in the E-W direction = 6
Seismic weight of the building (kN)	Both N-S moment-resisting frames at the ground level		5,217
	For the 1st level		5,521
	For the 2nd level to 19th level		5,413
	For the 20th level		5,727
The seismic weight the entire building above ground level (kN)			108,900

11.3.2 Evaluation Model for 20-Story Benchmark Building

The study focuses on an in-plane (2-D) analysis of the 20-story benchmark building. The frames considered in the development of the evaluation models are the N-S moment-resisting frames in the short (weak) direction of the building. Structural member nonlinearities are included in the in-plane finite element model to capture the inelastic behavior of the building during strong earthquakes. The beams and columns are modeled as plane-frame elements, and the mass and stiffness matrices for each element are formulated. A bilinear hysteresis model is used to characterize the nonlinear bending stiffness of the structural members. The damping matrix is formulated based on an assumption of Rayleigh damping. Total 138 nodes are located at the beam-column joints. Elements are created between the nodes to represent beams and columns in the structure. The beam members extend from c/c of columns, thus ignoring the column panel zone. The inertial loads are uniformly distributed at the nodes of each respective level assuming a lumped mass formulation. The building frame contains column splices. The column joint of the splice story is located 1.83 m above the centerline of the beam. For simplicity, the spliced columns are modeled as having uniform properties over the story height equal to the weighted average of the upper and lower column properties of that story. There is no node modeled at the splice. Each node has 3 DOFs; horizontal, vertical, and rotational. The DOFs for the structure are 414, i.e., (3 × 138), prior to the application of boundary conditions/constraints. Each element, modeled as a plane frame element, contains two nodes and six DOFs. For each element, the length, area, moment of inertia, modulus of elasticity, and mass density are predefined. The elemental lumped mass and stiffness matrices are functions of these properties. Global mass and stiffness matrices are assembled from the elemental mass and stiffness matrices by

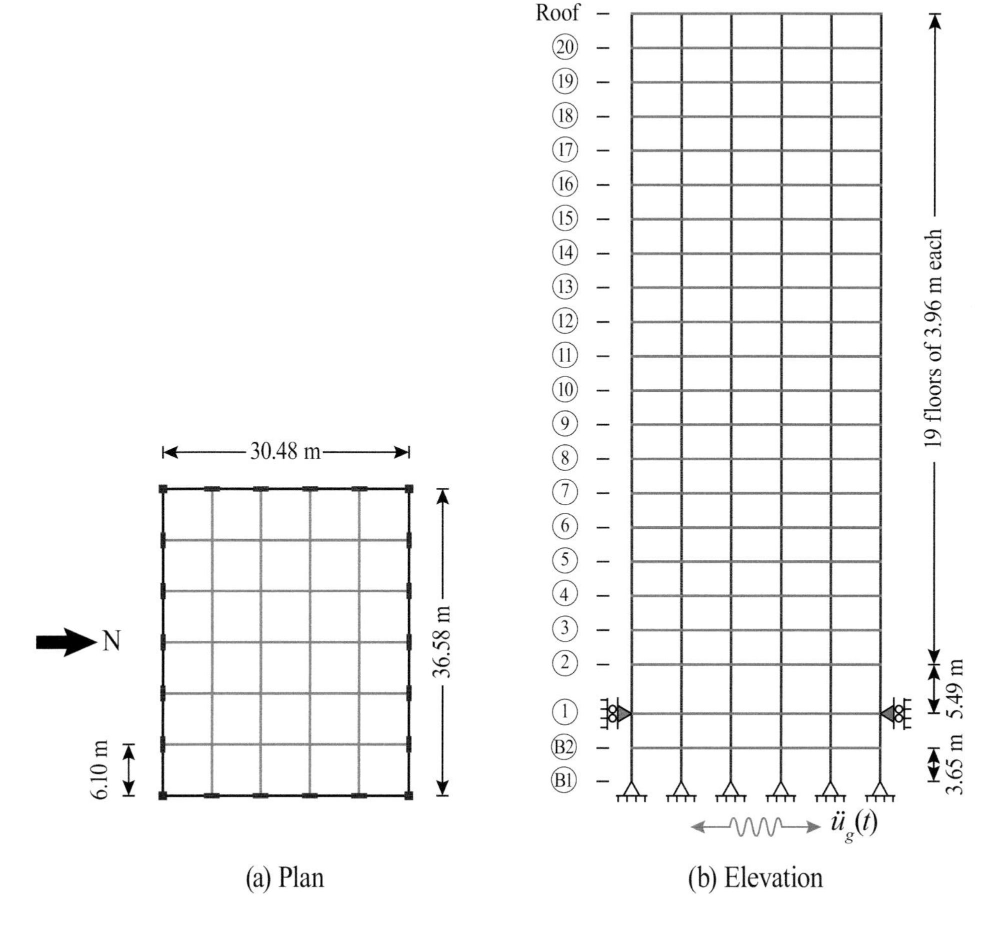

FIGURE 11.6 The 20-story N-S moment-resisting frame benchmark building: (a) plan and (b) elevation. (Spencer et al., 1998.)

summing the mass and stiffness associated with each DOF for each element of the entire structure. Rotational inertia is ignored; thus, rotational mass is assigned a very small value. The DOFs corresponding to fixed boundary conditions are then constrained by eliminating the rows and columns associated with these DOFs from the global mass and stiffness matrices. The first ten modal natural frequencies of the 20-story benchmark evaluation model are 0.261, 0.753, 1.30, 1.83, 2.40, 2.44, 2.92, 3.01, 3.63, and 3.68 Hz.

11.3.3 Governing Equations of Motion

During large seismic events, structural members can yield, resulting in nonlinear response behavior that may be significantly different than a linear approximation. For better representation of the nonlinear behavior, a bilinear hysteresis model, as shown in Figure 11.7, is used to model the plastic hinges, and the points of yielding of the structural members of the 20-story building. The plastic hinges are assumed to occur at the moment-resisting connections.

Newmark's time-step integration scheme is effectively implemented in MATLAB to determine the time domain response of the building. The incremental dynamic equations of motion for the nonlinear structural system take the following form.

$$[M]\{\ddot{u}\}+[C]\{\dot{u}\}+[K]\{u\}=-[M]\{r\}\ddot{u}_g+\{P\}\{f\} \tag{11.4}$$

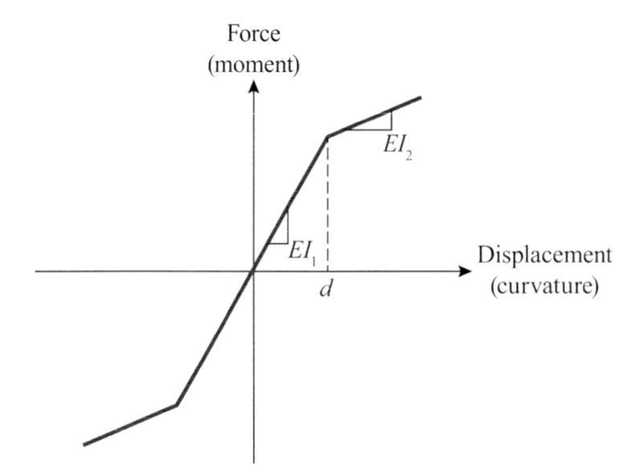

FIGURE 11.7 Bilinear hysteresis model.

where $[M]$,$[C]$, and $[K]$ are the mass, damping, and stiffness matrices of the finite element model, each of size 106 × 106; $\{u\}$ is the vector of floor displacement relative to ground; $\{\dot{u}\}$ and $\{\ddot{u}\}$ are the floor velocity and acceleration vectors. $\{f\} = [f_{d1}, f_{d2}, \ldots f_{dr}]^T$ is the vector of damper forces input to the structure and $\ddot{u}_g$ is the ground acceleration. Further, $\{r\}$ is the vector of zeros and ones defining loading of the ground acceleration to the structure, and $\{P\}$ is the vector defining how the forces produced by the control devices are transmitted to the structure. The unbalanced force is the difference between the restoring force evaluated using the hysteresis model and the restoring force assuming constant linear stiffness at any time during the time interval (t) to ($t + \Delta t$). This unbalanced force is handled at the next time step; i.e., the unbalanced force is added as an external force at the next time step. The floor slab is assumed to be rigid in the horizontal plane. Therefore, the nodes associated with each story level undergo the same horizontal displacement. Therefore, the dependent (slave) horizontal DOFs on each floor slab can be expressed in terms of a single active horizontal DOF. The nonlinear evaluation model requires a MATLAB program, implemented as a SIMULINK system function(S-function), to evaluate the nonlinear response of the model. This S-function performs the nonlinear dynamic analysis using Newmark's method.

11.3.4 Earthquake Ground Motions

In order to evaluate the proposed control strategies, two far-fault and two near-fault historical records are selected by a committee of ASCE working on structural control.

1. El Centro: The N-S component recorded at the Imperial Valley Irrigation District substation in El Centro, California, during the Imperial Valley, California earthquake of May 18, 1940.
2. Hachinohe: The N-S component recorded at Hachinohe City during the Tokachi-Oki earthquake of May 16, 1968.
3. Northridge: The N-S component recorded at Sylmar County Hospital parking lot in Sylmar, California, during the Northridge, California earthquake of January 17, 1994.
4. Kobe: The N-S component recorded at the Kobe Japanese Meteorological Agency (JMA) station during the Hyogo-ken Nanbu earthquake of January 17, 1995.

Additionally, this benchmark problem study considers various levels of each of the earthquake records, including 0.5, 1.0, and 1.5 times the magnitude of El Centro and Hachinohe earthquake ground motions; and 0.5 and 1.0 times the magnitude of Northridge and Kobe earthquake ground motions. Thus, a total of 10 earthquake records are considered in the evaluation of each control strategy. A summary of the earthquakes is provided in Table 11.11.

TABLE 11.11
Particulars of Earthquake Ground Motions

	Detail/Value of Parameter for Earthquake			
Parameter	**El Centro**	**Hachinohe**	**Northridge**	**Kobe**
Date of occurrence	May 18, 1940	May 16, 1968	January 17, 1994	January 17, 1995
Recording station	Imperial Valley Irrigation District substation, California	Hachinohe City station	Sylmar County Hospital parking lot	JMA (Japan Meteorological Agency) station
Magnitude	7.1	8.3	6.9	7.2
Distance from epicenter (km)	11.58	26	19	17
PGA (m/s^2)	3.414	2.453	8.270	8.182
PGA (g)	0.348	0.250	0.843	0.834
Duration (s)	50	36	60	150
Type	Major	Great	Strong	Major

Note: "Strong" refers to earthquake magnitude 6–6.9, "Major" refers to magnitude 7–7.9, and "Great" refers to magnitude 8 or more (based on US Geological Survey data).

11.3.5 Evaluation Criteria

The evaluation criteria are divided into four categories; viz., building responses, building damage, control devices, and control strategy requirements. The first three categories have both peak- and normed-based criteria. Small values of the evaluation criteria are more desirable as they indicate maximum response reduction.

1. The first category of the evaluation criteria is related to the building responses. The first three criteria are based on peak inter-story drift ratio (J_1), level acceleration (J_2), and base shear (J_3).

$$J_1 = \max_{\substack{El\,Centro\\Hachinohe\\Northgridge\\Kobe}} \left\{ \frac{\max\limits_{t,i} \dfrac{|d_i(t)|}{h_i}}{\delta^{\max}} \right\}$$

$$J_2 = \max_{\substack{El\,Centro\\Hachinohe\\Northgridge\\Kobe}} \left\{ \frac{\max\limits_{t,i} |\ddot{u}_{ai}(t)|}{\ddot{u}_a^{\max}} \right\}$$

$$J_3 = \max_{\substack{El\,Centro\\Hachinohe\\Northgridge\\Kobe}} \left\{ \frac{\max\limits_{t} \left| \sum\limits_i m_i \ddot{u}_{ai}(t) \right|}{F_b^{\max}} \right\}.$$

The criteria are defined over the range $i = [1, 20]$ where i is floor number. The next three criteria are based on normed building responses. The inter-story drift (J_4), level acceleration (J_5), and base shear (J_6) are defined in their normed based forms as,

$$J_4 = \max_{\substack{El\ Centro \\ Hachinohe \\ Northgridge \\ Kobe}} \left\{ \frac{\max\limits_{t,i} \dfrac{\|d_i(t)\|}{h_i}}{\|\delta^{\max}\|} \right\}$$

$$J_5 = \max_{\substack{El\ Centro \\ Hachinohe \\ Northgridge \\ Kobe}} \left\{ \frac{\max\limits_{t,i} \|\ddot{u}_{ai}(t)\|}{\|\ddot{u}_a^{\max}\|} \right\}$$

$$J_6 = \max_{6\substack{El\ Centro \\ Hachinohe \\ Northgridge \\ Kobe}} \left\{ \frac{\max\limits_{t} \left\| \sum\limits_{i} m_i \ddot{u}_{ai}(t) \right\|}{\|F_b^{\max}\|} \right\}.$$

The norm $\|\cdot\|$ is computed using the following equation,

$$\|\cdot\| = \sqrt{\frac{1}{t_f} \int_0^{t_f} [\cdot]^2\, dt}$$

where t_f is a sufficiently large time to allow the response of the structure to attenuate. In this benchmark study, the duration of 100 s is adopted for the El Centro, Hachinohe, and Northridge earthquakes; and 180 s for the Kobe earthquake.

2. The second category of the evaluation criteria assesses the building damage. These criteria have been added because of the nonlinear considerations of the benchmark study. Both ends of each element are considered in these criteria to assess the yielding. The 7th and 8th evaluation criteria are based on peak responses while the 9th and 10th are normed-based criteria. The evaluation criteria for the ductility factor (J_7) and dissipated energy of the curvatures at the end of members (J_8) are defined as

$$J_7 = \max_{\substack{El\ Centro \\ Hachinohe \\ Northgridge \\ Kobe}} \left\{ \frac{\max\limits_{t,j} \dfrac{|\varphi_j(t)|}{\varphi_{yj}}}{\varphi^{\max}} \right\}.$$

$$J_8 = \max_{\substack{El\ Centro \\ Hachinohe \\ Northgridge \\ Kobe}} \left\{ \frac{\max\limits_{t,j} \dfrac{\int dE_j}{F_{yj}, \varphi_{yj}}}{E^{\max}} \right\}.$$

The 9th evaluation criterion (J_9) is the ratio of the plastic connections sustained by the structure when controlled and uncontrolled and is given by

$$J_9 = \max_{\substack{El\,Centro\\Hachinohe\\Northgridge\\Kobe}} \left\{ \frac{N_d^c}{N_d} \right\}$$

The evaluation criteria (J_8) and (J_9) are meaningful only when the structure undergoes plastic deformations and, therefore, are not defined when the uncontrolled structure remains elastic. The 10th evaluation criterion (J_{10}) is the normed ductility factor, given by

$$J_{10} = \max_{\substack{El\,Centro\\Hachinohe\\Northgridge\\Kobe}} \left\{ \frac{\max_j \dfrac{\|\varphi_j(t)\|}{\varphi_{yj}}}{\|\varphi^{\max}\|} \right\}.$$

The curvature $\varphi_j(t)$, normed curvature $\left(\|\varphi_j(t)\|\right)$, dissipated energy $\left(\int dE_j\right)$, and information to determine the number of plastic hinges $\left(N_d^c\right)$ are determined within the nonlinear S-function during the simulation and are saved to a specified data file.

3. The third category of evaluation criteria is related to the control devices. This category assesses the required performance of the devices. Peak criteria (J_{11}), (J_{12}), and (J_{13}) indicate the control force, control device stroke, and power used for control.

$$J_{11} = \max_{\substack{El\,Centro\\Hachinohe\\Northgridge\\Kobe}} \left\{ \frac{\max_{t,\,1} |f_l(t)|}{W} \right\}$$

$$J_{12} = \max_{\substack{El\,Centro\\Hachinohe\\Northgridge\\Kobe}} \left\{ \frac{\max_{t,\,l} |y_l^a(t)|}{u^{\max}} \right\}$$

$$J_{13} = \max_{\substack{El\,Centro\\Hachinohe\\Northgridge\\Kobe}} \left\{ \frac{\max_{t,} \sum P_l(t)}{\dot{u}^{\max} W} \right\}$$

The 14th evaluation criterion (J_{14}) is a measure of the total power required for the control of the structure and is defined as

$$J_{14} = \max_{\substack{El\,Centro\\Hachinohe\\Northgridge\\Kobe}} \left\{ \frac{\sum \frac{1}{t_f} \int_0^{t_f} P_l(t)}{\dot{u}^{\max} W} \right\}$$

When passive control devices are installed, $J_{14} = 0$.

TABLE 11.12
Summary of Evaluation Criteria

Inter-story drift ratio	Level acceleration	Base shear
$J_1 = \max\limits_{\substack{El\,Centro\\Hachinohe\\Northridge\\Kobe}} \left\{ \dfrac{\max\limits_{t,i} \dfrac{\lvert d_i(t)\rvert}{h_i}}{\delta^{\max}} \right\}$	$J_2 = \max\limits_{\substack{El\,Centro\\Hachinohe\\Northridge\\Kobe}} \left\{ \dfrac{\max\limits_{t,i} \lvert \ddot{u}_{ai}(t)\rvert}{\ddot{u}_a^{\max}} \right\}$	$J_3 = \max\limits_{\substack{El\,Centro\\Hachinohe\\Northridge\\Kobe}} \left\{ \dfrac{\max\limits_{t} \left\lvert \sum_i m_i \ddot{u}_{ai}(t)\right\rvert}{F_b^{\max}} \right\}$
Normed inter-story drift ratio	Normed level acceleration	Normed base shear
$J_4 = \max\limits_{\substack{El\,Centro\\Hachinohe\\Northridge\\Kobe}} \left\{ \dfrac{\max\limits_{t,i} \dfrac{\lVert d_i(t)\rVert}{h_i}}{\lVert\delta^{\max}\rVert} \right\}$	$J_5 = \max\limits_{\substack{El\,Centro\\Hachinohe\\Northridge\\Kobe}} \left\{ \dfrac{\max\limits_{t,i} \lVert \ddot{u}_{ai}(t)\rVert}{\lVert\ddot{u}_a^{\max}\rVert} \right\}$	$J_6 = \max\limits_{\substack{6El\,Centro\\Hachinohe\\Northridge\\Kobe}} \left\{ \dfrac{\max\limits_{t} \left\lVert \sum_i m_i \ddot{u}_{ai}(t)\right\rVert}{\lVert F_b^{\max}\rVert} \right\}$
Ductility	Dissipated energy	Plastic connections
$J_7 = \max\limits_{\substack{El\,Centro\\Hachinohe\\Northridge\\Kobe}} \left\{ \dfrac{\max\limits_{t,j} \dfrac{\lvert \varphi_j(t)\rvert}{\varphi_{yj}}}{\varphi^{\max}} \right\}$	$J_8 = \max\limits_{\substack{El\,Centro\\Hachinohe\\Northridge\\Kobe}} \left\{ \dfrac{\max\limits_{t,j} \dfrac{\int dE_j}{F_{yj}, \varphi_{yj}}}{E^{\max}} \right\}.$	$J_9 = \max\limits_{\substack{El\,Centro\\Hachinohe\\Northridge\\Kobe}} \left\{ \dfrac{N_d^c}{N_d} \right\}$
Normed ductility	Control force	Control device stroke
$J_{10} = \max\limits_{\substack{El\,Centro\\Hachinohe\\Northridge\\Kobe}} \left\{ \dfrac{\max\limits_{j} \dfrac{\lVert \varphi_j(t)\rVert}{\varphi_{yj}}}{\varphi^{\max}} \right\}$	$J_{11} = \max\limits_{\substack{El\,Centro\\Hachinohe\\Northridge\\Kobe}} \left\{ \dfrac{\max\limits_{t,1} \lvert f_l(t)\rvert}{W} \right\}$	$J_{12} = \max\limits_{\substack{El\,Centro\\Hachinohe\\Northridge\\Kobe}} \left\{ \dfrac{\max\limits_{t,l} \lvert y_l^a(t)\rvert}{u^{\max}} \right\}$
Control power	Normed control power	J_{15} Number of control devices
$J_{13} = \max\limits_{\substack{El\,Centro\\Hachinohe\\Northridge\\Kobe}} \left\{ \dfrac{\max\limits_{t,} \sum P_l(t)}{\dot{u}^{\max} W} \right\}$	$J_{14} = \max\limits_{\substack{El\,Centro\\Hachinohe\\Northridge\\Kobe}} \left\{ \dfrac{\sum \dfrac{1}{t_f} \int_0^{t_f} P_l(t)}{\dot{u}^{\max} W} \right\}$	J_{16} Number of required sensors J_{17} Number of computational resources

4. The fourth category for evaluating the structural performance is related to the control strategy employed. The 15th (J_{15}), 16th (J_{16}), and 17th (J_{17}) evaluation criteria are J_{15} = number of control devices, J_{16} = number of required sensors, and J_{17} = number of computational resources. A summary of the evaluation criteria is presented in Table 11.12.

The response quantities for the uncontrolled 20-story nonlinear benchmark building subjected to four earthquakes are presented in Table 11.13.

11.3.6 Numerical Study

For seismic response evaluation of the 20-story benchmark building, passive control devices; viz., (i) shape memory alloy (SMA), (ii) viscous damping wall (VDW), (iii) friction damper (FD), and (iv) distributed tuned mass damper (d-TMD) are employed. Currently, hybrid isolation systems are

TABLE 11.13
Uncontrolled Response Quantities for the 20-Story Nonlinear Benchmark Building

Earthquake	El Centro			Hachinohe			Northridge		Kobe	
Intensity	0.5	1	1.5	0.5	1	1.5	0.5	1	0.5	1
δ^{max}	0.004	0.007	0.011	0.003	0.006	0.009	0.011	0.019	0.011	0.020
$\ddot{x}_a^{max}$ (m/s²)	2.670	5.339	7.786	1.829	3.658	4.715	6.008	8.511	7.494	9.974
F_b^{max} (N)	3.69E + 06	7.38E + 06	9.53E + 06	3.18E + 06	6.36E + 06	9.13E + 06	1.04E + 07	1.43E + 07	1.01E + 07	1.14E + 07
$\lVert\delta^{max}\rVert$	0.0007	0.0013	0.0020	0.0006	0.0013	0.0018	0.0019	0.0067	0.0015	0.0074
$\lVert\ddot{x}_a^{max}\rVert$(m/s²)	0.389	0.777	1.139	0.276	0.552	0.804	0.803	1.014	0.759	1.003
$\lVert F_b^{max}\rVert$(N)	7.64E + 05	1.53E + 06	2.27E + 06	7.06E + 05	1.41E + 06	2.08E + 06	2.15E + 06	2.39E + 06	1.63E + 06	2.03E + 06
φ^{max}	0.466	0.933	1.503	0.428	0.856	1.397	1.675	3.360	1.923	3.199
E^{max}	0.001	0.004	19.466	0.000	0.002	6.092	20.606	86.226	18.701	282.690
N_d	0	0	86	0	0	86	96	192	78	168
$\lVert\varphi^{max}\rVert$	0.099	0.197	0.331	0.094	0.187	0.278	0.363	1.406	0.242	1.423
x^{max} (m)	0.152	0.304	0.454	0.174	0.348	0.492	0.489	0.750	0.314	0.517
$\dot{x}^{max}$ (m/s)	0.458	0.916	1.301	0.451	0.902	1.268	1.662	2.039	1.315	1.846

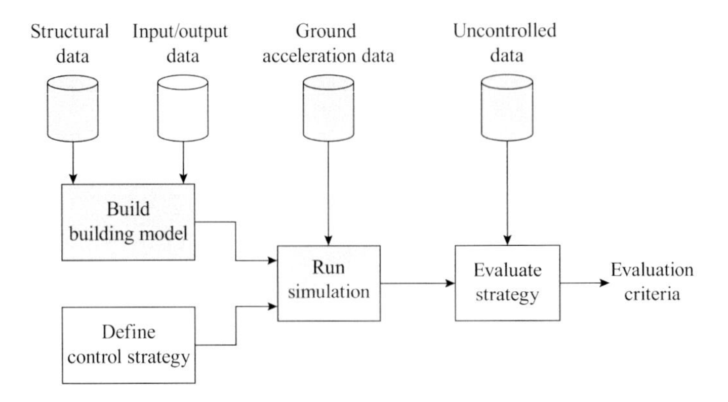

FIGURE 11.8 Flow diagram of the benchmark control problem.

being used to mitigate the seismic response of the structural system, e.g., combination of two passive devices or one passive device and one active or semi-active device. It is seen that the hybrid systems perform better than the traditional isolation systems.

11.3.6.1 Application of SMA Dampers

The SIMULINK models for the 20-story building with SMA damper are presented in Figures 11.8–11.13. Figure 11.8 shows the flow diagram of the benchmark control problem solution.

Figure 11.9 portrays the main SIMULINK model, showing the steps involved in finding out the response of control devices, when the building is subjected to earthquake ground motions. Further, Figure 11.10 shows the three subsystems used in the main model shown in Figure 11.9, whereas Figures 11.11–11.13 show the SIMULINK models of the three subsystems.

Six different configurations for SMA damper placement are selected as shown in Figures 11.14 and 11.15. These configurations are selected according to heuristic approach, in which various combinations of damper placement are studied for reduction of responses. Main criteria for placing

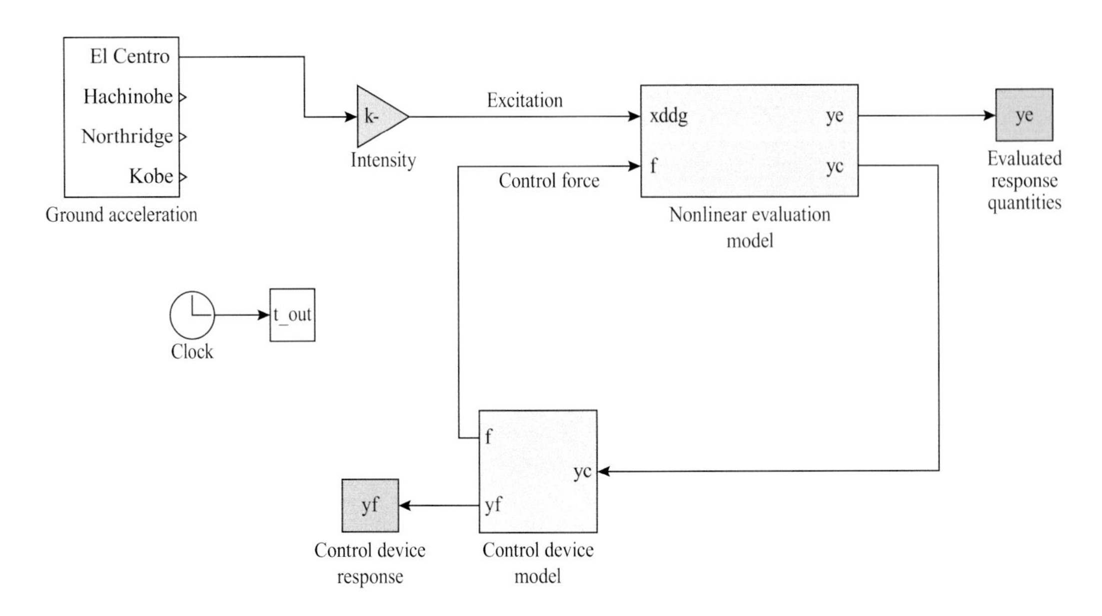

FIGURE 11.9 SIMULINK model for 20-story nonlinear benchmark building.

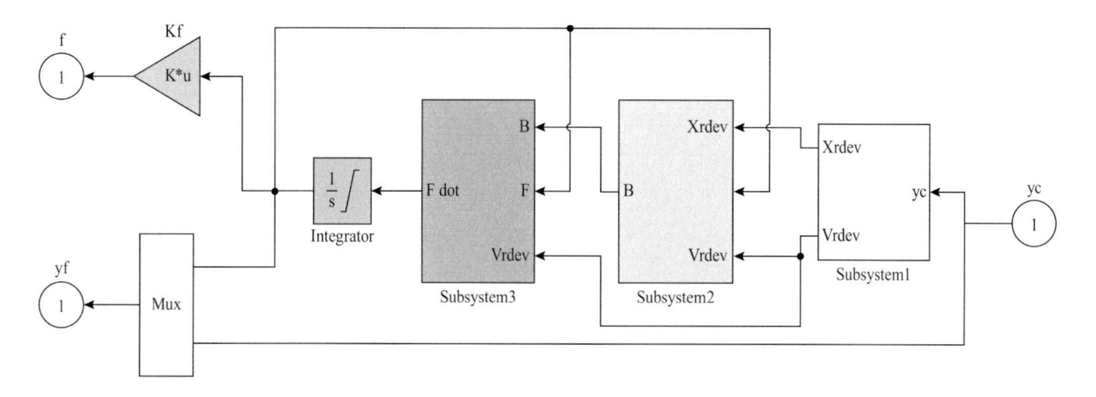

FIGURE 11.10 SIMULINK model for SMA control device.

the dampers are to reduce overall responses for maximum number of considered earthquakes. Evaluation criteria for the six cases of damper placement, shown in Figures 11.14 and 11.15, are presented in Tables 11.14–11.19.

Table 11.20 presents the maximum values of the evaluation criteria for the six cases of placement of SMA damper.

Figure 11.16 represents the top five members with the highest curvature of case 6 for Kobe 1.0 earthquake; the most severe near-fault earthquake among all earthquakes considered.

11.3.6.2 Application of Viscous Damping Walls

VDWs are placed on each floor of the 20-story benchmark building, to evaluate the performance criteria. At the ground level, four walls are installed, and two damping walls are installed on each of the first and second levels. Properties of the VDWs used in the study are presented in Table 11.21.

The evaluation criteria, after the installation of VDWs, are presented in Table 11.22.

11.3.6.3 Application of Friction Dampers

FDs are usually classified as one of the displacement-dependent energy dissipation devices and the damper force is independent of the velocity and frequency content of excitation. Raut and Jangid (2014) studied the effectiveness of the FD device (FDD) in order to mitigate the seismic

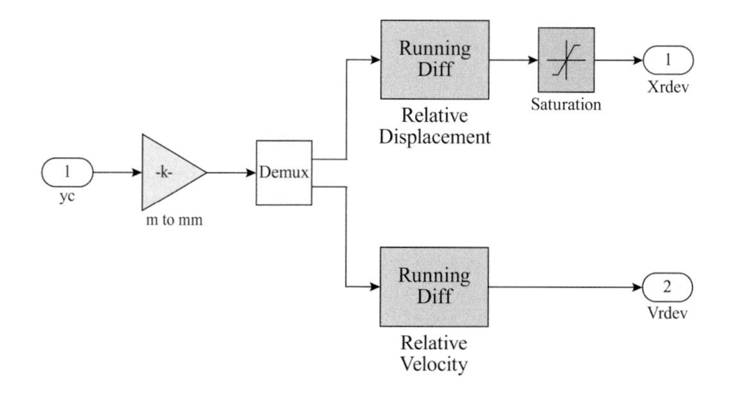

FIGURE 11.11 SIMULINK model for subsystem 1.

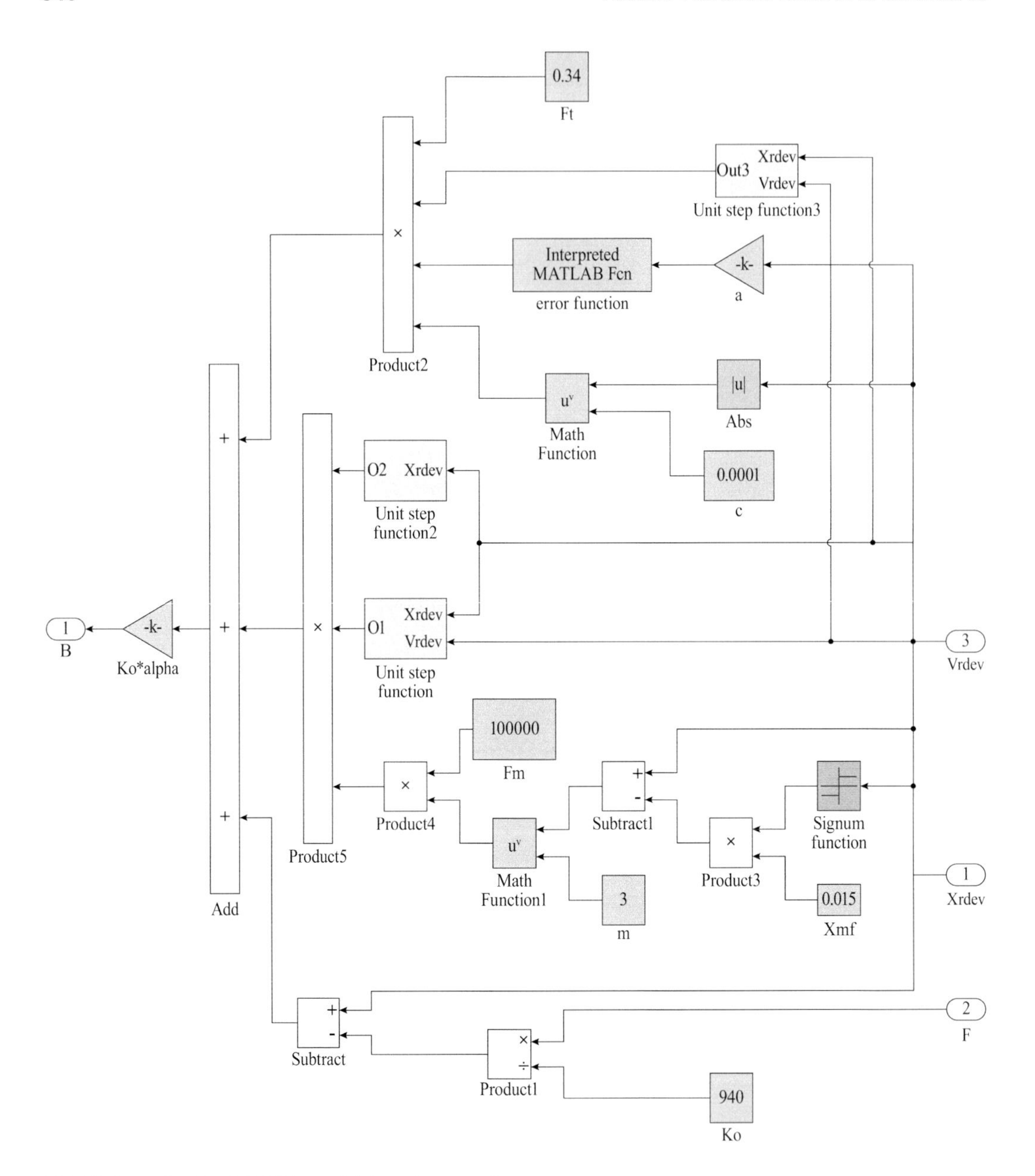

FIGURE 11.12 SIMULINK model for subsystem 2.

response of the 20-story benchmark building. Figure 11.17 shows the arrangement of the FDD installed in the benchmark building. Structural model of the building is shown according to the number of dampers employed and their location in the structure. Chevron bracing system is used to fix the FDs.

Slip force depicts the performance of the FD. Higher the slip force, inferior will be the performance of the damper because in this case, the FD will act as a brace, which reduces its effectiveness. However, if the slip force is low, the energy dissipating capacity of the damper is not utilized to its full capacity. If the slip force is zero, the building will act as an uncontrolled structure. Raut and Jangid (2014) carried out the work in two parts, benchmark building installed with FDs, (i) having equal slip force in all dampers and (ii) having different slip force in the dampers. Based on the change of dynamic properties of the building from before to after strong-motion earthquake,

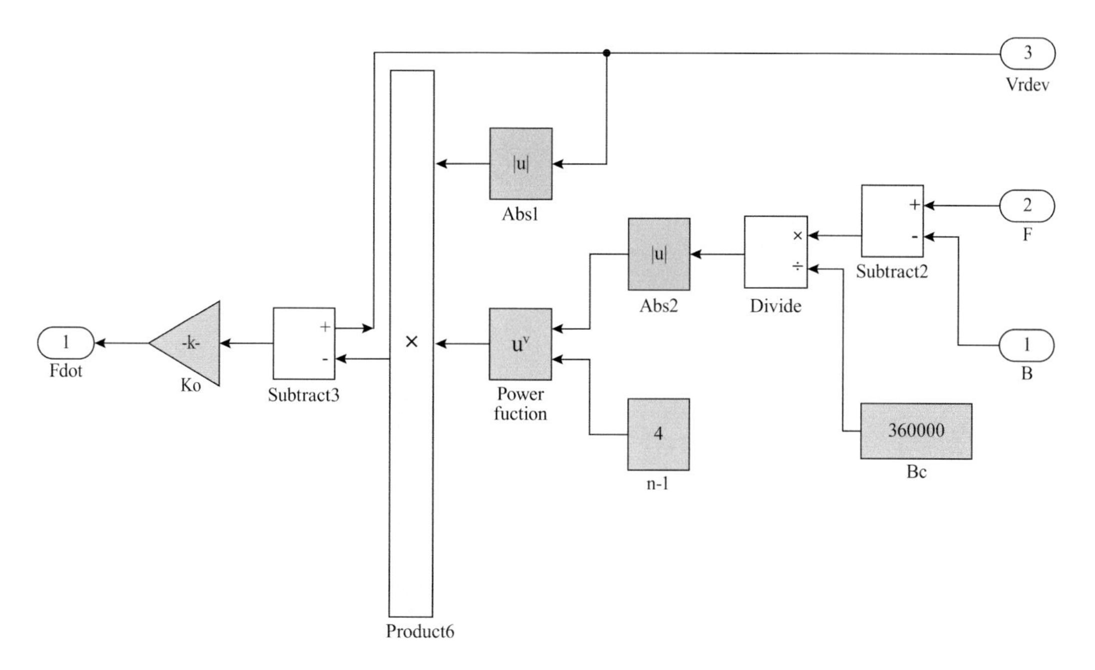

FIGURE 11.13 SIMULINK model for subsystem 3.

two models, viz., pre-earthquake evaluation model and post-earthquake evaluation model are proposed (Spencer et al., 1998). In the pre-earthquake evaluation model, the stiffness is increased proportionally such that first natural frequency of the model is 10% greater than that of the nominal model due to the presence of nonstructural elements such as partitions and cladding since the finite element model is derived without considering stiffness effect of the nonstructural elements. The damping is derived accordingly based on the increased stiffness. Similarly, the post-earthquake

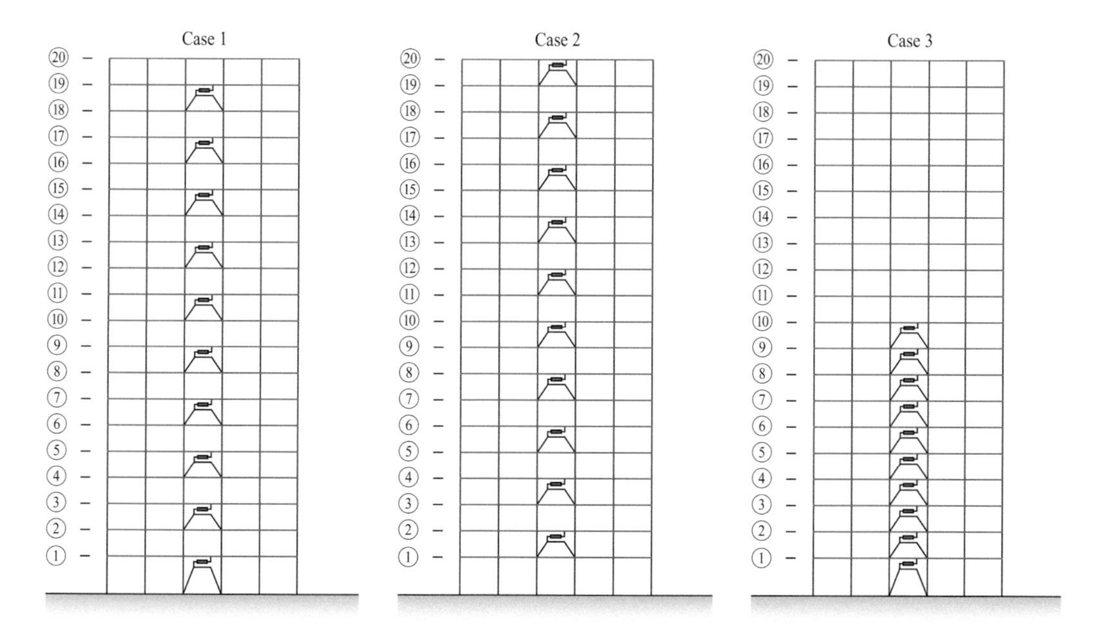

FIGURE 11.14 Case 1, case 2, and case 3 of damper placement.

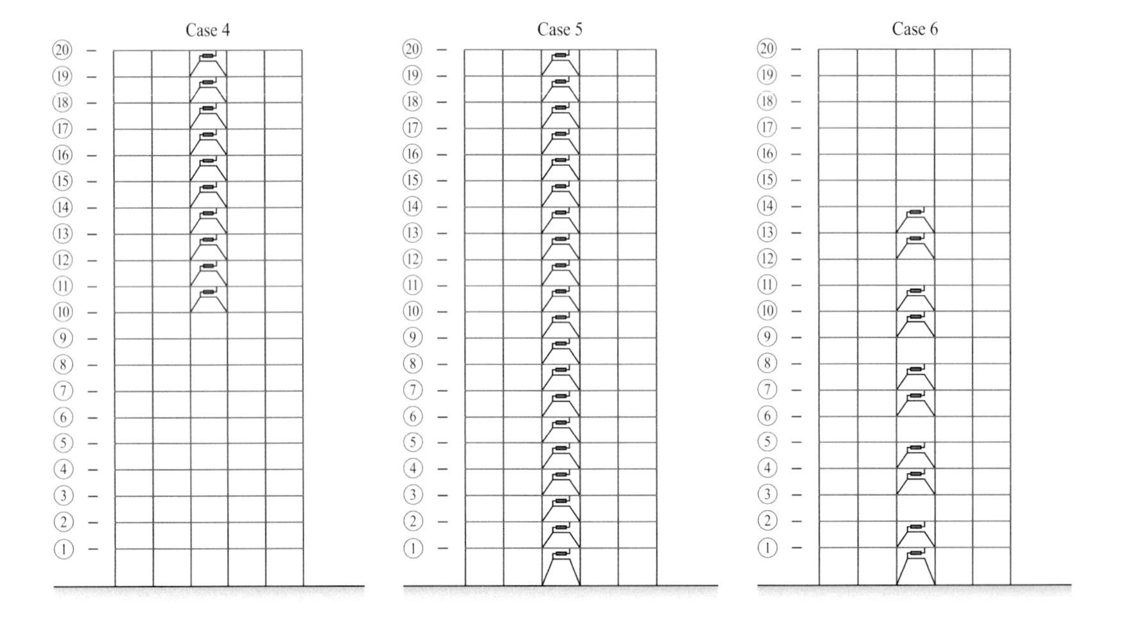

FIGURE 11.15 Case 4, case 5, and case 6 of damper placement.

evaluation model is arrived at by decreasing the stiffness of the nominal model on account of damage to nonstructural elements, such that natural frequency of the structure is decreased by 10%. The damping is derived from decreased stiffness.

Table 11.23 presents the natural frequency for pre-earthquake and post-earthquake models, for the first 10 modes of vibration. The post-earthquake models show reduced frequency of vibration.

TABLE 11.14
Evaluation Criteria for Case 1 of Placement of SMA Damper

Performance Criteria	Value of Performance Criteria under Earthquake Ground Motion of Specific Value of PGA										
	El Centro			Hachinohe			Northridge		Kobe		
	PGA = 0.5 g	PGA = 1 g	PGA = 1.5 g	0.5 g	1 g	1.5 g	0.5 g	1 g	0.5 g	1 g	Maximum
J_1	0.969	0.902	0.928	1.217	1.066	1.111	1.158	1.044	1.018	0.855	1.217
J_2	0.970	0.926	0.953	1.455	0.989	1.171	1.005	0.973	1.011	0.967	1.455
J_3	0.829	0.877	1.001	0.900	0.865	0.941	0.969	0.982	0.964	0.967	1.001
J_4	1.061	1.087	1.101	1.137	1.030	1.053	0.990	1.075	1.162	0.795	1.162
J_5	0.838	0.894	0.972	0.987	0.935	1.004	0.996	0.952	0.956	0.963	1.004
J_6	0.654	0.937	1.000	0.769	0.883	0.940	0.856	0.929	0.935	0.971	1.000
J_7	0.894	0.918	0.963	1.013	0.899	0.905	0.910	1.072	0.906	0.815	1.072
J_8	–	–	1.017	–	–	0.584	0.918	1.145	1.105	1.086	1.145
J_9	–	–	0.674	–	–	0.930	1.042	1.000	1.000	1.024	1.042
J_{10}	0.743	0.935	0.881	0.843	0.902	0.929	0.833	1.137	1.074	0.784	1.137
J_{11}	0.003	0.003	0.003	0.003	0.003	0.003	0.003	0.003	0.003	0.003	0.003
J_{13}	0.089	0.083	0.086	0.089	0.078	0.082	0.102	0.108	0.154	0.134	0.154
J_{17}	Number of dampers installed = 10										

TABLE 11.15
Evaluation Criteria for Case 2 of Placement of SMA Damper

Performance Criteria	Value of Performance Criteria under Earthquake Ground Motion of Specific Value of PGA										
	El Centro			Hachinohe			Northridge		Kobe		
	PGA = 0.5 g	PGA = 1 g	PGA = 1.5 g	PGA = 0.5 g	PGA = 1 g	PGA = 1.5 g	0.5 g	1 g	0.5 g	1 g	Maximum
J_1	1.057	0.961	0.962	1.142	1.039	1.105	1.125	1.071	1.073	0.942	1.142
J_2	1.086	0.904	0.952	1.175	1.046	1.215	1.045	0.958	0.997	1.103	1.215
J_3	0.752	0.871	0.978	0.898	0.905	0.961	0.952	0.998	0.926	0.963	0.998
J_4	1.213	1.052	1.054	1.266	1.023	0.995	0.936	1.131	1.310	1.002	1.310
J_5	0.778	0.885	0.964	0.877	0.931	0.998	0.975	0.955	0.945	0.961	0.998
J_6	0.604	0.899	1.005	0.663	0.877	0.926	0.830	0.934	0.916	0.970	1.005
J_7	0.863	0.899	0.947	0.906	0.896	0.878	0.913	1.094	0.857	0.940	1.094
J_8	–	–	0.926	–	–	0.529	0.946	1.192	1.483	1.073	1.483
J_9	–	–	0.581	–	–	0.907	1.021	1.010	1.000	1.013	1.021
J_{10}	0.743	0.965	0.932	0.754	0.933	0.975	0.775	1.134	1.311	1.075	1.311
J_{11}	0.003	0.003	0.003	0.003	0.003	0.003	0.003	0.003	0.003	0.003	0.003
J_{13}	0.097	0.089	0.089	0.083	0.076	0.082	0.099	0.111	0.146	0.148	0.148
J_{17}	Number of dampers installed = 10										

TABLE 11.16
Evaluation Criteria for Case 3 of Placement of SMA Damper

Performance Criteria	Earthquake Ground Motion and Intensity (PGA)										
	El Centro			Hachinohe			Northridge		Kobe		
	PGA = 0.5 g	PGA = 1 g	PGA = 1.5 g	PGA = 0.5 g	PGA = 1 g	PGA = 1.5 g	PGA = 0.5 g	PGA = 1 g	PGA = 0.5 g	PGA = 1 g	Maximum
J_1	0.982	0.943	0.977	0.893	0.940	0.982	1.068	1.017	1.029	0.949	1.068
J_2	1.284	0.953	1.007	1.341	1.052	1.071	1.022	0.976	1.049	1.019	1.341
J_3	0.726	0.816	0.936	0.913	0.902	0.934	0.990	0.985	0.961	0.935	0.990
J_4	0.789	0.965	0.981	1.006	0.951	0.962	0.920	1.134	0.960	1.000	1.134
J_5	0.889	0.980	0.965	1.009	0.974	0.979	1.004	1.022	0.963	0.978	1.022
J_6	0.682	0.956	0.990	0.793	0.883	0.909	0.864	0.997	0.918	0.939	0.997
J_7	0.896	0.946	0.979	0.995	0.946	0.951	0.948	1.029	0.943	0.960	1.029
J_8	–	–	1.071	–	–	0.998	0.719	1.049	1.110	0.965	1.110
J_9	–	–	0.744	–	–	0.814	0.979	0.979	0.974	1.013	1.013
J_{10}	0.821	1.010	1.088	0.984	0.957	0.969	0.936	1.165	1.047	1.109	1.165
J_{11}	0.003	0.003	0.003	0.003	0.003	0.003	0.003	0.003	0.003	0.003	0.003
J_{13}	0.091	0.087	0.091	0.066	0.069	0.073	0.095	0.107	0.158	0.149	0.158
J_{17}	Number of dampers installed = 10										

TABLE 11.17
Evaluation Criteria for Case 4 of Placement of SMA Damper

Performance Criteria	Earthquake Ground Motion and Intensity (PGA)										
	El Centro			Hachinohe			Northridge		Kobe		
	PGA = 0.5 g	PGA = 1 g	PGA = 1.5 g	PGA = 0.5 g	PGA = 1 g	PGA = 1.5 g	PGA = 0.5 g	PGA = 1 g	PGA = 0.5 g	PGA = 1 g	Maximum
J_1	1.254	0.996	0.933	1.571	1.076	1.162	1.207	1.106	1.098	0.824	1.571
J_2	1.181	0.905	0.905	2.179	0.980	1.255	1.024	1.026	1.022	1.026	2.179
J_3	0.902	0.928	1.036	0.965	0.900	0.944	0.981	0.995	0.944	0.978	1.036
J_4	1.483	1.085	1.148	1.578	1.091	1.083	1.015	1.231	1.296	0.613	1.578
J_5	0.895	0.859	0.967	1.043	0.985	1.041	0.981	0.978	0.961	0.946	1.043
J_6	0.676	0.850	0.991	0.765	0.856	0.929	0.829	0.931	0.905	0.972	0.991
J_7	1.031	0.947	0.931	1.315	0.947	0.918	0.945	1.091	0.772	0.759	1.315
J_8	–	–	0.905	–	–	0.504	1.454	1.240	2.917	1.209	2.917
J_9	–	–	0.791	–	–	0.953	1.000	1.000	1.026	0.988	1.026
J_{10}	1.107	0.856	0.930	1.106	0.854	0.965	0.929	1.140	1.114	0.661	1.140
J_{11}	0.003	0.003	0.003	0.003	0.003	0.003	0.003	0.003	0.003	0.003	0.003
J_{13}	0.116	0.092	0.089	0.115	0.079	0.086	0.107	0.110	0.150	0.139	0.150
J_{17}	Number of dampers installed = 10										

TABLE 11.18
Evaluation Criteria for Case 5 of Placement of SMA Damper

Performance Criteria	Earthquake Ground Motion and Intensity (PGA)										
	El Centro			Hachinohe			Northridge		Kobe		
	PGA = 0.5 g	PGA = 1 g	PGA = 1.5 g	PGA = 0.5 g	PGA = 1 g	PGA = 1.5 g	PGA = 0.5 g	PGA = 1 g	PGA = 0.5 g	PGA = 1 g	Maximum
J_1	1.276	1.075	0.885	1.211	1.200	1.152	1.346	1.085	1.070	0.832	1.346
J_2	1.327	0.990	0.885	1.603	1.223	1.259	1.150	0.985	0.980	1.068	1.603
J_3	0.685	0.901	0.988	0.894	0.863	0.919	0.982	0.966	0.907	0.994	0.994
J_4	1.383	1.008	1.036	1.573	1.006	1.066	1.139	1.227	1.158	0.513	1.573
J_5	0.736	0.814	0.904	0.899	0.897	0.984	0.968	0.947	0.936	0.936	0.984
J_6	0.513	0.651	0.845	0.550	0.735	0.928	0.744	0.909	0.891	0.929	0.929
J_7	0.914	0.919	0.908	0.928	1.001	0.839	1.098	1.153	0.796	0.738	1.153
J_8	–	–	0.765	–	–	0.535	1.788	1.354	2.189	1.275	2.189
J_9	–	–	0.581	–	–	0.814	1.000	1.010	1.000	1.024	1.024
J_{10}	1.015	0.773	0.790	1.072	0.826	0.974	0.975	1.219	1.159	0.509	1.219
J_{11}	0.003	0.003	0.003	0.003	0.003	0.003	0.003	0.003	0.003	0.003	0.003
J_{13}	0.118	0.099	0.082	0.088	0.088	0.085	0.119	0.108	0.146	0.131	0.146
J_{17}	Number of dampers installed = 20										

TABLE 11.19
Evaluation Criteria for Case 6 of SMA Damper

Performance Criteria	Earthquake Ground Motion and Intensity (PGA)										Maximum
		El Centro			Hachinohe		Northridge		Kobe		
	PGA = 0.5 g	PGA = 1 g	PGA = 1.5 g	PGA = 0.5 g	PGA = 1 g	PGA = 1.5 g	PGA = 0.5 g	PGA = 1 g	PGA = 0.5 g	PGA = 1 g	
J_1	0.971	0.946	0.947	0.910	0.985	0.997	1.079	0.999	1.026	0.893	1.079
J_2	1.097	0.982	0.994	1.369	0.969	1.058	1.018	0.931	1.053	1.022	1.369
J_3	0.733	0.814	0.929	0.866	0.903	0.936	0.986	0.991	0.969	0.986	0.991
J_4	0.839	0.991	0.962	1.048	1.033	1.023	0.938	1.130	0.955	0.750	1.130
J_5	0.895	0.942	0.957	0.989	0.965	0.965	0.990	0.989	0.962	0.963	0.990
J_6	0.681	0.964	0.983	0.773	0.897	0.927	0.869	0.957	0.943	0.963	0.983
J_7	0.872	0.920	0.935	0.936	0.927	0.933	0.944	1.016	0.997	0.859	1.016
J_8	–	–	0.826	–	–	0.662	0.799	0.993	0.866	0.965	0.993
J_9	–	–	0.651	–	–	0.837	0.958	0.969	0.949	1.000	1.000
J_{10}	0.744	0.990	0.949	0.940	0.934	0.926	0.809	1.138	1.147	0.718	1.147
J_{11}	0.003	0.003	0.003	0.003	0.003	0.003	0.003	0.003	0.003	0.003	0.003
J_{13}	0.089	0.087	0.088	0.066	0.072	0.074	0.095	0.106	0.163	0.140	0.163
J_{17}	Number of dampers installed = 10										

TABLE 11.20
Maximum Values Evaluation Criteria for the Six Cases of Placement of SMA Damper

Evaluation Criteria	Case 1	Case 2	Case 3	Case 4	Case 5	Case 6
J_1	1.217	1.142	1.068	1.571	1.346	1.079
J_2	1.455	1.215	1.341	2.179	1.603	1.369
J_3	1.001	0.998	0.990	1.036	0.994	0.991
J_4	1.162	1.310	1.134	1.578	1.573	1.130
J_5	1.004	0.998	1.022	1.043	0.984	0.990
J_6	1.000	1.005	0.997	0.991	0.929	0.983
J_7	1.072	1.094	1.029	1.315	1.153	1.016
J_8	1.145	1.483	1.110	2.917	2.189	0.993
J_9	1.042	1.021	1.012	1.026	1.024	1.000
J_{10}	1.137	1.311	1.165	1.140	1.219	1.147
J_{17}	10	10	10	10	20	10

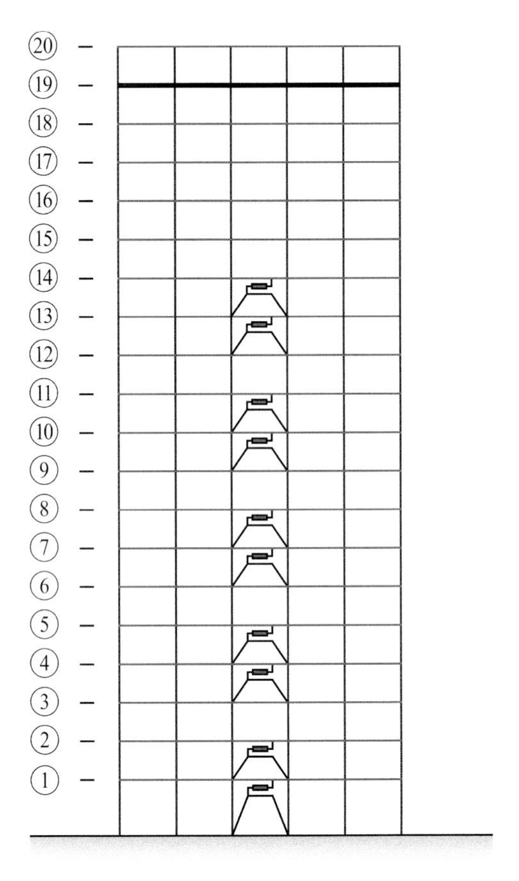

FIGURE 11.16 Members with the highest curvature for all cases.

TABLE 11.21
Properties of the Viscous Damping Wall

Variable	Symbol	Value
Width of wall (m)	W	2.0
Height of wall (m)	H	2.0
Effective shearing area (m^2)	S	8.0
Gap between the wall plates (m)	d	0.006
Temperature of viscous fluid (°C)	T	20
Coefficient (if $v_{di}/d \le 10$)	a	4.13
Coefficient (if $v_{di}/d > 10$)	a	6.37
Coefficient	b	−0.043

TABLE 11.22
Evaluation Criteria for 20-Story Benchmark Building Installed with Viscous Damping Walls

Evaluation Criteria	Intensity	El Centro	Hachinohe	Northridge	Kobe	Maximum
J_1	0.5	0.53034	0.64582	0.67781	0.42034	0.77714
Peak drift ratio	1.0	0.57280	0.66879	0.77714	0.58829	
	1.5	0.59599	0.72263	–	–	
J_2	0.5	0.50690	0.48766	0.55823	0.41942	0.79760
Peak level	1.0	0.48309	0.46251	0.79760	0.74962	
acceleration	1.5	0.49928	0.61813	–	–	
J_3	0.5	0.82265	1.0206	0.88483	0.72987	1.2137
Peak base shear	1.0	0.76154	1.0088	1.0723	1.2137	
	1.5	0.84403	1.0531	–	–	
J_4	0.5	0.45871	0.69904	0.47330	0.31705	0.91354
Normed drift ratio	1.0	0.47959	0.72709	0.91354	0.31320	
	1.5	0.50081	0.76285	–	–	
J_5	0.5	0.35133	0.37681	0.33688	0.29211	0.48117
Normed level	1.0	0.34216	0.40478	0.48117	0.45542	
acceleration	1.5	0.34825	0.43795	–	–	
J_6	0.5	0.53528	0.69148	0.48528	0.40906	0.73376
Normed base shear	1.0	0.54134	0.70481	0.73376	0.64962	
	1.5	0.5551	0.72888	–	–	
J_7	0.5	0.56418	0.7716	0.55805	0.37913	0.79977
Ductility	1.0	0.58473	0.79977	0.78902	0.65106	
	1.5	0.56010	0.75240	–	–	
J_{10}	0.5	0.47497	0.66678	0.35789	0.31709	0.92059
Normed ductility	1.0	0.50004	0.69438	0.92059	0.34817	
	1.5	0.46390	0.71907	–	–	
J_{11}	0.5	0.0042054	0.0034762	0.0069728	0.0071340	0.01181
Control force	1.0	0.0063514	0.0055208	0.0090864	0.01181	
	1.5	0.0077605	0.0067429	–	–	
J_{13}	0.5	0.048900	0.055291	0.060220	0.073177	0.10448
Device stroke	1.0	0.052822	0.06130	0.083885	0.10448	
	1.5	0.055398	0.067048	–	–	

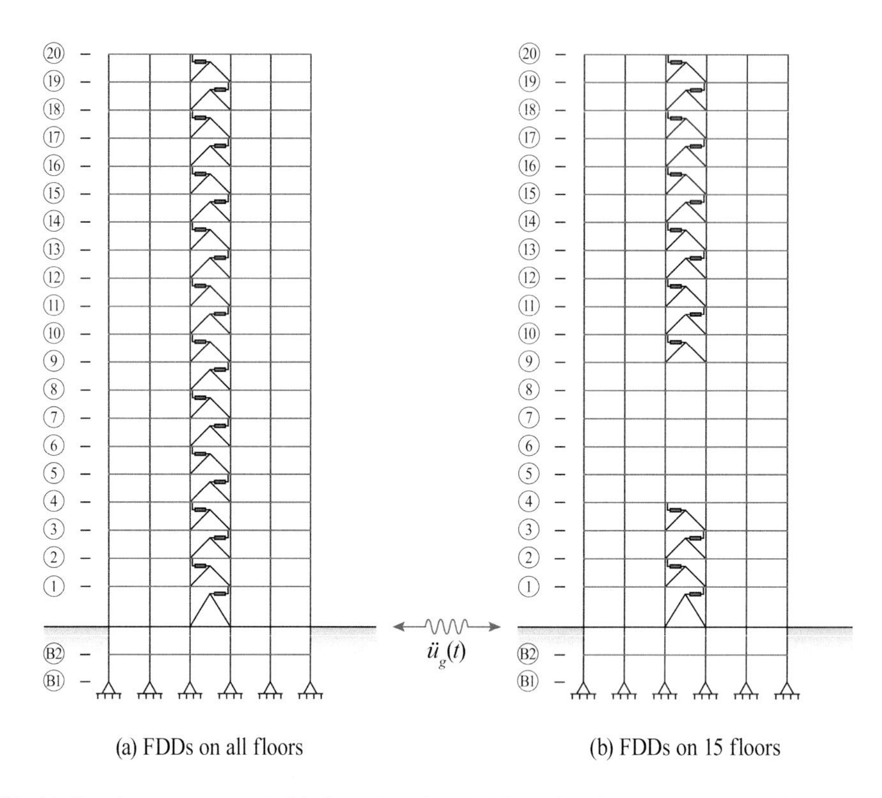

FIGURE 11.17 Arrangement of friction damping devices in the benchmark building. (Raut and Jangid, 2014)

Table 11.24 presents the peak uncontrolled response, and the corresponding response when installed with friction damping devices having the same slip force. Table 11.25 shows peak response quantities of the benchmark building installed with FDDs having the different slip force. Table 11.26 shows the performance criteria of the building installed with FDDs (pre-earthquake model). Table 11.27 shows the performance criteria of the building installed with FDDs (post-earthquake model). Table 11.28 shows the percentage variation of maximum relative displacement with respect to top floor (pre-earthquake model).

TABLE 11.23
Natural Frequency for Pre-Earthquake and Post-Earthquake Evaluation Models

Mode Number	Natural Frequency (Hz) Pre-Earthquake Model	Natural Frequency (Hz) Post-Earthquake Model
1	0.29	0.24
2	0.83	0.68
3	1.43	1.17
4	2.01	1.65
5	2.64	2.16
6	3.08	2.52
7	3.30	2.70
8	3.53	2.89
9	3.99	3.26
10	4.74	3.88

TABLE 11.24
Peak Response Quantities of the Benchmark Building Installed with FDDs Having the Same Slip Force

Earthquake	Evaluation Model	Top Floor Displacement (cm)			Top Floor Acceleration (g)			Normalized Base Shear		
		UC	Case (a)	Case (b)	UC	Case (a)	Case (b)	UC	Case (a)	Case (b)
El Centro, 1940	Pre-EQ	37.943	29.061	31.701	0.318	0.231	0.238	0.080	0.067	0.063
	Post-EQ	27.779	22.641	24.606	0.226	0.171	0.142	0.038	0.042	0.045
Hachinohe, 1968	Pre-EQ	51.705	27.555	33.503	0.290	0.259	0.256	0.099	0.066	0.075
	Post-EQ	29.828	25.349	28.374	0.198	0.152	0.153	0.049	0.052	0.055
Northridge, 1994	Pre-EQ	105.826	93.295	94.561	0.934	0.909	0.909	0.204	0.189	0.183
	Post-EQ	90.685	87.083	87.367	0.792	0.742	0.721	0.147	0.154	0.146
Kobe, 1995	Pre-EQ	56.862	49.043	49.582	0.910	0.828	0.864	0.174	0.139	0.134
	Post-EQ	65.610	48.047	49.569	0.738	0.576	0.643	0.115	0.108	0.112

Note: UC = Uncontrolled response, Case (a) = FDD installed on all floors, Case (b) = FDD installed on 15 floors, Pre-EQ = Pre-earthquake, Post-EQ = Post-earthquake.

11.3.6.4 Application of d-TMDs

The control effectiveness of using d-TMDs has also been investigated on the benchmark structure. Elias et al. (2019) installed distributed-multiple TMD (d-MTMD) at different levels of the benchmark building to investigate the performance criteria. Figure 11.18 shows the elevation of the 20-story benchmark building installed with various arrangements of TMD. The efficiency of the d-MTMD is established by comparing four cases:

- Placing single TMD at the topmost floor (STMD): Figure 11.18(d)
- Placing five TMDs all at the topmost floor (MTMDs-all.top): Figure 11.18(e)
- Placing five distributed TMDs on different floors (d-MTMDs): Figure 11.18(f)
- Placing five arbitrarily d-MTMDs on different floors (ad-MTMDs): Figure 11.18(g).

Table 11.29 shows the design parameters for the five tuned mass dampers, 5MTMDs-all.top, 5ad-MTMDs, 5d-MTMDs, and STMD with mass ratio (μ) of 5% and damping ratio (ξ_d) of 2%. Furthermore, Table 11.30 shows the reduction in the response quantities compared to the uncontrolled response under different real earthquake ground motions.

TABLE 11.25
Peak Response Quantities of the Benchmark Building Installed with FDDs Having the Different Slip Force

Earthquake	Evaluation Model	Top Floor Displacement (cm)			Top Floor Acceleration (g)			Normalized Base Shear		
		UC	Case (a)	Case (b)	UC	Case (a)	Case (b)	UC	Case (a)	Case (b)
El Centro, 1940	Pre-EQ	37.943	28.883	32.468	0.318	0.217	0.227	0.080	0.072	0.066
	Post-EQ	27.779	21.872	23.891	0.226	0.157	0.143	0.038	0.043	0.049
Hachinohe, 1968	Pre-EQ	51.705	25.442	31.970	0.290	0.264	0.248	0.099	0.062	0.074
	Post-EQ	29.828	23.326	27.707	0.198	0.181	0.186	0.049	0.051	0.053
Northridge, 1994	Pre-EQ	105.826	92.580	95.251	0.934	0.905	0.906	0.204	0.189	0.181
	Post-EQ	90.685	87.917	89.044	0.792	0.749	0.727	0.147	0.156	0.147
Kobe, 1995	Pre-EQ	56.862	49.363	49.999	0.910	0.777	0.842	0.174	0.139	0.143
	Post-EQ	65.610	48.669	50.567	0.738	0.569	0.622	0.115	0.113	0.116

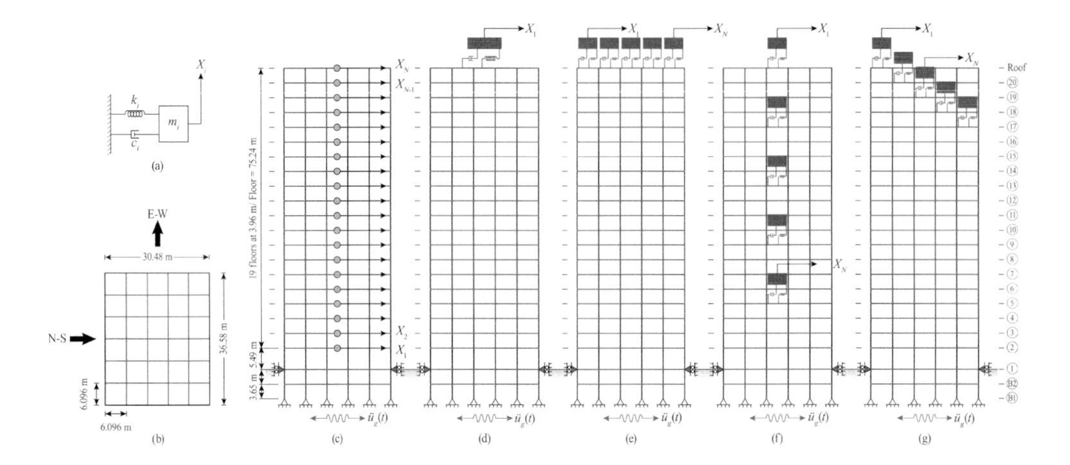

FIGURE 11.18 20-Story benchmark building (a) TMD, (b) plan of the building, (c) elevation of uncontrolled, (d) controlled with STMD, (e) controlled with MTMDs-all.top, (f) controlled with d-MTMDs, (g) controlled with ad-MTMDs.

TABLE 11.26
Performance Criteria of the Benchmark Building Installed with FDDs (Pre-Earthquake Model)

Evaluation	El Centro		Hachinohe		Northridge		Kobe		Maximum Value	
Criteria	Case (a)	Case (b)	Case (a)	Case (b)	Case (a)	Case (b)	Case (a)	Case (b)	Case (a)	Case (b)
J_1	0.7659	0.8355	0.5329	0.6480	0.8816	0.8936	0.8625	0.8720	0.8816	0.8936
J_2	0.7017	0.7976	0.5871	0.7159	0.9382	0.9344	0.7421	0.7902	0.9344	0.9344
J_3	0.9950	0.9795	0.8934	0.8831	0.9735	0.9740	0.9105	0.9502	0.9950	0.9795
J_4	0.8481	0.7951	0.6648	0.7562	0.9298	0.8970	0.7453	0.7725	0.9298	0.8970
J_5	0.4674	0.5542	0.4400	0.5079	0.6058	0.6623	0.6694	0.7184	0.6694	0.7184
J_6	0.5051	0.5768	0.4473	0.5392	0.5857	0.6425	0.6501	0.6735	0.6501	0.6735
J_7	0.5958	0.5765	0.4534	0.5296	0.6803	0.7137	0.6368	0.6541	0.6803	0.7137
J_8	0.5955	0.5860	0.4755	0.5616	0.5984	0.6562	0.6978	0.7097	0.6978	0.7097
J_9	0.0070	0.0070	0.0070	0.0070	0.0070	0.0070	0.0070	0.0070	0.0070	0.0070
J_{10}	0.0632	0.0644	0.0555	0.0555	0.0689	0.0664	0.0850	0.0878	0.0850	0.0878

TABLE 11.27
Performance Criteria of the Benchmark Building Installed with FDDs (Post-Earthquake Model)

Evaluation	El Centro		Hachinohe		Northridge		Kobe		Maximum Value	
Criteria	Case (a)	Case (b)	Case (a)	Case (b)	Case (a)	Case (b)	Case (a)	Case (b)	Case (a)	Case (b)
J_1	0.8150	0.8857	0.8498	0.9513	0.9603	0.9664	0.7323	0.7616	0.9603	0.9664
J_2	0.8096	0.9534	0.9480	1.0068	0.9029	0.8933	0.7457	0.7759	0.9480	1.0068
J_3	0.9629	0.9151	0.9632	0.9263	0.9372	0.9104	0.9086	0.9137	0.9632	0.9263
J_4	1.0849	1.1771	1.0694	1.1321	1.0477	0.9928	0.9470	0.9789	1.0849	1.1771
J_5	0.3730	0.4243	0.4378	0.4610	0.6571	0.6935	0.3453	0.4335	0.6571	0.6935
J_6	0.4875	0.5231	0.5530	0.5695	0.6221	0.6390	0.4088	0.4766	0.6221	0.6390
J_7	0.7027	0.6991	0.5991	0.6549	0.5868	0.5960	0.6399	0.6529	0.7027	0.6991
J_8	0.5837	0.5327	0.5839	0.5794	0.6287	0.6457	0.4496	0.5062	0.6287	0.6457
J_9	0.0070	0.0070	0.0070	0.0070	0.0070	0.0070	0.0070	0.0070	0.0070	0.0070
J_{10}	0.0778	0.0916	0.0954	0.1013	0.0952	0.0899	0.0887	0.0923	0.0954	0.1013

TABLE 11.28
Percentage Variation of Maximum Relative Displacement with Respect to Top Floor (Pre-Earthquake Model)

Floor Level	Percentage Variation under Earthquake				
	El Centro	Hachinohe	Northridge	Kobe	Average
1	8.32	8.09	7.95	12.09	9.11
2	13.94	13.77	13.35	19.52	15.14
3	19.28	19.50	18.55	26.48	20.95
4	24.42	25.29	23.62	32.75	26.52
5	29.41	31.07	28.45	38.31	31.81
6	34.18	36.75	32.96	42.96	36.71
7	38.77	42.29	37.08	46.91	41.26
8	43.24	47.61	40.86	50.37	45.52
9	47.64	52.67	44.31	53.36	49.50
10	52.05	57.50	49.13	55.84	53.63
11	56.67	62.24	54.42	57.85	57.79
13	61.49	66.87	59.86	59.21	61.86
13	66.32	71.25	65.33	59.90	65.70
14	71.17	75.72	70.86	59.93	69.42
15	76.00	80.09	76.33	65.08	74.38
16	80.61	84.31	81.45	71.95	79.58
17	85.37	88.64	86.46	79.15	84.90
18	90.31	92.77	91.28	86.08	90.11
19	95.39	96.64	95.89	92.92	95.21
20	100.00	100.00	100.00	100.00	100.00

TABLE 11.29
Design Parameters for 5MTMDs-All.Top, 5ad-MTMDs, and 5d-MTMDs with Mass Ratio (μ) 5% and Damping Ratio (ξ_d) 2%

Case	TMD Number	Tuning Frequency ω_i (rad/s)	Mass m_i (t)	Stiffness k_i (10^7 N/m)	Damping c_i (10^4 Ns/m)
5MTMDs-all.top	TMD-1	1.633	55.43	0.015	0.090
	TMD-2	4.710	55.43	0.133	0.261
	TMD-3	8.164	55.43	0.370	0.453
	TMD-4	11.492	55.43	0.733	0.637
	TMD-5	15.072	55.43	1.260	0.836
5ad-MTMDs	TMD-1	1.633	55.43	0.015	0.090
	TMD-2	4.710	55.43	0.133	0.261
	TMD-3	8.164	55.43	0.370	0.453
	TMD-4	11.492	55.43	0.733	0.637
	TMD-5	15.072	55.43	1.260	0.836
5d-MTMDs	TMD-1	1.633	55.43	0.015	0.090
	TMD-2	4.710	55.43	0.133	0.261
	TMD-3	8.164	55.43	0.370	0.453
	TMD-4	11.492	55.43	0.733	0.637
	TMD-5	15.072	55.43	1.260	0.836

TABLE 11.30
Evaluation Criteria of 20-Story Benchmark Building with Different TMD Schemes for the Four Earthquakes

Ground Motion	Evaluation Criteria	Case: STMD			5MTMDs-all.top			5ad-MTMDs			5d-MTMDs		
		$\mu = 0.02$	$\mu = 0.03$	$\mu = 0.05$	$\mu = 0.02$	$\mu = 0.03$	$\mu = 0.05$	$\mu = 0.02$	$\mu = 0.03$	$\mu = 0.05$	$\mu = 0.02$	$\mu = 0.03$	$\mu = 0.05$
Hachinohe, 1968	J_1	0.981	0.972	0.954	0.965	0.9	0.873	0.968	0.945	0.923	0.958	0.88	0.869
	J_2	1	1.025	1.058	0.891	0.889	0.884	0.913	0.92	0.925	0.901	0.918	0.924
	J_3	0.986	0.958	0.941	0.918	0.891	0.847	0.988	0.969	0.945	0.978	0.952	0.94
	J_4	1.222	1.352	1.422	0.751	0.759	0.762	0.858	0.888	0.893	0.825	0.845	0.867
	J_5	1.01	1.08	1.158	0.985	0.999	1.033	1.001	1.05	1.087	0.95	0.975	0.98
	J_6	0.965	0.948	0.902	0.758	0.715	0.65	0.787	0.759	0.73	0.761	0.735	0.681
	J_7	1.01	1.09	1.175	0.865	0.869	0.871	0.956	0.978	0.982	0.925	0.938	0.944
	J_8	1.135	1.255	1.351	0.795	0.777	0.76	0.935	0.916	0.905	0.888	0.826	0.804
Northridge, 1994	J_1	0.888	0.869	0.835	0.89	0.873	0.84	0.911	0.9	0.885	0.877	0.855	0.818
	J_2	0.868	0.878	0.887	1	1.085	1.155	0.878	0.888	0.894	0.855	0.861	0.879
	J_3	0.968	0.988	1	0.825	0.835	0.844	0.833	0.855	0.911	0.822	0.831	0.855
	J_4	0.935	0.948	0.961	1	1.033	1.045	0.926	0.936	0.966	0.911	0.928	0.954
	J_5	0.729	0.689	0.634	0.7	0.653	0.632	0.719	0.688	0.665	0.688	0.65	0.631
	J_6	0.656	0.666	0.677	0.633	0.629	0.613	0.677	0.666	0.657	0.62	0.615	0.61
	J_7	0.855	0.877	0.884	0.833	0.816	0.796	0.855	0.838	0.825	0.827	0.8	0.772
	J_8	0.811	0.822	0.832	0.809	0.768	0.729	0.811	0.777	0.709	0.789	0.7	0.679
Kobe, 1995	J_1	1.008	0.999	0.941	0.922	0.908	0.894	0.999	0.985	0.973	0.999	0.977	0.968
	J_2	0.8	0.836	0.856	0.958	0.933	0.9	0.808	0.811	0.815	0.899	0.855	0.802
	J_3	0.855	0.877	0.899	0.819	0.809	0.798	0.925	0.918	0.908	0.928	0.888	0.875
	J_4	0.875	0.888	0.892	0.889	0.868	0.851	0.888	0.859	0.832	0.908	0.865	0.826
	J_5	0.881	0.761	0.718	0.938	0.927	0.906	0.899	0.888	0.876	0.911	0.885	0.872
	J_6	0.652	0.659	0.661	0.833	0.818	0.792	0.857	0.801	0.783	0.888	0.822	0.78
	J_7	0.966	0.948	0.922	0.808	0.777	0.75	0.891	0.877	0.866	0.9	0.849	0.839
	J_8	0.7	0.666	0.658	0.855	0.824	0.802	0.837	0.799	0.786	0.836	0.788	0.779

11.4 WIND EXCITED 76-STORY BENCHMARK BUILDING

A benchmark problem on the wind response control of 76-story wind excited benchmark building has been developed and presented by Yang et al. (1998, 2004) as a test structure to evaluate response control effectiveness of new control devices and strategies. Passive control strategies have been developed and the building is evaluated with a broad set of carefully chosen parameters, the performance measures, and guidelines to the researchers, under dynamic wind excitation. The wind excited benchmark building is 306-m high RC office tower proposed for the city of Melbourne, Australia. The building is tall and slender with a height to width ratio of 7.3. Hence, it is considered as wind sensitive. Wind tunnel tests for the building model have been conducted at the University of Sydney, and the results of a crosswind data for a duration of 3,600 s are provided for the analysis of the benchmark problem (Samali et al., 2004). The building consists of RC core and RC frame. The RC frame was designed in order to primarily carry the gravitational loads and part of the wind loads. Concrete core was designed in order to resist the majority of wind loads. The building has square cross section having chamfer at two corners as shown in Figure 11.19. Figure 11.20 shows the elevation view of the benchmark building. Moreover, Table 11.31 presents the salient features of the wind excited 76-story benchmark building.

11.4.1 Model of the Wind Excited 76-Story Benchmark Building

Steel beams with a metal deck and 125-mm-thick slab make the floor light weight. The column sizes, core wall thickness, and mass of the floor vary along the height of the building. The building is modeled as a cantilever beam. The portion of the building between consecutive floors is assumed

TABLE 11.31
Salient Features of the Wind Excited 76-Story Benchmark Building

Building Feature		Detail of Feature
Type of building		RC square building consisting of concrete core and concrete frame
Floor height (m)	1st	10
	2nd and 3rd	4.5
	4th–38th	3.9
	39th and 40th	4.5
	41st–74th	3.9
	75th and 76th	4.5
Total height of building (m)		306.10
Total width of building (m)		42
Height-to-width ratio (aspect ratio)		7.3
Dimensions of the central RC core (m)		21 × 21
Total number of columns		24.6 on each side of the building
Spacing of columns		6.5 m apart, connected to a 900-mm-deep and 400-mm-wide spandrel beam on each floor
Total weight of the building (kN)		1,499,400
Total volume of the building (m^3)		510,000
Compressive strength of concrete (MPa)		60
Modulus of elasticity of concrete (GPa)		40
Natural frequencies of first five modes		0.16, 0.77, 1.99, 3.79, and 6.94 Hz
Structural damping		5%

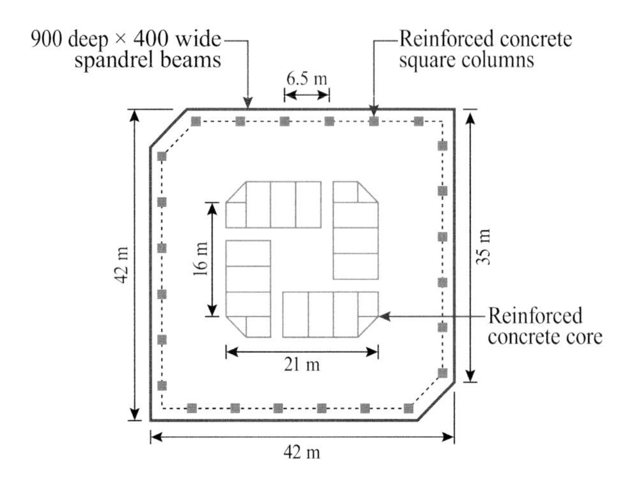

FIGURE 11.19 Plan view of the 76-story benchmark building (Yang et al., 2004).

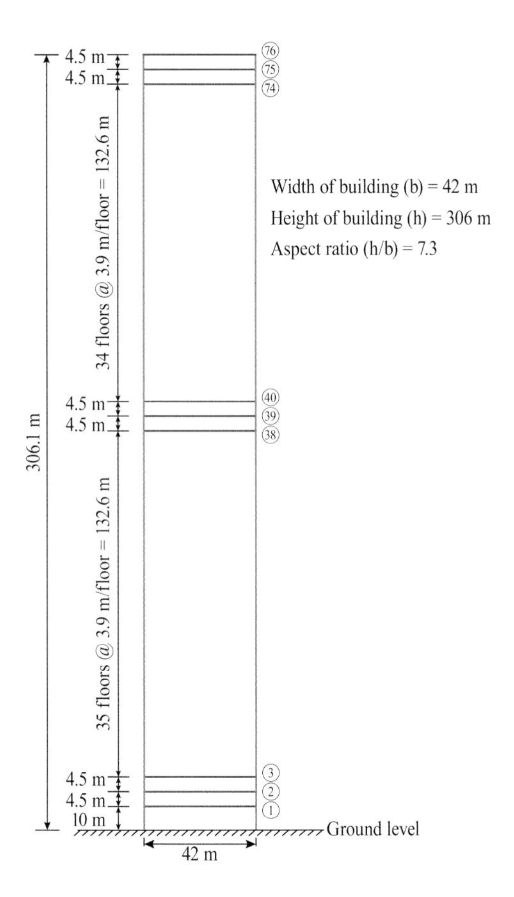

FIGURE 11.20 Elevation of the 76-story benchmark building (Yang et al., 2004).

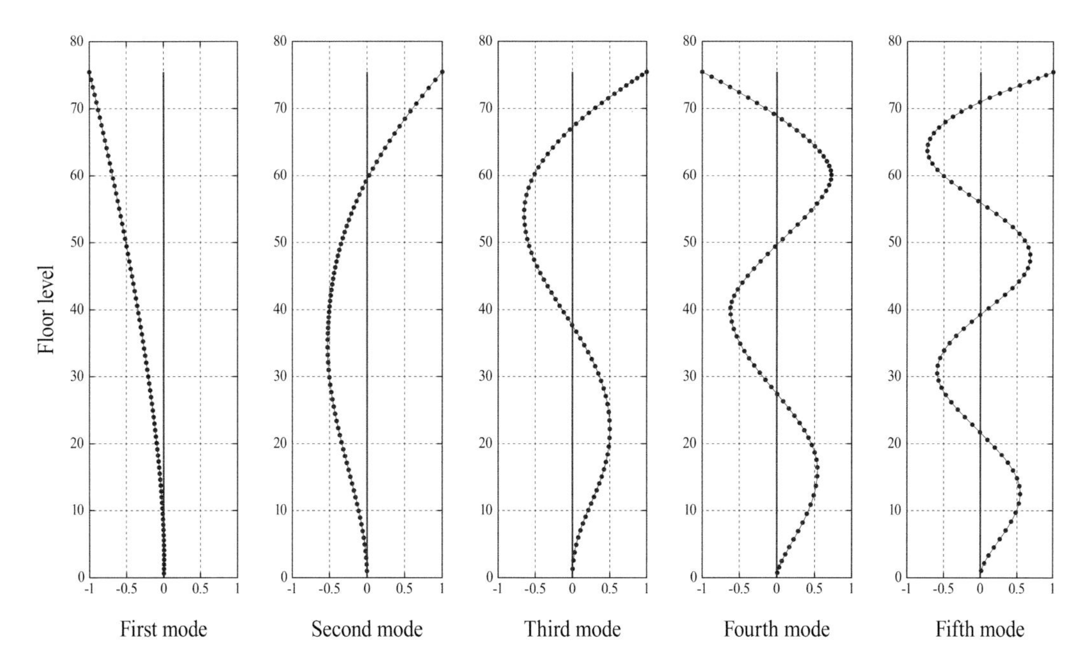

FIGURE 11.21 Mode shapes of the 76-story benchmark building.

to be a classical beam. The damping matrix of the building is constructed by assuming 1% damping ratio for the first five modes using Rayleigh's method. The performance criteria are evaluated considering 24 modes of vibration. The first five-mode shapes are presented in Figure 11.21.

The finite element model of the building is constructed assuming one DOF at each floor level. Thus, a total of 76 DOFs are considered. The rotational DOFs are eliminated by static condensation process.

11.4.2 Performance Criteria

Since the coupled lateral-torsional motion is neglected and crosswind and along-wind loads are uncorrelated, the building response quantities due to crosswinds and along-winds can be computed independently. Based on the wind tunnel test data, building response quantities due to crosswind loads are much higher than those due to along-wind loads. Consequently, only the design of controller using a crosswind loading is considered. From the response time histories, the peak response quantities can be obtained and the temporal RMS values can be computed. The main objective of installing control systems on the tall building is to reduce the absolute acceleration to alleviate the occupant's discomfort. However, no consideration is given to the frequency dependence of human perception to acceleration. The response quantities of the wind excited benchmark building are evaluated in terms of twelve dimensionless performance criteria. Criteria (J_1, J_2, J_7, J_8) are related to response in terms of acceleration, and criteria (J_3, J_4, J_9, J_{10}) are related to response in terms of displacement.

11.4.3 RMS Response

The first four evaluation criteria (J_1, J_2, J_3, J_4) are defined to measure the reduction in RMS response quantities of the wind excited benchmark building, evaluated by normalizing the response quantities by the corresponding response quantities of the uncontrolled building.

1. Reduction in the maximum floor RMS acceleration
   The first evaluation criterion for the controllers is their ability to reduce the maximum floor RMS acceleration.

$$J_1 = \max\left\{\frac{\sigma_{\ddot{x}1}}{\sigma_{\ddot{x}75o}}, \frac{\sigma_{\ddot{x}30}}{\sigma_{\ddot{x}75o}}, \frac{\sigma_{\ddot{x}50}}{\sigma_{\ddot{x}75o}}, \frac{\sigma_{\ddot{x}55}}{\sigma_{\ddot{x}75o}}, \frac{\sigma_{\ddot{x}60}}{\sigma_{\ddot{x}75o}}, \frac{\sigma_{\ddot{x}65}}{\sigma_{\ddot{x}75o}}, \frac{\sigma_{\ddot{x}70}}{\sigma_{\ddot{x}75o}}, \frac{\sigma_{\ddot{x}75}}{\sigma_{\ddot{x}75o}}\right\}$$

   In the performance criterion J_1, accelerations only up to 75th floor are considered because the 76th floor is the top of the building and is not used by the occupants.
2. Reduction in the average floor RMS acceleration for selected floors above the 49th floor

$$J_2 = \frac{1}{6}\sum_i \left[\sigma_{\ddot{x}i}/\sigma_{\ddot{x}io}\right]$$

   for $i = 50, 55, 60, 65, 70,$ and 75.
3. The ability of the controller to reduce the RMS top floor displacement compared to that of the uncontrolled structure

$$J_3 = \left\{\frac{\sigma_{x76}}{\sigma_{x76o}}\right\}$$

4. Ability of the controller to reduce the average RMS displacement of the selected floors

$$J_4 = \frac{1}{7}\sum_i \left[\sigma_{xi}/\sigma_{xio}\right]$$

   for $i = 50, 55, 60, 65, 70, 75,$ and 76

11.4.4 Peak Response

The evaluation criteria (J_7, J_8, J_9, J_{10}) are defined to measure the reduction in the peak responses, over the entire simulation time of the earthquake. These are calculated by normalizing the peak response quantities by the corresponding peak response quantities of the uncontrolled building.

5. Maximum peak acceleration normalized by the corresponding acceleration in the uncontrolled structure

$$J_7 = \max\left\{\frac{\ddot{x}_{p1}}{\ddot{x}_{p75o}}, \frac{\ddot{x}_{p30}}{\ddot{x}_{p75o}}, \frac{\ddot{x}_{p50}}{\ddot{x}_{p75o}}, \frac{\ddot{x}_{p55}}{\ddot{x}_{p75o}}, \frac{\ddot{x}_{p60}}{\ddot{x}_{p75o}}, \frac{\ddot{x}_{p65}}{\ddot{x}_{p75o}}, \frac{\ddot{x}_{p70}}{\ddot{x}_{p75o}}, \frac{\ddot{x}_{p75}}{\ddot{x}_{p75o}}\right\}$$

6. Normalized peak acceleration normalized by the corresponding acceleration in the uncontrolled structure

$$J_8 = \frac{1}{6}\sum_i \left[\ddot{x}_{pi}/\ddot{x}_{pio}\right]$$

TABLE 11.32
Summary of Evaluation Criteria for the 76-Story Benchmark Building

RMS Responses	Peak Responses
Maximum floor RMS acceleration	Maximum peak acceleration
$J_1 = \max\left\{\frac{\sigma_{\ddot{x}1}}{\sigma_{\ddot{x}75o}}, \frac{\sigma_{\ddot{x}30}}{\sigma_{\ddot{x}75o}}, \frac{\sigma_{\ddot{x}50}}{\sigma_{\ddot{x}75o}}, \frac{\sigma_{\ddot{x}55}}{\sigma_{\ddot{x}75o}}, \frac{\sigma_{\ddot{x}60}}{\sigma_{\ddot{x}75o}}, \frac{\sigma_{\ddot{x}65}}{\sigma_{\ddot{x}75o}}, \frac{\sigma_{\ddot{x}70}}{\sigma_{\ddot{x}75o}}, \frac{\sigma_{\ddot{x}75}}{\sigma_{\ddot{x}75o}}\right\}$	$J_7 = \max\left\{\frac{\ddot{x}_{p1}}{\ddot{x}_{p75o}}, \frac{\ddot{x}_{p30}}{\ddot{x}_{p75o}}, \frac{\ddot{x}_{p50}}{\ddot{x}_{p75o}}, \frac{\ddot{x}_{p55}}{\ddot{x}_{p75o}}, \frac{\ddot{x}_{p60}}{\ddot{x}_{p75o}}, \frac{\ddot{x}_{p65}}{\ddot{x}_{p75o}}, \frac{\ddot{x}_{p70}}{\ddot{x}_{p75o}}, \frac{\ddot{x}_{p75}}{\ddot{x}_{p75o}}\right\}$
Average floor acceleration	Normalized peak acceleration
$J_2 = \frac{1}{6}\sum_i \left[\sigma_{\ddot{x}i}/\sigma_{\ddot{x}io}\right]$	$J_8 = \frac{1}{6}\sum_i \left[\ddot{x}_{pi}/\ddot{x}_{pio}\right]$
Top floor displacement	Top peak displacement
$J_3 = \left\{\frac{\sigma_{x76}}{\sigma_{x76o}}\right\}$	$J_9 = \left\{\frac{x_{p76}}{x_{p76o}}\right\}$
Selected floor displacement	Normalized peak displacement
$J_4 = \frac{1}{7}\sum_i \left[\sigma_{xi}/\sigma_{xio}\right]$	$J_{10} = \frac{1}{7}\sum_i \left[x_{pi}/x_{pio}\right]$

7. Top peak displacement normalized by the corresponding displacement in the uncontrolled structure

$$J_9 = \left\{\frac{x_{p76}}{x_{p76o}}\right\}$$

8. Normalized peak displacement normalized by the corresponding displacement in the uncontrolled structure

$$J_{10} = \frac{1}{7}\sum_i \left[x_{pi}/x_{pio}\right]$$

In this study, the performance criteria related to only passive control systems are considered, i.e., J_1–J_4 and J_7–J_{10}. They are defined as shown in Table 11.32. The performance of dampers is studied for 900 s duration of wind load.

11.4.4.1 Application of Tuned Liquid Column Dampers (TLCD)

Min et al. (2005) studied the performance of TLCD for controlling the wind-induced responses of the benchmark building. The authors studied the robustness of multiple TLCD (MTLCD) and the effects of mass ratio, the number of dampers, central tuning frequency ratio, and the frequency range, on the performance of MTLCD. It is observed that the sensitivity of performance to the tuning frequency ratio is reduced with increasing mass ratio. However, a larger mass ratio does not always guarantee a superior performance, when the head loss coefficient is small. It is reported that the performance of MTLCD is almost equivalent to that of a single TLCD when there exists no uncertainty of stiffness; whereas MTLCD shows superior performance to a single TLCD when stiffness uncertainty exists.

11.4.4.2 Application of Friction Dampers

Patil and Jangid (2010) proposed and investigated a modified FD to attain enhanced performance when installed to the benchmark building. An additional plate is provided between the two sliding

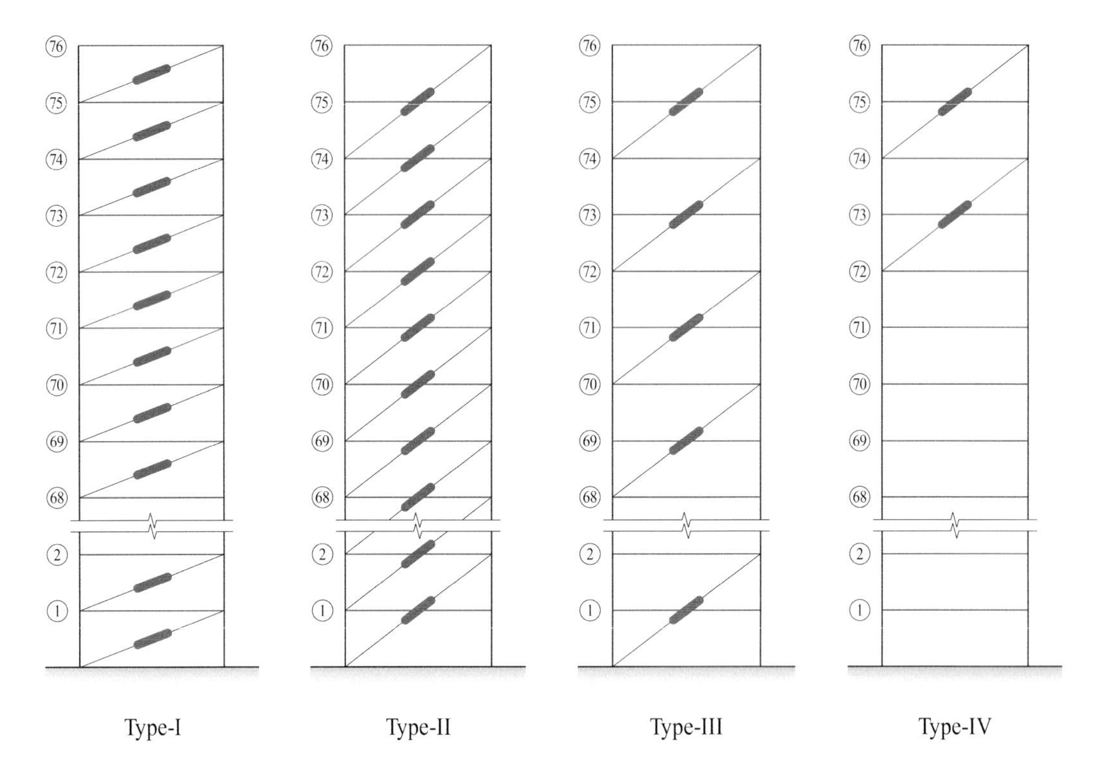

FIGURE 11.22 Various arrangements of fluid viscous dampers.

plates of a conventional FD, which results in an additional sliding interface with the same clamping force. Optimization of location and number of dampers are also carried out with the help of a controllability index which is obtained with the help of RMS value of the inter-story drift. Further, a parametric study of FDs is carried out by varying the slip force. At optimized locations, the proposed dampers give significant enhancement in the performance.

11.4.4.3 Application of Fluid Viscous Dampers

Patil and Jangid (2009) studied four different arrangements of fluid viscous dampers, as shown in Figure 11.22. The optimal location of dampers is identified with the help of a controllability index. To know the effect of damping ratio, a parametric study is carried out by varying the induced damping. The number of dampers installed in the four different arrangements is Type I (76), Type II (75), Type III (38), and Type IV (2).

There is a considerable reduction in the number of dampers by the optimization procedure carried out with the help of the controllability index, obtained with the help of RMS inter-story drift.

11.4.4.4 Application of TMD

Patil and Jangid (2011) proposed implementation of TMD or MTMDs at the top floor of the 76-story benchmark building. Optimum parameters of single TMD (STMD) and multiple TMD (MTMD) are obtained for the minimization of various performance criteria of the building by numerical search technique. The robustness of the MTMD is investigated by studying the effects of design parameters such as mass ratio, damping ratio, number of dampers, tuning frequency, and frequency bandwidth. From the trends of the numerical results obtained, it is found that by increasing the number of dampers in MTMD beyond 5 does not provide significant response reduction.

Table 11.31 presents the evaluation criteria when the 76-story benchmark building is installed with TLCD, MTLCD, double FD (DFD), linear fluid viscous dampers (LVD), TMD, and MTMD.

TABLE 11.33
Comparison of Evaluation Criteria for Passive Control Strategies

Performance Criterion	TLCD	MTLCD	DFD	LVD	TMD	MTMD
J_1	0.391	0.392	0.458	0.162	0.362	0.356
J_2	0.387	0.388	0.447	0.18	0.358	0.351
J_3	0.552	0.550	0.567	0.386	0.53	0.527
J_4	0.554	0.552	0.571	0.396	0.532	0.529
J_7	0.473	0.464	0.540	0.213	0.413	0.379
J_8	0.476	0.472	0.525	0.253	0.435	0.406
J_9	0.621	0.614	0.685	0.468	0.583	0.571
J_{10}	0.628	0.621	0.690	0.493	0.587	0.578

All the response quantitates are found to reduce significantly by installation of the designed passive control devices, with their optimum parameters.

11.5 SEISMICALLY EXCITED BENCHMARK HIGHWAY BRIDGE

A benchmark problem for seismically excited highway bridge has been developed and presented as a test bed structure for response control of highway bridges and development of control strategies (Agrawal et al., 2009). It provides a systematic and standardized means by which competing control strategies, including devices, control algorithms, and sensors, can be evaluated. Different passive control devices are examined analytically, to control the seismic response of the bridge under six historic earthquake ground motions. The evaluation criteria are developed under the coordination of the ASCE Task Committee on Structural Control Benchmarks. The benchmark structural control problem for a seismically excited highway bridge shown in Photo 11.1 is based on the 91/5 over-crossing, located in Orange County of Southern California. The salient features of the benchmark highway bridge are presented in Table 10.7.

Figure 10.48 shows the elevation and plan view and Figure 10.49 shows the cross-sectional elevation of the benchmark highway bridge.

The effect of soil-structure interaction at the end abutments and approach embankments is considered by modeling the interaction using equivalent spring and dashpot. A MATLAB-based nonlinear structural analysis tool is developed for the nonlinear dynamic analysis. The model is verified extensively through detailed comparisons with the nonlinear finite element model of the bridge in

PHOTO 11.1 91/5 overcrossing, located in Orange County of Southern California. (Agrawal et al., 2009.)

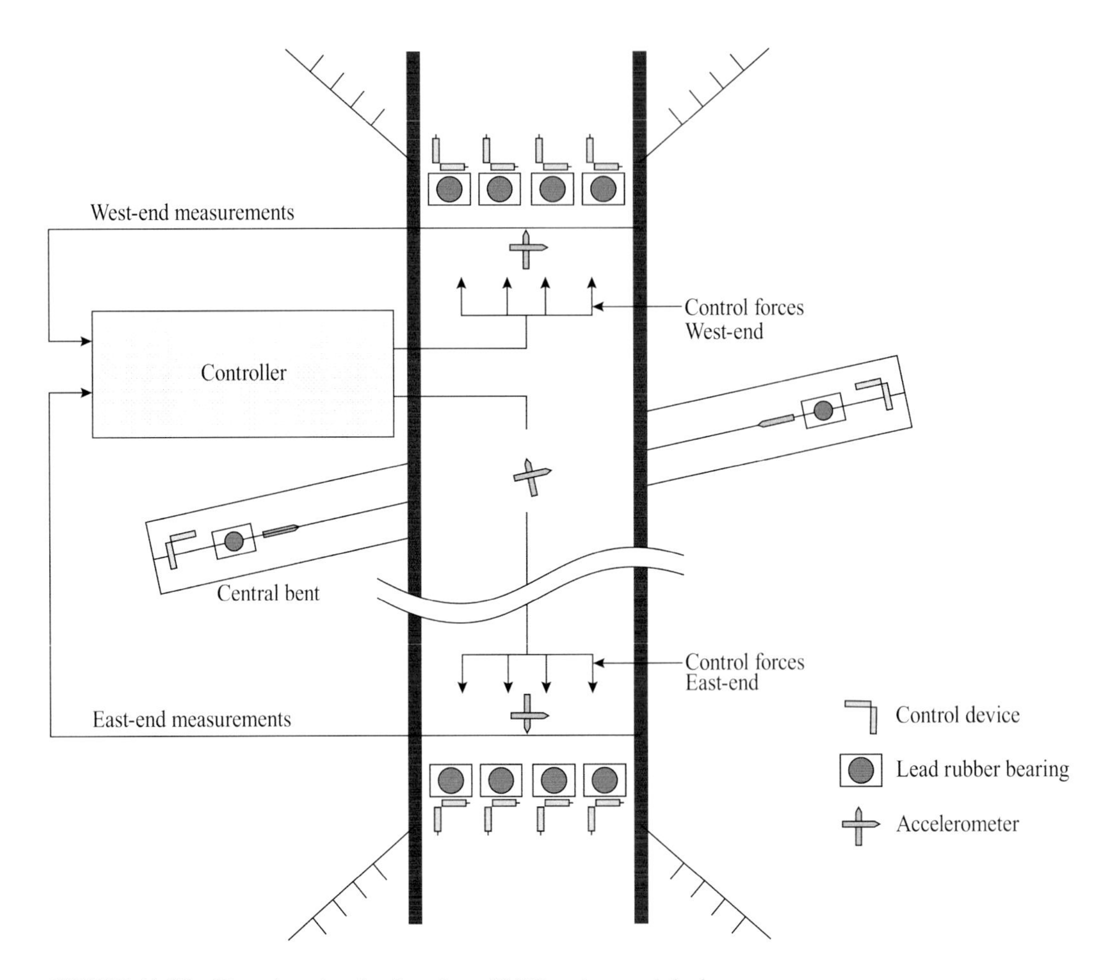

FIGURE 11.23 Plan view showing location of LRB and control devices.

ABAQUS. In the numerical simulations, a full order nonlinear model of the bridge is used as the evaluation model, to preserve the effects of column nonlinearity, and realistic implementation of the control strategies. The control devices are installed between the deck and the end abutments. Figure 11.23 shows the plan view, showing location of isolation bearings and control devices. Under dynamic loads, these devices are extremely stiff and are assumed to behave as rigid links. Since the bridge is located very close to two major faults, earthquake ground motions are applied simultaneously in the two horizontal directions. Six ground motions, representing soil classifications from A to D (according to NEHRP), are selected. Evaluation criteria and control constraints are specified for the design of controllers.

11.5.1 Evaluation Model

Agrawal et al. (2006, 2009) developed a 3D finite element evaluation model in MATLAB, which effectively represents the complex behavior of the full-scale benchmark bridge. The full-order nonlinear evaluation model is considered as a true representation of the real structural system. The model is used to assess the performance of applied controllers. The nonlinear behavior of the center columns and the isolation bearings is considered while formulating the bilinear force deformation relationship. Due to flexible soil at the embankment and foundation level, the bridge is likely to

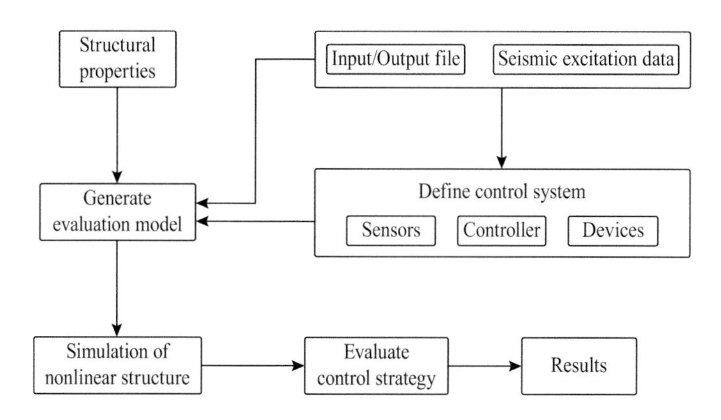

FIGURE 11.24 Flowchart of procedure to solve benchmark highway bridge problem.

undergo large displacements. Hence, soil-structure interaction is duly considered, since there is a possibility of extensive instability of ground by soil liquefaction, slope failure, and lateral spreading. In the actual bridge, four traditional non-seismic laminated rubber bearings (LRB) are provided at each abutment, and four nonlinear fluid viscous dampers are installed between the deck, and each end abutment.

However, in the evaluation model, lead rubber bearings replace the actually provided eight non-seismic elastomeric bearings, since lead rubber bearings provide improved seismic performance of an uncontrolled bridge. Figure 11.23 shows the location of control devices and accelerometers on the bridge. Eight accelerometers are installed to measure the absolute accelerations in two horizontal directions, and four displacement sensors measure the relative displacements.

The evaluation model is used to simulate the performance of control strategies. Figure 11.24 outlines the general procedure to solve the benchmark highway bridge problem. The resulting structural system has 430 (phase I) and 442 (phase II) DOF, representing the full order evaluation model.

11.5.2 Finite Element Model

A 3D finite element model developed in ABAQUS has a total of 108 nodes, 4 rigid links (to model the abutment and the deck-ends), 70 beam elements, 24 springs, 27 dashpots, and 8 user-defined bearing elements. The bearings installed between the bridge deck and abutments at both ends are idealized by bidirectional bilinear plasticity model. The bilinear force deformation relationship is hysteretic in both directions. The bearings are modeled as shear elements, with infinite vertical axial stiffness. The torsional rigidity, bending stiffness, mass, and damping of bearings are assumed to be negligible. The superstructure of the bridge, including the deck and the bent beam, is assumed to be elastic, and the deck structure elements are assumed to be linear. The element mass and stiffness matrices are derived from the finite element model and are assembled at the nodes using lumped mass and stiffness approximation. The nonlinear elements are added to the linear elastic deck elements and the resulting model is used for evaluating the structural responses. Both the abutments and deck-ends are assumed to be rigid. Structural member nonlinearities are included in the model to capture the inelastic moment-curvature behavior of columns and shear-displacement relationship of bearings. Rigid links with a skew angle of 33° are used to connect the isolation bearings and control devices, between the deck and end abutments. Soil-structure interaction at the end abutments/approach embankments is modeled by frequency-independent springs and dashpots. Dashpots are used

TABLE 11.34
Characteristics of Mode Shapes of the Benchmark Highway Bridge

Mode No	Natural Frequency (Hz)	Time Period (s)	Mode of Vibration
1	1.23	0.813	Torsional
2	1.28	0.781	Torsional coupled with vertical
3	1.55	0.645	First vertical
4	1.69	0.592	First transverse
5	1.77	0.565	Second vertical
6	3.26	0.307	Second transverse

to simulate soil radiation damping in the three translational directions. Each column of the bent is modeled as stiff column cap, elastic column segments, and nonlinear column segments in ABAQUS. The length of column segment is considered as 0.732 m. The yield moment of bent columns is 5.0822×10^7 Nm and; the per-yield flexure stiffness and post-yield flexural stiffness are 2.091×10^{10} N/m and 1.791×10^8 N/m, respectively. Approach embankments on each side of the bridge are massive deformable bodies, which have a tendency to amplify earthquake motions and interact strongly with the bridge structure. An accurate consideration of the approach embankments is obtained by evaluating the soil stiffness and damping coefficients. The pile foundations at both abutments are assumed to behave like an equivalent linear viscoelastic element and are modeled as springs and dashpots in the two horizontal directions. All element mass matrices and initial elastic element stiffness matrices, obtained in ABAQUS, are summed at nodal masses to assemble global stiffness and mass matrices in MATLAB. The nonlinear elements are then modeled for generating the evaluation model. Nodal masses of the deck and bent are assigned six dynamic DOF. The equations are partitioned into active and constrained DOF. Constraints are applied by eliminating the rows and columns associated with the fixed boundary conditions and condensing out rigid links by applying kinematic constraints. Combined spring (stiffness) and dashpot (damping) values are obtained to account for embankment and pile foundation at both ends of the bridge. Table 11.34 gives characteristics of mode shapes of the benchmark highway bridge.

11.5.3 Reduced-Order Controller Design Model

The model resulting from finite element formulation has a large number of DOF. Some assumptions are made regarding the behavior of the bridge, to make the model more manageable for dynamic simulation while retaining the fundamental behavior of the bridge. Static condensation is performed, by partitioning the mass and stiffness matrices corresponding to the structure DOF into active and dependent DOF. Application of static condensation to the full model of the bridge results in the evaluation model, which involves a large number of DOF and high-frequency dynamics. Hence, a reduced-order model of the system is developed from the initial elastic evaluation model. The reduced-order model has the same output as the evaluation model and is assumed to retain the dominant characteristics of the full-order evaluation system. This model is designated as the controller design model.

11.5.4 Governing Equations of Motion

The nonlinear evaluation model is evaluated to find the response of the bridge to the selected earthquake ground motions. It is assumed that the properties of the system remain constant during the time increment. Incremental dynamic equations of motion are solved using the Newmark-beta method, in combination with the modal reduction method. The input and output matrices are found using the state-space form of the equation. The global damping matrix is a combination of the distributed 5% inherent

Rayleigh damping in the first two modes and soil radiation damping. The basic equations of motion and the nonlinear analysis tool remain the same for all the controllers used in the study. A MATLAB-based nonlinear structural analysis tool is developed for vibration control simulations. This tool is written as S-function and is incorporated into the SIMULINK model. The S-function solves the incremental equations of motion and convergence technique at the overshooting regions. The inputs to the S-function block in SIMULINK are the seismic excitation and control forces provided by the control devices.

11.5.5 Evaluation Criteria

To facilitate direct and systematic comparison of various control strategies, a set of 21 normalized evaluation criteria is defined in the benchmark problem. The evaluation criteria consider the ability of the controller to reduce the peak responses, the normalized responses over the entire time record, and the control requirements. Since the earthquake is bidirectional, many of these criteria are evaluated in both horizontal directions. The merit of a controller is based on the criteria, given in terms of both, the RMS, and the peak response quantities. The RMS displacement provides a measure of the required physical size of the device. The RMS velocity provides a measure of the control power required, and RMS absolute acceleration, a measure of the magnitude of the control forces generated. For each passive control design, the 16 evaluation criteria are found for the six earthquake records. Since there are two piers, response quantities related to piers are normalized by the maximum of the response quantities for two piers. For highway bridges subjected to earthquake excitation, critical responses are related to the structural integrity of the bridge and the serviceability.

11.5.6 Peak Response

The first eight evaluation criteria are defined to measure ability of the controller to reduce peak response quantities of the benchmark highway bridge.

1. Peak base shear force in the controlled structure, normalized by the corresponding shear in the uncontrolled structure

$$J_1 = \max\left\{\frac{\max_{i,t}\left|F_{bi}(t)\right|}{F_{0b}^{\max}}\right\}$$

2. Peak overturning moment in the controlled structure, normalized by the corresponding moment in the uncontrolled structure

$$J_2 = \max\left\{\frac{\max_{i,t}\left|M_{bi}(t)\right|}{M_{0b}^{\max}}\right\}$$

3. Peak displacement at the midspan of the controlled structure, normalized by the corresponding midspan displacement of the uncontrolled structure

$$J_3 = \max\left\{\max_{i,t}\left|\frac{y_{mi}(t)}{y_{0m}^{\max}}\right|\right\}$$

Maximum relative displacement of the deck governs the width of expansion joints. Furthermore, it may result in pounding between deck segments and deck with piers and abutments, causing excessively large lateral force transfer from deck to abutments, and from deck to piers.

4. Peak acceleration at the midspan of the controlled structure, normalized by the corresponding peak midspan acceleration of the uncontrolled structure

$$J_4 = \max\left\{\max_{i,t}\left|\frac{\ddot{y}_{mi}(t)}{\ddot{y}_{0m}^{\max}}\right|\right\}$$

5. Peak deformation of the bearings in the controlled structure, normalized by the corresponding peak deformation of bearings in the uncontrolled structure

$$J_5 = \max\left\{\max_{i,t}\left|\frac{y_{bi}(t)}{y_{0b}^{\max}}\right|\right\}$$

Excessive increase in the time period due to seismic isolation results in large deck displacements. Bearing displacement is a quantity of prime interest in the isolated bridges, because if it exceeds certain limits, the bearing may fail and result in a collapse of the bridge.

6. Peak curvature at the bent column in the controlled structure, normalized by the corresponding curvature in the uncontrolled structure

$$J_6 = \max\left\{\max_{j,t}\frac{\Phi_j(t)}{\Phi^{\max}}\right\}$$

Ductility factors are useful indicators of the level of inelastic response.

7. Peak dissipated energy of curvature at the bent column in the controlled structure, normalized by the corresponding dissipated energy in the uncontrolled structure

$$J_7 = \max\left\{\frac{\max_{j,t}\int dE_j}{E^{\max}}\right\}$$

The hysteretic energy dissipation is due to deformation of piers in the inelastic range. The dissipation of energy in the yielding members reduces the seismic response of the bridge. The seismic energy is assumed to be dissipated by the inelastic pier behavior, by LRB, and by control devices.

8. The number of plastic connections with control, normalized by the corresponding number of plastic connections without control

$$J_8 = \max\left\{\frac{N_d^c}{N_d}\right\}$$

The primary element in dissipation of seismic energy in bridges is the pier, where a plastic hinge is expected to form in the nonlinear segments.

11.5.7 Normed Response

The second set of six evaluation criteria is based on normed responses, over the entire simulation time of the earthquake. The normed value of the response is denoted by $\|\bullet\|$, and is defined as

$$\|\bullet\| = \sqrt{\frac{1}{t_f}\int_0^{t_f}(\bullet)^2\,dt}$$

where t_f is the time required for the response to attenuate.

9. Normed base shear force in the controlled structure, normalized by the corresponding normed shear in the uncontrolled structure

$$J_9 = \max\left\{\frac{\max_t \|F_{bi}(t)\|}{\|F_{0b}^{\max}\|}\right\}$$

10. Normed overturning moment in the controlled structure, normalized by the corresponding normed overturning moment in the uncontrolled structure

$$J_{10} = \max\left\{\frac{\max_{t,i} \|M_{bi}(t)\|}{\|M_{0b}^{\max}\|}\right\}$$

11. Normed displacement at the midspan in the controlled structure, normalized by the corresponding normed displacement at the midspan in the uncontrolled structure

$$J_{11} = \max\left\{\max_i \frac{\|y_{mi}(t)\|}{\|y_{0m}^{\max}\|}\right\}$$

12. Normed acceleration at midspan in the controlled structure, normalized by the corresponding normed acceleration at midspan in the uncontrolled structure

$$J_{12} = \max\left\{\max_i \frac{\|\ddot{y}_{mi}(t)\|}{\|\ddot{y}_{0m}^{\max}\|}\right\}$$

13. Normed deformation of bearings in the controlled structure, normalized by the corresponding normed displacement of bearings in the uncontrolled structure

$$J_{13} = \max\left\{\max_i \left\|\frac{y_{bi}(t)}{y_{0b}^{\max}}\right\|\right\}$$

14. Normed curvature at the bent column in the controlled structure, normalized by the corresponding normed curvature at the bent column in the uncontrolled structure

$$J_{14} = \max\left\{\max_{j,t} \left\|\frac{\Phi_j(t)}{\Phi^{\max}}\right\|\right\}$$

11.5.8 Control Requirement

The third set of three evaluation criteria is defined for quantifying the measure of control resource required by the controller.

15. Peak control force generated by the control devices, normalized by the seismic weight of the bridge, based on the mass of the superstructure

$$J_{15} = \max\left\{\max_{l,t}\left(\frac{f_l(t)}{W}\right)\right\}$$

where $f_l(t)$ is the force generated by lth control device.

TABLE 11.35
Summary of Evaluation Criteria

Peak Responses	Normed Responses	Control Strategy
Peak base shear	Base shear	Peak force
$J_1 = \max\left\{\dfrac{\max\limits_{i,t}\left\|F_{bi}(t)\right\|}{F_{0b}^{\max}}\right\}$	$J_9 = \max\left\{\dfrac{\max\limits_{t}\left\|F_{bi}(t)\right\|}{\left\|F_{0b}^{\max}\right\|}\right\}$	$J_{15} = \max\left\{\max\limits_{l,t}\left(\dfrac{f_l(t)}{W}\right)\right\}$
Peak overturning moment	Overturning moment	Device stroke
$J_2 = \max\left\{\dfrac{\max\limits_{i,t}\left\|M_{bi}(t)\right\|}{M_{0b}^{\max}}\right\}$	$J_{10} = \max\left\{\dfrac{\max\limits_{t,i}\left\|M_{bi}(t)\right\|}{\left\|M_{0b}^{\max}\right\|}\right\}$	$J_{16} = \max\left\{\max\limits_{l,t}\left(\dfrac{d_l(t)}{x_{0m}^{\max}}\right)\right\}$
Displacement at midspan	Displacement at midspan	NA
$J_3 = \max\left\{\max\limits_{i,t}\left\|\dfrac{y_{mi}(t)}{y_{0m}^{\max}}\right\|\right\}$	$J_{11} = \max\left\{\max\limits_{i}\dfrac{\left\|y_{mi}(t)\right\|}{\left\|y_{0m}^{\max}\right\|}\right\}$	
Acceleration at midspan	Acceleration at midspan	NA
$J_4 = \max\left\{\max\limits_{i,t}\left\|\dfrac{\ddot{y}_{mi}(t)}{\ddot{y}_{0m}^{\max}}\right\|\right\}$	$J_{12} = \max\left\{\max\limits_{i}\left\|\dfrac{\ddot{y}_{mi}(t)}{\ddot{y}_{0m}^{\max}}\right\|\right\}$	
Bearing deformation	Displacement at abutment	Control devices J_{19} = number of control devices
$J_5 = \max\left\{\max\limits_{i,t}\left\|\dfrac{y_{bi}(t)}{y_{0b}^{\max}}\right\|\right\}$	$J_{13} = \max\left\{\max\limits_{i}\left\|\dfrac{y_{bi}(t)}{y_{0b}^{\max}}\right\|\right\}$	
Ductility	Ductility	NA
$J_6 = \max\left\{\max\limits_{j,t}\dfrac{\Phi_j(t)}{\Phi^{\max}}\right\}$	$J_{14} = \max\left\{\max\limits_{j,t}\left\|\dfrac{\Phi_j(t)}{\Phi^{\max}}\right\|\right\}$	
Dissipated energy	Plastic connections	NA
$J_7 = \max\left\{\dfrac{\max\limits_{j,t}\int dE_j}{E^{\max}}\right\}$	$J_8 = \max\left\{\dfrac{N_d^c}{N_d}\right\}$	

Note: NA = Not applicable for passive control devices.

16. Peak stroke of the control devices, normalized by the maximum deformation of bearings in the uncontrolled structure

$$J_{16} = \max\left\{\max_{l,t}\left(\frac{d_l(t)}{x_{0m}^{\max}}\right)\right\}$$

17. Total number of control devices required in the control system to control the response of the bridge

    J_{19} = number of control devices.

    For each control design, the 16 evaluation criteria are calculated for the six earthquakes, and the performance of the controller is analyzed. Table 11.35 summarizes the evaluation criteria.

11.5.9 Earthquake Ground Excitations

As a consequence of the strong ground motions recorded during Kobe, Northridge, and Turkey earthquakes, it is revealed that stronger design earthquakes control the seismic design of highway

TABLE 11.36
Properties of Selected Earthquake Records

Earthquake	Magnitude	Distance to Fault (km)	Peak Ground Acceleration (g)	
			EW	NS
North Palm Springs (1986)	6.0	7.3	0.492	0.613
Chi-Chi (1999)	7.6	10.39	1.157	0.417
El Centro (1940)	7.0	8.3	0.313	0.215
Northridge (1994)	6.7	7.1	0.838	0.472
Turkey (1999)	7.1	17.6	0.728	0.822
Kobe (1995)	6.9	11.1	0.509	0.503

bridges. These stronger earthquakes include the effects of near-fault pulses, fault-normal motions, and near-fault deep soil site motions. The benchmark highway bridge is subjected to six ground motions, which cover a large range of characteristics, namely, frequency content, duration, and peak ground acceleration (PGA). The fault-normal component is applied along the transverse direction, while the fault-parallel component is applied simultaneously, along the longitudinal direction. Following six historical earthquake records are considered as ground excitations for the numerical simulations of seismic protective systems installed in the bridge; North Palm Springs (1986) earthquake, TCU084 component of the 1999 Chi-Chi earthquake, El Centro component of 1940 Imperial Valley earthquake, Rinaldi component of the 1994 Northridge earthquake, Bolu component of the 1999 Duzce earthquake, and Nishi-Akashi component of the 1995 Kobe earthquake. Since the bridge is located very close to active faults, five near-fault motions are considered along with the El Centro record. Table 11.36 mentions the properties of the earthquake records.

11.5.10 Phase I and Phase II of the Benchmark Highway Bridge

The benchmark highway bridge model consists of two phases, phase I, in which the bridge deck is fixed to the outriggers and is isolated at the abutments, and phase II, in which the bridge deck is isolated from the outriggers, as well as at the abutments. In phase I, a total of 16 control devices are installed at two locations between the deck and end abutments. In phase II, four additional devices are installed at the center. Table 11.37 presents uncontrolled responses of North Palm Spring, Chi-Chi, and El Centro earthquake for phase I and phase II; and Table 11.38 presents uncontrolled responses of Northridge, Turkey, and Kobe earthquakes for phase I and phase II.

Table 11.39 presents the ratio of the uncontrolled response for phase II to the uncontrolled response for phase I. It is observed that provision of isolators at the center in phase II substantially reduces the pier base shear, pier base moment, midspan deck acceleration and their corresponding normed values. However, there is tremendous increase in the midspan deck displacement, isolator displacement and their corresponding normed values. This is due to increased lateral flexibility of the bridge arising from isolators installed at the center.

11.5.11 Sample Controllers for the Benchmark Highway Bridge

A sample passive control system using non-LVD is designed for the benchmark highway bridge. For control system, 16 control devices (phase I) are considered, placed orthogonally between the deck and end abutments. Four additional devices, placed at the center in phase II, link the bent beam to the piers. The non-LVD have damping coefficient 800 kNs/m (phase I) and 100 kNs/m (phase II), and nonlinearity constant $(\alpha) = 0.6$. Figure 11.25 shows the basic SIMULINK model for passive control devices.

TABLE 11.37
Uncontrolled Response of North Palm Spring, Chi-Chi, and El Centro Earthquakes

	North Palm Springs		Chi-Chi		El Centro	
Response	**Phase I**	**Phase II**	**Phase I**	**Phase II**	**Phase I**	**Phase II**
F_{0b}^{max} (N)	5,897,100	1,049,200	18,691,000	1,049,200	5,811,300	995,010
M_{0b}^{max} (Nm)	27,647,000	6,884,300	55,321,000	6,884,300	25,832,000	7,329,800
y_{0m}^{max} (m)	0.062	0.537	0.292	0.537	0.057	0.251
$\ddot{y}_{0m}^{max}$ (m/s^2)	5.715	1.614	13.079	1.614	3.777	1.137
y_{0b}^{max} (m)	0.094	0.544	0.299	0.544	0.101	0.249
Φ^{max}	0.001	0	0.024	0	0.001	0
E^{max}	0	0	21,445,000	0	0	0
$\lVert F_{0b}^{max} \rVert$ (N)	1,068,600	335,140	3,296,100	335,140	1,375,300	302,910
$\lVert M_{0b}^{max} \rVert$ (Nm)	5,294,200	2,453,600	17,637,000	2,453,600	6,609,700	2,211,300
$\lVert y_{0m}^{max} \rVert$ (m)	0.013	0.144	0.051	0.144	0.015	0.118
$\lVert \ddot{y}_{0m}^{max} \rVert$ (m/s^2)	1.067	0.286	2.492	0.286	0.903	0.305
$\lVert y_{0b}^{max} \rVert$ (m)	0.025	0.147	0.051	0.147	0.022	0.130
$\lVert \Phi^{max} \rVert$	0	0	0.005	0	0	0
x (m)	0.097351	0.54401	0.31354	0.54401	0.10933	0.24867
$\dot{x}^{max}$ (m/s)	1.131	1.3815	2.1552	1.3815	0.82198	0.40618

TABLE 11.38
Uncontrolled Response of Northridge, Turkey, and Kobe Earthquakes

	Northridge		Turkey		Kobe	
Response	**Phase I**	**Phase II**	**Phase I**	**Phase II**	**Phase I**	**Phase II**
F_{0b}^{max} (N)	19,403,000	1,326,100	13,732,000	1,107,800	6,898,700	783,250
M_{0b}^{max} (Nm)	56,094,000	8,032,700	52,548,000	8,001,000	33,262,000	5,067,700
y_{0m}^{max} (m)	0.310	0.543	0.181	0.324	0.075	0.276
$\ddot{y}_{0m}^{max}$ (m/s^2)	14.413	2.055	9.847	1.852	5.430	0.973
y_{0b}^{max}(m)	0.316	0.550	0.190	0.328	0.160	0.286
Φ^{max}	0.029	0	0.011	0	0.002	0
E^{max}	4,139,800	0	1,119,500	0	0	0
$\lVert F_{0b}^{max} \rVert$ (N)	3,913,900	427,320	1,409,000	586,550	1,316,500	334,060
$\lVert M_{0b}^{max} \rVert$ (Nm)	19,231,000	3,118,600	13,362,000	4,285,600	6,071,900	2,434,500
$\lVert y_{0m}^{max} \rVert$ (m)	0.058	0.150	0.023	0.246	0.014	0.133
$\lVert \ddot{y}_{0m}^{max} \rVert$ (m/s^2)	2.894	0.584	0.979	0.327	0.869	0.301
$\lVert y_{0b}^{max} \rVert$ (m)	0.058	0.149	0.033	0.252	0.034	0.141
$\lVert \Phi^{max} \rVert$	0.005	0	0.006	0	0	0
x (m)	0.34697	0.54982	0.19181	0.32842	0.16197	0.2858
$\dot{x}^{max}$ (m/s)	2.3611	1.5581	1.9672	1.029	1.4553	0.56092

TABLE 11.39
Comparison of Uncontrolled Response (Phase II/Phase I)

Evaluation Criteria	Value of Performance Criteria under Earthquake					
	North Palm Springs	Chi-Chi	El Centro	Northridge	Turkey	Kobe
J_1	0.118	0.056	0.171	0.063	0.087	0.114
J_2	0.118	0.134	0.284	0.143	0.152	0.152
J_3	1.557	1.837	4.417	1.749	1.788	3.670
J_4	0.176	0.133	0.298	0.143	0.188	0.179
J_5	1.096	1.818	2.473	1.740	1.727	1.790
J_6	0.118	0.013	0.284	0.013	0.034	0.152
J_7	0	0	0	0	0	0
J_8	0	0	0	0	0	0
J_9	0.183	0.102	0.220	0.109	0.416	0.254
J_{10}	0.258	0.139	0.335	0.162	0.347	0.401
J_{11}	4.217	2.833	7.961	2.590	10.573	9.713
J_{12}	0.180	0.115	0.338	0.202	0.334	0.346
J_{13}	2.111	2.893	5.524	2.544	7.596	4.133
J_{14}	0.258	0.022	0.335	0.031	0.033	0.401
Maximum displacement	1.055	1.741	2.275	1.585	1.713	1.765
Maximum velocity	0.413	0.641	0.494	0.660	0.523	0.385

11.5.11.1 Application of Sliding Isolators

In recent years, there have been important studies on the use of friction isolators for protection of bridges from severe earthquake attacks. In the present study, response of the benchmark highway bridge, isolated with VFPS, and variable frequency pendulum isolator (VFPI) is investigated under the six earthquake ground motions. In order to verify the effectiveness of VFPS and VFPI, response of the bridge isolated with these systems is compared to that, isolated with the conventional FPS. Parametric studies are carried out by evaluating various performance indices to critically

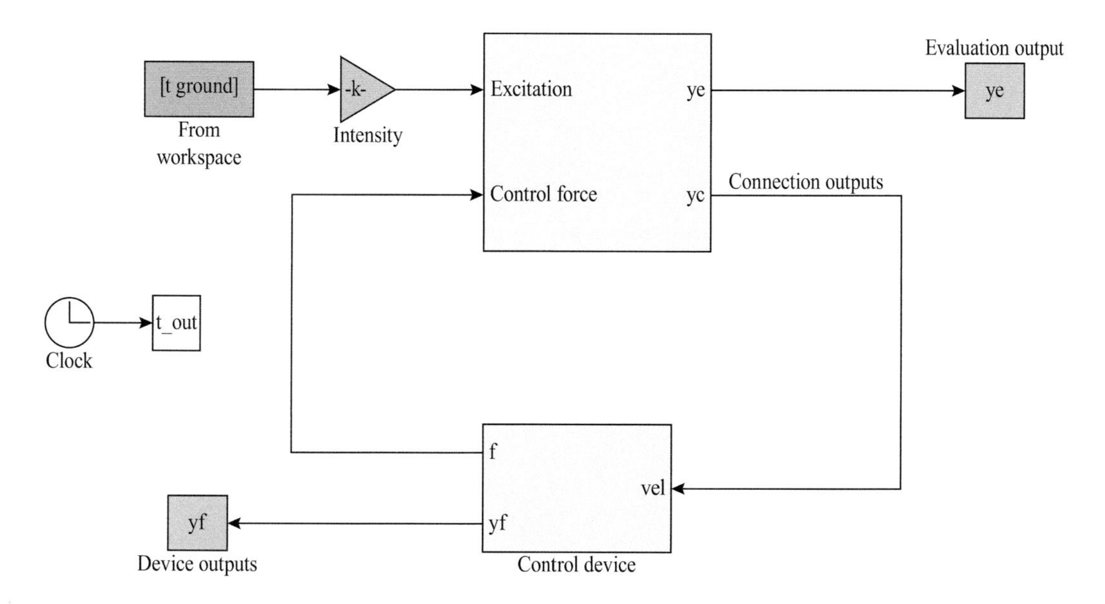

FIGURE 11.25 Basic SIMULINK model for functioning of a passive controller.

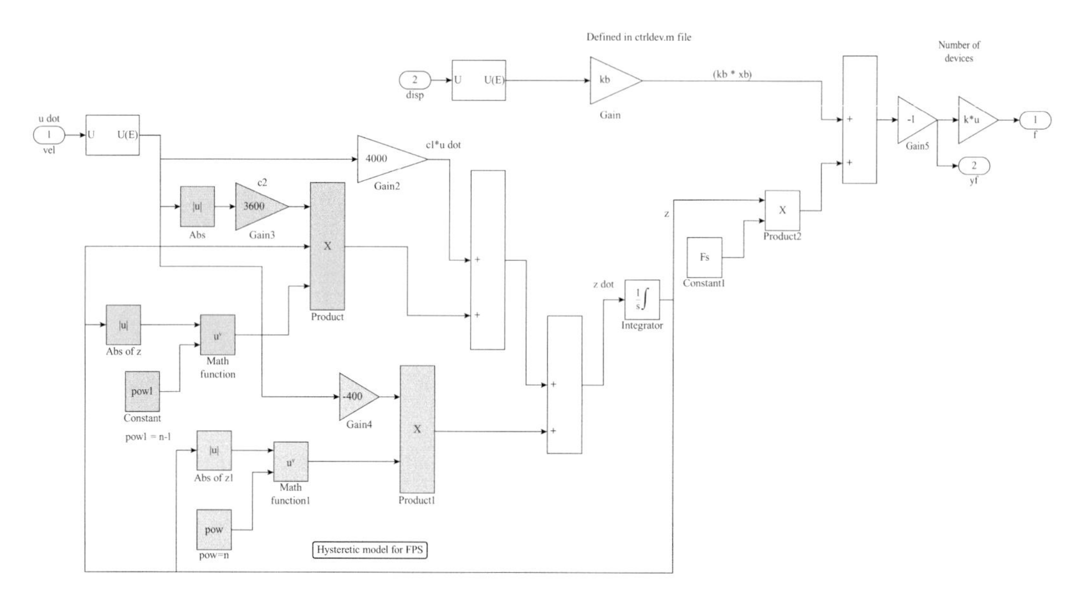

FIGURE 11.26 SIMULINK model for friction pendulum system.

examine the behavior of the bridge isolated with the FPS, VFPS and VFPI. Figure 11.26 shows the SIMULINK model for FPS.

Figure 11.27 shows the SIMULINK model for finding coefficient of friction of FPS, and Figure 11.28 depicts the SIMULINK model for VFPS. Figure 11.29 represents the combined SIMULINK model of subsystem to find coefficient of friction for VFPS and for VFPI.

11.5.12 Numerical Study

To implement the sliding isolation systems, the isolator parameters are selected such that, during the six seismic excitations, the restoring force of isolators does not exceed 1,000 kN. The values

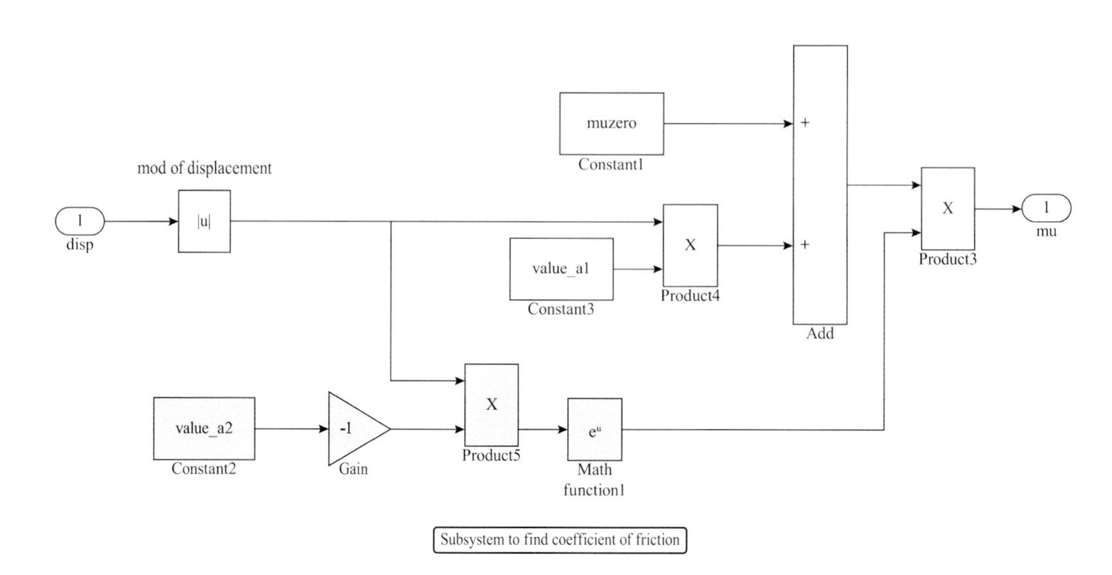

FIGURE 11.27 SIMULINK model for finding coefficient of friction of friction pendulum system.

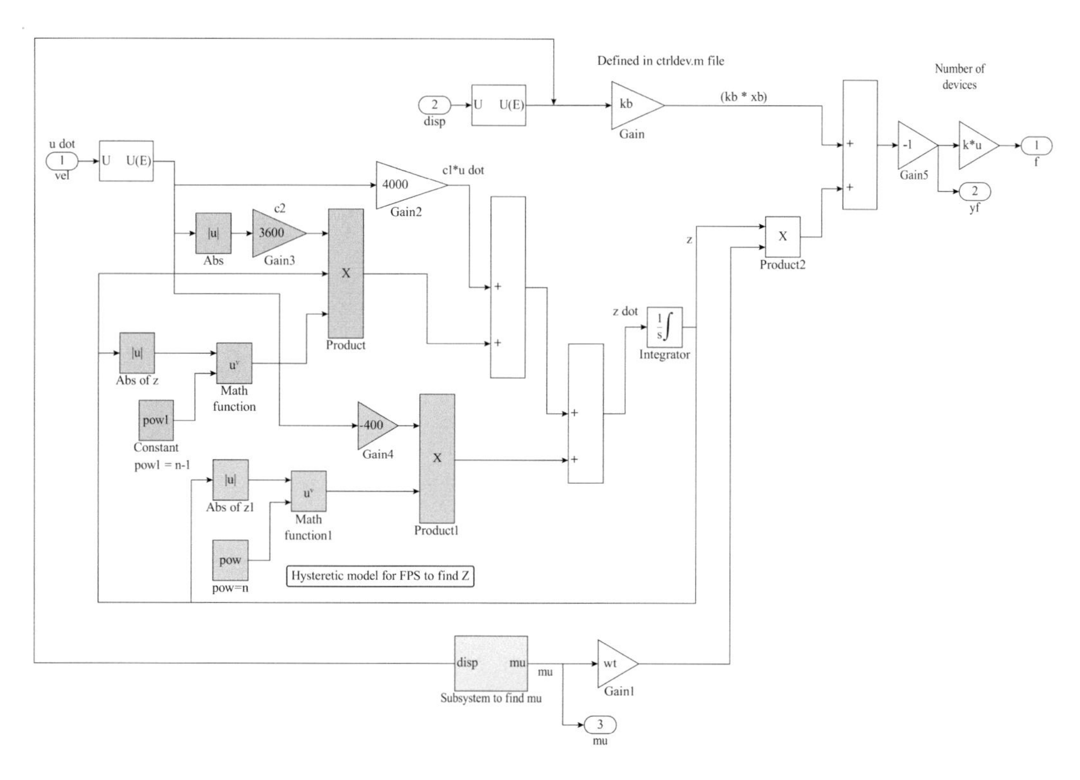

FIGURE 11.28 SIMULINK model for variable friction pendulum system.

of parameters are selected so as to give minimum response of the benchmark highway bridge. Table 11.40 presents the evaluation criteria for non-FVD, FPS, VFPS, and VFPI.

11.5.13 Seismic Response Control of Benchmark Highway Bridge: Phase II

In phase II of the benchmark highway bridge, the bridge is isolated at the central bent in addition to the deck-ends. A total of 16 devices are placed at the junction of the deck-ends and abutments (eight devices on each side) and four additional devices are installed at the junction of the bent beam and piers (two devices at each junction). The response of the bridge is evaluated with sliding isolators, viz., FPS, VFPS, and VFPI. The optimum parameters of the control devices are presented in Table 11.41. The parameters are selected so that the control force is limited to the capacity of the device, i.e., 1,000 kN.

11.5.14 Evaluation Criteria

Table 11.42 presents the comparison of evaluation criteria under the North Palm Springs (1986), Chi-Chi (1999), and El Centro (1940) Earthquakes; and Table 11.43 presents the comparison of evaluation criteria under the Northridge (1994), Turkey (1999), and Kobe (1995) earthquake for the passive control strategies. Table 11.44 presents the comparison of J_1 to J_{16} criteria for all earthquakes and all considered control strategies.

11.6 SEISMICALLY EXCITED BENCHMARK CABLE-STAYED BRIDGE

A benchmark problem on the seismic response control of cable-stayed bridge has been developed and presented as a testbed structure for the development of control strategies (Dyke et al., 2002a, 2002b; Caicedo et al., 2003). The benchmark cable-stayed bridge is Missouri 74–Illinois

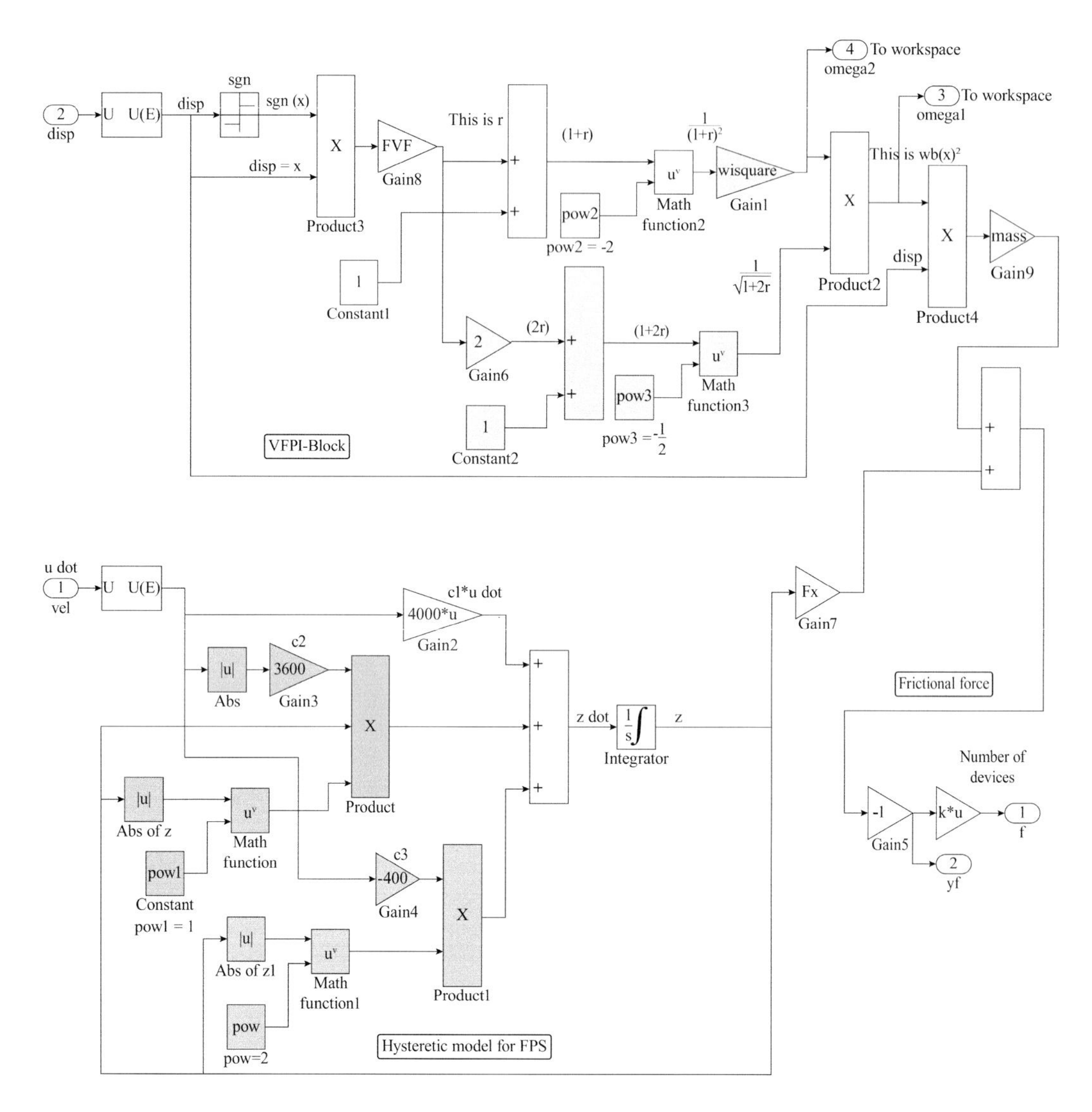

FIGURE 11.29 Combined SIMULINK model of subsystem to find coefficient of friction for VFPS and VFPI.

146 bridge, spanning the Mississippi River near Cape Girardeau, Missouri. It is designed according to AASHTO specifications, by HNTB Corporation, an American infrastructure design firm. The construction was completed in 2003. Different passive control devices are examined analytically, to control the seismic response under three historic earthquake ground motions. The salient features of benchmark cable-stayed bridge are presented in Table 11.45. The model of the benchmark bridge is based on the Bill Emerson Memorial Bridge, shown in Figure 11.30, and Figure 11.31 shows the cross-sectional elevation of towers of the cable-stayed bridge.

A MATLAB-based nonlinear structural analysis tool is developed for the nonlinear dynamic analysis. The model is verified extensively through detailed comparisons with the nonlinear finite element model of the bridge in ABAQUS. Evaluation criteria and control constraints are specified for the design of controllers. Damping for evaluation model is considered as 3%. The bridge is subjected to earthquake excitations in the longitudinal direction. The responses considered for evaluation of a cable-stayed bridge are (i) force responses of the tower, (ii) displacement responses

TABLE 11.40
Evaluation Criteria for FVD, FPS, VFPS, and VFPI

Evaluation Criteria	Control Type	Value of Evaluation Criteria						
		N. Palm Springs	Chi-Chi	El Centro	Northridge	Turkey	Kobe	Maximum
Peak base shear, J_1	FVD	1.224	0.763	0.636	0.775	0.776	0.860	1.224
	FPS	1.059	0.822	0.755	0.891	0.940	0.885	1.059
	VFPS	1.006	0.824	0.803	0.894	0.901	0.883	1.006
	VFPI	1.023	0.891	0.845	0.930	0.939	0.902	1.023
Peak overturning moment, J_2	FVD	0.633	0.959	0.576	0.960	0.879	0.549	0.960
	FPS	0.728	0.981	0.720	0.978	0.990	0.668	0.990
	VFPS	0.742	0.977	0.751	0.980	0.987	0.658	0.987
	VFPI	0.786	0.982	0.803	0.983	0.987	0.768	0.987
Peak midspan displacement, J_3	FVD	0.642	0.713	0.653	0.703	0.587	0.632	0.713
	FPS	0.761	0.939	0.798	0.865	0.847	0.753	0.939
	VFPS	0.773	0.935	0.826	0.848	0.818	0.745	0.935
	VFPI	0.815	0.964	0.878	0.915	0.897	0.764	0.964
Peak midspan acceleration, J_4	FVD	1.296	0.954	0.940	0.901	0.937	1.072	1.296
	FPS	1.147	1.044	1.049	0.996	1.000	1.039	1.147
	VFPS	1.280	1.001	1.108	0.938	0.958	1.086	1.280
	VFPI	1.018	0.997	1.025	0.969	0.984	1.034	1.034
Peak bearing displacement, J_5	FVD	0.397	0.671	0.344	0.661	0.542	0.275	0.671
	FPS	0.647	0.898	0.436	0.834	0.800	0.428	0.898
	VFPS	0.578	0.905	0.458	0.822	0.770	0.379	0.905
	VFPI	0.683	0.939	0.492	0.899	0.852	0.534	0.939
Peak ductility, J_6	FVD	0.633	0.598	0.576	0.585	0.194	0.549	0.633
	FPS	0.728	0.936	0.720	0.811	0.677	0.668	0.936
	VFPS	0.742	0.921	0.751	0.799	0.598	0.658	0.921
	VFPI	0.786	0.973	0.803	0.896	0.801	0.768	0.973
Peak dissipated energy, J_7	FVD	0	0.226	0	0.333	0	0	0.333
	FPS	0	0.681	0	0.696	0.307	0	0.696
	VFPS	0	0.634	0	0.663	0.243	0	0.663
	VFPI	0	0.774	0	0.787	0.508	0	0.787
Maximum plastic connections, J_8	FVD	0	0.667	0	0.750	0	0	0.750
	FPS	0	0.833	0	1	0.667	0	1
	VFPS	0	0.667	0	1	0.333	0	1
	VFPI	0	0.833	0	1	0.667	0	1
Normed base shear, J_9	FVD	1.031	0.747	0.518	0.709	0.738	0.746	1.031
	FPS	0.944	0.860	0.554	0.811	0.855	0.717	0.944
	VFPS	0.902	0.856	0.559	0.804	0.852	0.697	0.902
	VFPI	0.879	0.899	0.556	0.859	0.914	0.714	0.914
Normed overturning moment, J_{10}	FVD	0.529	0.747	0.393	0.718	0.366	0.515	0.747
	FPS	0.668	0.820	0.420	0.840	0.756	0.629	0.840
	VFPS	0.696	0.810	0.518	0.817	0.704	0.640	0.817
	VFPI	0.726	0.853	0.520	0.878	0.934	0.674	0.934
Normed midspan displacement, J_{11}	FVD	0.557	0.645	0.410	0.650	0.442	0.544	0.650
	FPS	0.692	0.821	0.436	0.807	0.659	0.656	0.821
	VFPS	0.713	0.805	0.537	0.791	0.601	0.668	0.805
	VFPI	0.748	0.876	0.535	0.866	0.725	0.699	0.876

(Continued)

TABLE 11.40 (*Continued*)
Evaluation Criteria for FVD, FPS, VFPS, and VFPI

		Value of Evaluation Criteria						
Evaluation Criteria	**Control Type**	**N. Palm Springs**	**Chi-Chi**	**El Centro**	**Northridge**	**Turkey**	**Kobe**	**Maximum**
Normed midspan acceleration, J_{12}	FVD	1.017	0.765	0.743	0.777	1.008	1.059	1.059
	FPS	1.009	0.937	0.824	0.923	0.995	1.014	1.014
	VFPS	0.977	0.916	0.749	0.897	0.981	1.010	1.010
	VFPI	0.977	0.940	0.795	0.924	0.959	0.933	0.977
Normed displacement at abutment, J_{13}	FVD	0.251	0.616	0.248	0.616	0.291	0.196	0.616
	FPS	0.531	0.794	0.272	0.780	0.476	0.257	0.794
	VFPS	0.601	0.781	0.340	0.767	0.431	0.265	0.781
	VFPI	0.460	0.860	0.337	0.851	0.522	0.317	0.860
Norm ductility, J_{14}	FVD	0.529	0.657	0.393	0.994	0.035	0.515	0.994
	FPS	0.668	0.621	0.420	0.850	0.705	0.629	0.850
	VFPS	0.696	0.565	0.518	0.796	0.581	0.640	0.796
	VFPI	0.726	0.634	0.520	1.003	0.896	0.674	1.003
Peak control force, J_{15}	FVD	0.013	0.024	0.010	0.022	0.017	0.013	0.024
	FPS	0.006	0.013	0.006	0.013	0.009	0.007	0.013
	VFPS	0.008	0.009	0.008	0.009	0.009	0.009	0.009
	VFPI	0.004	0.004	0.004	0.004	0.004	0.004	0.004
Maximum device stroke, J_{16}	FVD	0.382	0.642	0.317	0.602	0.537	0.271	0.642
	FPS	0.623	0.859	0.401	0.759	0.793	0.422	0.859
	VFPS	0.556	0.866	0.421	0.749	0.764	0.374	0.866
	VFPI	0.657	0.899	0.453	0.819	0.844	0.526	0.899

of the deck, and (iii) the variation of force in the stays, which varies between the 20% and 70% of the maximum tension in the cable.

11.6.1 Evaluation Model

Based on the description of the constructed cable-stayed bridge, Dyke et al. (2002a) developed a 3D finite element evaluation model in MATLAB, which effectively represents the complex behavior of the full-scale benchmark bridge. A full-order linear evaluation model is considered as a true representation of the real structural system. As the bridge is founded on bedrock, the effects of soil-structure

TABLE 11.41
Parameters of Control Devices for Phase I and Phase II

Control Device	Parameter	Phase I	Phase II
Friction pendulum system (FPS)	μ	0.10	0.10
	T_b (s)	2.5	3.5
Variable friction pendulum system (VFPS)	μ_{max}	0.16	0.16
	T_i (s)	1.2	1.25
	T_b (s)	3.0	3.5
Variable frequency pendulum isolator (VFPI)	μ	0.07	0.07
	FVF	5	4
	T_i (s)	2.25	1.25

TABLE 11.42
Evaluation Criteria for North Palm Springs (1986), Chi-Chi (1999), and El Centro (1940) Earthquakes

Evaluation	North Palm Springs (1986)			Chi-Chi (1999)			El Centro (1940)		
Criteria	FPS	VFPS	VFPI	FPS	VFPS	VFPI	FPS	VFPS	VFPI
J_1	0.925	1.098	0.945	1.301	1.221	1.366	0.674	0.773	0.859
J_2	1.072	1.311	1.189	1.458	1.296	1.502	0.662	0.771	0.846
J_3	0.847	0.753	0.726	1.117	1.008	1.103	0.317	0.284	0.295
J_4	2.106	2.969	2.644	1.863	1.647	1.559	1.613	1.799	1.657
J_5	0.802	0.700	0.680	1.100	0.999	1.104	0.299	0.263	0.275
J_6	1.072	1.311	1.189	1.458	1.296	1.502	0.662	0.771	0.846
J_7	0	0	0	0	0	0	0	0	0
J_8	0	0	0	0	0	0	0	0	0
J_9	0.975	1.155	0.882	1.072	0.729	1.507	0.651	0.708	0.681
J_{10}	0.959	1.109	0.815	1.058	0.693	1.490	0.636	0.692	0.661
J_{11}	0.650	0.473	0.247	0.654	0.477	0.611	0.195	0.213	0.138
J_{12}	1.768	2.448	2.049	1.831	2.009	1.851	1.343	1.473	1.506
J_{13}	0.627	0.454	0.230	0.676	0.469	0.645	0.199	0.219	0.133
J_{14}	0.959	1.109	0.815	1.058	0.693	1.490	0.636	0.692	0.661
J_{15}	0.004	0.007	0.006	0.013	0.007	0.007	0.004	0.007	0.006
J_{16}	0.802	0.700	0.680	1.100	0.999	1.104	0.299	0.263	0.275

TABLE 11.43
Evaluation Criteria for Northridge (1986), Turkey (1994), and Kobe (1995) Earthquakes

Evaluation	Northridge (1994)			Turkey (1999)			Kobe (1995)		
Criteria	FPS	VFPS	VFPI	FPS	VFPS	VFPI	FPS	VFPS	VFPI
J_1	1.055	1.025	1.246	1.041	0.933	1.369	0.834	1.130	0.901
J_2	1.110	1.063	1.239	1.093	0.964	1.406	0.948	1.206	1.074
J_3	0.709	0.642	0.634	0.532	0.543	0.495	0.295	0.320	0.295
J_4	1.386	1.847	1.572	1.269	1.758	1.414	2.043	2.422	2.505
J_5	0.709	0.649	0.640	0.510	0.509	0.472	0.284	0.307	0.279
J_6	1.110	1.063	1.239	1.093	0.964	1.406	0.948	1.206	1.074
J_7	0	0	0	0	0	0	0	0	0
J_8	0	0	0	0	0	0	0	0	0
J_9	0.884	0.965	1.296	0.313	0.444	0.620	0.559	0.592	0.575
J_{10}	0.858	0.924	1.276	0.300	0.415	0.609	0.544	0.573	0.555
J_{11}	0.666	0.663	0.643	0.132	0.136	0.113	0.154	0.238	0.116
J_{12}	1.444	1.626	1.713	1.174	1.564	1.443	1.390	1.540	1.607
J_{13}	0.670	0.669	0.633	0.117	0.137	0.133	0.153	0.250	0.098
J_{14}	0.858	0.924	1.276	0.300	0.415	0.609	0.544	0.573	0.555
J_{15}	0.009	0.007	0.007	0.005	0.007	0.007	0.004	0.007	0.006
J_{16}	0.709	0.649	0.640	1.100	0.999	1.104	0.284	0.307	0.279

TABLE 11.44
Comparison of J_1 to J_{16} Criteria for All Earthquakes and All Control Strategies

		Control Case		
Evaluation Criteria	**Earthquake**	**FPS**	**VFPS**	**VFPI**
J_1	North Palm Springs	0.925	1.098	0.945
	Chi-Chi	1.301	1.221	1.366
	El Centro	0.674	0.773	0.859
	Northridge	1.055	1.025	1.246
	Turkey	1.041	0.933	1.369
	Kobe	0.834	1.130	0.901
J_2	North Palm Springs	1.072	1.311	1.189
	Chi-Chi	1.458	1.296	1.502
	El Centro	0.662	0.771	0.846
	Northridge	1.110	1.063	1.239
	Turkey	1.093	0.964	1.406
	Kobe	0.948	1.206	1.074
J_3	North Palm Springs	0.847	0.753	0.726
	Chi-Chi	1.117	1.008	1.103
	El Centro	0.317	0.284	0.295
	Northridge	0.709	0.642	0.634
	Turkey	0.532	0.543	0.495
	Kobe	0.295	0.320	0.295
J_4	North Palm Springs	2.106	2.969	2.644
	Chi-Chi	1.863	1.647	1.559
	El Centro	1.613	1.799	1.657
	Northridge	1.386	1.847	1.572
	Turkey	1.269	1.758	1.414
	Kobe	2.043	2.422	2.505
J_5	North Palm Springs	0.802	0.700	0.680
	Chi-Chi	1.100	0.999	1.104
	El Centro	0.299	0.263	0.275
	Northridge	0.709	0.649	0.640
	Turkey	0.510	0.509	0.472
	Kobe	0.284	0.307	0.279
J_6	North Palm Springs	1.072	1.311	1.189
	Chi-Chi	1.458	1.296	1.502
	El Centro	0.662	0.771	0.846
	Northridge	1.110	1.063	1.239
	Turkey	1.093	0.964	1.406
	Kobe	0.948	1.206	1.074
J_9	North Palm Springs	0.975	1.155	0.882
	Chi-Chi	1.072	0.729	1.507
	El Centro	0.651	0.708	0.681
	Northridge	0.884	0.965	1.296
	Turkey	0.313	0.444	0.620
	Kobe	0.559	0.592	0.575
J_{10}	North Palm Springs	0.959	1.109	0.815
	Chi-Chi	1.058	0.693	1.490
	El Centro	0.636	0.692	0.661
	Northridge	0.858	0.924	1.276
	Turkey	0.300	0.415	0.609
	Kobe	0.544	0.573	0.555

(Continued)

TABLE 11.44 (*Continued*)
Comparison of J_1 to J_{16} Criteria for All Earthquakes and All Control Strategies

		Control Case		
Evaluation Criteria	**Earthquake**	**FPS**	**VFPS**	**VFPI**
J_{11}	North Palm Springs	0.650	0.473	0.247
	Chi-Chi	0.654	0.477	0.611
	El Centro	0.195	0.213	0.138
	Northridge	0.666	0.663	0.643
	Turkey	0.132	0.136	0.113
	Kobe	0.154	0.238	0.116
J_{12}	North Palm Springs	1.768	2.448	2.049
	Chi-Chi	1.831	2.009	1.851
	El Centro	1.343	1.473	1.506
	Northridge	1.444	1.626	1.713
	Turkey	1.174	1.564	1.443
	Kobe	1.390	1.540	1.607
J_{13}	North Palm Springs	0.627	0.454	0.230
	Chi-Chi	0.676	0.469	0.645
	El Centro	0.199	0.219	0.133
	Northridge	0.670	0.669	0.633
	Turkey	0.117	0.137	0.133
	Kobe	0.153	0.250	0.098
J_{14}	North Palm Springs	0.959	1.109	0.815
	Chi-Chi	1.058	0.693	1.490
	El Centro	0.636	0.692	0.661
	Northridge	0.858	0.924	1.276
	Turkey	0.300	0.415	0.609
	Kobe	0.544	0.573	0.555
J_{15}	North Palm Springs	0.004	0.007	0.006
	Chi-Chi	0.013	0.007	0.007
	El Centro	0.004	0.007	0.006
	Northridge	0.009	0.007	0.007
	Turkey	0.005	0.007	0.007
	Kobe	0.004	0.007	0.006
J_{16}	North Palm Springs	0.802	0.700	0.680
	Chi-Chi	1.100	0.999	1.104
	El Centro	0.299	0.263	0.275
	Northridge	0.709	0.649	0.640
	Turkey	0.510	0.509	0.472
	Kobe	0.284	0.307	0.279

interaction are ignored. Longitudinal direction is considered as the most destructive in cable-stayed bridges and ground acceleration is applied in this direction. The finite element model consists of beam elements, cable elements, and rigid links. ABAQUS is used to perform the nonlinear static analysis and the element mass and stiffness are output to MATLAB for assembly. Figure 11.32 shows the flowchart of the procedure used to develop the evaluation model of the cable-stayed bridge, and Figure 11.33 displays the chart demonstrating overall working of the SIMULINK models.

A modified evaluation model is formed to make feasible for the designers to place control devices longitudinally between the deck and tower. From Table 11.46, it is seen that the frequencies of the modified evaluation model are less than those of the basic evaluation model of the bridge.

TABLE 11.45
Salient Features of Benchmark Cable-Stayed Bridge

Property of Bridge	Value/Detail
Total length of bridge (m)	1205.8
Length of main span (m)	350.6
Length of side spans (m)	142.7
Total width of bridge (m)	29.3
Total number of lanes	Four lanes + two narrower bicycle lanes
Composition of deck	Steel beams and prestressed concrete + cables
Number of cables	128 Made of high-strength, low relaxation steel
RC towers	Two H-shaped towers
Height of tower (m)	102.4 @ pier 2
	108.5 @ pier 3
Number of piers	12 in the approach from the Illinois side

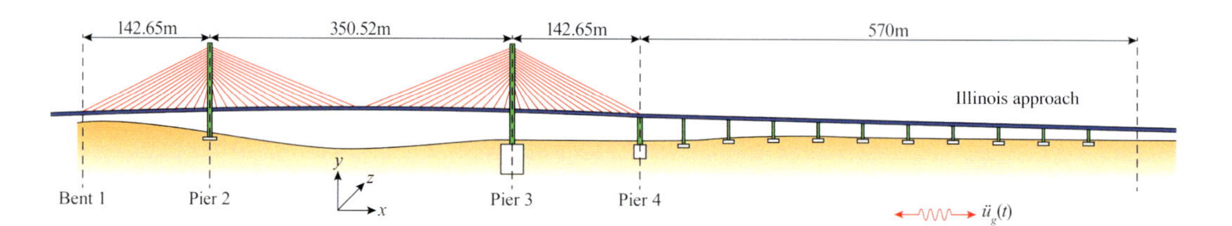

FIGURE 11.30 Elevation of benchmark cable-stayed bridge. (Caicedo et al., 2003)

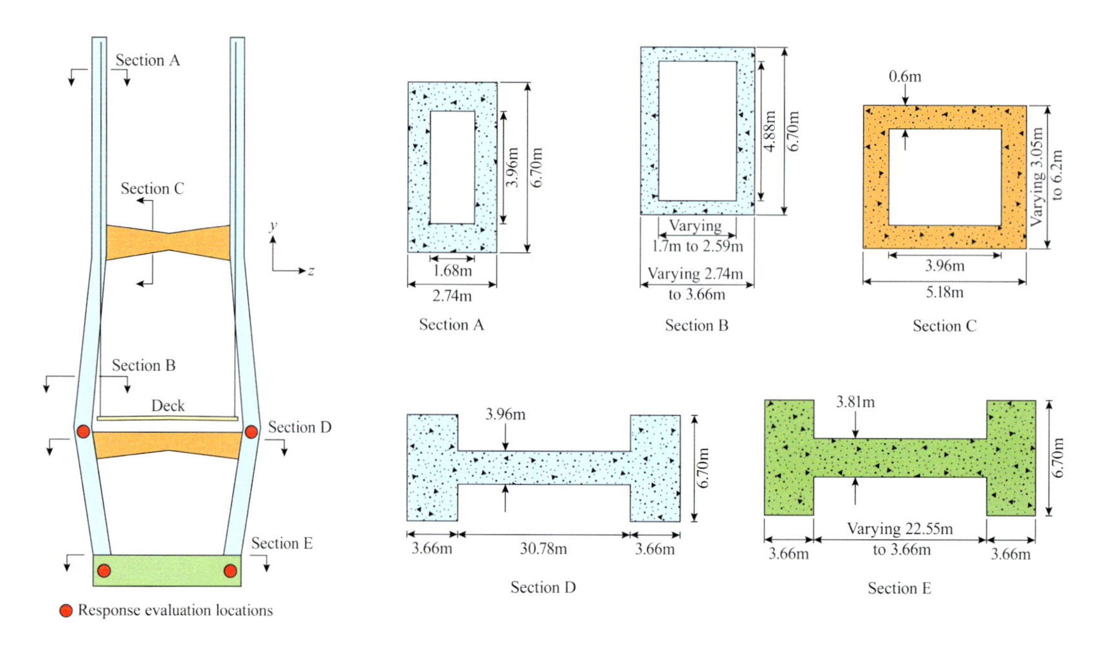

FIGURE 11.31 Cross-sectional elevation of tower. (Caicedo et al., 2003.)

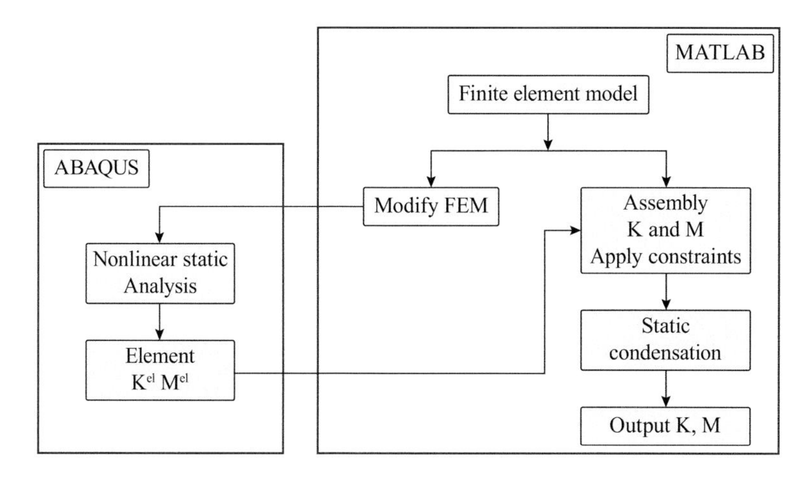

FIGURE 11.32 Flowchart to develop the evaluation model of the cable-stayed bridge.

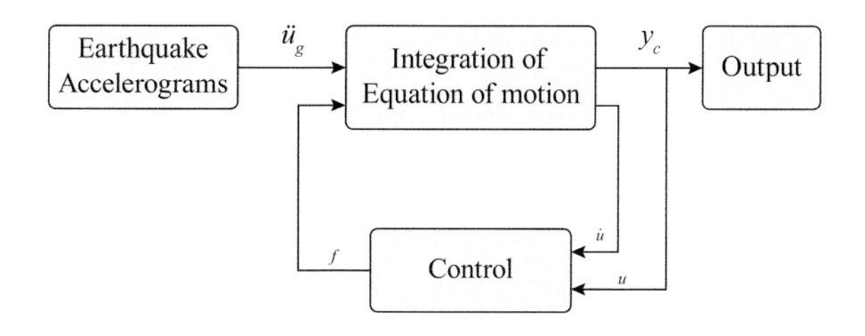

FIGURE 11.33 Working of the SIMULINK model.

TABLE 11.46
First Ten Frequencies of Basic Evaluation Model and Modified Evaluation Model

Mode No.	Frequency of Basic Evaluation Model (Hz)	Frequency of Modified Evaluation Model (Hz)
1	0.2899	0.1618
2	0.3699	0.2666
3	0.4683	0.3723
4	0.5158	0.4545
5	0.5812	0.5015
6	0.6490	0.5650
7	0.6687	0.6187
8	0.6970	0.6486
9	0.7102	0.6965
10	0.7203	0.7094

11.6.2 Finite Element Model

A 3D finite element model of the cable-stayed bridge is developed in ABAQUS, which consists of 579 nodes, 420 rigid links, to model the abutment and the deck-ends, 134 nodal masses, 162 beam elements, and 128 cable elements. The H-shaped tower consists of 50 nodes, 43 beam elements, and 74 rigid links. Constraints are applied, in order to restrict the lateral movement at piers 2, 3, and 4, shown in Figure 11.30. Only longitudinal displacement and rotation about the Z-axis are allowed at pier 1. As the attachment points of the cables to the deck are above the neutral axis of the deck, and the attachment points of the cables to the tower are outside the neutral axis of the tower, the cables are connected to the tower and to the deck by means of rigid links as shown in Figure 11.34.

The use of rigid links in the model gives assurance of the total length, and inclination angle of the cable with the actual drawings of the bridge. This FEM model is used only when the control devices are placed in the longitudinal direction between the deck and tower, i.e., in the direction of application of earthquake ground motion.

11.6.3 Nonlinear Static Analysis

Due to variations of the catenary shapes of the inclined cables of the cable-stayed bridge, it shows nonlinear behavior. Also, the cable tension induces compressive forces and large displacements in the deck and towers. Equivalent elastic modulus, given by Equation 11.5, is used to model the catenary shape and its variation with the axial force in the cable.

$$E_{eq} = \frac{E_c}{1 + \left[\dfrac{(wL_x)^2 A_c E_c}{12T_c^3}\right]} \tag{11.5}$$

where T_c, w, and L_x are the tension in cable, unit weight of cable, and the projected length in the X–Z plane, respectively. E_c is the modulus of elasticity of material of the cable. Lumped masses are

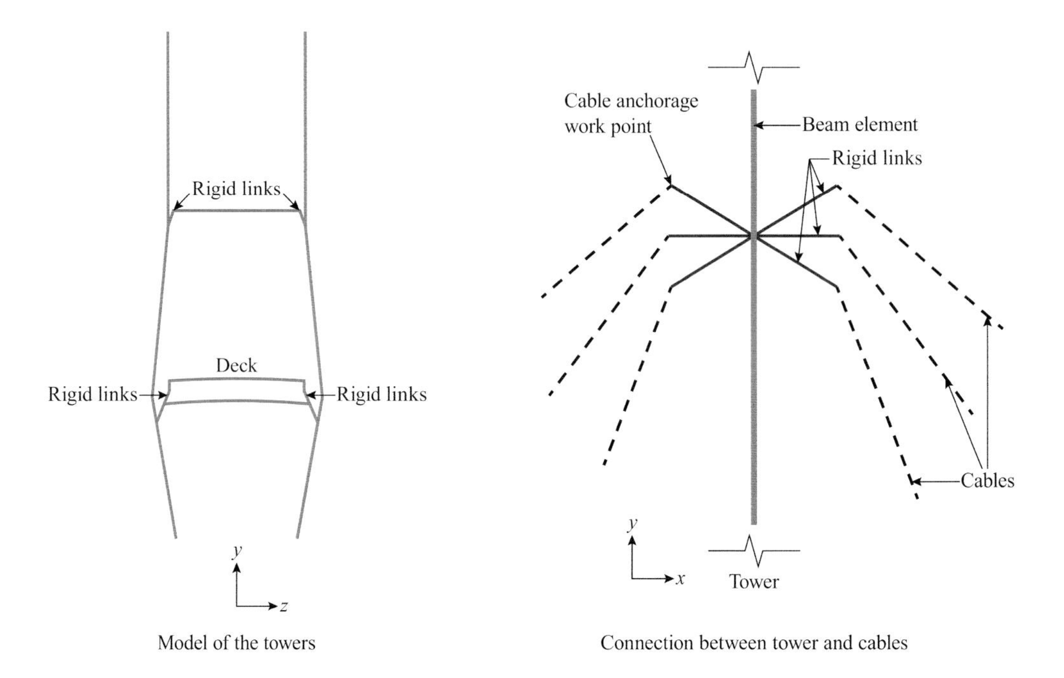

FIGURE 11.34 Finite element model of tower of cable-stayed bridge.

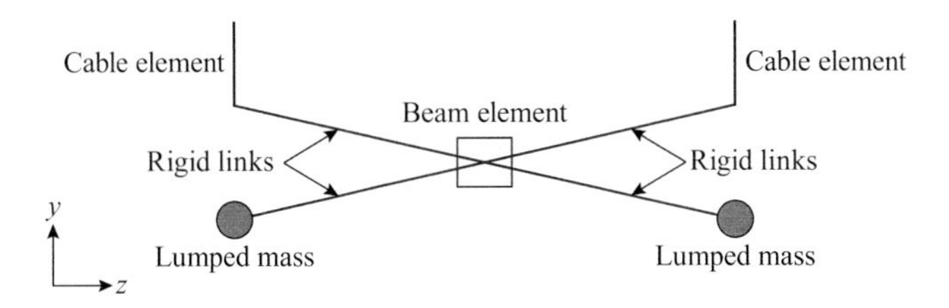

FIGURE 11.35 Finite element model of the cross section of the bridge deck.

used to model the mass of the deck, which in turn is connected to the central beam as shown in Figure 11.35. The deck is treated as a C-shaped section as it comprises two main steel girders along each longitudinal edge of the deck supporting the concrete slab as shown in Figure 11.36.

While calculating the torsional stiffness of the deck section, warping and pure torsional constants are taken into account. The warping and pure torsional constants are given in Equations 11.6 and 11.7 respectively,

$$J_t = \sum_{i=1}^{n} \frac{b_i t_i^3}{3} \tag{11.6}$$

$$\Gamma_w = \frac{d^2}{4}\left\{ I_{zz} + e^2 A\left(1 - \frac{d\ A}{4I_{yy}}\right)\right\} \tag{11.7}$$

where b_i and t_i are the length and thickness of the section, respectively, d is the distance between the webs of the steel beams located along the edges of the deck, e is the distance between the neutral axis and middle of the concrete slab, A is the equivalent cross-sectional area, and I_{zz} and I_{yy} are the moment of inertia of the deck about Z and Y axes, respectively. The torsional stiffness of the deck is given by

$$G_s J_{eq} = G_s\left[J + \frac{E_s \Gamma_w \pi^2}{G_s L^2} \right] \tag{11.8}$$

where G_s is the shear modulus of steel, J_{eq} is the equivalent torsional constant, and L is the length of the main span. The mass of the deck includes the mass of the steel beams, rigid concrete slab, barriers, and railings. The total mass of the deck is 264.57 kg/m. In order to include the behavior of C-shaped section, the deck is represented by two lumped mass system, as shown in Figure 11.33. Each lumped mass is 50% of the total deck mass. The lumped masses are joined by means of rigid links. The mass moment of inertia of the lumped masses with respect to the jth axis, i.e., either X, Y, or Z axis is

$$I_j = 2M_l r^2 \tag{11.9}$$

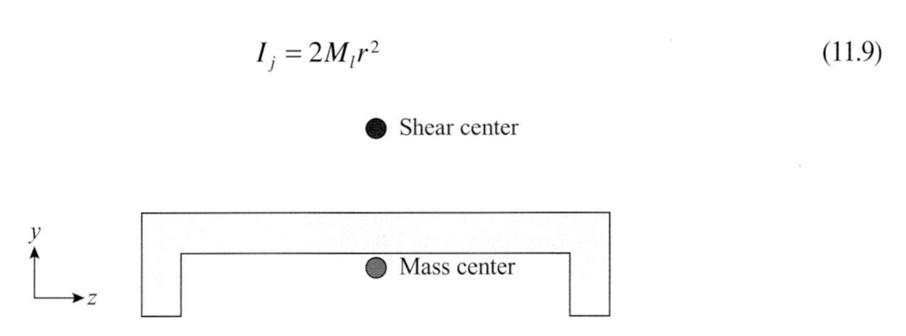

FIGURE 11.36 C-shaped section used to determine properties of deck.

where M_l is the magnitude of each lumped mass, and r is the perpendicular distance from the lumped mass to each axis. However, the actual moment of inertia of the deck with respect to the jth axis i.e., either X, Y, or Z axis is expressed as

$$I_{mj} = \sum_{i=1}^{n} \left(I_{mi} + m_l r_i^2 \right) \tag{11.10}$$

where m_i is the mass of each component, and r_i is the perpendicular distance between centroid of each component and the jth axis. Hence, the corrected mass moment of inertia of the section is

$$\Delta_j = \left(I_{mj} - I_j \right) \tag{11.11}$$

11.6.4 Reduced-Order Controller Design Model

The model resulting from finite element formulation has a large number of DOF and involves high-frequency dynamics. Appropriate assumptions are made regarding the behavior of the bridge, to make the model more manageable for dynamic simulation, while retaining the fundamental behavior of the bridge. A reduced-order model of the system is developed from the initial elastic evaluation model. Eigenmode reduction method is used to convert the evaluation model to the reduced-order model. This is a more general and accurate method for representing the dynamics of a full-order bridge model. Static condensation is performed by partitioning the mass and stiffness matrices corresponding to the DOF of structure into active and dependent DOF. The static transformation matrix is determined, and the transformed mass, stiffness, and input coefficient matrices are generated. The reduced-order model has the same output as the evaluation model and is assumed to retain the dominant characteristics of the full-order evaluation system. This model is designated as the controller design model. In the present study, the response due to vertical ground motion component is not taken into account. Accordingly, the corresponding DOFs are restrained without significantly affecting the expected response.

11.6.5 Evaluation Criteria

To facilitate direct and systematic comparison of various control strategies, a set of 18 dimensionless and normalized evaluation criteria is defined in the benchmark problem definition.

11.6.6 Peak Response

The first two evaluation criteria are the measures of the shear force at key location in the bridge to measure ability of the controller to reduce peak response quantities of the benchmark cable-stayed bridge.

1. Peak base shear force in the controlled structure, normalized by the corresponding base shear in the uncontrolled structure

$$J_1 = \max \left\{ \frac{\max_{i,t} \left| F_{bi}(t) \right|}{F_{0b}^{\max}} \right\}$$

2. Peak shear force at deck level of the controlled structure, normalized by the corresponding shear force of the uncontrolled structure

$$J_2 = \max \left\{ \frac{\max_{i,t} \left| F_{di}(t) \right|}{F_{0d}^{\max}} \right\}$$

The second set of evaluation criteria is measure of moments in the bridge at the key locations.

3. Peak overturning moment in the controlled structure, normalized by the corresponding moment in the uncontrolled structure

$$J_3 = \max\left\{\frac{\max\limits_{i,t}\left|M_{bi}(t)\right|}{M_{0b}^{\max}}\right\}$$

4. Peak moment at deck level of the controlled structure, normalized by the corresponding peak moment at deck level of the uncontrolled structure

$$J_4 = \max\left\{\frac{\max\limits_{i,t}\left|M_{di}(t)\right|}{M_{0d}^{\max}}\right\}$$

The fifth evaluation criterion is a measure of deviation of the tension in the stay cables from the nominal pretension.

5. Peak cable tension in the controlled structure, normalized by the corresponding peak cable tension in the uncontrolled structure

$$J_5 = \max\left\{\max_{i,t}\left|\frac{T_{ai}(t)-T_{0i}}{T_{0i}}\right|\right\}$$

The sixth evaluation criterion is a measure of the peak deck displacement at piers 1 and 4.

6. Peak deck displacement at abutment in the controlled structure, normalized by the corresponding peak deck displacement at abutment in the uncontrolled structure

$$J_6 = \max\left\{\max_{i,t}\left|\frac{x_{bi}(t)}{x_{0b}}\right|\right\}$$

11.6.7 Normed Response

The seventh and eighth evaluation criteria are measures of the normed values of the base shear and shear at the deck level in each of the towers, respectively.

7. Normed base shear force in the controlled structure, normalized by the corresponding normed base shear force in the uncontrolled structure

$$J_7 = \max\left\{\frac{\max\limits_{t}\left\|F_{bi}(t)\right\|}{\left\|F_{0b}^{\max}\right\|}\right\}$$

8. Normed shear force at deck level in the controlled structure, normalized by the corresponding normed shear force at deck level

$$J_8 = \max\left\{\frac{\max\limits_{i}\left\|F_{di}(t)\right\|}{\left\|F_{0d}^{\max}\right\|}\right\}$$

The ninth and tenth evaluation criteria are measures of the normed values of the overturning moment, and moment at the deck level in each of the towers, respectively.

9. Normed overturning moment in the controlled structure, normalized by the corresponding normed overturning moment in the uncontrolled structure

$$J_9 = \max\left\{ \frac{\max\limits_i \left\| M_{bi}(t) \right\|}{\left\| M_{0b}^{\max} \right\|} \right\}$$

10. Normed moment at deck level in the controlled structure, normalized by the corresponding normed moment at deck level in the uncontrolled structure

$$J_{10} = \max\left\{ \frac{\max\limits_i \left\| M_{di}(t) \right\|}{\left\| M_{0d}^{\max} \right\|} \right\}$$

The eleventh evaluation criterion is a measure of the normed value of the deviation of tension in the stay cables from the nominal pretension.

11. Normed cable tension in the controlled structure, normalized by the corresponding normed cable tension in the uncontrolled structure

$$J_{11} = \max\left\{ \max_i \frac{\left\| T_{ai}(t) - T_{0i} \right\|}{T_{0i}} \right\}$$

11.6.8 Control Requirement

The 12th evaluation criterion deals with the maximum force generated by the control devices. It is defined as the peak control force generated by the control devices, normalized by the seismic weight of the bridge.

$$J_{12} = \max\left\{ \max_{l,t} \left(\frac{f_l(t)}{W} \right) \right\}$$

The 13th criterion is based on the maximum stroke of the control devices.

12. Peak stroke of the control devices, normalized by the maximum deformation of bearings in the uncontrolled structure

$$J_{13} = \max\left\{ \max_i \left(\frac{y_i^d(t)}{x_0^{\max}} \right) \right\}$$

The 14th evaluation criterion is a measure of the maximum instantaneous power required to control the bridge.

13. Peak power of the control devices, normalized by the maximum deformation of bearings in the uncontrolled structure

$$J_{14} = \max\left\{ \max_t \frac{\left[\sum P_i(t) \right]}{x_0^{\max} W} \right\}$$

The 15th evaluation criterion is a measure of the total power required to control the bridge.

14. Peak total power of the control devices, normalized by the seismic weight of the bridge

$$J_{15} = \max \left\{ \frac{\left[\sum \int_0^{t_f} P_i(t) dt \right]}{x_0^{\max} W} \right\}$$

15. The 16th evaluation criterion is a measure of the total number of control devices required in the control system to control the response of the bridge

    J_{16} = number of control devices.

    A summary of the sixteen evaluation criteria is presented in Table 11.47.

TABLE 11.47
Summary of Evaluation Criteria

Peak Responses	Normed Responses	Control Strategy
Peak base shear	Normed base shear	Peak force
$J_1 = \max \left\{ \frac{\max_{i,t} \lvert F_{bi}(t) \rvert}{F_{0b}^{\max}} \right\}$	$J_7 = \max \left\{ \frac{\max_{t} \lVert F_{bi}(t) \rVert}{\lVert F_{0b}^{\max} \rVert} \right\}$	$J_{12} = \max \left\{ \max_{l,t} \left(\frac{f_l(t)}{W} \right) \right\}$
Peak shear at deck level	Normed shear at deck level	Device stroke
$J_2 = \max \left\{ \frac{\max_{i,t} \lvert F_{di}(t) \rvert}{F_{0d}^{\max}} \right\}$	$J_8 = \max \left\{ \frac{\max_{i} \lVert F_{di}(t) \rVert}{\lVert F_{0d}^{\max} \rVert} \right\}$	$J_{13} = \max \left\{ \max_{i} \left(\frac{y_i^d(t)}{x_0^{\max}} \right) \right\}$
Peak overturning moment	Normed overturning moment	Peak power
$J_3 = \max \left\{ \frac{\max_{i,t} \lvert M_{bi}(t) \rvert}{M_{0b}^{\max}} \right\}$	$J_9 = \max \left\{ \frac{\max_{i} \lVert M_{bi}(t) \rVert}{\lVert M_{0b}^{\max} \rVert} \right\}$	$J_{14} = \max \left\{ \max_{t} \frac{\left[\sum P_i(t) \right]}{x_0^{\max} W} \right\}$
Peak moment at deck level	Normed moment at deck level	Total power
$J_4 = \max \left\{ \frac{\max_{i,t} \lvert M_{di}(t) \rvert}{M_{0d}^{\max}} \right\}$	$J_{10} = \max \left\{ \frac{\max_{i} \lVert M_{di}(t) \rVert}{\lVert M_{0d}^{\max} \rVert} \right\}$	$J_{15} = \max \left\{ \frac{\left[\sum \int_0^{t_f} P_i(t) dt \right]}{x_0^{\max} W} \right\}$
Cable tension	Cable tension	
$J_5 = \max \left\{ \max_{i,t} \left\lvert \frac{T_{ai}(t) - T_{0i}}{T_{0i}} \right\rvert \right\}$	$J_{11} = \max \left\{ \max_{i} \frac{\lVert T_{ai}(t) - T_{0i} \rVert}{T_{0i}} \right\}$	J_{16} = number of control devices
	Deck displacement at abutment	
	$J_6 = \max \left\{ \max_{i,t} \left\lvert \frac{x_{bi}(t)}{x_{0b}} \right\rvert \right\}$	

TABLE 11.48
Parameters of Selected Earthquake Records

Earthquake	Magnitude	PGA (g)
El Centro NS, 1940	7.1	0.348
Mexico City, 1985	8.1	0.143
Gebze NS, 1999	6.8	0.265

11.6.9 Earthquake Ground Excitations

In order to evaluate the effectiveness of the controllers when installed in the benchmark cable-stayed bridge, three historical earthquakes are used to perform numerical simulation. Table 11.48 presents the parameters of the earthquake records.

1. Imperial Valley: The N-S component the Imperial Valley earthquake recorded at the Imperial Valley Irrigation District substation in El Centro, California, during the Imperial Valley, California earthquake of May 18, 1940
2. Mexico City: Recorded at the Caleta de Campos station with site Geology of Meta-Andesite Breccia on September 19, 1985
3. Gebze: The N-S component recorded at the Gebze Tubitak Marmara Arastirma Merkezi on August 17, 1999

11.6.10 Numerical Study

In recent years, there have been important studies on the use of friction isolators for protection of bridges from severe earthquake attacks. In this problem, FPS, high damping rubber bearing (HDRB), lead rubber bearing (LRB), and resilient-friction bearing isolator (R-FBI) are used for evaluation of the benchmark cable-stayed bridge. The optimum parameters of the control devices are presented in Table 11.49.

In the benchmark study, the performance of the hybrid passive isolation system, consisting of isolators, and linear and non-LVD, is investigated. The fundamental base isolation period for all

TABLE 11.49
Parameters of Control Devices

Control Device	Parameter	Value
Friction pendulum system (FPS)	μ	0.05
	T_b (s)	1.5
	ξ (%)	10
High damping rubber bearing (HDRB)	μ_{max}	0.16
	T_b (s)	1.5
	ξ (%)	15
Lead rubber bearing (LRB)	μ	0.07
	T_b (s)	1.5
	F_o	0.1
	ξ (%)	5
Resilient-friction bearing isolator (R-FBI)	μ	0.04
	T_b (s)	1.5
	ξ (%)	10

TABLE 11.50
Evaluation Criteria for El Centro Earthquake (1940)

Evaluation	Value of Evaluation Criteria for Control Cases											
Criteria	HDRB	LRB	FPS	RFBI	HDRB + LFVD	LRB + LFVD	FPS + LFVD	RFBI + LFVD	HDRB + NLFVD	LRB + NLFVD	FPS + NLFVD	RFBI + NLFVD
J_1	0.451	0.447	0.442	0.436	0.427	0.420	0.415	0.413	0.407	0.399	0.397	0.390
J_2	1.344	1.327	1.313	1.299	1.254	1.240	1.220	1.206	1.180	1.123	1.103	1.089
J_3	0.382	0.378	0.375	0.370	0.361	0.352	0.349	0.347	0.345	0.326	0.325	0.313
J_4	0.645	0.668	0.635	0.632	0.550	0.615	0.586	0.578	0.503	0.540	0.526	0.515
J_5	0.257	0.253	0.252	0.249	0.240	0.232	0.233	0.232	0.231	0.224	0.227	0.219
J_6	1.848	1.801	1.753	1.698	1.392	1.557	1.472	1.431	1.123	1.284	1.299	1.127
J_7	0.254	0.251	0.247	0.244	0.234	0.236	0.236	0.234	0.357	0.224	0.342	0.220
J_8	1.417	1.389	1.287	1.297	1.145	1.167	1.090	1.078	1.401	0.925	1.342	0.856
J_9	0.353	0.342	0.318	0.318	0.283	0.309	0.292	0.289	0.243	0.262	0.256	0.245
J_{10}	1.240	1.058	0.996	0.999	0.958	0.961	0.906	0.903	0.750	0.776	0.757	0.710
J_{11}	0.032	0.034	0.032	0.032	0.029	0.030	0.028	0.028	0.024	0.024	0.023	0.022
J_{12}	0.001	0.004	0.004	0.004	0.004	0.006	0.005	0.005	0.005	0.008	0.006	0.008
J_{13}	1.213	1.182	1.151	1.115	0.914	1.022	0.966	0.940	0.737	0.843	0.853	0.740
J_{16}	24	24	24	24	24 + 8	24 + 8	24 + 8	24 + 8	24 + 8	24 + 8	24 + 8	24 + 8

TABLE 11.51
Evaluation Criteria for Various Cases under the 1985 Mexico City Earthquake

Evaluation Criteria	Value of Evaluation Criteria for Control Cases											
	HDRB	LRB	FPS	RFBI	HDRB + LFVD	LRB + LFVD	FPS + LFVD	RFBI + LFVD	HDRB + NLFVD	LRB + NLFVD	FPS + NLFVD	RFBI + NLFVD
J_1	0.550	0.564	0.482	0.482	0.494	0.494	0.487	0.487	0.675	0.671	0.689	0.720
J_2	1.667	1.522	1.480	1.480	1.362	1.36	1.199	1.193	1.314	1.405	1.276	1.412
J_3	0.822	0.862	0.699	0.699	0.690	0.69	0.566	0.580	0.478	0.569	0.508	0.503
J_4	0.922	0.957	0.811	0.811	0.674	0.674	0.701	0.694	0.489	0.506	0.449	0.440
J_5	0.084	0.106	0.083	0.083	0.077	0.077	0.072	0.073	0.061	0.059	0.057	0.056
J_6	3.916	4.380	2.988	2.988	3.364	3.364	2.690	2.666	1.641	1.964	1.522	1.468
J_7	0.422	0.438	0.439	0.439	0.386	0.386	0.425	0.411	1.143	1.149	1.089	1.321
J_8	1.470	1.430	1.255	1.255	1.165	1.165	1.092	1.078	2.718	2.727	2.574	3.141
J_9	0.583	0.642	0.552	0.552	0.451	0.451	0.519	0.501	0.358	0.391	0.373	0.375
J_{10}	1.642	1.495	1.214	1.214	1.242	1.242	1.108	1.110	0.809	0.803	0.701	0.707
J_{11}	0.012	0.014	0.011	0.011	0.011	0.011	0.010	0.010	0.007	0.007	0.007	0.007
J_{12}	0.001	0.003	0.002	0.002	0.001	0.001	0.003	0.003	0.005	0.006	0.005	0.006
J_{13}	1.972	2.205	1.505	1.505	1.694	1.694	1.355	1.343	0.827	0.989	0.767	0.739
J_{16}	24	24	24	24	24 + 8	24 + 8	24 + 8	24 + 8	24 + 8	24 + 8	24 + 8	24 + 8

TABLE 11.52
Evaluation Criteria for Various Cases under the 1999 Gebze Earthquake

Evaluation	Value of Evaluation Criteria for Control Cases											
Criteria	HDRB	LRB	FPS	RFBI	HDRB + LFVD	LRB + LFVD	FPS + LFVD	RFBI + LFVD	HDRB + NLFVD	LRB + NLFVD	FPS + NLFVD	RFBI + NLFVD
J_1	0.385	0.498	0.495	0.489	0.412	0.475	0.475	0.472	0.420	0.451	0.440	0.445
J_2	1.900	1.808	1.813	1.753	1.444	1.477	1.479	1.444	1.271	1.302	1.288	1.257
J_3	0.774	0.765	0.775	0.743	0.584	0.618	0.614	0.606	0.504	0.524	0.510	0.505
J_4	3.123	2.249	2.276	2.180	2.103	1.679	1.691	1.645	1.597	1.256	1.244	1.209
J_5	0.307	0.288	0.300	0.286	0.228	0.195	0.200	0.193	0.173	0.151	0.158	0.156
J_6	9.609	7.125	7.490	6.909	6.607	5.201	5.312	5.176	4.856	3.793	3.846	3.723
J_7	0.476	0.557	0.589	0.517	0.327	0.393	0.400	0.388	0.326	0.705	0.667	0.787
J_8	3.429	2.269	2.427	2.181	1.791	1.539	1.552	1.502	1.498	2.473	2.371	2.770
J_9	1.429	1.174	1.269	1.084	0.669	0.667	0.674	0.650	0.511	0.506	0.497	0.485
J_{10}	6.041	3.593	3.955	3.414	2.675	1.975	2.018	1.940	1.918	1.388	1.404	1.370
J_{11}	0.042	0.040	0.043	0.037	0.023	0.022	0.022	0.022	0.018	0.017	0.017	0.016
J_{12}	0.005	0.012	0.012	0.011	0.006	0.010	0.010	0.010	0.006	0.009	0.009	0.009
J_{13}	5.268	3.907	4.107	3.788	3.622	2.852	2.912	2.838	2.662	2.080	2.108	2.041
J_{16}	24	24	24	24	24 + 8	24 + 8	24 + 8	24 + 8	24 + 8	24 + 8	24 + 8	24 + 8

hybrid strategies is taken as 2 s. A total number of twenty-four isolators and eight fluid viscous dampers are installed at eight locations between the deck and the pier; i.e., three isolators and one fluid viscous damper at each location. The evaluation criteria of the benchmark cable-stayed bridge considering isolation device parameters mentioned in Table 11.49 and fluid viscous dampers with damping coefficient 3,000 kNs/m, and $\alpha = 0.5$ are presented in Tables 11.50–11.52 for the three earthquakes.

REFERENCES

Agrawal, A. K., Tan, P., Nagarajaiah, S., and Zhang, J. (2006). Benchmark structural control problem for a seismically excited highway bridge. Structures Congress 2006.

Agrawal, A., Tan, P., Nagarajaiah, S., and Zhang, J. (2009). Benchmark structural control problem for a seismically excited highway bridge – Part I: Phase I problem definition. *Structural Control and Health Monitoring*, 16(5), 509–529.

Caicedo, J. M., Dyke, S. J., Moon, S. J., Bergman, L. A., Turan, G., and Hague, S. (2003). Phase II benchmark control problem for seismic response of cable-stayed bridges. *Journal of Structural Control*, 10(3–4), 137–168.

Dyke, S. J., Caicedo, J. M., Turan, G., Bergman, L. A., and Hague, S. (2002a). Introducing phase I of the benchmark problem on the seismic response control of cable-stayed bridges. 7th US National Conference on Earthquake Engineering. Boston, MA: Earthquake Engineering Research Institute (EERI).

Dyke, S. J., Caicedo, J. M., Turan, G., Bergman, L. A., and Hague, S. (2002b). Phase I benchmark control problem for seismic response of cable-stayed bridges. *Journal of Structural Engineering*, 129(7), 857–872.

Elias, S., Matsagar, V., and Datta, T. K. (2019). Distributed tuned mass dampers for multi-mode control of benchmark building under seismic excitations. *Journal of Earthquake Engineering*, 23(7), 1137–1172.

Min, K.-W., Kim, H.-S., Lee, S.-H., Kim, H., and Kyung Ahn, S. (2005). Performance evaluation of tuned liquid column dampers for response control of a 76-story benchmark building. *Engineering Structures*, 27(7), 1101–1112.

Narasimhan, S., Nagarajaiah, S., Johnson, E. A., and Gavin, H. P. (2006). Smart base-isolated benchmark building. Part I: Problem definition. *Structural Control and Health Monitoring*, 13(2–3), 573–588.

Patil, V. B., and Jangid, R. S. (2009). Response of wind-excited benchmark building installed with dampers. *The Structural Design of Tall and Special Buildings*, 20(4), 497–514.

Patil, V. B., and Jangid, R. S. (2010). Double friction dampers for wind excited benchmark building. *International Journal of Applied Science and Engineering*, 7(2), 95–114.

Patil, V. B., and Jangid, R. S. (2011). Optimum multiple tuned mass dampers for the wind excited benchmark building. *Journal of Civil Engineering and Management*, 17(4), 540–557.

Raut, B. R., and Jangid, R. S. (2014). Seismic analysis of benchmark building installed with friction dampers. *The IES Journal Part A: Civil and Structural Engineering*, 7(1), 20–37.

Samali, B., Kwok, K. C. S., Wood, G. S., and Yang, J. N. (2004). Wind tunnel tests for wind-excited benchmark building. *Journal of Engineering Mechanics*, 130(4), 447–450.

Sharma, A., and Jangid, R. S. (2010). Seismic response of base-isolated benchmark building with variable sliding isolators. *Journal of Earthquake Engineering*, 14(7), 1063–1091.

Spencer, B. F. Jr., Christenson, R. E., and Dyke, S. J. (1998). Next generation benchmark control problem for seismically excited buildings. Proceedings of the Second World Conference on Structural Control, 2, 1135–1360. Japan: Kyoto.

Yang, J. N., Agrawal, A. K., Samali, B., and Wu, J. C. (1998). A benchmark problem for response control of wind-excited tall buildings. Proceeding of the 2nd World Conference on Structural Control, 2, 151–157.

Yang, J. N., Agrawal, A. K., Samali, B., and Wu, J. C. (2004). Benchmark problem for response control of wind-excited tall buildings. *Journal of Engineering Mechanics*, 130(4), 437–446.

Index

T

U

V

W

X